# APEX Calculus III
Version 4.0

**Gregory Hartman, Ph.D.**
*Department of Applied Mathematics*
*Virginia Military Institute*

*Contributing Authors*

**Troy Siemers, Ph.D.**
*Department of Applied Mathematics*
*Virginia Military Institute*

**Brian Heinold, Ph.D.**
*Department of Mathematics and Computer Science*
*Mount Saint Mary's University*

**Dimplekumar Chalishajar, Ph.D.**
*Department of Applied Mathematics*
*Virginia Military Institute*

*Editor*

**Jennifer Bowen, Ph.D.**
*Department of Mathematics and Computer Science*
*The College of Wooster*

Copyright © 2018 Gregory Hartman
Licensed to the public under Creative Commons Attribution-Noncommercial 4.0 International Public License

# Contents

Table of Contents     iii

Preface     v

**9 Curves in the Plane**     **497**
    9.1 Conic Sections . . . . . . . . . . . . . . . . . . . . . 497
    9.2 Parametric Equations . . . . . . . . . . . . . . . . . 511
    9.3 Calculus and Parametric Equations . . . . . . . . . . 521
    9.4 Introduction to Polar Coordinates . . . . . . . . . . . 533
    9.5 Calculus and Polar Functions . . . . . . . . . . . . . 546

**10 Vectors**     **559**
    10.1 Introduction to Cartesian Coordinates in Space . . . . . . . . 559
    10.2 An Introduction to Vectors . . . . . . . . . . . . . . . 574
    10.3 The Dot Product . . . . . . . . . . . . . . . . . . . . 588
    10.4 The Cross Product . . . . . . . . . . . . . . . . . . . 601
    10.5 Lines . . . . . . . . . . . . . . . . . . . . . . . . . . 612
    10.6 Planes . . . . . . . . . . . . . . . . . . . . . . . . . 623

**11 Vector Valued Functions**     **631**
    11.1 Vector–Valued Functions . . . . . . . . . . . . . . . . 631
    11.2 Calculus and Vector–Valued Functions . . . . . . . . . 637
    11.3 The Calculus of Motion . . . . . . . . . . . . . . . . 651
    11.4 Unit Tangent and Normal Vectors . . . . . . . . . . . 664
    11.5 The Arc Length Parameter and Curvature . . . . . . . 673

**12 Functions of Several Variables**     **683**
    12.1 Introduction to Multivariable Functions . . . . . . . . 683
    12.2 Limits and Continuity of Multivariable Functions . . . . . . 690
    12.3 Partial Derivatives . . . . . . . . . . . . . . . . . . . 700
    12.4 Differentiability and the Total Differential . . . . . . . 712
    12.5 The Multivariable Chain Rule . . . . . . . . . . . . . 721
    12.6 Directional Derivatives . . . . . . . . . . . . . . . . . 729
    12.7 Tangent Lines, Normal Lines, and Tangent Planes . . . . . . 739
    12.8 Extreme Values . . . . . . . . . . . . . . . . . . . . 749

## 13 Multiple Integration — 759
- 13.1 Iterated Integrals and Area — 759
- 13.2 Double Integration and Volume — 769
- 13.3 Double Integration with Polar Coordinates — 780
- 13.4 Center of Mass — 787
- 13.5 Surface Area — 799
- 13.6 Volume Between Surfaces and Triple Integration — 806
- 13.7 Triple Integration with Cylindrical and Spherical Coordinates — 828

## 14 Vector Analysis — 839
- 14.1 Introduction to Line Integrals — 840
- 14.2 Vector Fields — 850
- 14.3 Line Integrals over Vector Fields — 859
- 14.4 Flow, Flux, Green's Theorem and the Divergence Theorem — 870
- 14.5 Parametrized Surfaces and Surface Area — 880
- 14.6 Surface Integrals — 891
- 14.7 The Divergence Theorem and Stokes' Theorem — 900

## A  Solutions To Selected Problems — A.1

## Index — A.19

# Preface
*A Note on Using this Text*

Thank you for reading this short preface. Allow us to share a few key points about the text so that you may better understand what you will find beyond this page.

This text is Part III of a three–text series on Calculus. The first part covers material taught in many "Calc 1" courses: limits, derivatives, and the basics of integration, found in Chapters 1 through 6.1. The second text covers material often taught in "Calc 2:" integration and its applications, along with an introduction to sequences, series and Taylor Polynomials, found in Chapters 5 through 8. The third text covers topics common in "Calc 3" or "multivariable calc:" parametric equations, polar coordinates, vector–valued functions, and functions of more than one variable, found in Chapters 9 through 14. All three are available separately for free at www.apexcalculus.com. These three texts are intended to work together and make one cohesive text, *APEX Calculus*, which can also be downloaded from the website.

Printing the entire text as one volume makes for a large, heavy, cumbersome book. One can certainly only print the pages they currently need, but some prefer to have a nice, bound copy of the text. Therefore this text has been split into these three manageable parts, each of which can be purchased for about $15 at Amazon.com.

A result of this splitting is that sometimes a concept is said to be explored in an "earlier section," though that section does not actually appear in this particular text. Also, the index makes reference to topics, and page numbers, that do not appear in this text. This is done intentionally to show the reader what topics are available for study. Downloading the .pdf of *APEX Calculus* will ensure that you have all the content.

## For Students: How to Read this Text

Mathematics textbooks have a reputation for being hard to read. High–level mathematical writing often seeks to say much with few words, and this style often seeps into texts of lower–level topics. This book was written with the goal of being easier to read than many other calculus textbooks, without becoming too verbose.

Each chapter and section starts with an introduction of the coming material, hopefully setting the stage for "why you should care," and ends with a look ahead to see how the just–learned material helps address future problems.

*Please read the text;* it is written to explain the concepts of Calculus. There are numerous examples to demonstrate the meaning of definitions, the truth of theorems, and the application of mathematical techniques. When you encounter a sentence you don't understand, read it again. If it still doesn't make sense, read on anyway, as sometimes confusing sentences are explained by later sentences.

*You don't have to read every equation.* The examples generally show "all" the steps needed to solve a problem. Sometimes reading through each step is helpful; sometimes it is confusing. When the steps are illustrating a new technique, one probably should follow each step closely to learn the new technique. When the steps are showing the mathematics needed to find a number to be used later, one can usually skip ahead and see how that number is being used, instead of getting bogged down in reading how the number was found.

*Most proofs have been omitted.* In mathematics, *proving* something is always true is extremely important, and entails much more than testing to see if it works twice. However, students often are confused by the details of a proof, or become concerned that they should have been able to construct this proof on their own. To alleviate this potential problem, we do not include the proofs to most theorems in the text. The interested reader is highly encouraged to find proofs online or from their instructor. In most cases, one is very capable of understanding what a theorem *means* and *how to apply it* without knowing fully *why* it is true.

### Interactive, 3D Graphics

New to Version 3.0 was the addition of interactive, 3D graphics in the .pdf version. Nearly all graphs of objects in space can be rotated, shifted, and zoomed in/out so the reader can better understand the object illustrated.

As of this writing, the only pdf viewers that support these 3D graphics are Adobe Reader & Acrobat (and only the versions for PC/Mac/Unix/Linux computers, not tablets or smartphones). To activate the interactive mode, click on the image. Once activated, one can click/drag to rotate the object and use the scroll wheel on a mouse to zoom in/out. (A great way to investigate an image is to first zoom in on the page of the pdf viewer so the graphic itself takes up much of the screen, then zoom inside the graphic itself.) A CTRL-click/drag pans the object left/right or up/down. By right-clicking on the graph one can access a menu of other options, such as changing the lighting scheme or perspective. One can also revert the graph back to its default view. If you wish to deactivate the interactivity, one can right-click and choose the "Disable Content" option.

### Thanks

There are many people who deserve recognition for the important role they have played in the development of this text. First, I thank Michelle for her support and encouragement, even as this "project from work" occupied my time and attention at home. Many thanks to Troy Siemers, whose most important contributions extend far beyond the sections he wrote or the 227 figures he coded in Asymptote for 3D interaction. He provided incredible support, advice and encouragement for which I am very grateful. My thanks to Brian Heinold and Dimplekumar Chalishajar for their contributions and to Jennifer Bowen for reading through so much material and providing great feedback early on. Thanks to Troy, Lee Dewald, Dan Joseph, Meagan Herald, Bill Lowe, John David, Vonda Walsh, Geoff Cox, Jessica Libertini and other faculty of VMI who have given me numerous suggestions and corrections based on their experience with teaching from the text. (Special thanks to Troy, Lee & Dan for their patience in teaching Calc III while I was still writing the Calc III material.) Thanks to Randy Cone for encouraging his tutors of VMI's Open Math Lab to read through the text and check the solutions, and thanks to the tutors for spending their time doing so. A very special thanks to Kristi Brown and Paul Janiczek who took this opportunity far above & beyond what I expected, meticulously checking every solution and carefully reading every example. Their comments have been extraordinarily helpful. I am also thankful for the support provided by Wane Schneiter, who as my Dean provided me with extra time to work on this project. Finally, a huge heap of thanks is to be bestowed on the numerous people I do not know who took the time to email me corrections and suggestions. I am blessed to have so many people give of their time to make this book better.

# APEX – Affordable Print and Electronic teXts

APEX is a consortium of authors who collaborate to produce high–quality, low–cost textbooks. The current textbook–writing paradigm is facing a potential revolution as desktop publishing and electronic formats increase in popularity. However, writing a good textbook is no easy task, as the time requirements alone are substantial. It takes countless hours of work to produce text, write examples and exercises, edit and publish. Through collaboration, however, the cost to any individual can be lessened, allowing us to create texts that we freely distribute electronically and sell in printed form for an incredibly low cost. Having said that, nothing is entirely free; someone always bears some cost. This text "cost" the authors of this book their time, and that was not enough. *APEX Calculus* would not exist had not the Virginia Military Institute, through a generous Jackson–Hope grant, given the lead author significant time away from teaching so he could focus on this text.

Each text is available as a free .pdf, protected by a Creative Commons Attribution - Noncommercial 4.0 copyright. That means you can give the .pdf to anyone you like, print it in any form you like, and even edit the original content and redistribute it. If you do the latter, you must clearly reference this work and you cannot sell your edited work for money.

We encourage others to adapt this work to fit their own needs. One might add sections that are "missing" or remove sections that your students won't need. The source files can be found at github.com/APEXCalculus.

You can learn more at www.vmi.edu/APEX.

## Version 4.0

Key changes from Version 3.0 to 4.0:

- Numerous typographical and "small" mathematical corrections (again, thanks to all my close readers!).

- "Large" mathematical corrections and adjustments. There were a number of places in Version 3.0 where a definition/theorem was not correct as stated. See www.apexcalculus.com for more information.

- More useful numbering of Examples, Theorems, etc. "Definition 11.4.2" refers to the second definition of Chapter 11, Section 4.

- The addition of Section 13.7: Triple Integration with Cylindrical and Spherical Coordinates

- The addition of Chapter 14: Vector Analysis.

# 9: Curves in the Plane

We have explored functions of the form $y = f(x)$ closely throughout this text. We have explored their limits, their derivatives and their antiderivatives; we have learned to identify key features of their graphs, such as relative maxima and minima, inflection points and asymptotes; we have found equations of their tangent lines, the areas between portions of their graphs and the x-axis, and the volumes of solids generated by revolving portions of their graphs about a horizontal or vertical axis.

Despite all this, the graphs created by functions of the form $y = f(x)$ are limited. Since each x-value can correspond to only 1 y-value, common shapes like circles cannot be fully described by a function in this form. Fittingly, the "vertical line test" excludes vertical lines from being functions of x, even though these lines are important in mathematics.

In this chapter we'll explore new ways of drawing curves in the plane. We'll still work within the framework of functions, as an input will still only correspond to one output. However, our new techniques of drawing curves will render the vertical line test pointless, and allow us to create important – and beautiful – new curves. Once these curves are defined, we'll apply the concepts of calculus to them, continuing to find equations of tangent lines and the areas of enclosed regions.

## 9.1 Conic Sections

The ancient Greeks recognized that interesting shapes can be formed by intersecting a plane with a *double napped* cone (i.e., two identical cones placed tip–to–tip as shown in the following figures). As these shapes are formed as sections of conics, they have earned the official name "conic sections."

The three "most interesting" conic sections are given in the top row of Figure 9.1.1. They are the parabola, the ellipse (which includes circles) and the hyperbola. In each of these cases, the plane does not intersect the tips of the cones (usually taken to be the origin).

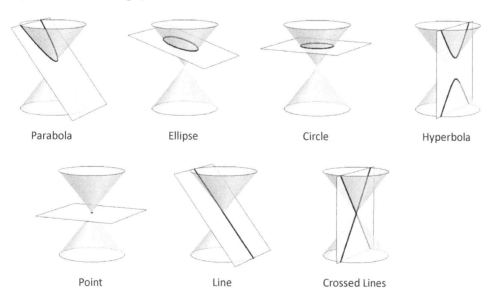

Figure 9.1.1: Conic Sections

# Chapter 9 Curves in the Plane

When the plane does contain the origin, three **degenerate** cones can be formed as shown the bottom row of Figure 9.1.1: a point, a line, and crossed lines. We focus here on the nondegenerate cases.

While the above geometric constructs define the conics in an intuitive, visual way, these constructs are not very helpful when trying to analyze the shapes algebraically or consider them as the graph of a function. It can be shown that all conics can be defined by the general second–degree equation

$$Ax^2 + Bxy + Cy^2 + Dx + Ey + F = 0.$$

While this algebraic definition has its uses, most find another geometric perspective of the conics more beneficial.

Each nondegenerate conic can be defined as the **locus**, or set, of points that satisfy a certain distance property. These distance properties can be used to generate an algebraic formula, allowing us to study each conic as the graph of a function.

## Parabolas

> **Definition 9.1.1**    **Parabola**
>
> A **parabola** is the locus of all points equidistant from a point (called a **focus**) and a line (called the **directrix**) that does not contain the focus.

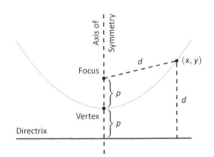

Figure 9.1.2: Illustrating the definition of the parabola and establishing an algebraic formula.

Figure 9.1.2 illustrates this definition. The point halfway between the focus and the directrix is the **vertex**. The line through the focus, perpendicular to the directrix, is the **axis of symmetry**, as the portion of the parabola on one side of this line is the mirror–image of the portion on the opposite side.

The definition leads us to an algebraic formula for the parabola. Let $P = (x, y)$ be a point on a parabola whose focus is at $F = (0, p)$ and whose directrix is at $y = -p$. (We'll assume for now that the focus lies on the y-axis; by placing the focus $p$ units above the x-axis and the directrix $p$ units below this axis, the vertex will be at $(0, 0)$.)

We use the Distance Formula to find the distance $d_1$ between $F$ and $P$:

$$d_1 = \sqrt{(x-0)^2 + (y-p)^2}.$$

The distance $d_2$ from $P$ to the directrix is more straightforward:

$$d_2 = y - (-p) = y + p.$$

---

Notes:

These two distances are equal. Setting $d_1 = d_2$, we can solve for $y$ in terms of $x$:

$$d_1 = d_2$$
$$\sqrt{x^2 + (y-p)^2} = y + p$$

Now square both sides.

$$x^2 + (y-p)^2 = (y+p)^2$$
$$x^2 + y^2 - 2yp + p^2 = y^2 + 2yp + p^2$$
$$x^2 = 4yp$$
$$y = \frac{1}{4p}x^2.$$

The geometric definition of the parabola has led us to the familiar quadratic function whose graph is a parabola with vertex at the origin. When we allow the vertex to not be at $(0,0)$, we get the following standard form of the parabola.

---

**Key Idea 9.1.1**     **General Equation of a Parabola**

1. **Vertical Axis of Symmetry:** The equation of the parabola with vertex at $(h, k)$ and directrix $y = k - p$ in standard form is

$$y = \frac{1}{4p}(x - h)^2 + k.$$

The focus is at $(h, k + p)$.

2. **Horizontal Axis of Symmetry:** The equation of the parabola with vertex at $(h, k)$ and directrix $x = h - p$ in standard form is

$$x = \frac{1}{4p}(y - k)^2 + h.$$

The focus is at $(h + p, k)$.

Note: $p$ is not necessarily a positive number.

---

**Example 9.1.1**     **Finding the equation of a parabola**

Give the equation of the parabola with focus at $(1, 2)$ and directrix at $y = 3$.

**SOLUTION**     The vertex is located halfway between the focus and directrix, so $(h, k) = (1, 2.5)$. This gives $p = -0.5$. Using Key Idea 9.1.1 we have the

Notes:

# Chapter 9  Curves in the Plane

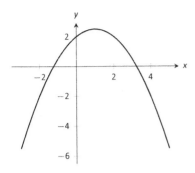

Figure 9.1.3: The parabola described in Example 9.1.1.

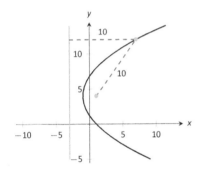

Figure 9.1.4: The parabola described in Example 9.1.2. The distances from a point on the parabola to the focus and directrix is given.

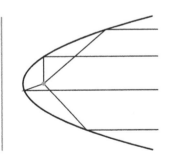

Figure 9.1.5: Illustrating the parabola's reflective property.

equation of the parabola as

$$y = \frac{1}{4(-0.5)}(x-1)^2 + 2.5 = -\frac{1}{2}(x-1)^2 + 2.5.$$

The parabola is sketched in Figure 9.1.3.

**Example 9.1.2  Finding the focus and directrix of a parabola**
Find the focus and directrix of the parabola $x = \frac{1}{8}y^2 - y + 1$. The point $(7, 12)$ lies on the graph of this parabola; verify that it is equidistant from the focus and directrix.

**Solution**  We need to put the equation of the parabola in its general form. This requires us to complete the square:

$$\begin{aligned} x &= \frac{1}{8}y^2 - y + 1 \\ &= \frac{1}{8}(y^2 - 8y + 8) \\ &= \frac{1}{8}(y^2 - 8y + 16 - 16 + 8) \\ &= \frac{1}{8}((y-4)^2 - 8) \\ &= \frac{1}{8}(y-4)^2 - 1. \end{aligned}$$

Hence the vertex is located at $(-1, 4)$. We have $\frac{1}{8} = \frac{1}{4p}$, so $p = 2$. We conclude that the focus is located at $(1, 4)$ and the directrix is $x = -3$. The parabola is graphed in Figure 9.1.4, along with its focus and directrix.

The point $(7, 12)$ lies on the graph and is $7 - (-3) = 10$ units from the directrix. The distance from $(7, 12)$ to the focus is:

$$\sqrt{(7-1)^2 + (12-4)^2} = \sqrt{100} = 10.$$

Indeed, the point on the parabola is equidistant from the focus and directrix.

## Reflective Property

One of the fascinating things about the nondegenerate conic sections is their reflective properties. Parabolas have the following reflective property:

> Any ray emanating from the focus that intersects the parabola reflects off along a line perpendicular to the directrix.

This is illustrated in Figure 9.1.5. The following theorem states this more rigorously.

Notes:

## 9.1 Conic Sections

> **Theorem 9.1.1**     **Reflective Property of the Parabola**
>
> Let $P$ be a point on a parabola. The tangent line to the parabola at $P$ makes equal angles with the following two lines:
>
> 1. The line containing $P$ and the focus $F$, and
> 2. The line perpendicular to the directrix through $P$.

Because of this reflective property, paraboloids (the 3D analogue of parabolas) make for useful flashlight reflectors as the light from the bulb, ideally located at the focus, is reflected along parallel rays. Satellite dishes also have paraboloid shapes. Signals coming from satellites effectively approach the dish along parallel rays. The dish then *focuses* these rays at the focus, where the sensor is located.

## Ellipses

> **Definition 9.1.2**     **Ellipse**
>
> An **ellipse** is the locus of all points whose sum of distances from two fixed points, each a **focus** of the ellipse, is constant.

An easy way to visualize this construction of an ellipse is to pin both ends of a string to a board. The pins become the foci. Holding a pencil tight against the string places the pencil on the ellipse; the sum of distances from the pencil to the pins is constant: the length of the string. See Figure 9.1.6.

We can again find an algebraic equation for an ellipse using this geometric definition. Let the foci be located along the $x$-axis, $c$ units from the origin. Let these foci be labeled as $F_1 = (-c, 0)$ and $F_2 = (c, 0)$. Let $P = (x, y)$ be a point on the ellipse. The sum of distances from $F_1$ to $P$ ($d_1$) and from $F_2$ to $P$ ($d_2$) is a constant $d$. That is, $d_1 + d_2 = d$. Using the Distance Formula, we have

$$\sqrt{(x+c)^2 + y^2} + \sqrt{(x-c)^2 + y^2} = d.$$

Using a fair amount of algebra can produce the following equation of an ellipse (note that the equation is an implicitly defined function; it has to be, as an ellipse fails the Vertical Line Test):

$$\frac{x^2}{\left(\frac{d}{2}\right)^2} + \frac{y^2}{\left(\frac{d}{2}\right)^2 - c^2} = 1.$$

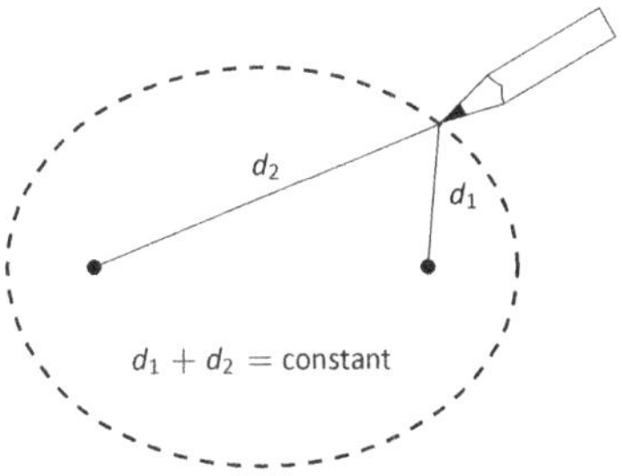

Figure 9.1.6: Illustrating the construction of an ellipse with pins, pencil and string.

Notes:

# Chapter 9 Curves in the Plane

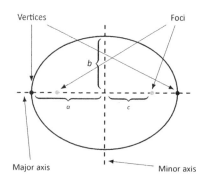

Figure 9.1.7: Labeling the significant features of an ellipse.

This is not particularly illuminating, but by making the substitution $a = d/2$ and $b = \sqrt{a^2 - c^2}$, we can rewrite the above equation as

$$\frac{x^2}{a^2} + \frac{y^2}{b^2} = 1.$$

This choice of $a$ and $b$ is not without reason; as shown in Figure 9.1.7, the values of $a$ and $b$ have geometric meaning in the graph of the ellipse.

In general, the two foci of an ellipse lie on the **major axis** of the ellipse, and the midpoint of the segment joining the two foci is the **center**. The major axis intersects the ellipse at two points, each of which is a **vertex**. The line segment through the center and perpendicular to the major axis is the **minor axis**. The "constant sum of distances" that defines the ellipse is the length of the major axis, i.e., $2a$.

Allowing for the shifting of the ellipse gives the following standard equations.

---

**Key Idea 9.1.2**  **Standard Equation of the Ellipse**

The equation of an ellipse centered at $(h, k)$ with major axis of length $2a$ and minor axis of length $2b$ in standard form is:

1. **Horizontal major axis:** $\dfrac{(x-h)^2}{a^2} + \dfrac{(y-k)^2}{b^2} = 1.$

2. **Vertical major axis:** $\dfrac{(x-h)^2}{b^2} + \dfrac{(y-k)^2}{a^2} = 1.$

The foci lie along the major axis, $c$ units from the center, where $c^2 = a^2 - b^2$.

---

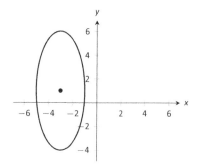

Figure 9.1.8: The ellipse used in Example 9.1.3.

**Example 9.1.3**  **Finding the equation of an ellipse**

Find the general equation of the ellipse graphed in Figure 9.1.8.

**SOLUTION**  The center is located at $(-3, 1)$. The distance from the center to a vertex is 5 units, hence $a = 5$. The minor axis seems to have length 4, so $b = 2$. Thus the equation of the ellipse is

$$\frac{(x+3)^2}{4} + \frac{(y-1)^2}{25} = 1.$$

**Example 9.1.4**  **Graphing an ellipse**

Graph the ellipse defined by $4x^2 + 9y^2 - 8x - 36y = -4$.

---

Notes:

**SOLUTION** It is simple to graph an ellipse once it is in standard form. In order to put the given equation in standard form, we must complete the square with both the x and y terms. We first rewrite the equation by regrouping:

$$4x^2 + 9y^2 - 8x - 36y = -4 \quad \Rightarrow \quad (4x^2 - 8x) + (9y^2 - 36y) = -4.$$

Now we complete the squares.

$$(4x^2 - 8x) + (9y^2 - 36y) = -4$$
$$4(x^2 - 2x) + 9(y^2 - 4y) = -4$$
$$4(x^2 - 2x + 1 - 1) + 9(y^2 - 4y + 4 - 4) = -4$$
$$4((x-1)^2 - 1) + 9((y-2)^2 - 4) = -4$$
$$4(x-1)^2 - 4 + 9(y-2)^2 - 36 = -4$$
$$4(x-1)^2 + 9(y-2)^2 = 36$$
$$\frac{(x-1)^2}{9} + \frac{(y-2)^2}{4} = 1.$$

We see the center of the ellipse is at $(1, 2)$. We have $a = 3$ and $b = 2$; the major axis is horizontal, so the vertices are located at $(-2, 2)$ and $(4, 2)$. We find $c = \sqrt{9-4} = \sqrt{5} \approx 2.24$. The foci are located along the major axis, approximately 2.24 units from the center, at $(1 \pm 2.24, 2)$. This is all graphed in Figure 9.1.9.

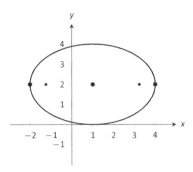

Figure 9.1.9: Graphing the ellipse in Example 9.1.4.

## Eccentricity

When $a = b$, we have a circle. The general equation becomes

$$\frac{(x-h)^2}{a^2} + \frac{(y-k)^2}{a^2} = 1 \quad \Rightarrow \quad (x-h)^2 + (y-k)^2 = a^2,$$

the familiar equation of the circle centered at $(h, k)$ with radius $a$. Since $a = b$, $c = \sqrt{a^2 - b^2} = 0$. The circle has "two" foci, but they lie on the same point, the center of the circle.

Consider Figure 9.1.10, where several ellipses are graphed with $a = 1$. In (a), we have $c = 0$ and the ellipse is a circle. As $c$ grows, the resulting ellipses look less and less circular. A measure of this "noncircularness" is *eccentricity*.

> **Definition 9.1.3**     **Eccentricity of an Ellipse**
>
> The eccentricity $e$ of an ellipse is $e = \dfrac{c}{a}$.

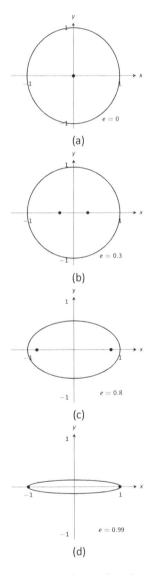

Figure 9.1.10: Understanding the eccentricity of an ellipse.

The eccentricity of a circle is 0; that is, a circle has no "noncircularness." As $c$ approaches $a$, $e$ approaches 1, giving rise to a very noncircular ellipse, as seen in Figure 9.1.10 (d).

It was long assumed that planets had circular orbits. This is known to be incorrect; the orbits are elliptical. Earth has an eccentricity of 0.0167 – it has a nearly circular orbit. Mercury's orbit is the most eccentric, with $e = 0.2056$. (Pluto's eccentricity is greater, at $e = 0.248$, the greatest of all the currently known dwarf planets.) The planet with the most circular orbit is Venus, with $e = 0.0068$. The Earth's moon has an eccentricity of $e = 0.0549$, also very circular.

## Reflective Property

The ellipse also possesses an interesting reflective property. Any ray emanating from one focus of an ellipse reflects off the ellipse along a line through the other focus, as illustrated in Figure 9.1.11. This property is given formally in the following theorem.

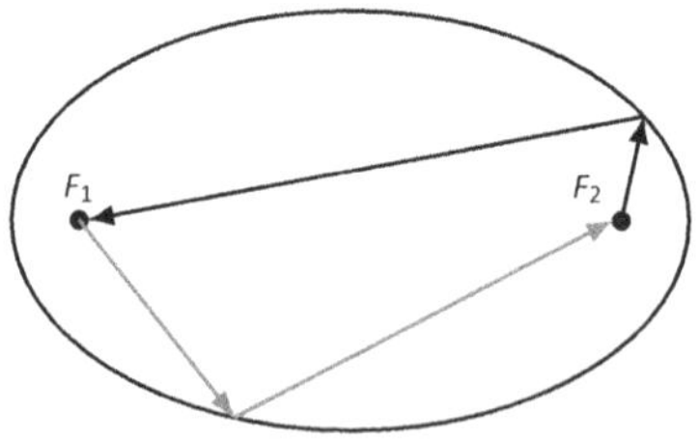

Figure 9.1.11: Illustrating the reflective property of an ellipse.

> **Theorem 9.1.2**     **Reflective Property of an Ellipse**
>
> Let $P$ be a point on a ellipse with foci $F_1$ and $F_2$. The tangent line to the ellipse at $P$ makes equal angles with the following two lines:
>
> 1. The line through $F_1$ and $P$, and
> 2. The line through $F_2$ and $P$.

This reflective property is useful in optics and is the basis of the phenomena experienced in whispering halls.

## Hyperbolas

The definition of a hyperbola is very similar to the definition of an ellipse; we essentially just change the word "sum" to "difference."

> **Definition 9.1.4**     **Hyperbola**
>
> A **hyperbola** is the locus of all points where the absolute value of difference of distances from two fixed points, each a focus of the hyperbola, is constant.

Notes:

We do not have a convenient way of visualizing the construction of a hyperbola as we did for the ellipse. The geometric definition does allow us to find an algebraic expression that describes it. It will be useful to define some terms first.

The two foci lie on the **transverse axis** of the hyperbola; the midpoint of the line segment joining the foci is the **center** of the hyperbola. The transverse axis intersects the hyperbola at two points, each a **vertex** of the hyperbola. The line through the center and perpendicular to the transverse axis is the **conjugate axis.** This is illustrated in Figure 9.1.12. It is easy to show that the constant difference of distances used in the definition of the hyperbola is the distance between the vertices, i.e., $2a$.

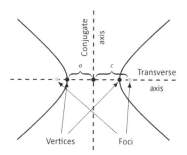

Figure 9.1.12: Labeling the significant features of a hyperbola.

---

**Key Idea 9.1.3    Standard Equation of a Hyperbola**

The equation of a hyperbola centered at $(h, k)$ in standard form is:

1. **Horizontal Transverse Axis:**  $\dfrac{(x-h)^2}{a^2} - \dfrac{(y-k)^2}{b^2} = 1.$

2. **Vertical Transverse Axis:**  $\dfrac{(y-k)^2}{a^2} - \dfrac{(x-h)^2}{b^2} = 1.$

The vertices are located $a$ units from the center and the foci are located $c$ units from the center, where $c^2 = a^2 + b^2$.

---

## Graphing Hyperbolas

Consider the hyperbola $\frac{x^2}{9} - \frac{y^2}{1} = 1$. Solving for $y$, we find $y = \pm\sqrt{x^2/9 - 1}$. As $x$ grows large, the "$-1$" part of the equation for $y$ becomes less significant and $y \approx \pm\sqrt{x^2/9} = \pm x/3$. That is, as $x$ gets large, the graph of the hyperbola looks very much like the lines $y = \pm x/3$. These lines are asymptotes of the hyperbola, as shown in Figure 9.1.13.

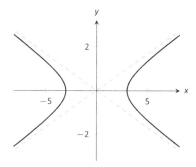

Figure 9.1.13: Graphing the hyperbola $\frac{x^2}{9} - \frac{y^2}{1} = 1$ along with its asymptotes, $y = \pm x/3$.

This is a valuable tool in sketching. Given the equation of a hyperbola in general form, draw a rectangle centered at $(h, k)$ with sides of length $2a$ parallel to the transverse axis and sides of length $2b$ parallel to the conjugate axis. (See Figure 9.1.14 for an example with a horizontal transverse axis.) The diagonals of the rectangle lie on the asymptotes.

These lines pass through $(h, k)$. When the transverse axis is horizontal, the slopes are $\pm b/a$; when the transverse axis is vertical, their slopes are $\pm a/b$. This gives equations:

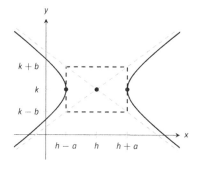

Figure 9.1.14: Using the asymptotes of a hyperbola as a graphing aid.

Notes:

Chapter 9  Curves in the Plane

|Horizontal Transverse Axis|Vertical Transverse Axis|
|---|---|
|$y = \pm\dfrac{b}{a}(x-h) + k$|$y = \pm\dfrac{a}{b}(x-h) + k.$|

**Example 9.1.5   Graphing a hyperbola**

Sketch the hyperbola given by $\dfrac{(y-2)^2}{25} - \dfrac{(x-1)^2}{4} = 1$.

**SOLUTION**   The hyperbola is centered at $(1,2)$; $a = 5$ and $b = 2$. In Figure 9.1.15 we draw the prescribed rectangle centered at $(1,2)$ along with the asymptotes defined by its diagonals. The hyperbola has a vertical transverse axis, so the vertices are located at $(1,7)$ and $(1,-3)$. This is enough to make a good sketch.

We also find the location of the foci: as $c^2 = a^2 + b^2$, we have $c = \sqrt{29} \approx 5.4$. Thus the foci are located at $(1, 2 \pm 5.4)$ as shown in the figure.

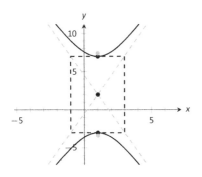

Figure 9.1.15: Graphing the hyperbola in Example 9.1.5.

**Example 9.1.6   Graphing a hyperbola**

Sketch the hyperbola given by $9x^2 - y^2 + 2y = 10$.

**SOLUTION**   We must complete the square to put the equation in general form. (We recognize this as a hyperbola since it is a general quadratic equation and the $x^2$ and $y^2$ terms have opposite signs.)

$$
\begin{aligned}
9x^2 - y^2 + 2y &= 10 \\
9x^2 - (y^2 - 2y) &= 10 \\
9x^2 - (y^2 - 2y + 1 - 1) &= 10 \\
9x^2 - \big((y-1)^2 - 1\big) &= 10 \\
9x^2 - (y-1)^2 &= 9 \\
x^2 - \dfrac{(y-1)^2}{9} &= 1
\end{aligned}
$$

We see the hyperbola is centered at $(0,1)$, with a horizontal transverse axis, where $a = 1$ and $b = 3$. The appropriate rectangle is sketched in Figure 9.1.16 along with the asymptotes of the hyperbola. The vertices are located at $(\pm 1, 1)$. We have $c = \sqrt{10} \approx 3.2$, so the foci are located at $(\pm 3.2, 1)$ as shown in the figure.

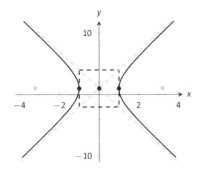

Figure 9.1.16: Graphing the hyperbola in Example 9.1.6.

Notes:

## Eccentricity

> **Definition 9.1.5**     **Eccentricity of a Hyperbola**
>
> The eccentricity of a hyperbola is $e = \dfrac{c}{a}$.

Note that this is the definition of eccentricity as used for the ellipse. When $c$ is close in value to $a$ (i.e., $e \approx 1$), the hyperbola is very narrow (looking almost like crossed lines). Figure 9.1.17 shows hyperbolas centered at the origin with $a = 1$. The graph in (a) has $c = 1.05$, giving an eccentricity of $e = 1.05$, which is close to 1. As $c$ grows larger, the hyperbola widens and begins to look like parallel lines, as shown in part (d) of the figure.

## Reflective Property

Hyperbolas share a similar reflective property with ellipses. However, in the case of a hyperbola, a ray emanating from a focus that intersects the hyperbola reflects along a line containing the other focus, but moving *away* from that focus. This is illustrated in Figure 9.1.19 (on the next page). Hyperbolic mirrors are commonly used in telescopes because of this reflective property. It is stated formally in the following theorem.

> **Theorem 9.1.3**     **Reflective Property of Hyperbolas**
>
> Let $P$ be a point on a hyperbola with foci $F_1$ and $F_2$. The tangent line to the hyperbola at $P$ makes equal angles with the following two lines:
>
> 1. The line through $F_1$ and $P$, and
> 2. The line through $F_2$ and $P$.

## Location Determination

Determining the location of a known event has many practical uses (locating the epicenter of an earthquake, an airplane crash site, the position of the person speaking in a large room, etc.).

To determine the location of an earthquake's epicenter, seismologists use *trilateration* (not to be confused with *triangulation*). A seismograph allows one

Notes:

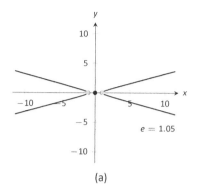

(a)

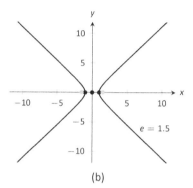

(b)

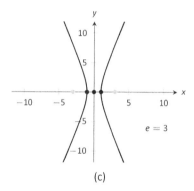

(c)

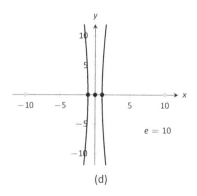

(d)

Figure 9.1.17: Understanding the eccentricity of a hyperbola.

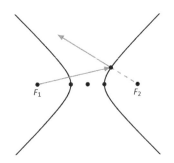

Figure 9.1.19: Illustrating the reflective property of a hyperbola.

to determine how far away the epicenter was; using three separate readings, the location of the epicenter can be approximated.

A key to this method is knowing distances. What if this information is not available? Consider three microphones at positions *A*, *B* and *C* which all record a noise (a person's voice, an explosion, etc.) created at unknown location *D*. The microphone does not "know" when the sound was *created*, only when the sound was *detected*. How can the location be determined in such a situation?

If each location has a clock set to the same time, hyperbolas can be used to determine the location. Suppose the microphone at position *A* records the sound at exactly 12:00, location *B* records the time exactly 1 second later, and location *C* records the noise exactly 2 seconds after that. We are interested in the *difference* of times. Since the speed of sound is approximately 340 m/s, we can conclude quickly that the sound was created 340 meters closer to position *A* than position *B*. If *A* and *B* are a known distance apart (as shown in Figure 9.1.18 (a)), then we can determine a hyperbola on which *D* must lie.

The "difference of distances" is 340; this is also the distance between vertices of the hyperbola. So we know $2a = 340$. Positions *A* and *B* lie on the foci, so $2c = 1000$. From this we can find $b \approx 470$ and can sketch the hyperbola, given in part (b) of the figure. We only care about the side closest to *A*. (Why?)

We can also find the hyperbola defined by positions *B* and *C*. In this case, $2a = 680$ as the sound traveled an extra 2 seconds to get to *C*. We still have $2c = 1000$, centering this hyperbola at $(-500, 500)$. We find $b \approx 367$. This hyperbola is sketched in part (c) of the figure. The intersection point of the two graphs is the location of the sound, at approximately $(188, -222.5)$.

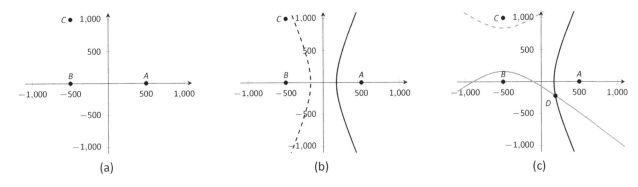

Figure 9.1.18: Using hyperbolas in location detection.

This chapter explores curves in the plane, in particular curves that cannot be described by functions of the form $y = f(x)$. In this section, we learned of ellipses and hyperbolas that are defined implicitly, not explicitly. In the following sections, we will learn completely new ways of describing curves in the plane, using *parametric equations* and *polar coordinates*, then study these curves using calculus techniques.

Notes:

# Exercises 9.1

## Terms and Concepts

1. What is the difference between degenerate and nondegenerate conics?

2. Use your own words to explain what the eccentricity of an ellipse measures.

3. What has the largest eccentricity: an ellipse or a hyperbola?

4. Explain why the following is true: "If the coefficient of the $x^2$ term in the equation of an ellipse in standard form is smaller than the coefficient of the $y^2$ term, then the ellipse has a horizontal major axis."

5. Explain how one can quickly look at the equation of a hyperbola in standard form and determine whether the transverse axis is horizontal or vertical.

6. Fill in the blank: It can be said that ellipses and hyperbolas share the *same* reflective property: "A ray emanating from one focus will reflect off the conic along a _____ that contains the other focus."

## Problems

In Exercises 7 – 14, find the equation of the parabola defined by the given information. Sketch the parabola.

7. Focus: $(3, 2)$; directrix: $y = 1$

8. Focus: $(-1, -4)$; directrix: $y = 2$

9. Focus: $(1, 5)$; directrix: $x = 3$

10. Focus: $(1/4, 0)$; directrix: $x = -1/4$

11. Focus: $(1, 1)$; vertex: $(1, 2)$

12. Focus: $(-3, 0)$; vertex: $(0, 0)$

13. Vertex: $(0, 0)$; directrix: $y = -1/16$

14. Vertex: $(2, 3)$; directrix: $x = 4$

In Exercises 15 – 16, the equation of a parabola and a point on its graph are given. Find the focus and directrix of the parabola, and verify that the given point is equidistant from the focus and directrix.

15. $y = \frac{1}{4}x^2$, $P = (2, 1)$

16. $x = \frac{1}{8}(y - 2)^2 + 3$, $P = (11, 10)$

In Exercises 17 – 18, sketch the ellipse defined by the given equation. Label the center, foci and vertices.

17. $\dfrac{(x-1)^2}{3} + \dfrac{(y-2)^2}{5} = 1$

18. $\dfrac{1}{25}x^2 + \dfrac{1}{9}(y+3)^2 = 1$

In Exercises 19 – 20, find the equation of the ellipse shown in the graph. Give the location of the foci and the eccentricity of the ellipse.

19.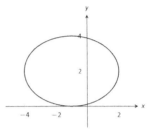

20.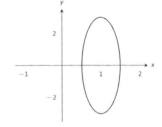

In Exercises 21 – 24, find the equation of the ellipse defined by the given information. Sketch the elllipse.

21. Foci: $(\pm 2, 0)$; vertices: $(\pm 3, 0)$

22. Foci: $(-1, 3)$ and $(5, 3)$; vertices: $(-3, 3)$ and $(7, 3)$

23. Foci: $(2, \pm 2)$; vertices: $(2, \pm 7)$

24. Focus: $(-1, 5)$; vertex: $(-1, -4)$; center: $(-1, 1)$

In Exercises 25 – 28, write the equation of the given ellipse in standard form.

25. $x^2 - 2x + 2y^2 - 8y = -7$

26. $5x^2 + 3y^2 = 15$

27. $3x^2 + 2y^2 - 12y + 6 = 0$

28. $x^2 + y^2 - 4x - 4y + 4 = 0$

**In Exercises 29 – 32, find the equation of the hyperbola shown in the graph.**

29.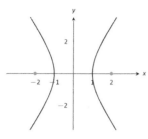

30.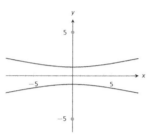

31.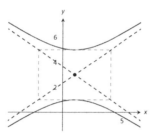

32.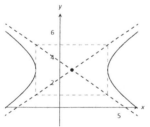

**In Exercises 33 – 34, sketch the hyperbola defined by the given equation. Label the center and foci.**

33. $\dfrac{(x-1)^2}{16} - \dfrac{(y+2)^2}{9} = 1$

34. $(y-4)^2 - \dfrac{(x+1)^2}{25} = 1$

**In Exercises 35 – 38, find the equation of the hyperbola defined by the given information. Sketch the hyperbola.**

35. Foci: $(\pm 3, 0)$; vertices: $(\pm 2, 0)$

36. Foci: $(0, \pm 3)$; vertices: $(0, \pm 2)$

37. Foci: $(-2, 3)$ and $(8, 3)$; vertices: $(-1, 3)$ and $(7, 3)$

38. Foci: $(3, -2)$ and $(3, 8)$; vertices: $(3, 0)$ and $(3, 6)$

**In Exercises 39 – 42, write the equation of the hyperbola in standard form.**

39. $3x^2 - 4y^2 = 12$

40. $3x^2 - y^2 + 2y = 10$

41. $x^2 - 10y^2 + 40y = 30$

42. $(4y - x)(4y + x) = 4$

43. Consider the ellipse given by $\dfrac{(x-1)^2}{4} + \dfrac{(y-3)^2}{12} = 1$.

    (a) Verify that the foci are located at $(1, 3 \pm 2\sqrt{2})$.

    (b) The points $P_1 = (2, 6)$ and $P_2 = (1 + \sqrt{2}, 3 + \sqrt{6}) \approx (2.414, 5.449)$ lie on the ellipse. Verify that the sum of distances from each point to the foci is the same.

44. Johannes Kepler discovered that the planets of our solar system have elliptical orbits with the Sun at one focus. The Earth's elliptical orbit is used as a standard unit of distance; the distance from the center of Earth's elliptical orbit to one vertex is 1 Astronomical Unit, or A.U.

    The following table gives information about the orbits of three planets.

    |  | Distance from center to vertex | eccentricity |
    |---|---|---|
    | Mercury | 0.387 A.U. | 0.2056 |
    | Earth | 1 A.U. | 0.0167 |
    | Mars | 1.524 A.U. | 0.0934 |

    (a) In an ellipse, knowing $c^2 = a^2 - b^2$ and $e = c/a$ allows us to find $b$ in terms of $a$ and $e$. Show $b = a\sqrt{1 - e^2}$.

    (b) For each planet, find equations of their elliptical orbit of the form $\dfrac{x^2}{a^2} + \dfrac{y^2}{b^2} = 1$. (This places the center at $(0, 0)$, but the Sun is in a different location for each planet.)

    (c) Shift the equations so that the Sun lies at the origin. Plot the three elliptical orbits.

45. A loud sound is recorded at three stations that lie on a line as shown in the figure below. Station A recorded the sound 1 second after Station B, and Station C recorded the sound 3 seconds after B. Using the speed of sound as 340m/s, determine the location of the sound's origination.

    A  1000m  B        2000m        C

## 9.2 Parametric Equations

We are familiar with sketching shapes, such as parabolas, by following this basic procedure:

Choose $x$ → Use a function $f$ to find $y$ ($y = f(x)$) → Plot point $(x, y)$

The **rectangular equation** $y = f(x)$ works well for some shapes like a parabola with a vertical axis of symmetry, but in the previous section we encountered several shapes that could not be sketched in this manner. (To plot an ellipse using the above procedure, we need to plot the "top" and "bottom" separately.)

In this section we introduce a new sketching procedure:

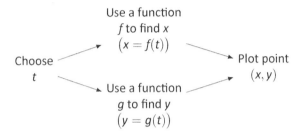

Here, $x$ and $y$ are found separately but then plotted together. This leads us to a definition.

---

**Definition 9.2.1**  **Parametric Equations and Curves**

Let $f$ and $g$ be continuous functions on an interval $I$. The set of all points $(x, y) = \big(f(t), g(t)\big)$ in the Cartesian plane, as $t$ varies over $I$, is the **graph** of the **parametric equations** $x = f(t)$ and $y = g(t)$, where $t$ is the **parameter**. A **curve** is a graph along with the parametric equations that define it.

---

This is a formal definition of the word *curve*. When a curve lies in a plane (such as the Cartesian plane), it is often referred to as a **plane curve**. Examples will help us understand the concepts introduced in the definition.

**Example 9.2.1**  **Plotting parametric functions**

Plot the graph of the parametric equations $x = t^2$, $y = t + 1$ for $t$ in $[-2, 2]$.

---

Notes:

# Chapter 9  Curves in the Plane

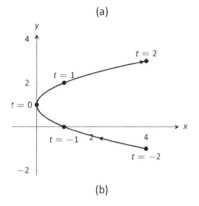

| t | x | y |
|---|---|---|
| −2 | 4 | −1 |
| −1 | 1 | 0 |
| 0 | 0 | 1 |
| 1 | 1 | 2 |
| 2 | 4 | 3 |

(a)

(b)

Figure 9.2.1: A table of values of the parametric equations in Example 9.2.1 along with a sketch of their graph.

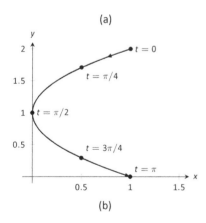

| t | x | y |
|---|---|---|
| 0 | 1 | 2 |
| $\pi/4$ | 1/2 | $1 + \sqrt{2}/2$ |
| $\pi/2$ | 0 | 1 |
| $3\pi/4$ | 1/2 | $1 - \sqrt{2}/2$ |
| $\pi$ | 1 | 0 |

(a)

(b)

Figure 9.2.2: A table of values of the parametric equations in Example 9.2.2 along with a sketch of their graph.

**SOLUTION** We plot the graphs of parametric equations in much the same manner as we plotted graphs of functions like $y = f(x)$: we make a table of values, plot points, then connect these points with a "reasonable" looking curve. Figure 9.2.1(a) shows such a table of values; note how we have 3 columns.

The points $(x, y)$ from the table are plotted in Figure 9.2.1(b). The points have been connected with a smooth curve. Each point has been labeled with its corresponding $t$-value. These values, along with the two arrows along the curve, are used to indicate the **orientation** of the graph. This information helps us determine the direction in which the graph is "moving."

We often use the letter $t$ as the parameter as we often regard $t$ as representing *time*. Certainly there are many contexts in which the parameter is not time, but it can be helpful to think in terms of time as one makes sense of parametric plots and their orientation (for instance, "At time $t = 0$ the position is $(1, 2)$ and at time $t = 3$ the position is $(5, 1)$.").

**Example 9.2.2  Plotting parametric functions**

Sketch the graph of the parametric equations $x = \cos^2 t$, $y = \cos t + 1$ for $t$ in $[0, \pi]$.

**SOLUTION** We again start by making a table of values in Figure 9.2.2(a), then plot the points $(x, y)$ on the Cartesian plane in Figure 9.2.2(b).

It is not difficult to show that the curves in Examples 9.2.1 and 9.2.2 are portions of the same parabola. While the *parabola* is the same, the *curves* are different. In Example 9.2.1; if we let $t$ vary over all real numbers, we'd obtain the entire parabola. In this example, letting $t$ vary over all real numbers would still produce the same graph; this portion of the parabola would be traced, and re-traced, infinitely many times. The orientation shown in Figure 9.2.2 shows the orientation on $[0, \pi]$, but this orientation is reversed on $[\pi, 2\pi]$.

These examples begin to illustrate the powerful nature of parametric equations. Their graphs are far more diverse than the graphs of functions produced by "$y = f(x)$" functions.

**Technology Note:** Most graphing utilities can graph functions given in parametric form. Often the word "parametric" is abbreviated as "PAR" or "PARAM" in the options. The user usually needs to determine the graphing window (i.e, the minimum and maximum $x$- and $y$-values), along with the values of $t$ that are to be plotted. The user is often prompted to give a $t$ minimum, a $t$ maximum, and a "$t$-step" or "$\Delta t$." Graphing utilities effectively plot parametric functions just as we've shown here: they plots lots of points. A smaller $t$-step plots more points, making for a smoother graph (but may take longer). In Figure 9.2.1, the $t$-step is

Notes:

1; in Figure 9.2.2, the *t*-step is $\pi/4$.

One nice feature of parametric equations is that their graphs are easy to shift. While this is not too difficult in the "$y = f(x)$" context, the resulting function can look rather messy. (Plus, to shift to the right by two, we replace *x* with $x - 2$, which is counter–intuitive.) The following example demonstrates this.

**Example 9.2.3**    **Shifting the graph of parametric functions**
Sketch the graph of the parametric equations $x = t^2 + t$, $y = t^2 - t$. Find new parametric equations that shift this graph to the right 3 places and down 2.

**Solution**    The graph of the parametric equations is given in Figure 9.2.3 (a). It is a parabola with a axis of symmetry along the line $y = x$; the vertex is at $(0, 0)$.

In order to shift the graph to the right 3 units, we need to increase the *x*-value by 3 for every point. The straightforward way to accomplish this is simply to add 3 to the function defining *x*: $x = t^2 + t + 3$. To shift the graph down by 2 units, we wish to decrease each *y*-value by 2, so we subtract 2 from the function defining *y*: $y = t^2 - t - 2$. Thus our parametric equations for the shifted graph are $x = t^2 + t + 3$, $y = t^2 - t - 2$. This is graphed in Figure 9.2.3 (b). Notice how the vertex is now at $(3, -2)$.

Because the *x*- and *y*-values of a graph are determined independently, the graphs of parametric functions often possess features not seen on "$y = f(x)$" type graphs. The next example demonstrates how such graphs can arrive at the same point more than once.

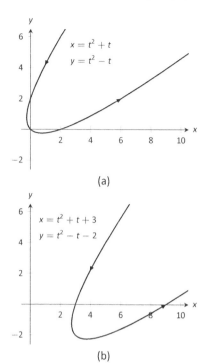

Figure 9.2.3: Illustrating how to shift graphs in Example 9.2.3.

**Example 9.2.4**    **Graphs that cross themselves**
Plot the parametric functions $x = t^3 - 5t^2 + 3t + 11$ and $y = t^2 - 2t + 3$ and determine the *t*-values where the graph crosses itself.

**Solution**    Using the methods developed in this section, we again plot points and graph the parametric equations as shown in Figure 9.2.4. It appears that the graph crosses itself at the point $(2, 6)$, but we'll need to analytically determine this.

We are looking for two different values, say, *s* and *t*, where $x(s) = x(t)$ and $y(s) = y(t)$. That is, the *x*-values are the same precisely when the *y*-values are the same. This gives us a system of 2 equations with 2 unknowns:

$$s^3 - 5s^2 + 3s + 11 = t^3 - 5t^2 + 3t + 11$$
$$s^2 - 2s + 3 = t^2 - 2t + 3$$

Solving this system is not trivial but involves only algebra. Using the quadratic

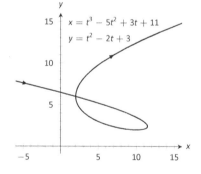

Figure 9.2.4: A graph of the parametric equations from Example 9.2.4.

Notes:

formula, one can solve for $t$ in the second equation and find that $t = 1 \pm \sqrt{s^2 - 2s + 1}$. This can be substituted into the first equation, revealing that the graph crosses itself at $t = -1$ and $t = 3$. We confirm our result by computing $x(-1) = x(3) = 2$ and $y(-1) = y(3) = 6$.

## Converting between rectangular and parametric equations

It is sometimes useful to rewrite equations in rectangular form (i.e., $y = f(x)$) into parametric form, and vice–versa. Converting from rectangular to parametric can be very simple: given $y = f(x)$, the parametric equations $x = t$, $y = f(t)$ produce the same graph. As an example, given $y = x^2$, the parametric equations $x = t$, $y = t^2$ produce the familiar parabola. However, other parametrizations can be used. The following example demonstrates one possible alternative.

**Example 9.2.5    Converting from rectangular to parametric**
Consider $y = x^2$. Find parametric equations $x = f(t)$, $y = g(t)$ for the parabola where $t = \frac{dy}{dx}$. That is, $t = a$ corresponds to the point on the graph whose tangent line has slope $a$.

**SOLUTION**    We start by computing $\frac{dy}{dx}$: $y' = 2x$. Thus we set $t = 2x$. We can solve for $x$ and find $x = t/2$. Knowing that $y = x^2$, we have $y = t^2/4$. Thus parametric equations for the parabola $y = x^2$ are

$$x = t/2 \quad y = t^2/4.$$

To find the point where the tangent line has a slope of $-2$, we set $t = -2$. This gives the point $(-1, 1)$. We can verify that the slope of the line tangent to the curve at this point indeed has a slope of $-2$.

We sometimes choose the parameter to accurately model physical behavior.

**Example 9.2.6    Converting from rectangular to parametric**
An object is fired from a height of 0ft and lands 6 seconds later, 192ft away. Assuming ideal projectile motion, the height, in feet, of the object can be described by $h(x) = -x^2/64 + 3x$, where $x$ is the distance in feet from the initial location. (Thus $h(0) = h(192) = 0$ft.) Find parametric equations $x = f(t)$, $y = g(t)$ for the path of the projectile where $x$ is the horizontal distance the object has traveled at time $t$ (in seconds) and $y$ is the height at time $t$.

**SOLUTION**    Physics tells us that the horizontal motion of the projectile is linear; that is, the horizontal speed of the projectile is constant. Since the object travels 192ft in 6s, we deduce that the object is moving horizontally at a rate of 32ft/s, giving the equation $x = 32t$. As $y = -x^2/64 + 3x$, we find

Notes:

$y = -16t^2 + 96t$. We can quickly verify that $y'' = -32\text{ft/s}^2$, the acceleration due to gravity, and that the projectile reaches its maximum at $t = 3$, halfway along its path.

These parametric equations make certain determinations about the object's location easy: 2 seconds into the flight the object is at the point $(x(2), y(2)) = (64, 128)$. That is, it has traveled horizontally 64ft and is at a height of 128ft, as shown in Figure 9.2.5.

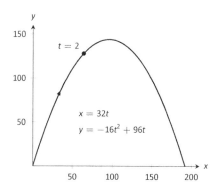

Figure 9.2.5: Graphing projectile motion in Example 9.2.6.

It is sometimes necessary to convert given parametric equations into rectangular form. This can be decidedly more difficult, as some "simple" looking parametric equations can have very "complicated" rectangular equations. This conversion is often referred to as "eliminating the parameter," as we are looking for a relationship between $x$ and $y$ that does not involve the parameter $t$.

**Example 9.2.7 Eliminating the parameter**
Find a rectangular equation for the curve described by

$$x = \frac{1}{t^2 + 1} \quad \text{and} \quad y = \frac{t^2}{t^2 + 1}.$$

**SOLUTION** There is not a set way to eliminate a parameter. One method is to solve for $t$ in one equation and then substitute that value in the second. We use that technique here, then show a second, simpler method.

Starting with $x = 1/(t^2 + 1)$, solve for $t$: $t = \pm\sqrt{1/x - 1}$. Substitute this value for $t$ in the equation for $y$:

$$y = \frac{t^2}{t^2 + 1}$$
$$= \frac{1/x - 1}{1/x - 1 + 1}$$
$$= \frac{1/x - 1}{1/x}$$
$$= \left(\frac{1}{x} - 1\right) \cdot x$$
$$= 1 - x.$$

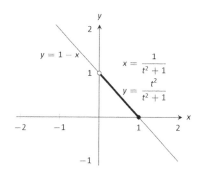

Figure 9.2.6: Graphing parametric and rectangular equations for a graph in Example 9.2.7.

Thus $y = 1 - x$. One may have recognized this earlier by manipulating the equation for $y$:

$$y = \frac{t^2}{t^2 + 1} = 1 - \frac{1}{t^2 + 1} = 1 - x.$$

This is a shortcut that is very specific to this problem; sometimes shortcuts exist and are worth looking for.

We should be careful to limit the domain of the function $y = 1 - x$. The parametric equations limit $x$ to values in $(0, 1]$, thus to produce the same graph we should limit the domain of $y = 1 - x$ to the same.

The graphs of these functions is given in Figure 9.2.6. The portion of the graph defined by the parametric equations is given in a thick line; the graph defined by $y = 1 - x$ with unrestricted domain is given in a thin line.

**Example 9.2.8    Eliminating the parameter**
Eliminate the parameter in $x = 4\cos t + 3$, $y = 2\sin t + 1$

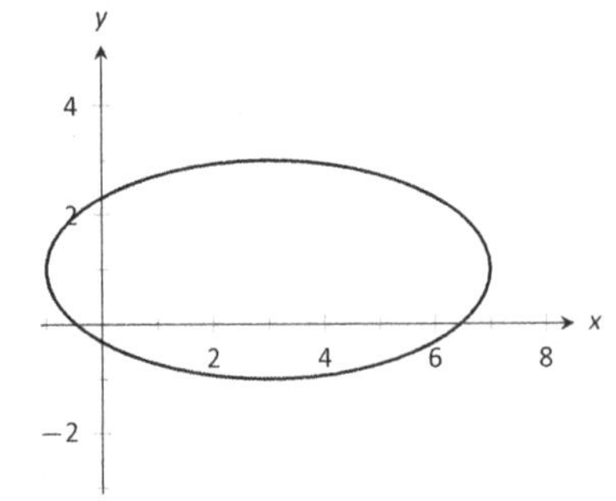

**SOLUTION**    We should not try to solve for $t$ in this situation as the resulting algebra/trig would be messy. Rather, we solve for $\cos t$ and $\sin t$ in each equation, respectively. This gives

$$\cos t = \frac{x-3}{4} \quad \text{and} \quad \sin t = \frac{y-1}{2}.$$

The Pythagorean Theorem gives $\cos^2 t + \sin^2 t = 1$, so:

$$\cos^2 t + \sin^2 t = 1$$
$$\left(\frac{x-3}{4}\right)^2 + \left(\frac{y-1}{2}\right)^2 = 1$$
$$\frac{(x-3)^2}{16} + \frac{(y-1)^2}{4} = 1$$

This final equation should look familiar — it is the equation of an ellipse! Figure 9.2.7 plots the parametric equations, demonstrating that the graph is indeed of an ellipse with a horizontal major axis and center at $(3, 1)$.

The Pythagorean Theorem can also be used to identify parametric equations for hyperbolas. We give the parametric equations for ellipses and hyperbolas in the following Key Idea.

Figure 9.2.7: Graphing the parametric equations $x = 4\cos t + 3$, $y = 2\sin t + 1$ in Example 9.2.8.

Notes:

## 9.2 Parametric Equations

> **Key Idea 9.2.1**      **Parametric Equations of Ellipses and Hyperbolas**
>
> - The parametric equations
>
>   $$x = a\cos t + h, \quad y = b\sin t + k$$
>
>   define an ellipse with horizontal axis of length $2a$ and vertical axis of length $2b$, centered at $(h, k)$.
>
> - The parametric equations
>
>   $$x = a\tan t + h, \quad y = \pm b\sec t + k$$
>
>   define a hyperbola with vertical transverse axis centered at $(h, k)$, and
>
>   $$x = \pm a\sec t + h, \quad y = b\tan t + k$$
>
>   defines a hyperbola with horizontal transverse axis. Each has asymptotes at $y = \pm b/a(x - h) + k$.

### Special Curves

Figure 9.2.8 gives a small gallery of "interesting" and "famous" curves along with parametric equations that produce them. Interested readers can begin learning more about these curves through internet searches.

One might note a feature shared by two of these graphs: "sharp corners," or **cusps**. We have seen graphs with cusps before and determined that such functions are not differentiable at these points. This leads us to a definition.

> **Definition 9.2.2**      **Smooth**
>
> A curve $C$ defined by $x = f(t), y = g(t)$ is **smooth** on an interval $I$ if $f'$ and $g'$ are continuous on $I$ and not simultaneously 0 (except possibly at the endpoints of $I$). A curve is **piecewise smooth** on $I$ if $I$ can be partitioned into subintervals where $C$ is smooth on each subinterval.

Consider the astroid, given by $x = \cos^3 t$, $y = \sin^3 t$. Taking derivatives, we have:

$$x' = -3\cos^2 t \sin t \quad \text{and} \quad y' = 3\sin^2 t \cos t.$$

It is clear that each is 0 when $t = 0, \pi/2, \pi, \ldots$. Thus the astroid is not smooth

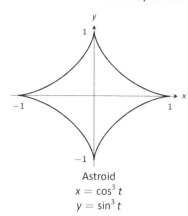

Astroid
$x = \cos^3 t$
$y = \sin^3 t$

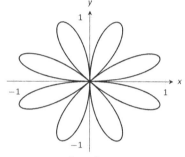

Rose Curve
$x = \cos(t)\sin(4t)$
$y = \sin(t)\sin(4t)$

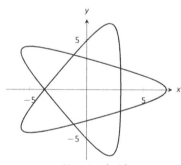

Hypotrochoid
$x = 2\cos(t) + 5\cos(2t/3)$
$y = 2\sin(t) - 5\sin(2t/3)$

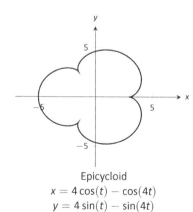

Epicycloid
$x = 4\cos(t) - \cos(4t)$
$y = 4\sin(t) - \sin(4t)$

Figure 9.2.8: A gallery of interesting planar curves.

at these points, corresponding to the cusps seen in the figure.

We demonstrate this once more.

**Example 9.2.9**     **Determine where a curve is not smooth**

Let a curve C be defined by the parametric equations $x = t^3 - 12t + 17$ and $y = t^2 - 4t + 8$. Determine the points, if any, where it is not smooth.

**SOLUTION**     We begin by taking derivatives.
$$x' = 3t^2 - 12, \quad y' = 2t - 4.$$

We set each equal to 0:
$$x' = 0 \Rightarrow 3t^2 - 12 = 0 \Rightarrow t = \pm 2$$
$$y' = 0 \Rightarrow 2t - 4 = 0 \Rightarrow t = 2$$

We see at $t = 2$ both $x'$ and $y'$ are 0; thus C is not smooth at $t = 2$, corresponding to the point $(1, 4)$. The curve is graphed in Figure 9.2.9, illustrating the cusp at $(1, 4)$.

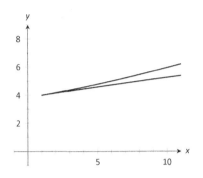

Figure 9.2.9: Graphing the curve in Example 9.2.9; note it is not smooth at $(1, 4)$.

If a curve is not smooth at $t = t_0$, it means that $x'(t_0) = y'(t_0) = 0$ as defined. This, in turn, means that rate of change of x (and y) is 0; that is, at that instant, neither x nor y is changing. If the parametric equations describe the path of some object, this means the object is at rest at $t_0$. An object at rest can make a "sharp" change in direction, whereas moving objects tend to change direction in a "smooth" fashion.

One should be careful to note that a "sharp corner" does not have to occur when a curve is not smooth. For instance, one can verify that $x = t^3$, $y = t^6$ produce the familiar $y = x^2$ parabola. However, in this parametrization, the curve is not smooth. A particle traveling along the parabola according to the given parametric equations comes to rest at $t = 0$, though no sharp point is created.

Our previous experience with cusps taught us that a function was not differentiable at a cusp. This can lead us to wonder about derivatives in the context of parametric equations and the application of other calculus concepts. Given a curve defined parametrically, how do we find the slopes of tangent lines? Can we determine concavity? We explore these concepts and more in the next section.

Notes:

# Exercises 9.2

## Terms and Concepts

1. T/F: When sketching the graph of parametric equations, the x and y values are found separately, then plotted together.

2. The direction in which a graph is "moving" is called the ___ of the graph.

3. An equation written as $y = f(x)$ is written in ___ form.

4. Create parametric equations $x = f(t), y = g(t)$ and sketch their graph. Explain any interesting features of your graph based on the functions $f$ and $g$.

## Problems

In Exercises 5 – 8, sketch the graph of the given parametric equations by hand, making a table of points to plot. Be sure to indicate the orientation of the graph.

5. $x = t^2 + t, \quad y = 1 - t^2, \quad -3 \leq t \leq 3$

6. $x = 1, \quad y = 5\sin t, \quad -\pi/2 \leq t \leq \pi/2$

7. $x = t^2, \quad y = 2, \quad -2 \leq t \leq 2$

8. $x = t^3 - t + 3, \quad y = t^2 + 1, \quad -2 \leq t \leq 2$

In Exercises 9 – 18, sketch the graph of the given parametric equations; using a graphing utility is advisable. Be sure to indicate the orientation of the graph.

9. $x = t^3 - 2t^2, \quad y = t^2, \quad -2 \leq t \leq 3$

10. $x = 1/t, \quad y = \sin t, \quad 0 < t \leq 10$

11. $x = 3\cos t, \quad y = 5\sin t, \quad 0 \leq t \leq 2\pi$

12. $x = 3\cos t + 2, \quad y = 5\sin t + 3, \quad 0 \leq t \leq 2\pi$

13. $x = \cos t, \quad y = \cos(2t), \quad 0 \leq t \leq \pi$

14. $x = \cos t, \quad y = \sin(2t), \quad 0 \leq t \leq 2\pi$

15. $x = 2\sec t, \quad y = 3\tan t, \quad -\pi/2 < t < \pi/2$

16. $x = \cosh t, \quad y = \sinh t, \quad -2 \leq t \leq 2$

17. $x = \cos t + \frac{1}{4}\cos(8t), \quad y = \sin t + \frac{1}{4}\sin(8t), \quad 0 \leq t \leq 2\pi$

18. $x = \cos t + \frac{1}{4}\sin(8t), \quad y = \sin t + \frac{1}{4}\cos(8t), \quad 0 \leq t \leq 2\pi$

In Exercises 19 – 20, four sets of parametric equations are given. Describe how their graphs are similar and different. Be sure to discuss orientation and ranges.

19. (a) $x = t \quad y = t^2, \quad -\infty < t < \infty$

    (b) $x = \sin t \quad y = \sin^2 t, \quad -\infty < t < \infty$

    (c) $x = e^t \quad y = e^{2t}, \quad -\infty < t < \infty$

    (d) $x = -t \quad y = t^2, \quad -\infty < t < \infty$

20. (a) $x = \cos t \quad y = \sin t, \quad 0 \leq t \leq 2\pi$

    (b) $x = \cos(t^2) \quad y = \sin(t^2), \quad 0 \leq t \leq 2\pi$

    (c) $x = \cos(1/t) \quad y = \sin(1/t), \quad 0 < t < 1$

    (d) $x = \cos(\cos t) \quad y = \sin(\cos t), \quad 0 \leq t \leq 2\pi$

In Exercises 21 – 30, eliminate the parameter in the given parametric equations.

21. $x = 2t + 5, \quad y = -3t + 1$

22. $x = \sec t, \quad y = \tan t$

23. $x = 4\sin t + 1, \quad y = 3\cos t - 2$

24. $x = t^2, \quad y = t^3$

25. $x = \dfrac{1}{t+1}, \quad y = \dfrac{3t+5}{t+1}$

26. $x = e^t, \quad y = e^{3t} - 3$

27. $x = \ln t, \quad y = t^2 - 1$

28. $x = \cot t, \quad y = \csc t$

29. $x = \cosh t, \quad y = \sinh t$

30. $x = \cos(2t), \quad y = \sin t$

In Exercises 31 – 34, eliminate the parameter in the given parametric equations. Describe the curve defined by the parametric equations based on its rectangular form.

31. $x = at + x_0, \quad y = bt + y_0$

32. $x = r\cos t, \quad y = r\sin t$

33. $x = a\cos t + h, \quad y = b\sin t + k$

34. $x = a\sec t + h, \quad y = b\tan t + k$

In Exercises 35 – 38, find parametric equations for the given rectangular equation using the parameter $t = \dfrac{dy}{dx}$. Verify that at $t = 1$, the point on the graph has a tangent line with slope of 1.

35. $y = 3x^2 - 11x + 2$

36. $y = e^x$

37. $y = \sin x$ on $[0, \pi]$

38. $y = \sqrt{x}$ on $[0, \infty)$

In Exercises 39 – 42, find the values of $t$ where the graph of the parametric equations crosses itself.

39. $x = t^3 - t + 3, \quad y = t^2 - 3$

40. $x = t^3 - 4t^2 + t + 7, \quad y = t^2 - t$

41. $x = \cos t, \quad y = \sin(2t)$ on $[0, 2\pi]$

42. $x = \cos t \cos(3t), \quad y = \sin t \cos(3t)$ on $[0, \pi]$

In Exercises 43 – 46, find the value(s) of $t$ where the curve defined by the parametric equations is not smooth.

43. $x = t^3 + t^2 - t, \quad y = t^2 + 2t + 3$

44. $x = t^2 - 4t, \quad y = t^3 - 2t^2 - 4t$

45. $x = \cos t, \quad y = 2\cos t$

46. $x = 2\cos t - \cos(2t), \quad y = 2\sin t - \sin(2t)$

In Exercises 47 – 55, find parametric equations that describe the given situation.

47. A projectile is fired from a height of 0ft, landing 16ft away in 4s.

48. A projectile is fired from a height of 0ft, landing 200ft away in 4s.

49. A projectile is fired from a height of 0ft, landing 200ft away in 20s.

50. A circle of radius 2, centered at the origin, that is traced clockwise once on $[0, 2\pi]$.

51. A circle of radius 3, centered at $(1, 1)$, that is traced once counter–clockwise on $[0, 1]$.

52. An ellipse centered at $(1, 3)$ with vertical major axis of length 6 and minor axis of length 2.

53. An ellipse with foci at $(\pm 1, 0)$ and vertices at $(\pm 5, 0)$.

54. A hyperbola with foci at $(5, -3)$ and $(-1, -3)$, and with vertices at $(1, -3)$ and $(3, -3)$.

55. A hyperbola with vertices at $(0, \pm 6)$ and asymptotes $y = \pm 3x$.

## 9.3 Calculus and Parametric Equations

The previous section defined curves based on parametric equations. In this section we'll employ the techniques of calculus to study these curves.

We are still interested in lines tangent to points on a curve. They describe how the *y*-values are changing with respect to the *x*-values, they are useful in making approximations, and they indicate instantaneous direction of travel.

The slope of the tangent line is still $\frac{dy}{dx}$, and the Chain Rule allows us to calculate this in the context of parametric equations. If $x = f(t)$ and $y = g(t)$, the Chain Rule states that

$$\frac{dy}{dt} = \frac{dy}{dx} \cdot \frac{dx}{dt}.$$

Solving for $\frac{dy}{dx}$, we get

$$\frac{dy}{dx} = \frac{dy}{dt} \bigg/ \frac{dx}{dt} = \frac{g'(t)}{f'(t)},$$

provided that $f'(t) \neq 0$. This is important so we label it a Key Idea.

---

**Key Idea 9.3.1**   Finding $\frac{dy}{dx}$ with Parametric Equations.

Let $x = f(t)$ and $y = g(t)$, where $f$ and $g$ are differentiable on some open interval $I$ and $f'(t) \neq 0$ on $I$. Then

$$\frac{dy}{dx} = \frac{g'(t)}{f'(t)}.$$

---

We use this to define the tangent line.

---

**Definition 9.3.1**   **Tangent and Normal Lines**

Let a curve $C$ be parametrized by $x = f(t)$ and $y = g(t)$, where $f$ and $g$ are differentiable functions on some interval $I$ containing $t = t_0$. The **tangent line** to $C$ at $t = t_0$ is the line through $\big(f(t_0), g(t_0)\big)$ with slope $m = g'(t_0)/f'(t_0)$, provided $f'(t_0) \neq 0$.

The **normal line** to $C$ at $t = t_0$ is the line through $\big(f(t_0), g(t_0)\big)$ with slope $m = -f'(t_0)/g'(t_0)$, provided $g'(t_0) \neq 0$.

---

The definition leaves two special cases to consider. When the tangent line is horizontal, the normal line is undefined by the above definition as $g'(t_0) = 0$.

---

Notes:

Likewise, when the normal line is horizontal, the tangent line is undefined. It seems reasonable that these lines be defined (one can draw a line tangent to the "right side" of a circle, for instance), so we add the following to the above definition.

1. If the tangent line at $t = t_0$ has a slope of 0, the normal line to $C$ at $t = t_0$ is the line $x = f(t_0)$.

2. If the normal line at $t = t_0$ has a slope of 0, the tangent line to $C$ at $t = t_0$ is the line $x = f(t_0)$.

**Example 9.3.1    Tangent and Normal Lines to Curves**
Let $x = 5t^2 - 6t + 4$ and $y = t^2 + 6t - 1$, and let $C$ be the curve defined by these equations.

1. Find the equations of the tangent and normal lines to $C$ at $t = 3$.

2. Find where $C$ has vertical and horizontal tangent lines.

**SOLUTION**

1. We start by computing $f'(t) = 10t - 6$ and $g'(t) = 2t + 6$. Thus
$$\frac{dy}{dx} = \frac{2t+6}{10t-6}.$$

   Make note of something that might seem unusual: $\frac{dy}{dx}$ is a function of $t$, not $x$. Just as points on the curve are found in terms of $t$, so are the slopes of the tangent lines.

   The point on $C$ at $t = 3$ is $(31, 26)$. The slope of the tangent line is $m = 1/2$ and the slope of the normal line is $m = -2$. Thus,

   - the equation of the tangent line is $y = \frac{1}{2}(x - 31) + 26$, and
   - the equation of the normal line is $y = -2(x - 31) + 26$.

   This is illustrated in Figure 9.3.1.

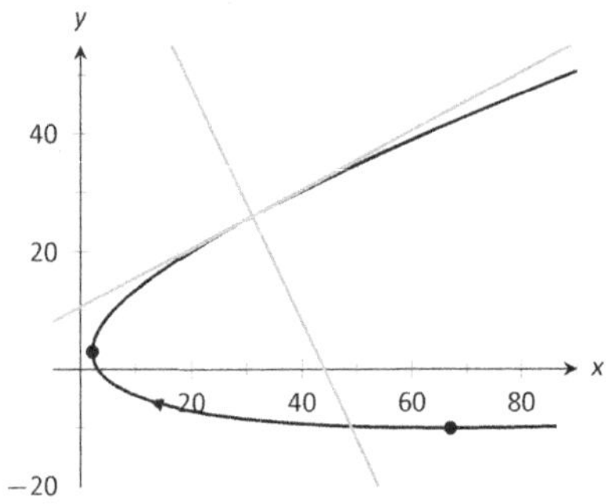

2. To find where $C$ has a horizontal tangent line, we set $\frac{dy}{dx} = 0$ and solve for $t$. In this case, this amounts to setting $g'(t) = 0$ and solving for $t$ (and making sure that $f'(t) \neq 0$).
$$g'(t) = 0 \quad \Rightarrow \quad 2t + 6 = 0 \quad \Rightarrow \quad t = -3.$$

   The point on $C$ corresponding to $t = -3$ is $(67, -10)$; the tangent line at that point is horizontal (hence with equation $y = -10$).

Figure 9.3.1: Graphing tangent and normal lines in Example 9.3.1.

Notes:

## 9.3 Calculus and Parametric Equations

To find where $C$ has a vertical tangent line, we find where it has a horizontal normal line, and set $-\frac{f'(t)}{g'(t)} = 0$. This amounts to setting $f'(t) = 0$ and solving for $t$ (and making sure that $g'(t) \neq 0$).

$$f'(t) = 0 \quad \Rightarrow \quad 10t - 6 = 0 \quad \Rightarrow \quad t = 0.6.$$

The point on $C$ corresponding to $t = 0.6$ is $(2.2, 2.96)$. The tangent line at that point is $x = 2.2$.

The points where the tangent lines are vertical and horizontal are indicated on the graph in Figure 9.3.1.

### Example 9.3.2 Tangent and Normal Lines to a Circle

1. Find where the unit circle, defined by $x = \cos t$ and $y = \sin t$ on $[0, 2\pi]$, has vertical and horizontal tangent lines.

2. Find the equation of the normal line at $t = t_0$.

**SOLUTION**

1. We compute the derivative following Key Idea 9.3.1:

$$\frac{dy}{dx} = \frac{g'(t)}{f'(t)} = -\frac{\cos t}{\sin t}.$$

The derivative is 0 when $\cos t = 0$; that is, when $t = \pi/2, 3\pi/2$. These are the points $(0, 1)$ and $(0, -1)$ on the circle.

The normal line is horizontal (and hence, the tangent line is vertical) when $\sin t = 0$; that is, when $t = 0, \pi, 2\pi$, corresponding to the points $(-1, 0)$ and $(0, 1)$ on the circle. These results should make intuitive sense.

2. The slope of the normal line at $t = t_0$ is $m = \dfrac{\sin t_0}{\cos t_0} = \tan t_0$. This normal line goes through the point $(\cos t_0, \sin t_0)$, giving the line

$$y = \frac{\sin t_0}{\cos t_0}(x - \cos t_0) + \sin t_0$$
$$= (\tan t_0)x,$$

as long as $\cos t_0 \neq 0$. It is an important fact to recognize that the normal lines to a circle pass through its center, as illustrated in Figure 9.3.2. Stated in another way, any line that passes through the center of a circle intersects the circle at right angles.

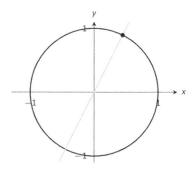

Figure 9.3.2: Illustrating how a circle's normal lines pass through its center.

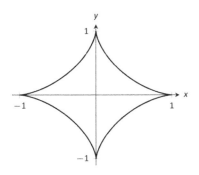

Figure 9.3.3: A graph of an astroid.

**Example 9.3.3**     **Tangent lines when $\frac{dy}{dx}$ is not defined**

Find the equation of the tangent line to the astroid $x = \cos^3 t$, $y = \sin^3 t$ at $t = 0$, shown in Figure 9.3.3.

**SOLUTION**     We start by finding $x'(t)$ and $y'(t)$:

$$x'(t) = -3 \sin t \cos^2 t, \qquad y'(t) = 3 \cos t \sin^2 t.$$

Note that both of these are 0 at $t = 0$; the curve is not smooth at $t = 0$ forming a cusp on the graph. Evaluating $\frac{dy}{dx}$ at this point returns the indeterminate form of "0/0".

We can, however, examine the slopes of tangent lines near $t = 0$, and take the limit as $t \to 0$.

$$\lim_{t \to 0} \frac{y'(t)}{x'(t)} = \lim_{t \to 0} \frac{3 \cos t \sin^2 t}{-3 \sin t \cos^2 t} \qquad \text{(We can cancel as } t \neq 0.\text{)}$$

$$= \lim_{t \to 0} -\frac{\sin t}{\cos t}$$

$$= 0.$$

We have accomplished something significant. When the derivative $\frac{dy}{dx}$ returns an indeterminate form at $t = t_0$, we can define its value by setting it to be $\lim\limits_{t \to t_0} \frac{dy}{dx}$, if that limit exists. This allows us to find slopes of tangent lines at cusps, which can be very beneficial.

We found the slope of the tangent line at $t = 0$ to be 0; therefore the tangent line is $y = 0$, the $x$-axis.

## Concavity

We continue to analyze curves in the plane by considering their concavity; that is, we are interested in $\frac{d^2y}{dx^2}$, "the second derivative of $y$ with respect to $x$." To find this, we need to find the derivative of $\frac{dy}{dx}$ with respect to $x$; that is,

$$\frac{d^2y}{dx^2} = \frac{d}{dx}\left[\frac{dy}{dx}\right],$$

but recall that $\frac{dy}{dx}$ is a function of $t$, not $x$, making this computation not straightforward.

To make the upcoming notation a bit simpler, let $h(t) = \frac{dy}{dx}$. We want $\frac{d}{dx}[h(t)]$; that is, we want $\frac{dh}{dx}$. We again appeal to the Chain Rule. Note:

$$\frac{dh}{dt} = \frac{dh}{dx} \cdot \frac{dx}{dt} \quad \Rightarrow \quad \frac{dh}{dx} = \frac{dh}{dt} \bigg/ \frac{dx}{dt}.$$

Notes:

## 9.3 Calculus and Parametric Equations

In words, to find $\frac{d^2y}{dx^2}$, we first take the derivative of $\frac{dy}{dx}$ *with respect to t*, then divide by $x'(t)$. We restate this as a Key Idea.

---

**Key Idea 9.3.2**     **Finding $\frac{d^2y}{dx^2}$ with Parametric Equations**

Let $x = f(t)$ and $y = g(t)$ be twice differentiable functions on an open interval $I$, where $f'(t) \neq 0$ on $I$. Then

$$\frac{d^2y}{dx^2} = \frac{d}{dt}\left[\frac{dy}{dx}\right] \bigg/ \frac{dx}{dt} = \frac{d}{dt}\left[\frac{dy}{dx}\right] \bigg/ f'(t).$$

---

Examples will help us understand this Key Idea.

**Example 9.3.4**     **Concavity of Plane Curves**
Let $x = 5t^2 - 6t + 4$ and $y = t^2 + 6t - 1$ as in Example 9.3.1. Determine the $t$-intervals on which the graph is concave up/down.

**Solution**     Concavity is determined by the second derivative of $y$ with respect to $x$, $\frac{d^2y}{dx^2}$, so we compute that here following Key Idea 9.3.2.

In Example 9.3.1, we found $\frac{dy}{dx} = \frac{2t+6}{10t-6}$ and $f'(t) = 10t - 6$. So:

$$\frac{d^2y}{dx^2} = \frac{d}{dt}\left[\frac{2t+6}{10t-6}\right] \bigg/ (10t-6)$$

$$= -\frac{72}{(10t-6)^2} \bigg/ (10t-6)$$

$$= -\frac{72}{(10t-6)^3}$$

$$= -\frac{9}{(5t-3)^3}$$

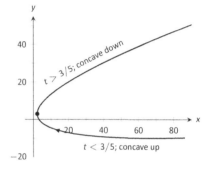

Figure 9.3.4: Graphing the parametric equations in Example 9.3.4 to demonstrate concavity.

The graph of the parametric functions is concave up when $\frac{d^2y}{dx^2} > 0$ and concave down when $\frac{d^2y}{dx^2} < 0$. We determine the intervals when the second derivative is greater/less than 0 by first finding when it is 0 or undefined.

As the numerator of $-\frac{9}{(5t-3)^3}$ is never 0, $\frac{d^2y}{dx^2} \neq 0$ for all $t$. It is undefined when $5t - 3 = 0$; that is, when $t = 3/5$. Following the work established in Section 3.4, we look at values of $t$ greater/less than $3/5$ on a number line:

---

Notes:

# Chapter 9 Curves in the Plane

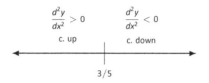

Reviewing Example 9.3.1, we see that when $t = 3/5 = 0.6$, the graph of the parametric equations has a vertical tangent line. This point is also a point of inflection for the graph, illustrated in Figure 9.3.4.

**Example 9.3.5    Concavity of Plane Curves**

Find the points of inflection of the graph of the parametric equations $x = \sqrt{t}$, $y = \sin t$, for $0 \leq t \leq 16$.

**Solution**    We need to compute $\frac{dy}{dx}$ and $\frac{d^2y}{dx^2}$.

$$\frac{dy}{dx} = \frac{y'(t)}{x'(t)} = \frac{\cos t}{1/(2\sqrt{t})} = 2\sqrt{t}\cos t.$$

$$\frac{d^2y}{dx^2} = \frac{\frac{d}{dt}\left[\frac{dy}{dx}\right]}{x'(t)} = \frac{\cos t/\sqrt{t} - 2\sqrt{t}\sin t}{1/(2\sqrt{t})} = 2\cos t - 4t\sin t.$$

The points of inflection are found by setting $\frac{d^2y}{dx^2} = 0$. This is not trivial, as equations that mix polynomials and trigonometric functions generally do not have "nice" solutions.

In Figure 9.3.5(a) we see a plot of the second derivative. It shows that it has zeros at approximately $t = 0.5, 3.5, 6.5, 9.5, 12.5$ and $16$. These approximations are not very good, made only by looking at the graph. Newton's Method provides more accurate approximations. Accurate to 2 decimal places, we have:

$$t = 0.65,\ 3.29,\ 6.36,\ 9.48,\ 12.61 \text{ and } 15.74.$$

The corresponding points have been plotted on the graph of the parametric equations in Figure 9.3.5(b). Note how most occur near the $x$-axis, but not exactly on the axis.

## Arc Length

We continue our study of the features of the graphs of parametric equations by computing their arc length.

Recall in Section 7.4 we found the arc length of the graph of a function, from $x = a$ to $x = b$, to be

$$L = \int_a^b \sqrt{1 + \left(\frac{dy}{dx}\right)^2}\, dx.$$

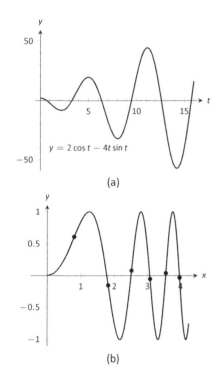

Figure 9.3.5: In (a), a graph of $\frac{d^2y}{dx^2}$, showing where it is approximately 0. In (b), graph of the parametric equations in Example 9.3.5 along with the points of inflection.

Notes:

We can use this equation and convert it to the parametric equation context. Letting $x = f(t)$ and $y = g(t)$, we know that $\frac{dy}{dx} = g'(t)/f'(t)$. It will also be useful to calculate the differential of $x$:

$$dx = f'(t)dt \quad \Rightarrow \quad dt = \frac{1}{f'(t)} \cdot dx.$$

Starting with the arc length formula above, consider:

$$L = \int_a^b \sqrt{1 + \left(\frac{dy}{dx}\right)^2}\, dx$$

$$= \int_a^b \sqrt{1 + \frac{g'(t)^2}{f'(t)^2}}\, dx.$$

Factor out the $f'(t)^2$:

$$= \int_a^b \sqrt{f'(t)^2 + g'(t)^2} \cdot \underbrace{\frac{1}{f'(t)}\, dx}_{=dt}$$

$$= \int_{t_1}^{t_2} \sqrt{f'(t)^2 + g'(t)^2}\, dt.$$

Note the new bounds (no longer "$x$" bounds, but "$t$" bounds). They are found by finding $t_1$ and $t_2$ such that $a = f(t_1)$ and $b = f(t_2)$. This formula is important, so we restate it as a theorem.

---

**Theorem 9.3.1**     **Arc Length of Parametric Curves**

Let $x = f(t)$ and $y = g(t)$ be parametric equations with $f'$ and $g'$ continuous on $[t_1, t_2]$, on which the graph traces itself only once. The arc length of the graph, from $t = t_1$ to $t = t_2$, is

$$L = \int_{t_1}^{t_2} \sqrt{f'(t)^2 + g'(t)^2}\, dt.$$

---

**Note:** Theorem 9.3.1 makes use of differentiability on closed intervals, just as was done in Section 7.4.

As before, these integrals are often not easy to compute. We start with a simple example, then give another where we approximate the solution.

Notes:

**Example 9.3.6**     **Arc Length of a Circle**

Find the arc length of the circle parametrized by $x = 3\cos t$, $y = 3\sin t$ on $[0, 3\pi/2]$.

**SOLUTION**      By direct application of Theorem 9.3.1, we have

$$L = \int_0^{3\pi/2} \sqrt{(-3\sin t)^2 + (3\cos t)^2}\, dt.$$

Apply the Pythagorean Theorem.

$$= \int_0^{3\pi/2} 3\, dt$$

$$= 3t\Big|_0^{3\pi/2} = 9\pi/2.$$

This should make sense; we know from geometry that the circumference of a circle with radius 3 is $6\pi$; since we are finding the arc length of $3/4$ of a circle, the arc length is $3/4 \cdot 6\pi = 9\pi/2$.

**Example 9.3.7**     **Arc Length of a Parametric Curve**

The graph of the parametric equations $x = t(t^2 - 1)$, $y = t^2 - 1$ crosses itself as shown in Figure 9.3.6, forming a "teardrop." Find the arc length of the teardrop.

**SOLUTION**      We can see by the parametrizations of $x$ and $y$ that when $t = \pm 1$, $x = 0$ and $y = 0$. This means we'll integrate from $t = -1$ to $t = 1$. Applying Theorem 9.3.1, we have

$$L = \int_{-1}^{1} \sqrt{(3t^2 - 1)^2 + (2t)^2}\, dt$$

$$= \int_{-1}^{1} \sqrt{9t^4 - 2t^2 + 1}\, dt.$$

Unfortunately, the integrand does not have an antiderivative expressible by elementary functions. We turn to numerical integration to approximate its value. Using 4 subintervals, Simpson's Rule approximates the value of the integral as 2.65051. Using a computer, more subintervals are easy to employ, and $n = 20$ gives a value of 2.71559. Increasing $n$ shows that this value is stable and a good approximation of the actual value.

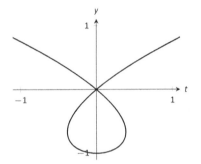

Figure 9.3.6: A graph of the parametric equations in Example 9.3.7, where the arc length of the teardrop is calculated.

## 9.3 Calculus and Parametric Equations

### Surface Area of a Solid of Revolution

Related to the formula for finding arc length is the formula for finding surface area. We can adapt the formula found in Theorem 7.4.2 from Section 7.4 in a similar way as done to produce the formula for arc length done before.

---

**Theorem 9.3.2**  **Surface Area of a Solid of Revolution**

Consider the graph of the parametric equations $x = f(t)$ and $y = g(t)$, where $f'$ and $g'$ are continuous on an open interval $I$ containing $t_1$ and $t_2$ on which the graph does not cross itself.

1. The surface area of the solid formed by revolving the graph about the x-axis is (where $g(t) \geq 0$ on $[t_1, t_2]$):

$$\text{Surface Area} = 2\pi \int_{t_1}^{t_2} g(t) \sqrt{f'(t)^2 + g'(t)^2}\, dt.$$

2. The surface area of the solid formed by revolving the graph about the y-axis is (where $f(t) \geq 0$ on $[t_1, t_2]$):

$$\text{Surface Area} = 2\pi \int_{t_1}^{t_2} f(t) \sqrt{f'(t)^2 + g'(t)^2}\, dt.$$

---

**Example 9.3.8**  **Surface Area of a Solid of Revolution**

Consider the teardrop shape formed by the parametric equations $x = t(t^2 - 1)$, $y = t^2 - 1$ as seen in Example 9.3.7. Find the surface area if this shape is rotated about the x-axis, as shown in Figure 9.3.7.

**SOLUTION**   The teardrop shape is formed between $t = -1$ and $t = 1$. Using Theorem 9.3.2, we see we need for $g(t) \geq 0$ on $[-1, 1]$, and this is not the case. To fix this, we simplify replace $g(t)$ with $-g(t)$, which flips the whole graph about the x-axis (and does not change the surface area of the resulting solid). The surface area is:

$$\text{Area } S = 2\pi \int_{-1}^{1} (1 - t^2)\sqrt{(3t^2 - 1)^2 + (2t)^2}\, dt$$
$$= 2\pi \int_{-1}^{1} (1 - t^2)\sqrt{9t^4 - 2t^2 + 1}\, dt.$$

Once again we arrive at an integral that we cannot compute in terms of ele-

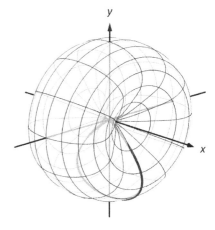

Figure 9.3.7: Rotating a teardrop shape about the x-axis in Example 9.3.8.

mentary functions. Using Simpson's Rule with $n = 20$, we find the area to be $S = 9.44$. Using larger values of $n$ shows this is accurate to 2 places after the decimal.

After defining a new way of creating curves in the plane, in this section we have applied calculus techniques to the parametric equation defining these curves to study their properties. In the next section, we define another way of forming curves in the plane. To do so, we create a new coordinate system, called *polar coordinates*, that identifies points in the plane in a manner different than from measuring distances from the *y*- and *x*- axes.

Notes:

# Exercises 9.3

## Terms and Concepts

1. T/F: Given parametric equations $x = f(t)$ and $y = g(t)$, $\frac{dy}{dx} = f'(t)/g'(t)$, as long as $g'(t) \neq 0$.

2. Given parametric equations $x = f(t)$ and $y = g(t)$, the derivative $\frac{dy}{dx}$ as given in Key Idea 9.3.1 is a function of _____?

3. T/F: Given parametric equations $x = f(t)$ and $y = g(t)$, to find $\frac{d^2y}{dx^2}$, one simply computes $\frac{d}{dt}\left(\frac{dy}{dx}\right)$.

4. T/F: If $\frac{dy}{dx} = 0$ at $t = t_0$, then the normal line to the curve at $t = t_0$ is a vertical line.

## Problems

In Exercises 5 – 12, parametric equations for a curve are given.

(a) Find $\frac{dy}{dx}$.

(b) Find the equations of the tangent and normal line(s) at the point(s) given.

(c) Sketch the graph of the parametric functions along with the found tangent and normal lines.

5. $x = t, y = t^2; \quad t = 1$

6. $x = \sqrt{t}, y = 5t + 2; \quad t = 4$

7. $x = t^2 - t, y = t^2 + t; \quad t = 1$

8. $x = t^2 - 1, y = t^3 - t; \quad t = 0$ and $t = 1$

9. $x = \sec t, y = \tan t$ on $(-\pi/2, \pi/2); \quad t = \pi/4$

10. $x = \cos t, y = \sin(2t)$ on $[0, 2\pi]; \quad t = \pi/4$

11. $x = \cos t \sin(2t), y = \sin t \sin(2t)$ on $[0, 2\pi]; \quad t = 3\pi/4$

12. $x = e^{t/10} \cos t, y = e^{t/10} \sin t; \quad t = \pi/2$

In Exercises 13 – 20, find $t$-values where the curve defined by the given parametric equations has a horizontal tangent line. Note: these are the same equations as in Exercises 5 – 12.

13. $x = t, y = t^2$

14. $x = \sqrt{t}, y = 5t + 2$

15. $x = t^2 - t, y = t^2 + t$

16. $x = t^2 - 1, y = t^3 - t$

17. $x = \sec t, y = \tan t$ on $(-\pi/2, \pi/2)$

18. $x = \cos t, y = \sin(2t)$ on $[0, 2\pi]$

19. $x = \cos t \sin(2t), y = \sin t \sin(2t)$ on $[0, 2\pi]$

20. $x = e^{t/10} \cos t, y = e^{t/10} \sin t$

In Exercises 21 – 24, find $t = t_0$ where the graph of the given parametric equations is not smooth, then find $\lim\limits_{t \to t_0} \frac{dy}{dx}$.

21. $x = \dfrac{1}{t^2 + 1}, \quad y = t^3$

22. $x = -t^3 + 7t^2 - 16t + 13, \quad y = t^3 - 5t^2 + 8t - 2$

23. $x = t^3 - 3t^2 + 3t - 1, \quad y = t^2 - 2t + 1$

24. $x = \cos^2 t, \quad y = 1 - \sin^2 t$

In Exercises 25 – 32, parametric equations for a curve are given. Find $\frac{d^2y}{dx^2}$, then determine the intervals on which the graph of the curve is concave up/down. Note: these are the same equations as in Exercises 5 – 12.

25. $x = t, \quad y = t^2$

26. $x = \sqrt{t}, \quad y = 5t + 2$

27. $x = t^2 - t, \quad y = t^2 + t$

28. $x = t^2 - 1, \quad y = t^3 - t$

29. $x = \sec t, \quad y = \tan t$ on $(-\pi/2, \pi/2)$

30. $x = \cos t, \quad y = \sin(2t)$ on $[0, 2\pi]$

31. $x = \cos t \sin(2t), \quad y = \sin t \sin(2t)$ on $[-\pi/2, \pi/2]$

32. $x = e^{t/10} \cos t, \quad y = e^{t/10} \sin t$

In Exercises 33 – 36, find the arc length of the graph of the parametric equations on the given interval(s).

33. $x = -3\sin(2t), \quad y = 3\cos(2t)$ on $[0, \pi]$

34. $x = e^{t/10} \cos t, \quad y = e^{t/10} \sin t$ on $[0, 2\pi]$ and $[2\pi, 4\pi]$

35. $x = 5t + 2, \quad y = 1 - 3t$ on $[-1, 1]$

36. $x = 2t^{3/2}, \quad y = 3t$ on $[0, 1]$

In Exercises 37 – 40, numerically approximate the given arc length.

37. Approximate the arc length of one petal of the rose curve $x = \cos t \cos(2t), \quad y = \sin t \cos(2t)$ using Simpson's Rule and $n = 4$.

38. Approximate the arc length of the "bow tie curve" $x = \cos t$, $y = \sin(2t)$ using Simpson's Rule and $n = 6$.

39. Approximate the arc length of the parabola $x = t^2 - t$, $y = t^2 + t$ on $[-1, 1]$ using Simpson's Rule and $n = 4$.

40. A common approximate of the circumference of an ellipse given by $x = a\cos t$, $y = b\sin t$ is $C \approx 2\pi\sqrt{\dfrac{a^2 + b^2}{2}}$. Use this formula to approximate the circumference of $x = 5\cos t$, $y = 3\sin t$ and compare this to the approximation given by Simpson's Rule and $n = 6$.

**In Exercises 41 – 44, a solid of revolution is described. Find or approximate its surface area as specified.**

41. Find the surface area of the sphere formed by rotating the circle $x = 2\cos t$, $y = 2\sin t$ about:

    (a) the x-axis and

    (b) the y-axis.

42. Find the surface area of the torus (or "donut") formed by rotating the circle $x = \cos t + 2$, $y = \sin t$ about the y-axis.

43. Approximate the surface area of the solid formed by rotating the "upper right half" of the bow tie curve $x = \cos t$, $y = \sin(2t)$ on $[0, \pi/2]$ about the x-axis, using Simpson's Rule and $n = 4$.

44. Approximate the surface area of the solid formed by rotating the one petal of the rose curve $x = \cos t \cos(2t)$, $y = \sin t \cos(2t)$ on $[0, \pi/4]$ about the x-axis, using Simpson's Rule and $n = 4$.

## 9.4 Introduction to Polar Coordinates

We are generally introduced to the idea of graphing curves by relating x-values to y-values through a function f. That is, we set $y = f(x)$, and plot lots of point pairs $(x, y)$ to get a good notion of how the curve looks. This method is useful but has limitations, not least of which is that curves that "fail the vertical line test" cannot be graphed without using multiple functions.

The previous two sections introduced and studied a new way of plotting points in the $x, y$-plane. Using parametric equations, $x$ and $y$ values are computed independently and then plotted together. This method allows us to graph an extraordinary range of curves. This section introduces yet another way to plot points in the plane: using **polar coordinates**.

### Polar Coordinates

Start with a point $O$ in the plane called the **pole** (we will always identify this point with the origin). From the pole, draw a ray, called the **initial ray** (we will always draw this ray horizontally, identifying it with the positive x-axis). A point $P$ in the plane is determined by the distance $r$ that $P$ is from $O$, and the angle $\theta$ formed between the initial ray and the segment $\overline{OP}$ (measured counter-clockwise). We record the distance and angle as an ordered pair $(r, \theta)$. To avoid confusion with rectangular coordinates, we will denote polar coordinates with the letter $P$, as in $P(r, \theta)$. This is illustrated in Figure 9.4.1

Practice will make this process more clear.

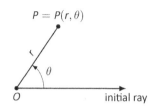

Figure 9.4.1: Illustrating polar coordinates.

### Example 9.4.1 Plotting Polar Coordinates
Plot the following polar coordinates:

$$A = P(1, \pi/4) \quad B = P(1.5, \pi) \quad C = P(2, -\pi/3) \quad D = P(-1, \pi/4)$$

**Solution**  To aid in the drawing, a polar grid is provided at the bottom of this page. To place the point $A$, go out 1 unit along the initial ray (putting you on the inner circle shown on the grid), then rotate counter-clockwise $\pi/4$ radians (or 45°). Alternately, one can consider the rotation first: think about the ray from $O$ that forms an angle of $\pi/4$ with the initial ray, then move out 1 unit along this ray (again placing you on the inner circle of the grid).

To plot $B$, go out 1.5 units along the initial ray and rotate $\pi$ radians (180°).

To plot $C$, go out 2 units along the initial ray then rotate *clockwise* $\pi/3$ radians, as the angle given is negative.

To plot $D$, move along the initial ray "$-1$" units – in other words, "back up" 1 unit, then rotate counter-clockwise by $\pi/4$. The results are given in Figure 9.4.2.

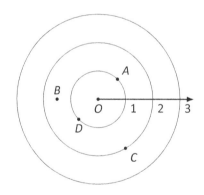

Figure 9.4.2: Plotting polar points in Example 9.4.1.

Notes:

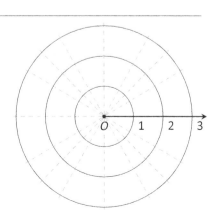

Consider the following two points: $A = P(1, \pi)$ and $B = P(-1, 0)$. To locate A, go out 1 unit on the initial ray then rotate $\pi$ radians; to locate B, go out $-1$ units on the initial ray and don't rotate. One should see that A and B are located at the same point in the plane. We can also consider $C = P(1, 3\pi)$, or $D = P(1, -\pi)$; all four of these points share the same location.

This ability to identify a point in the plane with multiple polar coordinates is both a "blessing" and a "curse." We will see that it is beneficial as we can plot beautiful functions that intersect themselves (much like we saw with parametric functions). The unfortunate part of this is that it can be difficult to determine when this happens. We'll explore this more later in this section.

## Polar to Rectangular Conversion

It is useful to recognize both the rectangular (or, Cartesian) coordinates of a point in the plane and its polar coordinates. Figure 9.4.3 shows a point P in the plane with rectangular coordinates $(x, y)$ and polar coordinates $P(r, \theta)$. Using trigonometry, we can make the identities given in the following Key Idea.

> **Key Idea 9.4.1   Converting Between Rectangular and Polar Coordinates**
>
> Given the polar point $P(r, \theta)$, the rectangular coordinates are determined by
> $$x = r\cos\theta \qquad y = r\sin\theta.$$
>
> Given the rectangular coordinates $(x, y)$, the polar coordinates are determined by
> $$r^2 = x^2 + y^2 \qquad \tan\theta = \frac{y}{x}.$$

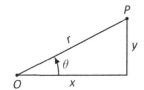

Figure 9.4.3: Converting between rectangular and polar coordinates.

**Example 9.4.2   Converting Between Polar and Rectangular Coordinates**

1. Convert the polar coordinates $P(2, 2\pi/3)$ and $P(-1, 5\pi/4)$ to rectangular coordinates.

2. Convert the rectangular coordinates $(1, 2)$ and $(-1, 1)$ to polar coordinates.

**SOLUTION**

Notes:

1. (a) We start with $P(2, 2\pi/3)$. Using Key Idea 9.4.1, we have
$$x = 2\cos(2\pi/3) = -1 \qquad y = 2\sin(2\pi/3) = \sqrt{3}.$$
So the rectangular coordinates are $(-1, \sqrt{3}) \approx (-1, 1.732)$.

   (b) The polar point $P(-1, 5\pi/4)$ is converted to rectangular with:
$$x = -1\cos(5\pi/4) = \sqrt{2}/2 \qquad y = -1\sin(5\pi/4) = \sqrt{2}/2.$$
So the rectangular coordinates are $(\sqrt{2}/2, \sqrt{2}/2) \approx (0.707, 0.707)$.

   These points are plotted in Figure 9.4.4 (a). The rectangular coordinate system is drawn lightly under the polar coordinate system so that the relationship between the two can be seen.

2. (a) To convert the rectangular point $(1, 2)$ to polar coordinates, we use the Key Idea to form the following two equations:
$$1^2 + 2^2 = r^2 \qquad \tan\theta = \frac{2}{1}.$$
The first equation tells us that $r = \sqrt{5}$. Using the inverse tangent function, we find
$$\tan\theta = 2 \quad \Rightarrow \quad \theta = \tan^{-1} 2 \approx 1.11 \approx 63.43°.$$
Thus polar coordinates of $(1, 2)$ are $P(\sqrt{5}, 1.11)$.

   (b) To convert $(-1, 1)$ to polar coordinates, we form the equations
$$(-1)^2 + 1^2 = r^2 \qquad \tan\theta = \frac{1}{-1}.$$
Thus $r = \sqrt{2}$. We need to be careful in computing $\theta$: using the inverse tangent function, we have
$$\tan\theta = -1 \quad \Rightarrow \quad \theta = \tan^{-1}(-1) = -\pi/4 = -45°.$$
This is not the angle we desire. The range of $\tan^{-1} x$ is $(-\pi/2, \pi/2)$; that is, it returns angles that lie in the 1st and 4th quadrants. To find locations in the 2nd and 3rd quadrants, add $\pi$ to the result of $\tan^{-1} x$. So $\pi + (-\pi/4)$ puts the angle at $3\pi/4$. Thus the polar point is $P(\sqrt{2}, 3\pi/4)$.

   An alternate method is to use the angle $\theta$ given by arctangent, but change the sign of $r$. Thus we could also refer to $(-1, 1)$ as $P(-\sqrt{2}, -\pi/4)$.

   These points are plotted in Figure 9.4.4 (b). The polar system is drawn lightly under the rectangular grid with rays to demonstrate the angles used.

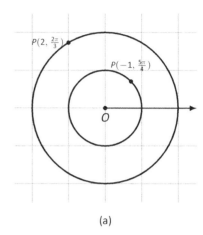

(a)

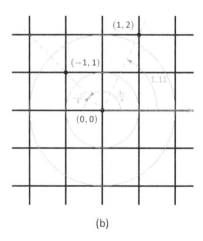

(b)

Figure 9.4.4: Plotting rectangular and polar points in Example 9.4.2.

## Polar Functions and Polar Graphs

Defining a new coordinate system allows us to create a new kind of function, a **polar function**. Rectangular coordinates lent themselves well to creating functions that related $x$ and $y$, such as $y = x^2$. Polar coordinates allow us to create functions that relate $r$ and $\theta$. Normally these functions look like $r = f(\theta)$, although we can create functions of the form $\theta = f(r)$. The following examples introduce us to this concept.

**Example 9.4.3    Introduction to Graphing Polar Functions**
Describe the graphs of the following polar functions.

1. $r = 1.5$
2. $\theta = \pi/4$

**SOLUTION**

1. The equation $r = 1.5$ describes all points that are 1.5 units from the pole; as the angle is not specified, any $\theta$ is allowable. All points 1.5 units from the pole describes a circle of radius 1.5.

   We can consider the rectangular equivalent of this equation; using $r^2 = x^2 + y^2$, we see that $1.5^2 = x^2 + y^2$, which we recognize as the equation of a circle centered at $(0, 0)$ with radius 1.5. This is sketched in Figure 9.4.5.

2. The equation $\theta = \pi/4$ describes all points such that the line through them and the pole make an angle of $\pi/4$ with the initial ray. As the radius $r$ is not specified, it can be any value (even negative). Thus $\theta = \pi/4$ describes the line through the pole that makes an angle of $\pi/4 = 45°$ with the initial ray.

   We can again consider the rectangular equivalent of this equation. Combine $\tan\theta = y/x$ and $\theta = \pi/4$:

   $$\tan \pi/4 = y/x \quad \Rightarrow x \tan \pi/4 = y \quad \Rightarrow y = x.$$

   This graph is also plotted in Figure 9.4.5.

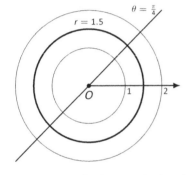

Figure 9.4.5: Plotting standard polar plots.

The basic rectangular equations of the form $x = h$ and $y = k$ create vertical and horizontal lines, respectively; the basic polar equations $r = h$ and $\theta = \alpha$ create circles and lines through the pole, respectively. With this as a foundation, we can create more complicated polar functions of the form $r = f(\theta)$. The input is an angle; the output is a length, how far in the direction of the angle to go out.

Notes:

We sketch these functions much like we sketch rectangular and parametric functions: we plot lots of points and "connect the dots" with curves. We demonstrate this in the following example.

**Example 9.4.4    Sketching Polar Functions**
Sketch the polar function $r = 1 + \cos\theta$ on $[0, 2\pi]$ by plotting points.

**Solution**    A common question when sketching curves by plotting points is "Which points should I plot?" With rectangular equations, we often choose "easy" values – integers, then add more if needed. When plotting polar equations, start with the "common" angles – multiples of $\pi/6$ and $\pi/4$. Figure 9.4.6 gives a table of just a few values of $\theta$ in $[0, \pi]$.

Consider the point $P(2, 0)$ determined by the first line of the table. The angle is 0 radians – we do not rotate from the initial ray – then we go out 2 units from the pole. When $\theta = \pi/6$, $r = 1.866$ (actually, it is $1 + \sqrt{3}/2$); so rotate by $\pi/6$ radians and go out 1.866 units.

The graph shown uses more points, connected with straight lines. (The points on the graph that correspond to points in the table are signified with larger dots.) Such a sketch is likely good enough to give one an idea of what the graph looks like.

**Technology Note:** Plotting functions in this way can be tedious, just as it was with rectangular functions. To obtain very accurate graphs, technology is a great aid. Most graphing calculators can plot polar functions; in the menu, set the plotting mode to something like `polar` or `POL`, depending on one's calculator. As with plotting parametric functions, the viewing "window" no longer determines the x-values that are plotted, so additional information needs to be provided. Often with the "window" settings are the settings for the beginning and ending $\theta$ values (often called $\theta_{\min}$ and $\theta_{\max}$) as well as the $\theta_{\text{step}}$ – that is, how far apart the $\theta$ values are spaced. The smaller the $\theta_{\text{step}}$ value, the more accurate the graph (which also increases plotting time). Using technology, we graphed the polar function $r = 1 + \cos\theta$ from Example 9.4.4 in Figure 9.4.7.

**Example 9.4.5    Sketching Polar Functions**
Sketch the polar function $r = \cos(2\theta)$ on $[0, 2\pi]$ by plotting points.

**Solution**    We start by making a table of $\cos(2\theta)$ evaluated at common angles $\theta$, as shown in Figure 9.4.8. These points are then plotted in Figure 9.4.9 (a). This particular graph "moves" around quite a bit and one can easily forget which points should be connected to each other. To help us with this, we numbered each point in the table and on the graph.

| $\theta$ | $r = 1 + \cos\theta$ |
|---|---|
| 0 | 2 |
| $\pi/6$ | 1.86603 |
| $\pi/2$ | 1 |
| $4\pi/3$ | 0.5 |
| $7\pi/4$ | 1.70711 |

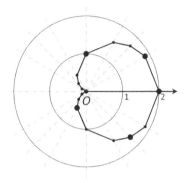

Figure 9.4.6: Graphing a polar function in Example 9.4.4 by plotting points.

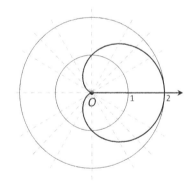

Figure 9.4.7: Using technology to graph a polar function.

Notes:

## Chapter 9  Curves in the Plane

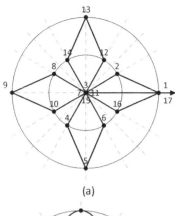

(a)

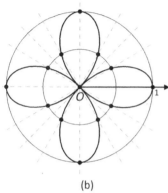

(b)

Figure 9.4.9: Polar plots from Example 9.4.5.

| Pt. | $\theta$ | $\cos(2\theta)$ | Pt. | $\theta$ | $\cos(2\theta)$ |
|---|---|---|---|---|---|
| 1 | 0 | 1. | 10 | $7\pi/6$ | 0.5 |
| 2 | $\pi/6$ | 0.5 | 11 | $5\pi/4$ | 0. |
| 3 | $\pi/4$ | 0. | 12 | $4\pi/3$ | $-0.5$ |
| 4 | $\pi/3$ | $-0.5$ | 13 | $3\pi/2$ | $-1.$ |
| 5 | $\pi/2$ | $-1.$ | 14 | $5\pi/3$ | $-0.5$ |
| 6 | $2\pi/3$ | $-0.5$ | 15 | $7\pi/4$ | 0. |
| 7 | $3\pi/4$ | 0. | 16 | $11\pi/6$ | 0.5 |
| 8 | $5\pi/6$ | 0.5 | 17 | $2\pi$ | 1. |
| 9 | $\pi$ | 1. | | | |

Figure 9.4.8: Tables of points for plotting a polar curve.

Using more points (and the aid of technology) a smoother plot can be made as shown in Figure 9.4.9 (b). This plot is an example of a *rose curve*.

It is sometimes desirable to refer to a graph via a polar equation, and other times by a rectangular equation. Therefore it is necessary to be able to convert between polar and rectangular functions, which we practice in the following example. We will make frequent use of the identities found in Key Idea 9.4.1.

**Example 9.4.6**  **Converting between rectangular and polar equations.**

Convert from rectangular to polar.

1. $y = x^2$
2. $xy = 1$

Convert from polar to rectangular.

3. $r = \dfrac{2}{\sin\theta - \cos\theta}$
4. $r = 2\cos\theta$

**SOLUTION**

1. Replace $y$ with $r\sin\theta$ and replace $x$ with $r\cos\theta$, giving:

$$y = x^2$$
$$r\sin\theta = r^2\cos^2\theta$$
$$\frac{\sin\theta}{\cos^2\theta} = r$$

We have found that $r = \sin\theta/\cos^2\theta = \tan\theta\sec\theta$. The domain of this polar function is $(-\pi/2, \pi/2)$; plot a few points to see how the familiar parabola is traced out by the polar equation.

Notes:

2. We again replace $x$ and $y$ using the standard identities and work to solve for $r$:

$$xy = 1$$
$$r\cos\theta \cdot r\sin\theta = 1$$
$$r^2 = \frac{1}{\cos\theta \sin\theta}$$
$$r = \frac{1}{\sqrt{\cos\theta \sin\theta}}$$

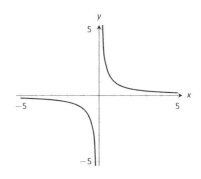

Figure 9.4.10: Graphing $xy = 1$ from Example 9.4.6.

This function is valid only when the product of $\cos\theta \sin\theta$ is positive. This occurs in the first and third quadrants, meaning the domain of this polar function is $(0, \pi/2) \cup (\pi, 3\pi/2)$.

We can rewrite the original rectangular equation $xy = 1$ as $y = 1/x$. This is graphed in Figure 9.4.10; note how it only exists in the first and third quadrants.

3. There is no set way to convert from polar to rectangular; in general, we look to form the products $r\cos\theta$ and $r\sin\theta$, and then replace these with $x$ and $y$, respectively. We start in this problem by multiplying both sides by $\sin\theta - \cos\theta$:

$$r = \frac{2}{\sin\theta - \cos\theta}$$
$$r(\sin\theta - \cos\theta) = 2$$
$$r\sin\theta - r\cos\theta = 2. \quad \text{Now replace with } y \text{ and } x:$$
$$y - x = 2$$
$$y = x + 2.$$

The original polar equation, $r = 2/(\sin\theta - \cos\theta)$ does not easily reveal that its graph is simply a line. However, our conversion shows that it is. The upcoming gallery of polar curves gives the general equations of lines in polar form.

4. By multiplying both sides by $r$, we obtain both an $r^2$ term and an $r\cos\theta$ term, which we replace with $x^2 + y^2$ and $x$, respectively.

$$r = 2\cos\theta$$
$$r^2 = 2r\cos\theta$$
$$x^2 + y^2 = 2x.$$

Notes:

# Chapter 9 Curves in the Plane

We recognize this as a circle; by completing the square we can find its radius and center.

$$x^2 - 2x + y^2 = 0$$
$$(x-1)^2 + y^2 = 1.$$

The circle is centered at $(1,0)$ and has radius 1. The upcoming gallery of polar curves gives the equations of *some* circles in polar form; circles with arbitrary centers have a complicated polar equation that we do not consider here.

Some curves have very simple polar equations but rather complicated rectangular ones. For instance, the equation $r = 1 + \cos\theta$ describes a *cardioid* (a shape important to the sensitivity of microphones, among other things; one is graphed in the gallery in the Limaçon section). It's rectangular form is not nearly as simple; it is the implicit equation $x^4 + y^4 + 2x^2y^2 - 2xy^2 - 2x^3 - y^2 = 0$. The conversion is not "hard," but takes several steps, and is left as a problem in the Exercise section.

## Gallery of Polar Curves

There are a number of basic and "classic" polar curves, famous for their beauty and/or applicability to the sciences. This section ends with a small gallery of some of these graphs. We encourage the reader to understand how these graphs are formed, and to investigate with technology other types of polar functions.

---

### Lines

**Through the origin:**

$\theta = \alpha$

**Horizontal line:**

$r = a \csc\theta$

**Vertical line:**

$r = a \sec\theta$

**Not through origin:**

$r = \dfrac{b}{\sin\theta - m\cos\theta}$

---

Notes:

## Circles

**Centered on x-axis:**
$r = a\cos\theta$

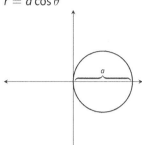

**Centered on y-axis:**
$r = a\sin\theta$

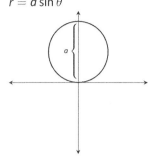

**Centered on origin:**
$r = a$

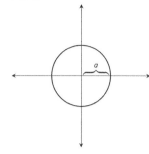

## Spiral

**Archimedean spiral**
$r = \theta$

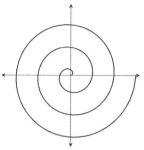

## Limaçons

Symmetric about x-axis: $r = a \pm b\cos\theta$; Symmetric about y-axis: $r = a \pm b\sin\theta$; $a, b > 0$

**With inner loop:**
$\dfrac{a}{b} < 1$

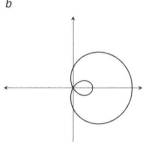

**Cardioid:**
$\dfrac{a}{b} = 1$

**Dimpled:**
$1 < \dfrac{a}{b} < 2$

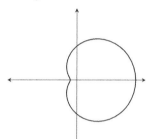

**Convex:**
$\dfrac{a}{b} > 2$

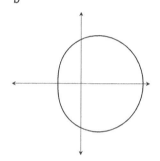

## Rose Curves

Symmetric about x-axis: $r = a\cos(n\theta)$; Symmetric about y-axis: $r = a\sin(n\theta)$
Curve contains 2n petals when n is even and n petals when n is odd.

$r = a\cos(2\theta)$

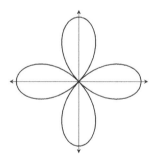

$r = a\sin(2\theta)$

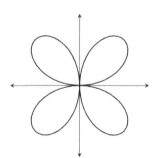

$r = a\cos(3\theta)$

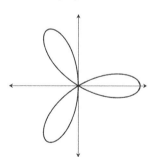

$r = a\sin(3\theta)$

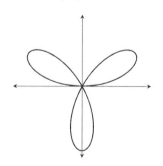

## Special Curves

**Rose curves**

$r = a\sin(\theta/5)$

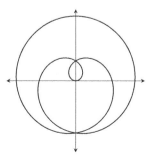

$r = a\sin(2\theta/5)$

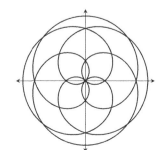

**Lemniscate:**
$r^2 = a^2\cos(2\theta)$

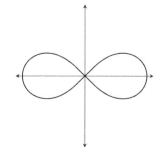

**Eight Curve:**
$r^2 = a^2\sec^4\theta\cos(2\theta)$

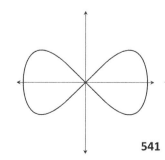

## Chapter 9 Curves in the Plane

Earlier we discussed how each point in the plane does not have a unique representation in polar form. This can be a "good" thing, as it allows for the beautiful and interesting curves seen in the preceding gallery. However, it can also be a "bad" thing, as it can be difficult to determine where two curves intersect.

**Example 9.4.7      Finding points of intersection with polar curves**
Determine where the graphs of the polar equations $r = 1 + 3\cos\theta$ and $r = \cos\theta$ intersect.

**SOLUTION**    As technology is generally readily available, it is usually a good idea to start with a graph. We have graphed the two functions in Figure 9.4.11(a); to better discern the intersection points, part (b) of the figure zooms in around the origin. We start by setting the two functions equal to each other and solving for $\theta$:

$$1 + 3\cos\theta = \cos\theta$$
$$2\cos\theta = -1$$
$$\cos\theta = -\frac{1}{2}$$
$$\theta = \frac{2\pi}{3}, \frac{4\pi}{3}.$$

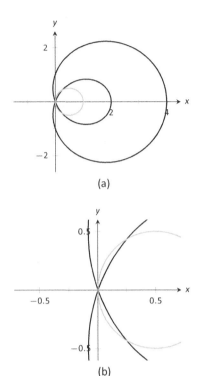

Figure 9.4.11: Graphs to help determine the points of intersection of the polar functions given in Example 9.4.7.

(There are, of course, infinite solutions to the equation $\cos\theta = -1/2$; as the limaçon is traced out once on $[0, 2\pi]$, we restrict our solutions to this interval.)

We need to analyze this solution. When $\theta = 2\pi/3$ we obtain the point of intersection that lies in the 4$^{\text{th}}$ quadrant. When $\theta = 4\pi/3$, we get the point of intersection that lies in the 2$^{\text{nd}}$ quadrant. There is more to say about this second intersection point, however. The circle defined by $r = \cos\theta$ is traced out once on $[0, \pi]$, meaning that this point of intersection occurs while tracing out the circle a second time. It seems strange to pass by the point once and then recognize it as a point of intersection only when arriving there a "second time." The first time the circle arrives at this point is when $\theta = \pi/3$. It is key to understand that these two points are the same: $(\cos\pi/3, \pi/3)$ and $(\cos 4\pi/3, 4\pi/3)$.

To summarize what we have done so far, we have found two points of intersection: when $\theta = 2\pi/3$ and when $\theta = 4\pi/3$. When referencing the circle $r = \cos\theta$, the latter point is better referenced as when $\theta = \pi/3$.

There is yet another point of intersection: the pole (or, the origin). We did not recognize this intersection point using our work above as each graph arrives at the pole at a different $\theta$ value.

A graph intersects the pole when $r = 0$. Considering the circle $r = \cos\theta$, $r = 0$ when $\theta = \pi/2$ (and odd multiples thereof, as the circle is repeatedly

Notes:

traced). The limaçon intersects the pole when $1 + 3\cos\theta = 0$; this occurs when $\cos\theta = -1/3$, or for $\theta = \cos^{-1}(-1/3)$. This is a nonstandard angle, approximately $\theta = 1.9106 = 109.47°$. The limaçon intersects the pole twice in $[0, 2\pi]$; the other angle at which the limaçon is at the pole is the reflection of the first angle across the $x$-axis. That is, $\theta = 4.3726 = 250.53°$.

If all one is concerned with is the $(x, y)$ coordinates at which the graphs intersect, much of the above work is extraneous. We know they intersect at $(0, 0)$; we might not care at what $\theta$ value. Likewise, using $\theta = 2\pi/3$ and $\theta = 4\pi/3$ can give us the needed rectangular coordinates. However, in the next section we apply calculus concepts to polar functions. When computing the area of a region bounded by polar curves, understanding the nuances of the points of intersection becomes important.

Notes:

# Exercises 9.4

## Terms and Concepts

1. In your own words, describe how to plot the polar point $P(r, \theta)$.

2. T/F: When plotting a point with polar coordinate $P(r, \theta)$, $r$ must be positive.

3. T/F: Every point in the Cartesian plane can be represented by a polar coordinate.

4. T/F: Every point in the Cartesian plane can be represented uniquely by a polar coordinate.

## Problems

5. Plot the points with the given polar coordinates.

   (a) $A = P(2, 0)$
   (b) $B = P(1, \pi)$
   (c) $C = P(-2, \pi/2)$
   (d) $D = P(1, \pi/4)$

6. Plot the points with the given polar coordinates.

   (a) $A = P(2, 3\pi)$
   (b) $B = P(1, -\pi)$
   (c) $C = P(1, 2)$
   (d) $D = P(1/2, 5\pi/6)$

7. For each of the given points give two sets of polar coordinates that identify it, where $0 \leq \theta \leq 2\pi$.

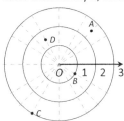

8. For each of the given points give two sets of polar coordinates that identify it, where $-\pi \leq \theta \leq \pi$.

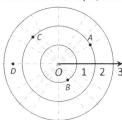

9. Convert each of the following polar coordinates to rectangular, and each of the following rectangular coordinates to polar.

   (a) $A = P(2, \pi/4)$
   (b) $B = P(2, -\pi/4)$
   (c) $C = (2, -1)$
   (d) $D = (-2, 1)$

10. Convert each of the following polar coordinates to rectangular, and each of the following rectangular coordinates to polar.

    (a) $A = P(3, \pi)$
    (b) $B = P(1, 2\pi/3)$
    (c) $C = (0, 4)$
    (d) $D = (1, -\sqrt{3})$

**In Exercises 11 – 30, graph the polar function on the given interval.**

11. $r = 2, \quad 0 \leq \theta \leq \pi/2$

12. $\theta = \pi/6, \quad -1 \leq r \leq 2$

13. $r = 1 - \cos\theta, \quad [0, 2\pi]$

14. $r = 2 + \sin\theta, \quad [0, 2\pi]$

15. $r = 2 - \sin\theta, \quad [0, 2\pi]$

16. $r = 1 - 2\sin\theta, \quad [0, 2\pi]$

17. $r = 1 + 2\sin\theta, \quad [0, 2\pi]$

18. $r = \cos(2\theta), \quad [0, 2\pi]$

19. $r = \sin(3\theta), \quad [0, \pi]$

20. $r = \cos(\theta/3), \quad [0, 3\pi]$

21. $r = \cos(2\theta/3), \quad [0, 6\pi]$

22. $r = \theta/2, \quad [0, 4\pi]$

23. $r = 3\sin(\theta), \quad [0, \pi]$

24. $r = 2\cos(\theta), \quad [0, \pi/2]$

25. $r = \cos\theta \sin\theta, \quad [0, 2\pi]$

26. $r = \theta^2 - (\pi/2)^2, \quad [-\pi, \pi]$

27. $r = \dfrac{3}{5\sin\theta - \cos\theta}, \quad [0, 2\pi]$

28. $r = \dfrac{-2}{3\cos\theta - 2\sin\theta}, \quad [0, 2\pi]$

29. $r = 3\sec\theta, \quad (-\pi/2, \pi/2)$

30. $r = 3\csc\theta, \quad (0, \pi)$

**In Exercises 31 – 40, convert the polar equation to a rectangular equation.**

31. $r = 6\cos\theta$

32. $r = -4\sin\theta$

33. $r = \cos\theta + \sin\theta$

34. $r = \dfrac{7}{5\sin\theta - 2\cos\theta}$

35. $r = \dfrac{3}{\cos\theta}$

36. $r = \dfrac{4}{\sin\theta}$

37. $r = \tan\theta$

38. $r = \cot\theta$

39. $r = 2$

40. $\theta = \pi/6$

**In Exercises 41 – 48, convert the rectangular equation to a polar equation.**

41. $y = x$

42. $y = 4x + 7$

43. $x = 5$

44. $y = 5$

45. $x = y^2$

46. $x^2 y = 1$

47. $x^2 + y^2 = 7$

48. $(x+1)^2 + y^2 = 1$

**In Exercises 49 – 56, find the points of intersection of the polar graphs.**

49. $r = \sin(2\theta)$ and $r = \cos\theta$ on $[0, \pi]$

50. $r = \cos(2\theta)$ and $r = \cos\theta$ on $[0, \pi]$

51. $r = 2\cos\theta$ and $r = 2\sin\theta$ on $[0, \pi]$

52. $r = \sin\theta$ and $r = \sqrt{3} + 3\sin\theta$ on $[0, 2\pi]$

53. $r = \sin(3\theta)$ and $r = \cos(3\theta)$ on $[0, \pi]$

54. $r = 3\cos\theta$ and $r = 1 + \cos\theta$ on $[-\pi, \pi]$

55. $r = 1$ and $r = 2\sin(2\theta)$ on $[0, 2\pi]$

56. $r = 1 - \cos\theta$ and $r = 1 + \sin\theta$ on $[0, 2\pi]$

57. Pick a integer value for $n$, where $n \neq 2, 3$, and use technology to plot $r = \sin\left(\dfrac{m}{n}\theta\right)$ for three different integer values of $m$. Sketch these and determine a minimal interval on which the entire graph is shown.

58. Create your own polar function, $r = f(\theta)$ and sketch it. Describe why the graph looks as it does.

## 9.5 Calculus and Polar Functions

The previous section defined polar coordinates, leading to polar functions. We investigated plotting these functions and solving a fundamental question about their graphs, namely, where do two polar graphs intersect?

We now turn our attention to answering other questions, whose solutions require the use of calculus. A basis for much of what is done in this section is the ability to turn a polar function $r = f(\theta)$ into a set of parametric equations. Using the identities $x = r\cos\theta$ and $y = r\sin\theta$, we can create the parametric equations $x = f(\theta)\cos\theta$, $y = f(\theta)\sin\theta$ and apply the concepts of Section 9.3.

### Polar Functions and $\dfrac{dy}{dx}$

We are interested in the lines tangent to a given graph, regardless of whether that graph is produced by rectangular, parametric, or polar equations. In each of these contexts, the slope of the tangent line is $\frac{dy}{dx}$. Given $r = f(\theta)$, we are generally *not* concerned with $r' = f'(\theta)$; that describes how fast $r$ changes with respect to $\theta$. Instead, we will use $x = f(\theta)\cos\theta$, $y = f(\theta)\sin\theta$ to compute $\frac{dy}{dx}$.

Using Key Idea 9.3.1 we have

$$\frac{dy}{dx} = \frac{dy}{d\theta} \bigg/ \frac{dx}{d\theta}.$$

Each of the two derivatives on the right hand side of the equality requires the use of the Product Rule. We state the important result as a Key Idea.

---

**Key Idea 9.5.1**     **Finding $\frac{dy}{dx}$ with Polar Functions**

Let $r = f(\theta)$ be a polar function. With $x = f(\theta)\cos\theta$ and $y = f(\theta)\sin\theta$,

$$\frac{dy}{dx} = \frac{f'(\theta)\sin\theta + f(\theta)\cos\theta}{f'(\theta)\cos\theta - f(\theta)\sin\theta}.$$

---

**Example 9.5.1**     **Finding $\frac{dy}{dx}$ with polar functions.**
Consider the limaçon $r = 1 + 2\sin\theta$ on $[0, 2\pi]$.

1. Find the equations of the tangent and normal lines to the graph at $\theta = \pi/4$.

2. Find where the graph has vertical and horizontal tangent lines.

---

Notes:

## 9.5 Calculus and Polar Functions

**SOLUTION**

1. We start by computing $\frac{dy}{dx}$. With $f'(\theta) = 2\cos\theta$, we have

$$\frac{dy}{dx} = \frac{2\cos\theta\sin\theta + \cos\theta(1 + 2\sin\theta)}{2\cos^2\theta - \sin\theta(1 + 2\sin\theta)}$$
$$= \frac{\cos\theta(4\sin\theta + 1)}{2(\cos^2\theta - \sin^2\theta) - \sin\theta}.$$

When $\theta = \pi/4$, $\frac{dy}{dx} = -2\sqrt{2} - 1$ (this requires a bit of simplification). In rectangular coordinates, the point on the graph at $\theta = \pi/4$ is $(1 + \sqrt{2}/2, 1 + \sqrt{2}/2)$. Thus the rectangular equation of the line tangent to the limaçon at $\theta = \pi/4$ is

$$y = (-2\sqrt{2} - 1)(x - (1 + \sqrt{2}/2)) + 1 + \sqrt{2}/2 \approx -3.83x + 8.24.$$

The limaçon and the tangent line are graphed in Figure 9.5.1.

The normal line has the opposite–reciprocal slope as the tangent line, so its equation is

$$y \approx \frac{1}{3.83}x + 1.26.$$

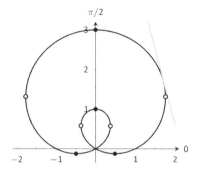

Figure 9.5.1: The limaçon in Example 9.5.1 with its tangent line at $\theta = \pi/4$ and points of vertical and horizontal tangency.

2. To find the horizontal lines of tangency, we find where $\frac{dy}{dx} = 0$; thus we find where the numerator of our equation for $\frac{dy}{dx}$ is 0.

$$\cos\theta(4\sin\theta + 1) = 0 \quad\Rightarrow\quad \cos\theta = 0 \quad\text{or}\quad 4\sin\theta + 1 = 0.$$

On $[0, 2\pi]$, $\cos\theta = 0$ when $\theta = \pi/2, 3\pi/2$.

Setting $4\sin\theta + 1 = 0$ gives $\theta = \sin^{-1}(-1/4) \approx -0.2527 = -14.48°$. We want the results in $[0, 2\pi]$; we also recognize there are two solutions, one in the $3^{\text{rd}}$ quadrant and one in the $4^{\text{th}}$. Using reference angles, we have our two solutions as $\theta = 3.39$ and $6.03$ radians. The four points we obtained where the limaçon has a horizontal tangent line are given in Figure 9.5.1 with black–filled dots.

To find the vertical lines of tangency, we set the denominator of $\frac{dy}{dx} = 0$.

$$2(\cos^2\theta - \sin^2\theta) - \sin\theta = 0.$$

Convert the $\cos^2\theta$ term to $1 - \sin^2\theta$:

$$2(1 - \sin^2\theta - \sin^2\theta) - \sin\theta = 0$$
$$4\sin^2\theta + \sin\theta - 2 = 0.$$

Notes:

Recognize this as a quadratic in the variable $\sin\theta$. Using the quadratic formula, we have

$$\sin\theta = \frac{-1\pm\sqrt{33}}{8}.$$

We solve $\sin\theta = \frac{-1+\sqrt{33}}{8}$ and $\sin\theta = \frac{-1-\sqrt{33}}{8}$:

$$\sin\theta = \frac{-1+\sqrt{33}}{8} \qquad \sin\theta = \frac{-1-\sqrt{33}}{8}$$
$$\theta = \sin^{-1}\left(\frac{-1+\sqrt{33}}{8}\right) \qquad \theta = \sin^{-1}\left(\frac{-1-\sqrt{33}}{8}\right)$$
$$\theta = 0.6349 \qquad \theta = -1.0030$$

In each of the solutions above, we only get one of the possible two solutions as $\sin^{-1}x$ only returns solutions in $[-\pi/2, \pi/2]$, the 4$^{\text{th}}$ and 1$^{\text{st}}$ quadrants. Again using reference angles, we have:

$$\sin\theta = \frac{-1+\sqrt{33}}{8} \quad\Rightarrow\quad \theta = 0.6349,\ 2.5067 \text{ radians}$$

and

$$\sin\theta = \frac{-1-\sqrt{33}}{8} \quad\Rightarrow\quad \theta = 4.1446,\ 5.2802 \text{ radians}.$$

These points are also shown in Figure 9.5.1 with white–filled dots.

When the graph of the polar function $r = f(\theta)$ intersects the pole, it means that $f(\alpha) = 0$ for some angle $\alpha$. Thus the formula for $\frac{dy}{dx}$ in such instances is very simple, reducing simply to

$$\frac{dy}{dx} = \tan\alpha.$$

This equation makes an interesting point. It tells us the slope of the tangent line at the pole is $\tan\alpha$; some of our previous work (see, for instance, Example 9.4.3) shows us that the line through the pole with slope $\tan\alpha$ has polar equation $\theta = \alpha$. Thus when a polar graph touches the pole at $\theta = \alpha$, the equation of the tangent line at the pole is $\theta = \alpha$.

**Example 9.5.2    Finding tangent lines at the pole.**
Let $r = 1 + 2\sin\theta$, a limaçon. Find the equations of the lines tangent to the graph at the pole.

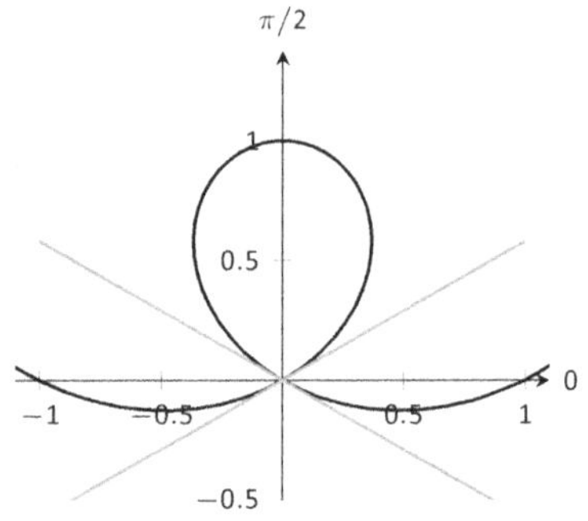

Figure 9.5.2: Graphing the tangent lines at the pole in Example 9.5.2.

Notes:

## 9.5 Calculus and Polar Functions

**SOLUTION**   We need to know when $r = 0$.

$$1 + 2\sin\theta = 0$$
$$\sin\theta = -1/2$$
$$\theta = \frac{7\pi}{6}, \frac{11\pi}{6}.$$

Thus the equations of the tangent lines, in polar, are $\theta = 7\pi/6$ and $\theta = 11\pi/6$. In rectangular form, the tangent lines are $y = \tan(7\pi/6)x$ and $y = \tan(11\pi/6)x$. The full limaçon can be seen in Figure 9.5.1; we zoom in on the tangent lines in Figure 9.5.2.

**Note:** Recall that the area of a sector of a circle with radius $r$ subtended by an angle $\theta$ is $A = \frac{1}{2}\theta r^2$.

## Area

When using rectangular coordinates, the equations $x = h$ and $y = k$ defined vertical and horizontal lines, respectively, and combinations of these lines create rectangles (hence the name "rectangular coordinates"). It is then somewhat natural to use rectangles to approximate area as we did when learning about the definite integral.

When using polar coordinates, the equations $\theta = \alpha$ and $r = c$ form lines through the origin and circles centered at the origin, respectively, and combinations of these curves form sectors of circles. It is then somewhat natural to calculate the area of regions defined by polar functions by first approximating with sectors of circles.

Consider Figure 9.5.3 (a) where a region defined by $r = f(\theta)$ on $[\alpha, \beta]$ is given. (Note how the "sides" of the region are the lines $\theta = \alpha$ and $\theta = \beta$, whereas in rectangular coordinates the "sides" of regions were often the vertical lines $x = a$ and $x = b$.)

Partition the interval $[\alpha, \beta]$ into $n$ equally spaced subintervals as $\alpha = \theta_1 < \theta_2 < \cdots < \theta_{n+1} = \beta$. The length of each subinterval is $\Delta\theta = (\beta - \alpha)/n$, representing a small change in angle. The area of the region defined by the $i^{th}$ subinterval $[\theta_i, \theta_{i+1}]$ can be approximated with a sector of a circle with radius $f(c_i)$, for some $c_i$ in $[\theta_i, \theta_{i+1}]$. The area of this sector is $\frac{1}{2}f(c_i)^2\Delta\theta$. This is shown in part (b) of the figure, where $[\alpha, \beta]$ has been divided into 4 subintervals. We approximate the area of the whole region by summing the areas of all sectors:

$$\text{Area} \approx \sum_{i=1}^{n} \frac{1}{2}f(c_i)^2 \Delta\theta.$$

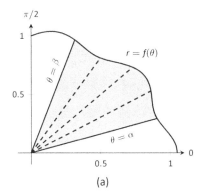

(a)

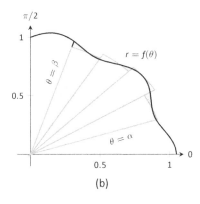

(b)

Figure 9.5.3: Computing the area of a polar region.

This is a Riemann sum. By taking the limit of the sum as $n \to \infty$, we find the

---

Notes:

## Chapter 9  Curves in the Plane

exact area of the region in the form of a definite integral.

> **Theorem 9.5.1**  **Area of a Polar Region**
>
> Let $f$ be continuous and non-negative on $[\alpha, \beta]$, where $0 \leq \beta - \alpha \leq 2\pi$. The area $A$ of the region bounded by the curve $r = f(\theta)$ and the lines $\theta = \alpha$ and $\theta = \beta$ is
>
> $$A = \frac{1}{2}\int_\alpha^\beta f(\theta)^2\, d\theta = \frac{1}{2}\int_\alpha^\beta r^2\, d\theta$$

**Note:** Example 9.5.3 requires the use of the integral $\int \cos^2\theta\, d\theta$. This is handled well by using the power reducing formula as found at the end of this text. Due to the nature of the area formula, integrating $\cos^2\theta$ and $\sin^2\theta$ is required often. We offer here these indefinite integrals as a time–saving measure.

$$\int \cos^2\theta\, d\theta = \frac{1}{2}\theta + \frac{1}{4}\sin(2\theta) + C$$

$$\int \sin^2\theta\, d\theta = \frac{1}{2}\theta - \frac{1}{4}\sin(2\theta) + C$$

The theorem states that $0 \leq \beta - \alpha \leq 2\pi$. This ensures that region does not overlap itself, which would give a result that does not correspond directly to the area.

**Example 9.5.3**   **Area of a polar region**
Find the area of the circle defined by $r = \cos\theta$. (Recall this circle has radius $1/2$.)

**Solution**   This is a direct application of Theorem 9.5.1. The circle is traced out on $[0, \pi]$, leading to the integral

$$\begin{aligned}
\text{Area} &= \frac{1}{2}\int_0^\pi \cos^2\theta\, d\theta \\
&= \frac{1}{2}\int_0^\pi \frac{1+\cos(2\theta)}{2}\, d\theta \\
&= \frac{1}{4}\left(\theta + \frac{1}{2}\sin(2\theta)\right)\Big|_0^\pi \\
&= \frac{1}{4}\pi.
\end{aligned}$$

Of course, we already knew the area of a circle with radius $1/2$. We did this example to demonstrate that the area formula is correct.

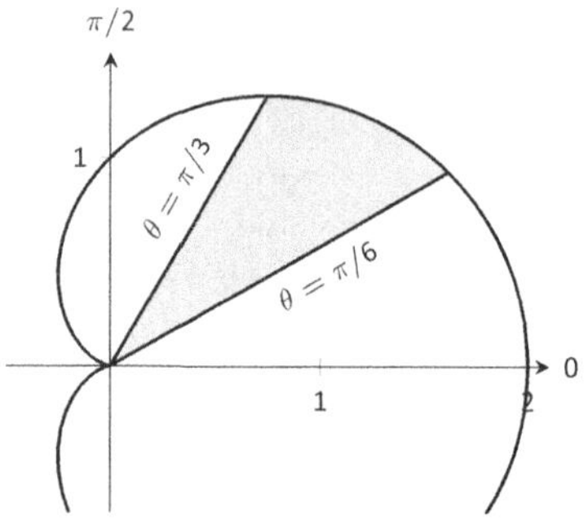

**Example 9.5.4**   **Area of a polar region**
Find the area of the cardioid $r = 1+\cos\theta$ bound between $\theta = \pi/6$ and $\theta = \pi/3$, as shown in Figure 9.5.4.

Figure 9.5.4: Finding the area of the shaded region of a cardioid in Example 9.5.4.

Notes:

## 9.5 Calculus and Polar Functions

**SOLUTION** This is again a direct application of Theorem 9.5.1.

$$\text{Area} = \frac{1}{2} \int_{\pi/6}^{\pi/3} (1+\cos\theta)^2 \, d\theta$$

$$= \frac{1}{2} \int_{\pi/6}^{\pi/3} (1 + 2\cos\theta + \cos^2\theta) \, d\theta$$

$$= \frac{1}{2} \left( \theta + 2\sin\theta + \frac{1}{2}\theta + \frac{1}{4}\sin(2\theta) \right) \Big|_{\pi/6}^{\pi/3}$$

$$= \frac{1}{8}\left(\pi + 4\sqrt{3} - 4\right) \approx 0.7587.$$

### Area Between Curves

Our study of area in the context of rectangular functions led naturally to finding area bounded between curves. We consider the same in the context of polar functions.

Consider the shaded region shown in Figure 9.5.5. We can find the area of this region by computing the area bounded by $r_2 = f_2(\theta)$ and subtracting the area bounded by $r_1 = f_1(\theta)$ on $[\alpha, \beta]$. Thus

$$\text{Area} = \frac{1}{2}\int_\alpha^\beta r_2^2 \, d\theta - \frac{1}{2}\int_\alpha^\beta r_1^2 \, d\theta = \frac{1}{2}\int_\alpha^\beta \left(r_2^2 - r_1^2\right) d\theta.$$

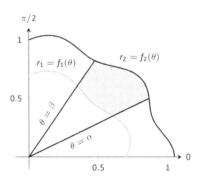

Figure 9.5.5: Illustrating area bound between two polar curves.

---

**Key Idea 9.5.2   Area Between Polar Curves**

The area $A$ of the region bounded by $r_1 = f_1(\theta)$ and $r_2 = f_2(\theta)$, $\theta = \alpha$ and $\theta = \beta$, where $f_1(\theta) \leq f_2(\theta)$ on $[\alpha, \beta]$, is

$$A = \frac{1}{2} \int_\alpha^\beta \left(r_2^2 - r_1^2\right) d\theta.$$

---

**Example 9.5.5   Area between polar curves**

Find the area bounded between the curves $r = 1 + \cos\theta$ and $r = 3\cos\theta$, as shown in Figure 9.5.6.

**SOLUTION** We need to find the points of intersection between these

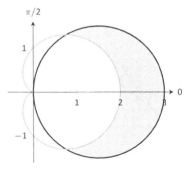

Figure 9.5.6: Finding the area between polar curves in Example 9.5.5.

Notes:

two functions. Setting them equal to each other, we find:

$$1 + \cos\theta = 3\cos\theta$$
$$\cos\theta = 1/2$$
$$\theta = \pm\pi/3$$

Thus we integrate $\frac{1}{2}\left((3\cos\theta)^2 - (1+\cos\theta)^2\right)$ on $[-\pi/3, \pi/3]$.

$$\text{Area} = \frac{1}{2}\int_{-\pi/3}^{\pi/3} \left((3\cos\theta)^2 - (1+\cos\theta)^2\right) d\theta$$

$$= \frac{1}{2}\int_{-\pi/3}^{\pi/3} \left(8\cos^2\theta - 2\cos\theta - 1\right) d\theta$$

$$= \frac{1}{2}\left(2\sin(2\theta) - 2\sin\theta + 3\theta\right)\Big|_{-\pi/3}^{\pi/3}$$

$$= \pi.$$

Amazingly enough, the area between these curves has a "nice" value.

**Example 9.5.6    Area defined by polar curves**
Find the area bounded between the polar curves $r = 1$ and $r = 2\cos(2\theta)$, as shown in Figure 9.5.7 (a).

**SOLUTION**    We need to find the point of intersection between the two curves. Setting the two functions equal to each other, we have

$$2\cos(2\theta) = 1 \quad \Rightarrow \quad \cos(2\theta) = \frac{1}{2} \quad \Rightarrow \quad 2\theta = \pi/3 \quad \Rightarrow \quad \theta = \pi/6.$$

In part (b) of the figure, we zoom in on the region and note that it is not really bounded *between* two polar curves, but rather *by* two polar curves, along with $\theta = 0$. The dashed line breaks the region into its component parts. Below the dashed line, the region is defined by $r = 1$, $\theta = 0$ and $\theta = \pi/6$. (Note: the dashed line lies on the line $\theta = \pi/6$.) Above the dashed line the region is bounded by $r = 2\cos(2\theta)$ and $\theta = \pi/6$. Since we have two separate regions, we find the area using two separate integrals.

Call the area below the dashed line $A_1$ and the area above the dashed line $A_2$. They are determined by the following integrals:

$$A_1 = \frac{1}{2}\int_0^{\pi/6} (1)^2 \, d\theta \qquad A_2 = \frac{1}{2}\int_{\pi/6}^{\pi/4} \left(2\cos(2\theta)\right)^2 d\theta.$$

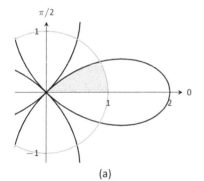

(a)

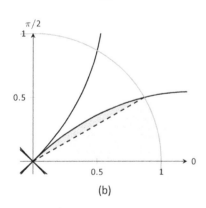

(b)

Figure 9.5.7: Graphing the region bounded by the functions in Example 9.5.6.

Notes:

(The upper bound of the integral computing $A_2$ is $\pi/4$ as $r = 2\cos(2\theta)$ is at the pole when $\theta = \pi/4$.)

We omit the integration details and let the reader verify that $A_1 = \pi/12$ and $A_2 = \pi/12 - \sqrt{3}/8$; the total area is $A = \pi/6 - \sqrt{3}/8$.

## Arc Length

As we have already considered the arc length of curves defined by rectangular and parametric equations, we now consider it in the context of polar equations. Recall that the arc length $L$ of the graph defined by the parametric equations $x = f(t)$, $y = g(t)$ on $[a, b]$ is

$$L = \int_a^b \sqrt{f'(t)^2 + g'(t)^2}\, dt = \int_a^b \sqrt{x'(t)^2 + y'(t)^2}\, dt. \qquad (9.1)$$

Now consider the polar function $r = f(\theta)$. We again use the identities $x = f(\theta)\cos\theta$ and $y = f(\theta)\sin\theta$ to create parametric equations based on the polar function. We compute $x'(\theta)$ and $y'(\theta)$ as done before when computing $\frac{dy}{dx}$, then apply Equation (9.1).

The expression $x'(\theta)^2 + y'(\theta)^2$ can be simplified a great deal; we leave this as an exercise and state that

$$x'(\theta)^2 + y'(\theta)^2 = f'(\theta)^2 + f(\theta)^2.$$

This leads us to the arc length formula.

---

**Theorem 9.5.2    Arc Length of Polar Curves**

Let $r = f(\theta)$ be a polar function with $f'$ continuous on $[\alpha, \beta]$, on which the graph traces itself only once. The arc length $L$ of the graph on $[\alpha, \beta]$ is

$$L = \int_\alpha^\beta \sqrt{f'(\theta)^2 + f(\theta)^2}\, d\theta = \int_\alpha^\beta \sqrt{(r')^2 + r^2}\, d\theta.$$

---

**Example 9.5.7    Arc length of a limaçon**
Find the arc length of the limaçon $r = 1 + 2\sin t$.

**SOLUTION**    With $r = 1 + 2\sin t$, we have $r' = 2\cos t$. The limaçon is traced out once on $[0, 2\pi]$, giving us our bounds of integration. Applying Theo-

Notes:

## Chapter 9 Curves in the Plane

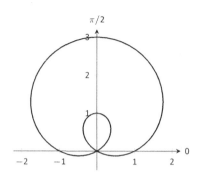

Figure 9.5.8: The limaçon in Example 9.5.7 whose arc length is measured.

rem 9.5.2, we have

$$L = \int_0^{2\pi} \sqrt{(2\cos\theta)^2 + (1+2\sin\theta)^2}\, d\theta$$

$$= \int_0^{2\pi} \sqrt{4\cos^2\theta + 4\sin^2\theta + 4\sin\theta + 1}\, d\theta$$

$$= \int_0^{2\pi} \sqrt{4\sin\theta + 5}\, d\theta$$

$$\approx 13.3649.$$

The final integral cannot be solved in terms of elementary functions, so we resorted to a numerical approximation. (Simpson's Rule, with $n = 4$, approximates the value with 13.0608. Using $n = 22$ gives the value above, which is accurate to 4 places after the decimal.)

### Surface Area

The formula for arc length leads us to a formula for surface area. The following Theorem is based on Theorem 9.3.2.

---

**Theorem 9.5.3    Surface Area of a Solid of Revolution**

Consider the graph of the polar equation $r = f(\theta)$, where $f'$ is continuous on $[\alpha, \beta]$, on which the graph does not cross itself.

1. The surface area of the solid formed by revolving the graph about the initial ray ($\theta = 0$) is:

$$\text{Surface Area} = 2\pi \int_\alpha^\beta f(\theta)\sin\theta \sqrt{f'(\theta)^2 + f(\theta)^2}\, d\theta.$$

2. The surface area of the solid formed by revolving the graph about the line $\theta = \pi/2$ is:

$$\text{Surface Area} = 2\pi \int_\alpha^\beta f(\theta)\cos\theta \sqrt{f'(\theta)^2 + f(\theta)^2}\, d\theta.$$

---

Notes:

## 9.5 Calculus and Polar Functions

**Example 9.5.8  Surface area determined by a polar curve**

Find the surface area formed by revolving one petal of the rose curve $r = \cos(2\theta)$ about its central axis (see Figure 9.5.9).

**SOLUTION** We choose, as implied by the figure, to revolve the portion of the curve that lies on $[0, \pi/4]$ about the initial ray. Using Theorem 9.5.3 and the fact that $f'(\theta) = -2\sin(2\theta)$, we have

$$\text{Surface Area} = 2\pi \int_0^{\pi/4} \cos(2\theta)\sin(\theta)\sqrt{\bigl(-2\sin(2\theta)\bigr)^2 + \bigl(\cos(2\theta)\bigr)^2}\, d\theta$$
$$\approx 1.36707.$$

The integral is another that cannot be evaluated in terms of elementary functions. Simpson's Rule, with $n = 4$, approximates the value at 1.36751.

This chapter has been about curves in the plane. While there is great mathematics to be discovered in the two dimensions of a plane, we live in a three dimensional world and hence we should also look to do mathematics in 3D – that is, in *space*. The next chapter begins our exploration into space by introducing the topic of *vectors*, which are incredibly useful and powerful mathematical objects.

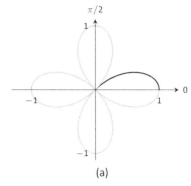

(a)

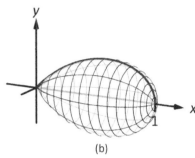

(b)

Figure 9.5.9: Finding the surface area of a rose-curve petal that is revolved around its central axis.

Notes:

# Exercises 9.5

## Terms and Concepts

1. Given polar equation $r = f(\theta)$, how can one create parametric equations of the same curve?

2. With rectangular coordinates, it is natural to approximate area with _____; with polar coordinates, it is natural to approximate area with _____.

## Problems

**In Exercises 3 – 10, find:**

(a) $\dfrac{dy}{dx}$

(b) the equation of the tangent and normal lines to the curve at the indicated $\theta$–value.

3. $r = 1;\quad \theta = \pi/4$

4. $r = \cos\theta;\quad \theta = \pi/4$

5. $r = 1 + \sin\theta;\quad \theta = \pi/6$

6. $r = 1 - 3\cos\theta;\quad \theta = 3\pi/4$

7. $r = \theta;\quad \theta = \pi/2$

8. $r = \cos(3\theta);\quad \theta = \pi/6$

9. $r = \sin(4\theta);\quad \theta = \pi/3$

10. $r = \dfrac{1}{\sin\theta - \cos\theta};\quad \theta = \pi$

**In Exercises 11 – 14, find the values of $\theta$ in the given interval where the graph of the polar function has horizontal and vertical tangent lines.**

11. $r = 3;\quad [0, 2\pi]$

12. $r = 2\sin\theta;\quad [0, \pi]$

13. $r = \cos(2\theta);\quad [0, 2\pi]$

14. $r = 1 + \cos\theta;\quad [0, 2\pi]$

**In Exercises 15 – 16, find the equation of the lines tangent to the graph at the pole.**

15. $r = \sin\theta;\quad [0, \pi]$

16. $r = \sin(3\theta);\quad [0, \pi]$

**In Exercises 17 – 28, find the area of the described region.**

17. Enclosed by the circle: $r = 4\sin\theta$

18. Enclosed by the circle $r = 5$

19. Enclosed by one petal of $r = \sin(3\theta)$

20. Enclosed by one petal of the rose curve $r = \cos(n\theta)$, where $n$ is a positive integer.

21. Enclosed by the cardioid $r = 1 - \sin\theta$

22. Enclosed by the inner loop of the limaçon $r = 1 + 2\cos\theta$

23. Enclosed by the outer loop of the limaçon $r = 1 + 2\cos\theta$ (including area enclosed by the inner loop)

24. Enclosed between the inner and outer loop of the limaçon $r = 1 + 2\cos\theta$

25. Enclosed by $r = 2\cos\theta$ and $r = 2\sin\theta$, as shown:

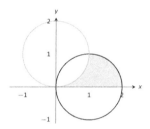

26. Enclosed by $r = \cos(3\theta)$ and $r = \sin(3\theta)$, as shown:

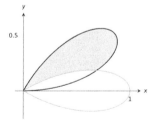

27. Enclosed by $r = \cos\theta$ and $r = \sin(2\theta)$, as shown:

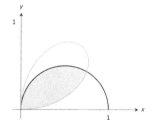

28. Enclosed by $r = \cos\theta$ and $r = 1 - \cos\theta$, as shown:

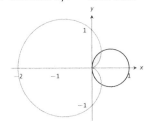

**In Exercises 29 – 34, answer the questions involving arc length.**

29. Use the arc length formula to compute the arc length of the circle $r = 2$.

30. Use the arc length formula to compute the arc length of the circle $r = 4\sin\theta$.

31. Use the arc length formula to compute the arc length of $r = \cos\theta + \sin\theta$.

32. Use the arc length formula to compute the arc length of the cardioid $r = 1 + \cos\theta$. (Hint: apply the formula, simplify, then use a Power–Reducing Formula to convert $1 + \cos\theta$ into a square.)

33. Approximate the arc length of one petal of the rose curve $r = \sin(3\theta)$ with Simpson's Rule and $n = 4$.

34. Let $x(\theta) = f(\theta)\cos\theta$ and $y(\theta) = f(\theta)\sin\theta$. Show, as suggested by the text, that
$$x'(\theta)^2 + y'(\theta)^2 = f'(\theta)^2 + f(\theta)^2.$$

**In Exercises 35 – 40, answer the questions involving surface area.**

35. Find the surface area of the sphere formed by revolving the circle $r = 2$ about the initial ray.

36. Find the surface area of the sphere formed by revolving the circle $r = 2\cos\theta$ about the initial ray.

37. Find the surface area of the solid formed by revolving the cardioid $r = 1 + \cos\theta$ about the initial ray.

38. Find the surface area of the solid formed by revolving the circle $r = 2\cos\theta$ about the line $\theta = \pi/2$.

39. Find the surface area of the solid formed by revolving the line $r = 3\sec\theta$, $-\pi/4 \leq \theta \leq \pi/4$, about the line $\theta = \pi/2$.

40. Find the surface area of the solid formed by revolving the line $r = 3\sec\theta$, $0 \leq \theta \leq \pi/4$, about the initial ray.

# 10: Vectors

This chapter introduces a new mathematical object, the **vector**. Defined in Section 10.2, we will see that vectors provide a powerful language for describing quantities that have magnitude and direction aspects. A simple example of such a quantity is force: when applying a force, one is generally interested in how much force is applied (i.e., the magnitude of the force) and the direction in which the force was applied. Vectors will play an important role in many of the subsequent chapters in this text.

This chapter begins with moving our mathematics out of the plane and into "space." That is, we begin to think mathematically not only in two dimensions, but in three. With this foundation, we can explore vectors both in the plane and in space.

## 10.1 Introduction to Cartesian Coordinates in Space

Up to this point in this text we have considered mathematics in a 2–dimensional world. We have plotted graphs on the $x$-$y$ plane using rectangular and polar coordinates and found the area of regions in the plane. We have considered properties of *solid* objects, such as volume and surface area, but only by first defining a curve in the plane and then rotating it out of the plane.

While there is wonderful mathematics to explore in "2D," we live in a "3D" world and eventually we will want to apply mathematics involving this third dimension. In this section we introduce Cartesian coordinates in space and explore basic surfaces. This will lay a foundation for much of what we do in the remainder of the text.

Each point $P$ in space can be represented with an ordered triple, $P = (a, b, c)$, where $a$, $b$ and $c$ represent the relative position of $P$ along the $x$-, $y$- and $z$-axes, respectively. Each axis is perpendicular to the other two.

Visualizing points in space on paper can be problematic, as we are trying to represent a 3-dimensional concept on a 2–dimensional medium. We cannot draw three lines representing the three axes in which each line is perpendicular to the other two. Despite this issue, standard conventions exist for plotting shapes in space that we will discuss that are more than adequate.

One convention is that the axes must conform to the **right hand rule**. This rule states that when the index finger of the right hand is extended in the direction of the positive $x$-axis, and the middle finger (bent "inward" so it is perpendicular to the palm) points along the positive $y$-axis, then the extended thumb will point in the direction of the positive $z$-axis. (It may take some thought to

# Chapter 10 Vectors

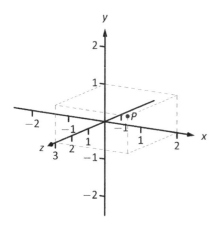

Figure 10.1.1: Plotting the point $P = (2, 1, 3)$ in space.

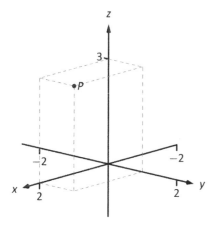

Figure 10.1.2: Plotting the point $P = (2, 1, 3)$ in space with a perspective used in this text.

verify this, but this system is inherently different from the one created by using the "left hand rule.")

As long as the coordinate axes are positioned so that they follow this rule, it does not matter how the axes are drawn on paper. There are two popular methods that we briefly discuss.

In Figure 10.1.1 we see the point $P = (2, 1, 3)$ plotted on a set of axes. The basic convention here is that the x-y plane is drawn in its standard way, with the z-axis down to the left. The perspective is that the paper represents the x-y plane and the positive z axis is coming up, off the page. This method is preferred by many engineers. Because it can be hard to tell where a single point lies in relation to all the axes, dashed lines have been added to let one see how far along each axis the point lies.

One can also consider the x-y plane as being a horizontal plane in, say, a room, where the positive z-axis is pointing up. When one steps back and looks at this room, one might draw the axes as shown in Figure 10.1.2. The same point $P$ is drawn, again with dashed lines. This point of view is preferred by most mathematicians, and is the convention adopted by this text.

Just as the x- and y-axes divide the plane into four *quadrants*, the x-, y-, and z-coordinate planes divide space into eight *octants*. The octant in which x, y, and z are positive is called the **first octant**. We do not name the other seven octants in this text.

## Measuring Distances

It is of critical importance to know how to measure distances between points in space. The formula for doing so is based on measuring distance in the plane, and is known (in both contexts) as the Euclidean measure of distance.

> **Definition 10.1.1**     **Distance In Space**
>
> Let $P = (x_1, y_1, z_1)$ and $Q = (x_2, y_2, z_2)$ be points in space. The distance $D$ between $P$ and $Q$ is
>
> $$D = \sqrt{(x_2 - x_1)^2 + (y_2 - y_1)^2 + (z_2 - z_1)^2}.$$

We refer to the line segment that connects points $P$ and $Q$ in space as $\overline{PQ}$, and refer to the length of this segment as $||\overline{PQ}||$. The above distance formula allows us to compute the length of this segment.

**Example 10.1.1**     **Length of a line segment**
Let $P = (1, 4, -1)$ and let $Q = (2, 1, 1)$. Draw the line segment $\overline{PQ}$ and find its length.

Notes:

SOLUTION  The points P and Q are plotted in Figure 10.1.3; no special consideration need be made to draw the line segment connecting these two points; simply connect them with a straight line. One *cannot* actually measure this line on the page and deduce anything meaningful; its true length must be measured analytically. Applying Definition 10.1.1, we have

$$\|\overrightarrow{PQ}\| = \sqrt{(2-1)^2 + (1-4)^2 + (1-(-1))^2} = \sqrt{14} \approx 3.74.$$

## Spheres

Just as a circle is the set of all points in the *plane* equidistant from a given point (its center), a sphere is the set of all points in *space* that are equidistant from a given point. Definition 10.1.1 allows us to write an equation of the sphere.

We start with a point $C = (a, b, c)$ which is to be the center of a sphere with radius $r$. If a point $P = (x, y, z)$ lies on the sphere, then $P$ is $r$ units from $C$; that is,

$$\|\overrightarrow{PC}\| = \sqrt{(x-a)^2 + (y-b)^2 + (z-c)^2} = r.$$

Squaring both sides, we get the standard equation of a sphere in space with center at $C = (a, b, c)$ with radius $r$, as given in the following Key Idea.

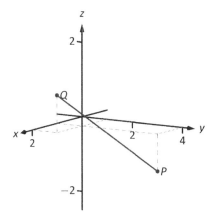

Figure 10.1.3: Plotting points P and Q in Example 10.1.1.

---

**Key Idea 10.1.1  Standard Equation of a Sphere in Space**

The standard equation of the sphere with radius $r$, centered at $C = (a, b, c)$, is

$$(x-a)^2 + (y-b)^2 + (z-c)^2 = r^2.$$

---

**Example 10.1.2  Equation of a sphere**
Find the center and radius of the sphere defined by $x^2 + 2x + y^2 - 4y + z^2 - 6z = 2$.

SOLUTION  To determine the center and radius, we must put the equation in standard form. This requires us to complete the square (three times).

$$x^2 + 2x + y^2 - 4y + z^2 - 6z = 2$$
$$(x^2 + 2x + 1) + (y^2 - 4y + 4) + (z^2 - 6z + 9) - 14 = 2$$
$$(x+1)^2 + (y-2)^2 + (z-3)^2 = 16$$

The sphere is centered at $(-1, 2, 3)$ and has a radius of 4.

The equation of a sphere is an example of an implicit function defining a surface in space. In the case of a sphere, the variables $x$, $y$ and $z$ are all used. We

Notes:

## Chapter 10 Vectors

now consider situations where surfaces are defined where one or two of these variables are absent.

### Introduction to Planes in Space

The coordinate axes naturally define three planes (shown in Figure 10.1.4), the **coordinate planes**: the x-y plane, the y-z plane and the x-z plane. The x-y plane is characterized as the set of all points in space where the z-value is 0. This, in fact, gives us an equation that describes this plane: $z = 0$. Likewise, the x-z plane is all points where the y-value is 0, characterized by $y = 0$.

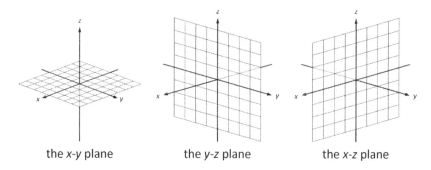

the x-y plane　　　the y-z plane　　　the x-z plane

Figure 10.1.4: The coordinate planes.

The equation $x = 2$ describes all points in space where the x-value is 2. This is a plane, parallel to the y-z coordinate plane, shown in Figure 10.1.5.

Figure 10.1.5: The plane $x = 2$.

**Example 10.1.3　　Regions defined by planes**
Sketch the region defined by the inequalities $-1 \leq y \leq 2$.

**Solution**　　The region is all points between the planes $y = -1$ and $y = 2$. These planes are sketched in Figure 10.1.6, which are parallel to the x-z plane. Thus the region extends infinitely in the x and z directions, and is bounded by planes in the y direction.

### Cylinders

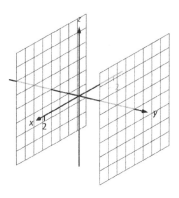

Figure 10.1.6: Sketching the boundaries of a region in Example 10.1.3.

The equation $x = 1$ obviously lacks the y and z variables, meaning it defines points where the y and z coordinates can take on any value. Now consider the equation $x^2 + y^2 = 1$ *in space*. In *the plane*, this equation describes a circle of radius 1, centered at the origin. In space, the z coordinate is not specified, meaning it can take on any value. In Figure 10.1.8 (a), we show part of the graph of the equation $x^2 + y^2 = 1$ by sketching 3 circles: the bottom one has a constant z-value of $-1.5$, the middle one has a z-value of 0 and the top circle has a z-value of 1. By plotting *all* possible z-values, we get the surface shown in Figure

Notes:

10.1.8(b). This surface looks like a "tube," or a "cylinder"; mathematicians call this surface a **cylinder** for an entirely different reason.

> **Definition 10.1.2**     **Cylinder**
>
> Let C be a curve in a plane and let L be a line not parallel to C. A **cylinder** is the set of all lines parallel to L that pass through C. The curve C is the **directrix** of the cylinder, and the lines are the **rulings**.

In this text, we consider curves C that lie in planes parallel to one of the coordinate planes, and lines L that are perpendicular to these planes, forming **right cylinders**. Thus the directrix can be defined using equations involving 2 variables, and the rulings will be parallel to the axis of the 3$^{rd}$ variable.

In the example preceding the definition, the curve $x^2 + y^2 = 1$ in the x-y plane is the directrix and the rulings are lines parallel to the z-axis. (Any circle shown in Figure 10.1.8 can be considered a directrix; we simply choose the one where $z = 0$.) Sample rulings can also be viewed in part (b) of the figure. More examples will help us understand this definition.

**Example 10.1.4**     **Graphing cylinders**
Graph the following cylinders.
 1. $z = y^2$     2. $x = \sin z$

**SOLUTION**

1. We can view the equation $z = y^2$ as a parabola in the y-z plane, as illustrated in Figure 10.1.7(a). As x does not appear in the equation, the rulings are lines through this parabola parallel to the x-axis, shown in (b). These rulings give an idea as to what the surface looks like, drawn in (c).

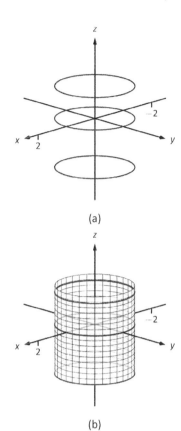

(a)

(b)

Figure 10.1.8: Sketching $x^2 + y^2 = 1$.

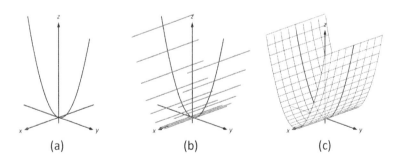

(a)         (b)         (c)

Figure 10.1.7: Sketching the cylinder defined by $z = y^2$.

Notes:

2. We can view the equation $x = \sin z$ as a sine curve that exists in the x-z plane, as shown in Figure 10.1.9 (a). The rules are parallel to the y axis as the variable y does not appear in the equation $x = \sin z$; some of these are shown in part (b). The surface is shown in part (c) of the figure.

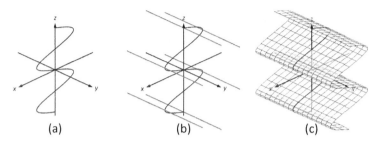

Figure 10.1.9: Sketching the cylinder defined by $x = \sin z$.

## Surfaces of Revolution

One of the applications of integration we learned previously was to find the volume of solids of revolution – solids formed by revolving a curve about a horizontal or vertical axis. We now consider how to find the equation of the surface of such a solid.

Consider the surface formed by revolving $y = \sqrt{x}$ about the x-axis. Cross–sections of this surface parallel to the y-z plane are circles, as shown in Figure 10.1.10(a). Each circle has equation of the form $y^2 + z^2 = r^2$ for some radius $r$. The radius is a function of x; in fact, it is $r(x) = \sqrt{x}$. Thus the equation of the surface shown in Figure 10.1.10b is $y^2 + z^2 = (\sqrt{x})^2$.

We generalize the above principles to give the equations of surfaces formed by revolving curves about the coordinate axes.

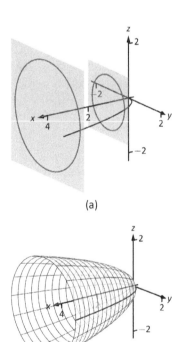

Figure 10.1.10: Introducing surfaces of revolution.

> **Key Idea 10.1.2**     **Surfaces of Revolution, Part 1**
>
> Let $r$ be a radius function.
>
> 1. The equation of the surface formed by revolving $y = r(x)$ or $z = r(x)$ about the x-axis is $y^2 + z^2 = r(x)^2$.
>
> 2. The equation of the surface formed by revolving $x = r(y)$ or $z = r(y)$ about the y-axis is $x^2 + z^2 = r(y)^2$.
>
> 3. The equation of the surface formed by revolving $x = r(z)$ or $y = r(z)$ about the z-axis is $x^2 + y^2 = r(z)^2$.

Notes:

## 10.1 Introduction to Cartesian Coordinates in Space

**Example 10.1.5    Finding equation of a surface of revolution**
Let $y = \sin z$ on $[0, \pi]$. Find the equation of the surface of revolution formed by revolving $y = \sin z$ about the $z$-axis.

**SOLUTION**    Using Key Idea 10.1.2, we find the surface has equation $x^2 + y^2 = \sin^2 z$. The curve is sketched in Figure 10.1.11(a) and the surface is drawn in Figure 10.1.11(b).

Note how the surface (and hence the resulting equation) is the same if we began with the curve $x = \sin z$, which is also drawn in Figure 10.1.11(a).

This particular method of creating surfaces of revolution is limited. For instance, in Example 7.3.4 of Section 7.3 we found the volume of the solid formed by revolving $y = \sin x$ about the $y$-axis. Our current method of forming surfaces can only rotate $y = \sin x$ about the $x$-axis. Trying to rewrite $y = \sin x$ as a function of $y$ is not trivial, as simply writing $x = \sin^{-1} y$ only gives part of the region we desire.

What we desire is a way of writing the surface of revolution formed by rotating $y = f(x)$ about the $y$-axis. We start by first recognizing this surface is the same as revolving $z = f(x)$ about the $z$-axis. This will give us a more natural way of viewing the surface.

A value of $x$ is a measurement of distance from the $z$-axis. At the distance $r$, we plot a $z$-height of $f(r)$. When rotating $f(x)$ about the $z$-axis, we want all points a distance of $r$ from the $z$-axis in the $x$-$y$ plane to have a $z$-height of $f(r)$. All such points satisfy the equation $r^2 = x^2 + y^2$; hence $r = \sqrt{x^2 + y^2}$. Replacing $r$ with $\sqrt{x^2 + y^2}$ in $f(r)$ gives $z = f(\sqrt{x^2 + y^2})$. This is the equation of the surface.

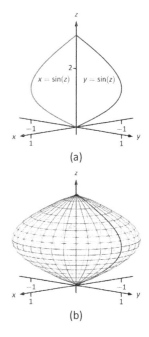

Figure 10.1.11: Revolving $y = \sin z$ about the $z$-axis in Example 10.1.5.

> **Key Idea 10.1.3    Surfaces of Revolution, Part 2**
>
> Let $z = f(x)$, $x \geq 0$, be a curve in the $x$-$z$ plane. The surface formed by revolving this curve about the $z$-axis has equation $z = f(\sqrt{x^2 + y^2})$.

**Example 10.1.6    Finding equation of surface of revolution**
Find the equation of the surface found by revolving $z = \sin x$ about the $z$-axis.

**SOLUTION**    Using Key Idea 10.1.3, the surface has equation $z = \sin(\sqrt{x^2 + y^2})$. The curve and surface are graphed in Figure 10.1.12.

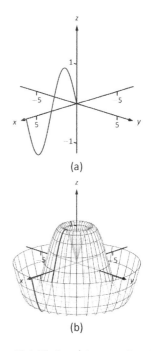

Figure 10.1.12: Revolving $z = \sin x$ about the $z$-axis in Example 10.1.6.

Notes:

## Quadric Surfaces

Spheres, planes and cylinders are important surfaces to understand. We now consider one last type of surface, a **quadric surface**. The definition may look intimidating, but we will show how to analyze these surfaces in an illuminating way.

> **Definition 10.1.3**     **Quadric Surface**
>
> A **quadric surface** is the graph of the general second–degree equation in three variables:
>
> $$Ax^2 + By^2 + Cz^2 + Dxy + Exz + Fyz + Gx + Hy + Iz + J = 0.$$

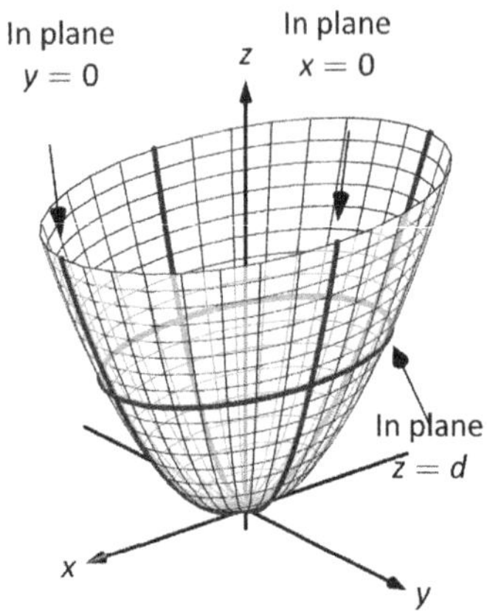

When the coefficients $D$, $E$ or $F$ are not zero, the basic shapes of the quadric surfaces are rotated in space. We will focus on quadric surfaces where these coefficients are 0; we will not consider rotations. There are six basic quadric surfaces: the elliptic paraboloid, elliptic cone, ellipsoid, hyperboloid of one sheet, hyperboloid of two sheets, and the hyperbolic paraboloid.

We study each shape by considering **traces**, that is, intersections of each surface with a plane parallel to a coordinate plane. For instance, consider the elliptic paraboloid $z = x^2/4 + y^2$, shown in Figure 10.1.13. If we intersect this shape with the plane $z = d$ (i.e., replace $z$ with $d$), we have the equation:

$$d = \frac{x^2}{4} + y^2.$$

Divide both sides by $d$:

$$1 = \frac{x^2}{4d} + \frac{y^2}{d}.$$

This describes an ellipse – so cross sections parallel to the $x$-$y$ coordinate plane are ellipses. This ellipse is drawn in the figure.

Now consider cross sections parallel to the $x$-$z$ plane. For instance, letting $y = 0$ gives the equation $z = x^2/4$, clearly a parabola. Intersecting with the plane $x = 0$ gives a cross section defined by $z = y^2$, another parabola. These parabolas are also sketched in the figure.

Thus we see where the elliptic paraboloid gets its name: some cross sections are ellipses, and others are parabolas.

Such an analysis can be made with each of the quadric surfaces. We give a sample equation of each, provide a sketch with representative traces, and describe these traces.

Figure 10.1.13: The elliptic paraboloid $z = x^2/4 + y^2$.

Notes:

**Elliptic Paraboloid,** $z = \dfrac{x^2}{a^2} + \dfrac{y^2}{b^2}$

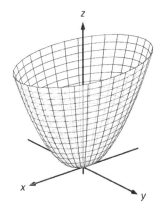

| Plane | Trace |
|---|---|
| $x = d$ | Parabola |
| $y = d$ | Parabola |
| $z = d$ | Ellipse |

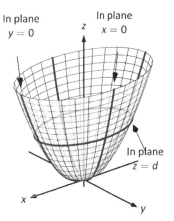

One variable in the equation of the elliptic paraboloid will be raised to the first power; above, this is the $z$ variable. The paraboloid will "open" in the direction of this variable's axis. Thus $x = y^2/a^2 + z^2/b^2$ is an elliptic paraboloid that opens along the $x$-axis.

Multiplying the right hand side by $(-1)$ defines an elliptic paraboloid that "opens" in the opposite direction.

---

**Elliptic Cone,** $z^2 = \dfrac{x^2}{a^2} + \dfrac{y^2}{b^2}$

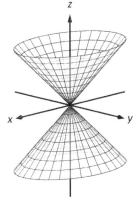

| Plane | Trace |
|---|---|
| $x = 0$ | Crossed Lines |
| $y = 0$ | Crossed Lines |
| $x = d$ | Hyperbola |
| $y = d$ | Hyperbola |
| $z = d$ | Ellipse |

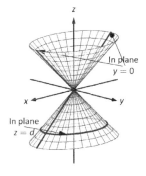 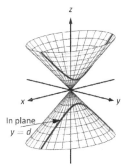

One can rewrite the equation as $z^2 - x^2/a^2 - y^2/b^2 = 0$. The one variable with a positive coefficient corresponds to the axis that the cones "open" along.

**Ellipsoid,** $\dfrac{x^2}{a^2} + \dfrac{y^2}{b^2} + \dfrac{z^2}{c^2} = 1$

| Plane | Trace |
|---|---|
| $x = d$ | Ellipse |
| $y = d$ | Ellipse |
| $z = d$ | Ellipse |

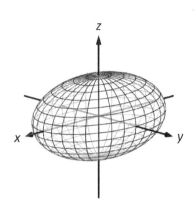
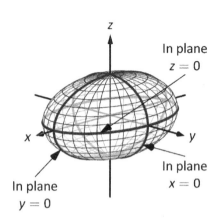

If $a = b = c \neq 0$, the ellipsoid is a sphere with radius $a$; compare to Key Idea 10.1.1.

**Hyperboloid of One Sheet,** $\dfrac{x^2}{a^2} + \dfrac{y^2}{b^2} - \dfrac{z^2}{c^2} = 1$

| Plane | Trace |
|---|---|
| $x = d$ | Hyperbola |
| $y = d$ | Hyperbola |
| $z = d$ | Ellipse |

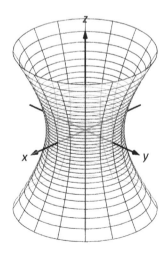
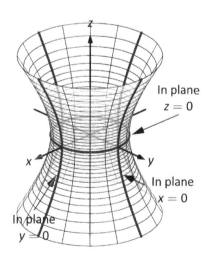

The one variable with a negative coefficient corresponds to the axis that the hyperboloid "opens" along.

**Hyperboloid of Two Sheets,** $\dfrac{z^2}{c^2} - \dfrac{x^2}{a^2} - \dfrac{y^2}{b^2} = 1$

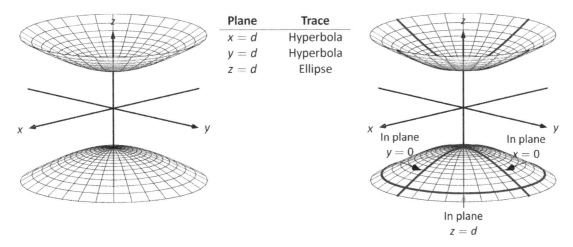

| Plane | Trace |
|---|---|
| $x = d$ | Hyperbola |
| $y = d$ | Hyperbola |
| $z = d$ | Ellipse |

The one variable with a positive coefficient corresponds to the axis that the hyperboloid "opens" along. In the case illustrated, when $|d| < |c|$, there is no trace.

---

**Hyperbolic Paraboloid,** $z = \dfrac{x^2}{a^2} - \dfrac{y^2}{b^2}$

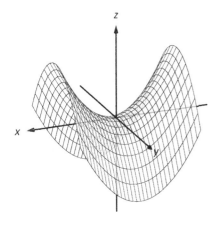

| Plane | Trace |
|---|---|
| $x = d$ | Parabola |
| $y = d$ | Parabola |
| $z = d$ | Hyperbola |

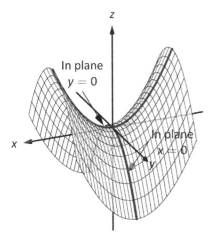
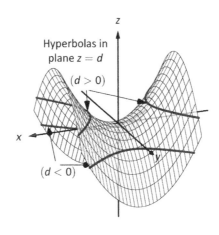

The parabolic traces will open along the axis of the one variable that is raised to the first power.

## Chapter 10 Vectors

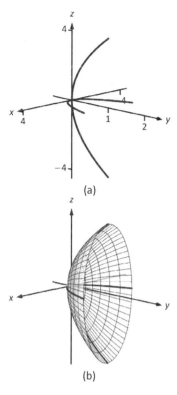

Figure 10.1.14: Sketching an elliptic paraboloid.

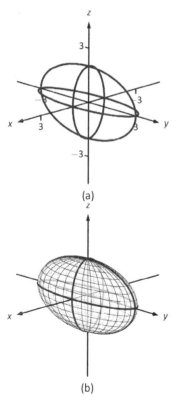

Figure 10.1.15: Sketching an ellipsoid.

**Example 10.1.7    Sketching quadric surfaces**

Sketch the quadric surface defined by the given equation.

1. $y = \dfrac{x^2}{4} + \dfrac{z^2}{16}$
2. $x^2 + \dfrac{y^2}{9} + \dfrac{z^2}{4} = 1$.
3. $z = y^2 - x^2$.

**Solution**

1. $y = \dfrac{x^2}{4} + \dfrac{z^2}{16}$:

    We first identify the quadric by pattern-matching with the equations given previously. Only two surfaces have equations where one variable is raised to the first power, the elliptic paraboloid and the hyperbolic paraboloid. In the latter case, the other variables have different signs, so we conclude that this describes a hyperbolic paraboloid. As the variable with the first power is $y$, we note the paraboloid opens along the $y$-axis.

    To make a decent sketch by hand, we need only draw a few traces. In this case, the traces $x = 0$ and $z = 0$ form parabolas that outline the shape.

    $x = 0$: The trace is the parabola $y = z^2/16$

    $z = 0$: The trace is the parabola $y = x^2/4$.

    Graphing each trace in the respective plane creates a sketch as shown in Figure 10.1.14(a). This is enough to give an idea of what the paraboloid looks like. The surface is filled in in (b).

2. $x^2 + \dfrac{y^2}{9} + \dfrac{z^2}{4} = 1$:

    This is an ellipsoid. We can get a good idea of its shape by drawing the traces in the coordinate planes.

    $x = 0$: The trace is the ellipse $\dfrac{y^2}{9} + \dfrac{z^2}{4} = 1$. The major axis is along the $y$-axis with length 6 (as $b = 3$, the length of the axis is 6); the minor axis is along the $z$-axis with length 4.

    $y = 0$: The trace is the ellipse $x^2 + \dfrac{z^2}{4} = 1$. The major axis is along the $z$-axis, and the minor axis has length 2 along the $x$-axis.

    $z = 0$: The trace is the ellipse $x^2 + \dfrac{y^2}{9} = 1$, with major axis along the $y$-axis.

    Graphing each trace in the respective plane creates a sketch as shown in Figure 10.1.15(a). Filling in the surface gives Figure 10.1.15(b).

3. $z = y^2 - x^2$:

Notes:

This defines a hyperbolic paraboloid, very similar to the one shown in the gallery of quadric sections. Consider the traces in the $y-z$ and $x-z$ planes:

$x = 0$: The trace is $z = y^2$, a parabola opening up in the $y - z$ plane.

$y = 0$: The trace is $z = -x^2$, a parabola opening down in the $x - z$ plane.

Sketching these two parabolas gives a sketch like that in Figure 10.1.16(a), and filling in the surface gives a sketch like (b).

**Example 10.1.8    Identifying quadric surfaces**
Consider the quadric surface shown in Figure 10.1.17. Which of the following equations best fits this surface?

(a)  $x^2 - y^2 - \dfrac{z^2}{9} = 0$    (c)  $z^2 - x^2 - y^2 = 1$

(b)  $x^2 - y^2 - z^2 = 1$    (d)  $4x^2 - y^2 - \dfrac{z^2}{9} = 1$

**Solution**    The image clearly displays a hyperboloid of two sheets. The gallery informs us that the equation will have a form similar to $\dfrac{z^2}{c^2} - \dfrac{x^2}{a^2} - \dfrac{y^2}{b^2} = 1$.

We can immediately eliminate option (a), as the constant in that equation is not 1.

The hyperboloid "opens" along the $x$-axis, meaning $x$ must be the only variable with a positive coefficient, eliminating (c).

The hyperboloid is wider in the $z$-direction than in the $y$-direction, so we need an equation where $c > b$. This eliminates (b), leaving us with (d). We should verify that the equation given in (d), $4x^2 - y^2 - \frac{z^2}{9} = 1$, fits.

We already established that this equation describes a hyperboloid of two sheets that opens in the $x$-direction and is wider in the $z$-direction than in the $y$. Now note the coefficient of the $x$-term. Rewriting $4x^2$ in standard form, we have: $4x^2 = \dfrac{x^2}{(1/2)^2}$. Thus when $y = 0$ and $z = 0$, $x$ must be $1/2$; i.e., each hyperboloid "starts" at $x = 1/2$. This matches our figure.

We conclude that $4x^2 - y^2 - \dfrac{z^2}{9} = 1$ best fits the graph.

This section has introduced points in space and shown how equations can describe surfaces. The next sections explore *vectors*, an important mathematical object that we'll use to explore curves in space.

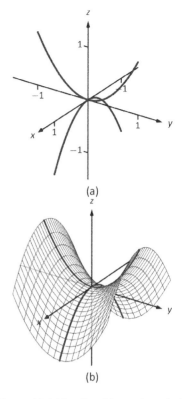

Figure 10.1.16: Sketching a hyperbolic paraboloid.

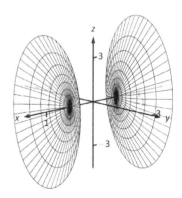

Figure 10.1.17: A possible equation of this quadric surface is found in Example 10.1.8.

# Exercises 10.1

## Terms and Concepts

1. Axes drawn in space must conform to the _____ _____ rule.

2. In the plane, the equation $x = 2$ defines a _____; in space, $x = 2$ defines a _____.

3. In the plane, the equation $y = x^2$ defines a _____; in space, $y = x^2$ defines a _____.

4. Which quadric surface looks like a Pringles® chip?

5. Consider the hyperbola $x^2 - y^2 = 1$ in the plane. If this hyperbola is rotated about the $x$-axis, what quadric surface is formed?

6. Consider the hyperbola $x^2 - y^2 = 1$ in the plane. If this hyperbola is rotated about the $y$-axis, what quadric surface is formed?

## Problems

7. The points $A = (1, 4, 2)$, $B = (2, 6, 3)$ and $C = (4, 3, 1)$ form a triangle in space. Find the distances between each pair of points and determine if the triangle is a right triangle.

8. The points $A = (1, 1, 3)$, $B = (3, 2, 7)$, $C = (2, 0, 8)$ and $D = (0, -1, 4)$ form a quadrilateral $ABCD$ in space. Is this a parallelogram?

9. Find the center and radius of the sphere defined by
$x^2 - 8x + y^2 + 2y + z^2 + 8 = 0$.

10. Find the center and radius of the sphere defined by
$x^2 + y^2 + z^2 + 4x - 2y - 4z + 4 = 0$.

In Exercises 11 – 14, describe the region in space defined by the inequalities.

11. $x^2 + y^2 + z^2 < 1$

12. $0 \leq x \leq 3$

13. $x \geq 0$, $y \geq 0$, $z \geq 0$

14. $y \geq 3$

In Exercises 15 – 18, sketch the cylinder in space.

15. $z = x^3$

16. $y = \cos z$

17. $\dfrac{x^2}{4} + \dfrac{y^2}{9} = 1$

18. $y = \dfrac{1}{x}$

In Exercises 19 – 22, give the equation of the surface of revolution described.

19. Revolve $z = \dfrac{1}{1 + y^2}$ about the $y$-axis.

20. Revolve $y = x^2$ about the $x$-axis.

21. Revolve $z = x^2$ about the $z$-axis.

22. Revolve $z = 1/x$ about the $z$-axis.

In Exercises 23 – 26, a quadric surface is sketched. Determine which of the given equations best fits the graph.

23.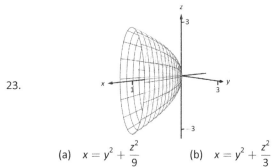

(a) $x = y^2 + \dfrac{z^2}{9}$   (b) $x = y^2 + \dfrac{z^2}{3}$

24.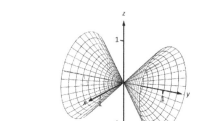

(a) $x^2 - y^2 - z^2 = 0$   (b) $x^2 - y^2 + z^2 = 0$

25.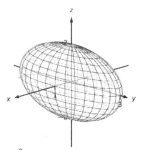

(a) $x^2 + \dfrac{y^2}{3} + \dfrac{z^2}{2} = 1$   (b) $x^2 + \dfrac{y^2}{9} + \dfrac{z^2}{4} = 1$

26.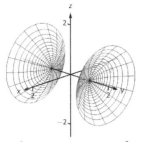

(a) $y^2 - x^2 - z^2 = 1$  (b) $y^2 + x^2 - z^2 = 1$

**In Exercises 27 – 32, sketch the quadric surface.**

27. $z - y^2 + x^2 = 0$

28. $z^2 = x^2 + \dfrac{y^2}{4}$

29. $x = -y^2 - z^2$

30. $16x^2 - 16y^2 - 16z^2 = 1$

31. $\dfrac{x^2}{9} - y^2 + \dfrac{z^2}{25} = 1$

32. $4x^2 + 2y^2 + z^2 = 4$

## 10.2 An Introduction to Vectors

Many quantities we think about daily can be described by a single number: temperature, speed, cost, weight and height. There are also many other concepts we encounter daily that cannot be described with just one number. For instance, a weather forecaster often describes wind with its speed and its direction ("... with winds from the southeast gusting up to 30 mph ..."). When applying a force, we are concerned with both the magnitude and direction of that force. In both of these examples, *direction* is important. Because of this, we study *vectors*, mathematical objects that convey both magnitude and direction information.

One "bare–bones" definition of a vector is based on what we wrote above: "a vector is a mathematical object with magnitude and direction parameters." This definition leaves much to be desired, as it gives no indication as to how such an object is to be used. Several other definitions exist; we choose here a definition rooted in a geometric visualization of vectors. It is very simplistic but readily permits further investigation.

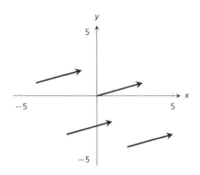

Figure 10.2.1: Drawing the same vector with different initial points.

> **Definition 10.2.1**     **Vector**
>
> A **vector** is a directed line segment.
>
> Given points $P$ and $Q$ (either in the plane or in space), we denote with $\overrightarrow{PQ}$ the vector *from $P$ to $Q$*. The point $P$ is said to be the **initial point** of the vector, and the point $Q$ is the **terminal point**.
>
> The **magnitude**, **length** or **norm** of $\overrightarrow{PQ}$ is the length of the line segment $\overline{PQ}$: $||\overrightarrow{PQ}|| = ||\overline{PQ}||$.
>
> Two vectors are **equal** if they have the same magnitude and direction.

Figure 10.2.1 shows multiple instances of the same vector. Each directed line segment has the same direction and length (magnitude), hence each is the same vector.

We use $\mathbb{R}^2$ (pronounced "r two") to represent all the vectors in the plane, and use $\mathbb{R}^3$ (pronounced "r three") to represent all the vectors in space.

Consider the vectors $\overrightarrow{PQ}$ and $\overrightarrow{RS}$ as shown in Figure 10.2.2. The vectors look to be equal; that is, they seem to have the same length and direction. Indeed, they are. Both vectors move 2 units to the right and 1 unit up from the initial point to reach the terminal point. One can analyze this movement to measure the

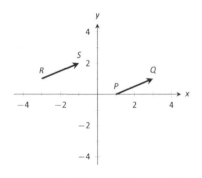

Figure 10.2.2: Illustrating how equal vectors have the same displacement.

Notes:

magnitude of the vector, and the movement itself gives direction information (one could also measure the slope of the line passing through P and Q or R and S). Since they have the same length and direction, these two vectors are equal.

This demonstrates that inherently all we care about is *displacement*; that is, how far in the x, y and possibly z directions the terminal point is from the initial point. Both the vectors $\vec{PQ}$ and $\vec{RS}$ in Figure 10.2.2 have an x-displacement of 2 and a y-displacement of 1. This suggests a standard way of describing vectors in the plane. A vector whose x-displacement is $a$ and whose y-displacement is $b$ will have terminal point $(a, b)$ when the initial point is the origin, $(0, 0)$. This leads us to a definition of a standard and concise way of referring to vectors.

---

**Definition 10.2.2**     **Component Form of a Vector**

1. The **component form** of a vector $\vec{v}$ in $\mathbb{R}^2$, whose terminal point is $(a, b)$ when its initial point is $(0, 0)$, is $\langle a, b \rangle$.

2. The **component form** of a vector $\vec{v}$ in $\mathbb{R}^3$, whose terminal point is $(a, b, c)$ when its initial point is $(0, 0, 0)$, is $\langle a, b, c \rangle$.

The numbers $a$, $b$ (and $c$, respectively) are the **components** of $\vec{v}$.

---

It follows from the definition that the component form of the vector $\vec{PQ}$, where $P = (x_1, y_1)$ and $Q = (x_2, y_2)$ is

$$\vec{PQ} = \langle x_2 - x_1, y_2 - y_1 \rangle\,;$$

in space, where $P = (x_1, y_1, z_1)$ and $Q = (x_2, y_2, z_2)$, the component form of $\vec{PQ}$ is

$$\vec{PQ} = \langle x_2 - x_1, y_2 - y_1, z_2 - z_1 \rangle\,.$$

We practice using this notation in the following example.

**Example 10.2.1**     **Using component form notation for vectors**

1. Sketch the vector $\vec{v} = \langle 2, -1 \rangle$ starting at $P = (3, 2)$ and find its magnitude.

2. Find the component form of the vector $\vec{w}$ whose initial point is $R = (-3, -2)$ and whose terminal point is $S = (-1, 2)$.

3. Sketch the vector $\vec{u} = \langle 2, -1, 3 \rangle$ starting at the point $Q = (1, 1, 1)$ and find its magnitude.

Notes:

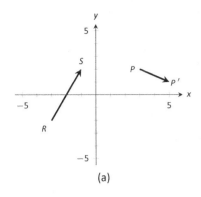

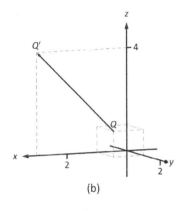

Figure 10.2.3: Graphing vectors in Example 10.2.1.

**SOLUTION**

1. Using $P$ as the initial point, we move 2 units in the positive $x$-direction and $-1$ units in the positive $y$-direction to arrive at the terminal point $P' = (5, 1)$, as drawn in Figure 10.2.3(a).

    The magnitude of $\vec{v}$ is determined directly from the component form:
    $$\|\vec{v}\| = \sqrt{2^2 + (-1)^2} = \sqrt{5}.$$

2. Using the note following Definition 10.2.2, we have
    $$\vec{RS} = \langle -1 - (-3), 2 - (-2) \rangle = \langle 2, 4 \rangle.$$

    One can readily see from Figure 10.2.3(a) that the $x$- and $y$-displacement of $\vec{RS}$ is 2 and 4, respectively, as the component form suggests.

3. Using $Q$ as the initial point, we move 2 units in the positive $x$-direction, $-1$ unit in the positive $y$-direction, and 3 units in the positive $z$-direction to arrive at the terminal point $Q' = (3, 0, 4)$, illustrated in Figure 10.2.3(b).

    The magnitude of $\vec{u}$ is:
    $$\|\vec{u}\| = \sqrt{2^2 + (-1)^2 + 3^2} = \sqrt{14}.$$

Now that we have defined vectors, and have created a nice notation by which to describe them, we start considering how vectors interact with each other. That is, we define an *algebra* on vectors.

Notes:

## 10.2 An Introduction to Vectors

**Definition 10.2.3  Vector Algebra**

1. Let $\vec{u} = \langle u_1, u_2 \rangle$ and $\vec{v} = \langle v_1, v_2 \rangle$ be vectors in $\mathbb{R}^2$, and let $c$ be a scalar.

    (a) The addition, or sum, of the vectors $\vec{u}$ and $\vec{v}$ is the vector
    $$\vec{u} + \vec{v} = \langle u_1 + v_1, u_2 + v_2 \rangle.$$

    (b) The scalar product of $c$ and $\vec{v}$ is the vector
    $$c\vec{v} = c\langle v_1, v_2 \rangle = \langle cv_1, cv_2 \rangle.$$

2. Let $\vec{u} = \langle u_1, u_2, u_3 \rangle$ and $\vec{v} = \langle v_1, v_2, v_3 \rangle$ be vectors in $\mathbb{R}^3$, and let $c$ be a scalar.

    (a) The addition, or sum, of the vectors $\vec{u}$ and $\vec{v}$ is the vector
    $$\vec{u} + \vec{v} = \langle u_1 + v_1, u_2 + v_2, u_3 + v_3 \rangle.$$

    (b) The scalar product of $c$ and $\vec{v}$ is the vector
    $$c\vec{v} = c\langle v_1, v_2, v_3 \rangle = \langle cv_1, cv_2, cv_3 \rangle.$$

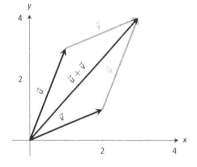

Figure 10.2.5: Illustrating how to add vectors using the Head to Tail Rule and Parallelogram Law.

In short, we say addition and scalar multiplication are computed "component–wise."

**Example 10.2.2  Adding vectors**

Sketch the vectors $\vec{u} = \langle 1, 3 \rangle$, $\vec{v} = \langle 2, 1 \rangle$ and $\vec{u} + \vec{v}$ all with initial point at the origin.

**Solution**   We first compute $\vec{u} + \vec{v}$.
$$\vec{u} + \vec{v} = \langle 1, 3 \rangle + \langle 2, 1 \rangle$$
$$= \langle 3, 4 \rangle.$$

These are all sketched in Figure 10.2.4.

As vectors convey magnitude and direction information, the sum of vectors also convey length and magnitude information. Adding $\vec{u} + \vec{v}$ suggests the following idea:

"Starting at an initial point, go out $\vec{u}$, then go out $\vec{v}$."

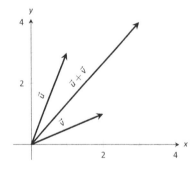

Figure 10.2.4: Graphing the sum of vectors in Example 10.2.2.

Notes:

577

This idea is sketched in Figure 10.2.5, where the initial point of $\vec{v}$ is the terminal point of $\vec{u}$. This is known as the "Head to Tail Rule" of adding vectors. Vector addition is very important. For instance, if the vectors $\vec{u}$ and $\vec{v}$ represent forces acting on a body, the sum $\vec{u}+\vec{v}$ gives the resulting force. Because of various physical applications of vector addition, the sum $\vec{u}+\vec{v}$ is often referred to as the **resultant vector**, or just the "resultant."

Analytically, it is easy to see that $\vec{u}+\vec{v}=\vec{v}+\vec{u}$. Figure 10.2.5 also gives a graphical representation of this, using gray vectors. Note that the vectors $\vec{u}$ and $\vec{v}$, when arranged as in the figure, form a parallelogram. Because of this, the Head to Tail Rule is also known as the Parallelogram Law: the vector $\vec{u}+\vec{v}$ is defined by forming the parallelogram defined by the vectors $\vec{u}$ and $\vec{v}$; the initial point of $\vec{u}+\vec{v}$ is the common initial point of parallelogram, and the terminal point of the sum is the common terminal point of the parallelogram.

While not illustrated here, the Head to Tail Rule and Parallelogram Law hold for vectors in $\mathbb{R}^3$ as well.

It follows from the properties of the real numbers and Definition 10.2.3 that

$$\vec{u}-\vec{v}=\vec{u}+(-1)\vec{v}.$$

The Parallelogram Law gives us a good way to visualize this subtraction. We demonstrate this in the following example.

**Example 10.2.3    Vector Subtraction**
Let $\vec{u}=\langle 3,1\rangle$ and $\vec{v}=\langle 1,2\rangle$. Compute and sketch $\vec{u}-\vec{v}$.

**SOLUTION**    The computation of $\vec{u}-\vec{v}$ is straightforward, and we show all steps below. Usually the formal step of multiplying by $(-1)$ is omitted and we "just subtract."

$$\begin{aligned}\vec{u}-\vec{v}&=\vec{u}+(-1)\vec{v}\\&=\langle 3,1\rangle+\langle -1,-2\rangle\\&=\langle 2,-1\rangle.\end{aligned}$$

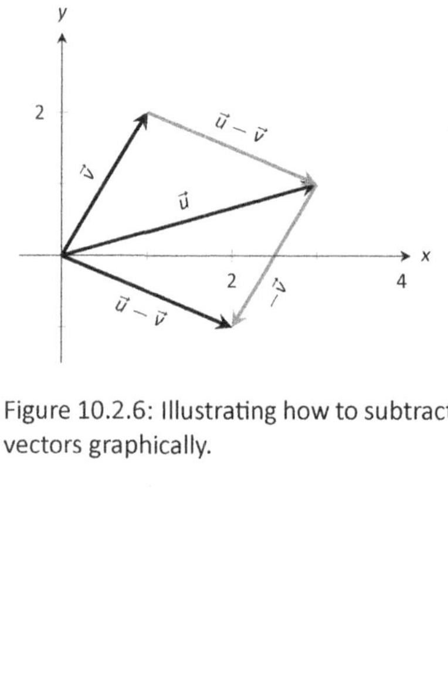

Figure 10.2.6 illustrates, using the Head to Tail Rule, how the subtraction can be viewed as the sum $\vec{u}+(-\vec{v})$. The figure also illustrates how $\vec{u}-\vec{v}$ can be obtained by looking only at the terminal points of $\vec{u}$ and $\vec{v}$ (when their initial points are the same).

**Example 10.2.4    Scaling vectors**

1. Sketch the vectors $\vec{v}=\langle 2,1\rangle$ and $2\vec{v}$ with initial point at the origin.

2. Compute the magnitudes of $\vec{v}$ and $2\vec{v}$.

Figure 10.2.6: Illustrating how to subtract vectors graphically.

Notes:

**SOLUTION**

1. We compute $2\vec{v}$:
$$2\vec{v} = 2\langle 2, 1\rangle$$
$$= \langle 4, 2\rangle.$$

Both $\vec{v}$ and $2\vec{v}$ are sketched in Figure 10.2.7. Make note that $2\vec{v}$ does not start at the terminal point of $\vec{v}$; rather, its initial point is also the origin.

2. The figure suggests that $2\vec{v}$ is twice as long as $\vec{v}$. We compute their magnitudes to confirm this.
$$\|\vec{v}\| = \sqrt{2^2 + 1^2}$$
$$= \sqrt{5}.$$
$$\|2\vec{v}\| = \sqrt{4^2 + 2^2}$$
$$= \sqrt{20}$$
$$= \sqrt{4 \cdot 5} = 2\sqrt{5}.$$

As we suspected, $2\vec{v}$ is twice as long as $\vec{v}$.

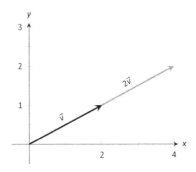

Figure 10.2.7: Graphing vectors $\vec{v}$ and $2\vec{v}$ in Example 10.2.4.

The **zero vector** is the vector whose initial point is also its terminal point. It is denoted by $\vec{0}$. Its component form, in $\mathbb{R}^2$, is $\langle 0, 0\rangle$; in $\mathbb{R}^3$, it is $\langle 0, 0, 0\rangle$. Usually the context makes is clear whether $\vec{0}$ is referring to a vector in the plane or in space.

Our examples have illustrated key principles in vector algebra: how to add and subtract vectors and how to multiply vectors by a scalar. The following theorem states formally the properties of these operations.

Notes:

> **Theorem 10.2.1**     **Properties of Vector Operations**
>
> The following are true for all scalars $c$ and $d$, and for all vectors $\vec{u}$, $\vec{v}$ and $\vec{w}$, where $\vec{u}$, $\vec{v}$ and $\vec{w}$ are all in $\mathbb{R}^2$ or where $\vec{u}$, $\vec{v}$ and $\vec{w}$ are all in $\mathbb{R}^3$:
>
> 1. $\vec{u} + \vec{v} = \vec{v} + \vec{u}$             Commutative Property
> 2. $(\vec{u} + \vec{v}) + \vec{w} = \vec{u} + (\vec{v} + \vec{w})$    Associative Property
> 3. $\vec{v} + \vec{0} = \vec{v}$                Additive Identity
> 4. $(cd)\vec{v} = c(d\vec{v})$
> 5. $c(\vec{u} + \vec{v}) = c\vec{u} + c\vec{v}$       Distributive Property
> 6. $(c + d)\vec{v} = c\vec{v} + d\vec{v}$        Distributive Property
> 7. $0\vec{v} = \vec{0}$
> 8. $\|c\vec{v}\| = |c| \cdot \|\vec{v}\|$
> 9. $\|\vec{u}\| = 0$ if, and only if, $\vec{u} = \vec{0}$.

As stated before, each nonvector $\vec{v}$ conveys magnitude and direction information. We have a method of extracting the magnitude, which we write as $\|\vec{v}\|$. *Unit vectors* are a way of extracting just the direction information from a vector.

> **Definition 10.2.4**     **Unit Vector**
>
> A **unit vector** is a vector $\vec{v}$ with a magnitude of 1; that is,
>
> $$\|\vec{v}\| = 1.$$

Consider this scenario: you are given a vector $\vec{v}$ and are told to create a vector of length 10 in the direction of $\vec{v}$. How does one do that? If we knew that $\vec{u}$ was the unit vector in the direction of $\vec{v}$, the answer would be easy: $10\vec{u}$. So how do we find $\vec{u}$?

Property 8 of Theorem 10.2.1 holds the key. If we divide $\vec{v}$ by its magnitude, it becomes a vector of length 1. Consider:

$$\left\| \frac{1}{\|\vec{v}\|}\vec{v} \right\| = \frac{1}{\|\vec{v}\|}\|\vec{v}\| \quad \text{(we can pull out } \frac{1}{\|\vec{v}\|} \text{ as it is a positive scalar)}$$
$$= 1.$$

Notes:

So the vector of length 10 in the direction of $\vec{v}$ is $10\dfrac{1}{||\vec{v}||}\vec{v}$. An example will make this more clear.

**Example 10.2.5    Using Unit Vectors**
Let $\vec{v} = \langle 3, 1 \rangle$ and let $\vec{w} = \langle 1, 2, 2 \rangle$.

1. Find the unit vector in the direction of $\vec{v}$.
2. Find the unit vector in the direction of $\vec{w}$.
3. Find the vector in the direction of $\vec{v}$ with magnitude 5.

**Solution**

1. We find $||\vec{v}|| = \sqrt{10}$. So the unit vector $\vec{u}$ in the direction of $\vec{v}$ is
$$\vec{u} = \frac{1}{\sqrt{10}}\vec{v} = \left\langle \frac{3}{\sqrt{10}}, \frac{1}{\sqrt{10}} \right\rangle.$$

2. We find $||\vec{w}|| = 3$, so the unit vector $\vec{z}$ in the direction of $\vec{w}$ is
$$\vec{u} = \frac{1}{3}\vec{w} = \left\langle \frac{1}{3}, \frac{2}{3}, \frac{2}{3} \right\rangle.$$

3. To create a vector with magnitude 5 in the direction of $\vec{v}$, we multiply the unit vector $\vec{u}$ by 5. Thus $5\vec{u} = \langle 15/\sqrt{10}, 5/\sqrt{10}\rangle$ is the vector we seek. This is sketched in Figure 10.2.8.

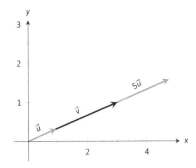

Figure 10.2.8: Graphing vectors in Example 10.2.5. All vectors shown have their initial point at the origin.

The basic formation of the unit vector $\vec{u}$ in the direction of a vector $\vec{v}$ leads to a interesting equation. It is:
$$\vec{v} = ||\vec{v}||\frac{1}{||\vec{v}||}\vec{v}.$$

We rewrite the equation with parentheses to make a point:
$$\vec{v} = \underbrace{||\vec{v}||}_{\text{magnitude}} \cdot \underbrace{\left(\frac{1}{||\vec{v}||}\vec{v}\right)}_{\text{direction}}.$$

This equation illustrates the fact that a nonzero vector has both magnitude and direction, where we view a unit vector as supplying *only* direction information. Identifying unit vectors with direction allows us to define **parallel vectors**.

Notes:

# Chapter 10  Vectors

**Note:** $\vec{0}$ is directionless; because $\|\vec{0}\| = 0$, there is no unit vector in the "direction" of $\vec{0}$.

Some texts define two vectors as being parallel if one is a scalar multiple of the other. By this definition, $\vec{0}$ is parallel to all vectors as $\vec{0} = 0\vec{v}$ for all $\vec{v}$.

We define what it means for two vectors to be perpendicular in Definition 10.3.2, which is written to exclude $\vec{0}$. It could be written to include $\vec{0}$; by such a definition, $\vec{0}$ is perpendicular to all vectors. While counter-intuitive, it is mathematically sound to allow $\vec{0}$ to be both parallel and perpendicular to all vectors.

We prefer the given definition of parallel as it is grounded in the fact that unit vectors provide direction information. One may adopt the convention that $\vec{0}$ is parallel to all vectors if they desire. (See also the marginal note on page 604.)

---

**Definition 10.2.5     Parallel Vectors**

1. Unit vectors $\vec{u}_1$ and $\vec{u}_2$ are **parallel** if $\vec{u}_1 = \pm\vec{u}_2$.

2. Nonzero vectors $\vec{v}_1$ and $\vec{v}_2$ are **parallel** if their respective unit vectors are parallel.

---

It is equivalent to say that vectors $\vec{v}_1$ and $\vec{v}_2$ are parallel if there is a scalar $c \neq 0$ such that $\vec{v}_1 = c\vec{v}_2$ (see marginal note).

If one graphed all unit vectors in $\mathbb{R}^2$ with the initial point at the origin, then the terminal points would all lie on the unit circle. Based on what we know from trigonometry, we can then say that the component form of all unit vectors in $\mathbb{R}^2$ is $\langle \cos\theta, \sin\theta \rangle$ for some angle $\theta$.

A similar construction in $\mathbb{R}^3$ shows that the terminal points all lie on the unit sphere. These vectors also have a particular component form, but its derivation is not as straightforward as the one for unit vectors in $\mathbb{R}^2$. Important concepts about unit vectors are given in the following Key Idea.

---

**Key Idea 10.2.1     Unit Vectors**

1. The unit vector in the direction of a nonzero vector $\vec{v}$ is
$$\vec{u} = \frac{1}{\|\vec{v}\|}\vec{v}.$$

2. A vector $\vec{u}$ in $\mathbb{R}^2$ is a unit vector if, and only if, its component form is $\langle \cos\theta, \sin\theta \rangle$ for some angle $\theta$.

3. A vector $\vec{u}$ in $\mathbb{R}^3$ is a unit vector if, and only if, its component form is $\langle \sin\theta\cos\varphi, \sin\theta\sin\varphi, \cos\theta \rangle$ for some angles $\theta$ and $\varphi$.

---

These formulas can come in handy in a variety of situations, especially the formula for unit vectors in the plane.

**Example 10.2.6     Finding Component Forces**
Consider a weight of 50lb hanging from two chains, as shown in Figure 10.2.9. One chain makes an angle of 30° with the vertical, and the other an angle of 45°. Find the force applied to each chain.

**Solution**     Knowing that gravity is pulling the 50lb weight straight down,

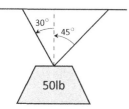

Figure 10.2.9: A diagram of a weight hanging from 2 chains in Example 10.2.6.

Notes:

we can create a vector $\vec{F}$ to represent this force.

$$\vec{F} = 50\,\langle 0, -1\rangle = \langle 0, -50\rangle\,.$$

We can view each chain as "pulling" the weight up, preventing it from falling. We can represent the force from each chain with a vector. Let $\vec{F}_1$ represent the force from the chain making an angle of $30°$ with the vertical, and let $\vec{F}_2$ represent the force form the other chain. Convert all angles to be measured from the horizontal (as shown in Figure 10.2.10), and apply Key Idea 10.2.1. As we do not yet know the magnitudes of these vectors, (that is the problem at hand), we use $m_1$ and $m_2$ to represent them.

$$\vec{F}_1 = m_1\,\langle \cos 120°, \sin 120°\rangle$$

$$\vec{F}_2 = m_2\,\langle \cos 45°, \sin 45°\rangle$$

As the weight is not moving, we know the sum of the forces is $\vec{0}$. This gives:

$$\vec{F} + \vec{F}_1 + \vec{F}_2 = \vec{0}$$
$$\langle 0, -50\rangle + m_1\,\langle \cos 120°, \sin 120°\rangle + m_2\,\langle \cos 45°, \sin 45°\rangle = \vec{0}$$

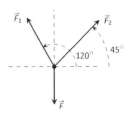

Figure 10.2.10: A diagram of the force vectors from Example 10.2.6.

The sum of the entries in the first component is 0, and the sum of the entries in the second component is also 0. This leads us to the following two equations:

$$m_1 \cos 120° + m_2 \cos 45° = 0$$
$$m_1 \sin 120° + m_2 \sin 45° = 50$$

This is a simple 2-equation, 2-unkown system of linear equations. We leave it to the reader to verify that the solution is

$$m_1 = 50(\sqrt{3} - 1) \approx 36.6; \qquad m_2 = \frac{50\sqrt{2}}{1 + \sqrt{3}} \approx 25.88.$$

It might seem odd that the sum of the forces applied to the chains is more than 50lb. We leave it to a physics class to discuss the full details, but offer this short explanation. Our equations were established so that the *vertical* components of each force sums to 50lb, thus supporting the weight. Since the chains are at an angle, they also pull against each other, creating an "additional" horizontal force while holding the weight in place.

Unit vectors were very important in the previous calculation; they allowed us to define a vector in the proper direction but with an unknown magnitude. Our computations were then computed component–wise. Because such calculations are often necessary, the *standard unit vectors* can be useful.

Notes:

## Chapter 10  Vectors

> **Definition 10.2.6  Standard Unit Vectors**
>
> 1. In $\mathbb{R}^2$, the standard unit vectors are
>
> $$\vec{i} = \langle 1, 0 \rangle \quad \text{and} \quad \vec{j} = \langle 0, 1 \rangle.$$
>
> 2. In $\mathbb{R}^3$, the standard unit vectors are
>
> $$\vec{i} = \langle 1, 0, 0 \rangle \quad \text{and} \quad \vec{j} = \langle 0, 1, 0 \rangle \quad \text{and} \quad \vec{k} = \langle 0, 0, 1 \rangle.$$

**Example 10.2.7  Using standard unit vectors**

1. Rewrite $\vec{v} = \langle 2, -3 \rangle$ using the standard unit vectors.

2. Rewrite $\vec{w} = 4\vec{i} - 5\vec{j} + 2\vec{k}$ in component form.

**Solution**

1. 
$$\begin{aligned}\vec{v} &= \langle 2, -3 \rangle \\ &= \langle 2, 0 \rangle + \langle 0, -3 \rangle \\ &= 2 \langle 1, 0 \rangle - 3 \langle 0, 1 \rangle \\ &= 2\vec{i} - 3\vec{j}\end{aligned}$$

2. 
$$\begin{aligned}\vec{w} &= 4\vec{i} - 5\vec{j} + 2\vec{k} \\ &= \langle 4, 0, 0 \rangle + \langle 0, -5, 0 \rangle + \langle 0, 0, 2 \rangle \\ &= \langle 4, -5, 2 \rangle\end{aligned}$$

These two examples demonstrate that converting between component form and the standard unit vectors is rather straightforward. Many mathematicians prefer component form, and it is the preferred notation in this text. Many engineers prefer using the standard unit vectors, and many engineering text use that notation.

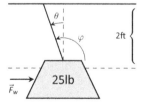

Figure 10.2.11: A figure of a weight being pushed by the wind in Example 10.2.8.

**Example 10.2.8  Finding Component Force**

A weight of 25lb is suspended from a chain of length 2ft while a wind pushes the weight to the right with constant force of 5lb as shown in Figure 10.2.11. What angle will the chain make with the vertical as a result of the wind's pushing? How much higher will the weight be?

Notes:

**Solution**    The force of the wind is represented by the vector $\vec{F}_w = 5\vec{i}$. The force of gravity on the weight is represented by $\vec{F}_g = -25\vec{j}$. The direction and magnitude of the vector representing the force on the chain are both unknown. We represent this force with

$$\vec{F}_c = m\langle \cos\varphi, \sin\varphi\rangle = m\cos\varphi\,\vec{i} + m\sin\varphi\,\vec{j}$$

for some magnitude $m$ and some angle with the horizontal $\varphi$. (Note: $\theta$ is the angle the chain makes with the *vertical*; $\varphi$ is the angle with the *horizontal*.)

As the weight is at equilibrium, the sum of the forces is $\vec{0}$:

$$\vec{F}_c + \vec{F}_w + \vec{F}_g = \vec{0}$$
$$m\cos\varphi\,\vec{i} + m\sin\varphi\,\vec{j} + 5\vec{i} - 25\vec{j} = \vec{0}$$

Thus the sum of the $\vec{i}$ and $\vec{j}$ components are 0, leading us to the following system of equations:

$$5 + m\cos\varphi = 0$$
$$-25 + m\sin\varphi = 0 \tag{10.1}$$

This is enough to determine $\vec{F}_c$ already, as we know $m\cos\varphi = -5$ and $m\sin\varphi = 25$. Thus $\vec{F}_c = \langle -5, 25\rangle$. We can use this to find the magnitude $m$:

$$m = \sqrt{(-5)^2 + 25^2} = 5\sqrt{26} \approx 25.5\text{lb}.$$

We can then use either equality from Equation (10.1) to solve for $\varphi$. We choose the first equality as using arccosine will return an angle in the 2nd quadrant:

$$5 + 5\sqrt{26}\cos\varphi = 0 \quad\Rightarrow\quad \varphi = \cos^{-1}\left(\frac{-5}{5\sqrt{26}}\right) \approx 1.7682 \approx 101.31°.$$

Subtracting 90° from this angle gives us an angle of 11.31° with the vertical.

We can now use trigonometry to find out how high the weight is lifted. The diagram shows that a right triangle is formed with the 2ft chain as the hypotenuse with an interior angle of 11.31°. The length of the adjacent side (in the diagram, the dashed vertical line) is $2\cos 11.31° \approx 1.96$ft. Thus the weight is lifted by about 0.04ft, almost 1/2in.

The algebra we have applied to vectors is already demonstrating itself to be very useful. There are two more fundamental operations we can perform with vectors, the *dot product* and the *cross product*. The next two sections explore each in turn.

Notes:

# Exercises 10.2

## Terms and Concepts

1. Name two different things that cannot be described with just one number, but rather need 2 or more numbers to fully describe them.

2. What is the difference between $(1, 2)$ and $\langle 1, 2 \rangle$?

3. What is a unit vector?

4. Unit vectors can be thought of as conveying what type of information?

5. What does it mean for two vectors to be parallel?

6. What effect does multiplying a vector by $-2$ have?

## Problems

In Exercises 7 – 10, points $P$ and $Q$ are given. Write the vector $\vec{PQ}$ in component form and using the standard unit vectors.

7. $P = (2, -1)$, $Q = (3, 5)$

8. $P = (3, 2)$, $Q = (7, -2)$

9. $P = (0, 3, -1)$, $Q = (6, 2, 5)$

10. $P = (2, 1, 2)$, $Q = (4, 3, 2)$

11. Let $\vec{u} = \langle 1, -2 \rangle$ and $\vec{v} = \langle 1, 1 \rangle$.
    (a) Find $\vec{u} + \vec{v}$, $\vec{u} - \vec{v}$, $2\vec{u} - 3\vec{v}$.
    (b) Sketch the above vectors on the same axes, along with $\vec{u}$ and $\vec{v}$.
    (c) Find $\vec{x}$ where $\vec{u} + \vec{x} = 2\vec{v} - \vec{x}$.

12. Let $\vec{u} = \langle 1, 1, -1 \rangle$ and $\vec{v} = \langle 2, 1, 2 \rangle$.
    (a) Find $\vec{u} + \vec{v}$, $\vec{u} - \vec{v}$, $\pi\vec{u} - \sqrt{2}\vec{v}$.
    (b) Sketch the above vectors on the same axes, along with $\vec{u}$ and $\vec{v}$.
    (c) Find $\vec{x}$ where $\vec{u} + \vec{x} = \vec{v} + 2\vec{x}$.

In Exercises 13 – 16, sketch $\vec{u}$, $\vec{v}$, $\vec{u} + \vec{v}$ and $\vec{u} - \vec{v}$ on the same axes.

13.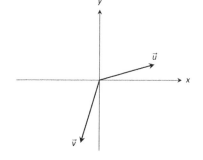

14.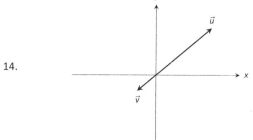

15.

16.

In Exercises 17 – 20, find $||\vec{u}||$, $||\vec{v}||$, $||\vec{u} + \vec{v}||$ and $||\vec{u} - \vec{v}||$.

17. $\vec{u} = \langle 2, 1 \rangle$, $\vec{v} = \langle 3, -2 \rangle$

18. $\vec{u} = \langle -3, 2, 2 \rangle$, $\vec{v} = \langle 1, -1, 1 \rangle$

19. $\vec{u} = \langle 1, 2 \rangle$, $\vec{v} = \langle -3, -6 \rangle$

20. $\vec{u} = \langle 2, -3, 6 \rangle$, $\vec{v} = \langle 10, -15, 30 \rangle$

21. Under what conditions is $||\vec{u}|| + ||\vec{v}|| = ||\vec{u} + \vec{v}||$?

In Exercises 22 – 25, find the unit vector $\vec{u}$ in the direction of $\vec{v}$.

22. $\vec{v} = \langle 3, 7 \rangle$

23. $\vec{v} = \langle 6, 8 \rangle$

24. $\vec{v} = \langle 1, -2, 2 \rangle$

25. $\vec{v} = \langle 2, -2, 2 \rangle$

26. Find the unit vector in the first quadrant of $\mathbb{R}^2$ that makes a 50° angle with the x-axis.

27. Find the unit vector in the second quadrant of $\mathbb{R}^2$ that makes a 30° angle with the y-axis.

28. Verify, from Key Idea 10.2.1, that
$$\vec{u} = \langle \sin\theta\cos\varphi, \sin\theta\sin\varphi, \cos\theta \rangle$$
is a unit vector for all angles $\theta$ and $\varphi$.

**A weight of 100lb is suspended from two chains, making angles with the vertical of $\theta$ and $\varphi$ as shown in the figure below.**

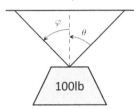

**In Exercises 29 – 32, angles $\theta$ and $\varphi$ are given. Find the magnitude of the force applied to each chain.**

29. $\theta = 30°$, $\varphi = 30°$

30. $\theta = 60°$, $\varphi = 60°$

31. $\theta = 20°$, $\varphi = 15°$

32. $\theta = 0°$, $\varphi = 0°$

**A weight of $p$lb is suspended from a chain of length $\ell$ while a constant force of $\vec{F}_w$ pushes the weight to the right, making an angle of $\theta$ with the vertical, as shown in the figure below.**

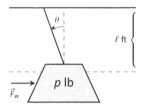

**In Exercises 33 – 36, a force $\vec{F}_w$ and length $\ell$ are given. Find the angle $\theta$ and the height the weight is lifted as it moves to the right.**

33. $\vec{F}_w = 1$lb, $\ell = 1$ft, $p = 1$lb

34. $\vec{F}_w = 1$lb, $\ell = 1$ft, $p = 10$lb

35. $\vec{F}_w = 1$lb, $\ell = 10$ft, $p = 1$lb

36. $\vec{F}_w = 10$lb, $\ell = 10$ft, $p = 1$lb

## 10.3 The Dot Product

The previous section introduced vectors and described how to add them together and how to multiply them by scalars. This section introduces *a* multiplication on vectors called the **dot product**.

> **Definition 10.3.1**     **Dot Product**
>
> 1. Let $\vec{u} = \langle u_1, u_2 \rangle$ and $\vec{v} = \langle v_1, v_2 \rangle$ in $\mathbb{R}^2$. The **dot product** of $\vec{u}$ and $\vec{v}$, denoted $\vec{u} \cdot \vec{v}$, is
>
> $$\vec{u} \cdot \vec{v} = u_1 v_1 + u_2 v_2.$$
>
> 2. Let $\vec{u} = \langle u_1, u_2, u_3 \rangle$ and $\vec{v} = \langle v_1, v_2, v_3 \rangle$ in $\mathbb{R}^3$. The **dot product** of $\vec{u}$ and $\vec{v}$, denoted $\vec{u} \cdot \vec{v}$, is
>
> $$\vec{u} \cdot \vec{v} = u_1 v_1 + u_2 v_2 + u_3 v_3.$$

Note how this product of vectors returns a *scalar*, not another vector. We practice evaluating a dot product in the following example, then we will discuss why this product is useful.

**Example 10.3.1**     **Evaluating dot products**

1. Let $\vec{u} = \langle 1, 2 \rangle$, $\vec{v} = \langle 3, -1 \rangle$ in $\mathbb{R}^2$. Find $\vec{u} \cdot \vec{v}$.

2. Let $\vec{x} = \langle 2, -2, 5 \rangle$ and $\vec{y} = \langle -1, 0, 3 \rangle$ in $\mathbb{R}^3$. Find $\vec{x} \cdot \vec{y}$.

**SOLUTION**

1. Using Definition 10.3.1, we have
$$\vec{u} \cdot \vec{v} = 1(3) + 2(-1) = 1.$$

2. Using the definition, we have
$$\vec{x} \cdot \vec{y} = 2(-1) - 2(0) + 5(3) = 13.$$

The dot product, as shown by the preceding example, is very simple to evaluate. It is only the sum of products. While the definition gives no hint as to why

---

Notes:

## 10.3 The Dot Product

we would care about this operation, there is an amazing connection between the dot product and angles formed by the vectors. Before stating this connection, we give a theorem stating some of the properties of the dot product.

---

**Theorem 10.3.1****Properties of the Dot Product**

Let $\vec{u}$, $\vec{v}$ and $\vec{w}$ be vectors in $\mathbb{R}^2$ or $\mathbb{R}^3$ and let $c$ be a scalar.

1. $\vec{u} \cdot \vec{v} = \vec{v} \cdot \vec{u}$      Commutative Property
2. $\vec{u} \cdot (\vec{v} + \vec{w}) = \vec{u} \cdot \vec{v} + \vec{u} \cdot \vec{w}$      Distributive Property
3. $c(\vec{u} \cdot \vec{v}) = (c\vec{u}) \cdot \vec{v} = \vec{u} \cdot (c\vec{v})$
4. $\vec{0} \cdot \vec{v} = 0$
5. $\vec{v} \cdot \vec{v} = \|\vec{v}\|^2$

---

The last statement of the theorem makes a handy connection between the magnitude of a vector and the dot product with itself. Our definition and theorem give properties of the dot product, but we are still likely wondering "What does the dot product *mean*?" It is helpful to understand that the dot product of a vector with itself is connected to its magnitude.

The next theorem extends this understanding by connecting the dot product to magnitudes and angles. Given vectors $\vec{u}$ and $\vec{v}$ in the plane, an angle $\theta$ is clearly formed when $\vec{u}$ and $\vec{v}$ are drawn with the same initial point as illustrated in Figure 10.3.1(a). (We always take $\theta$ to be the angle in $[0, \pi]$ as two angles are actually created.)

The same is also true of 2 vectors in space: given $\vec{u}$ and $\vec{v}$ in $\mathbb{R}^3$ with the same initial point, there is a plane that contains both $\vec{u}$ and $\vec{v}$. (When $\vec{u}$ and $\vec{v}$ are co-linear, there are infinitely many planes that contain both vectors.) In that plane, we can again find an angle $\theta$ between them (and again, $0 \leq \theta \leq \pi$). This is illustrated in Figure 10.3.1(b).

The following theorem connects this angle $\theta$ to the dot product of $\vec{u}$ and $\vec{v}$.

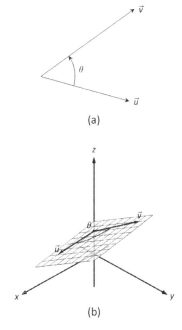

Figure 10.3.1: Illustrating the angle formed by two vectors with the same initial point.

---

Notes:

# Chapter 10 Vectors

> **Theorem 10.3.2    The Dot Product and Angles**
>
> Let $\vec{u}$ and $\vec{v}$ be nonzero vectors in $\mathbb{R}^2$ or $\mathbb{R}^3$. Then
>
> $$\vec{u} \cdot \vec{v} = \|\vec{u}\| \|\vec{v}\| \cos\theta,$$
>
> where $\theta$, $0 \leq \theta \leq \pi$, is the angle between $\vec{u}$ and $\vec{v}$.

Using Theorem 10.3.1, we can rewrite this theorem as

$$\frac{\vec{u}}{\|\vec{u}\|} \cdot \frac{\vec{v}}{\|\vec{v}\|} = \cos\theta.$$

Note how on the left hand side of the equation, we are computing the dot product of two unit vectors. Recalling that unit vectors essentially only provide direction information, we can informally restate Theorem 10.3.2 as saying "The dot product of two directions gives the cosine of the angle between them."

When $\theta$ is an acute angle (i.e., $0 \leq \theta < \pi/2$), $\cos\theta$ is positive; when $\theta = \pi/2$, $\cos\theta = 0$; when $\theta$ is an obtuse angle ($\pi/2 < \theta \leq \pi$), $\cos\theta$ is negative. Thus the sign of the dot product gives a general indication of the angle between the vectors, illustrated in Figure 10.3.2.

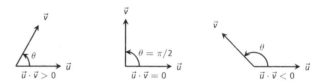

Figure 10.3.2: Illustrating the relationship between the angle between vectors and the sign of their dot product.

We *can* use Theorem 10.3.2 to compute the dot product, but generally this theorem is used to find the angle between known vectors (since the dot product is generally easy to compute). To this end, we rewrite the theorem's equation as

$$\cos\theta = \frac{\vec{u} \cdot \vec{v}}{\|\vec{u}\|\|\vec{v}\|} \quad \Leftrightarrow \quad \theta = \cos^{-1}\left(\frac{\vec{u} \cdot \vec{v}}{\|\vec{u}\|\|\vec{v}\|}\right).$$

We practice using this theorem in the following example.

**Example 10.3.2    Using the dot product to find angles**

Let $\vec{u} = \langle 3, 1 \rangle$, $\vec{v} = \langle -2, 6 \rangle$ and $\vec{w} = \langle -4, 3 \rangle$, as shown in Figure 10.3.3. Find the angles $\alpha$, $\beta$ and $\theta$.

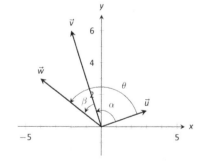

Figure 10.3.3: Vectors used in Example 10.3.2.

Notes:

**SOLUTION** We start by computing the magnitude of each vector.

$$\|\vec{u}\| = \sqrt{10}; \quad \|\vec{v}\| = 2\sqrt{10}; \quad \|\vec{w}\| = 5.$$

We now apply Theorem 10.3.2 to find the angles.

$$\alpha = \cos^{-1}\left(\frac{\vec{u}\cdot\vec{v}}{(\sqrt{10})(2\sqrt{10})}\right)$$
$$= \cos^{-1}(0) = \frac{\pi}{2} = 90°.$$

$$\beta = \cos^{-1}\left(\frac{\vec{v}\cdot\vec{w}}{(2\sqrt{10})(5)}\right)$$
$$= \cos^{-1}\left(\frac{26}{10\sqrt{10}}\right)$$
$$\approx 0.6055 \approx 34.7°.$$

$$\theta = \cos^{-1}\left(\frac{\vec{u}\cdot\vec{w}}{(\sqrt{10})(5)}\right)$$
$$= \cos^{-1}\left(\frac{-9}{5\sqrt{10}}\right)$$
$$\approx 2.1763 \approx 124.7°$$

We see from our computation that $\alpha + \beta = \theta$, as indicated by Figure 10.3.3. While we knew this should be the case, it is nice to see that this non-intuitive formula indeed returns the results we expected.

We do a similar example next in the context of vectors in space.

**Example 10.3.3   Using the dot product to find angles**
Let $\vec{u} = \langle 1, 1, 1\rangle$, $\vec{v} = \langle -1, 3, -2\rangle$ and $\vec{w} = \langle -5, 1, 4\rangle$, as illustrated in Figure 10.3.4. Find the angle between each pair of vectors.

**SOLUTION**

1. Between $\vec{u}$ and $\vec{v}$:

$$\theta = \cos^{-1}\left(\frac{\vec{u}\cdot\vec{v}}{\|\vec{u}\|\|\vec{v}\|}\right)$$
$$= \cos^{-1}\left(\frac{0}{\sqrt{3}\sqrt{14}}\right)$$
$$= \frac{\pi}{2}.$$

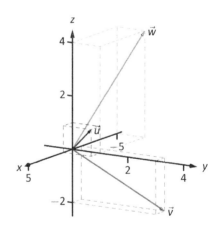

Figure 10.3.4: Vectors used in Example 10.3.3.

2. Between $\vec{u}$ and $\vec{w}$:

$$\theta = \cos^{-1}\left(\frac{\vec{u}\cdot\vec{w}}{\|\vec{u}\|\|\vec{w}\|}\right)$$
$$= \cos^{-1}\left(\frac{0}{\sqrt{3}\sqrt{42}}\right)$$
$$= \frac{\pi}{2}.$$

3. Between $\vec{v}$ and $\vec{w}$:

$$\theta = \cos^{-1}\left(\frac{\vec{v}\cdot\vec{w}}{\|\vec{v}\|\|\vec{w}\|}\right)$$
$$= \cos^{-1}\left(\frac{0}{\sqrt{14}\sqrt{42}}\right)$$
$$= \frac{\pi}{2}.$$

While our work shows that each angle is $\pi/2$, i.e., 90°, none of these angles looks to be a right angle in Figure 10.3.4. Such is the case when drawing three-dimensional objects on the page.

All three angles between these vectors was $\pi/2$, or 90°. We know from geometry and everyday life that 90° angles are "nice" for a variety of reasons, so it should seem significant that these angles are all $\pi/2$. Notice the common feature in each calculation (and also the calculation of $\alpha$ in Example 10.3.2): the dot products of each pair of angles was 0. We use this as a basis for a definition of the term **orthogonal**, which is essentially synonymous to *perpendicular*.

> **Definition 10.3.2**     **Orthogonal**
>
> Nonzero vectors $\vec{u}$ and $\vec{v}$ are **orthogonal** if their dot product is 0.

**Example 10.3.4**     **Finding orthogonal vectors**
Let $\vec{u} = \langle 3, 5\rangle$ and $\vec{v} = \langle 1, 2, 3\rangle$.

1. Find two vectors in $\mathbb{R}^2$ that are orthogonal to $\vec{u}$.

2. Find two non–parallel vectors in $\mathbb{R}^3$ that are orthogonal to $\vec{v}$.

**SOLUTION**

**Note:** The term *perpendicular* originally referred to lines. As mathematics progressed, the concept of "being at right angles to" was applied to other objects, such as vectors and planes, and the term *orthogonal* was introduced. It is especially used when discussing objects that are hard, or impossible, to visualize: two vectors in 5-dimensional space are orthogonal if their dot product is 0. It is not wrong to say they are *perpendicular*, but common convention gives preference to the word *orthogonal*.

Notes:

1. Recall that a line perpendicular to a line with slope $m$ has slope $-1/m$, the "opposite reciprocal slope." We can think of the slope of $\vec{u}$ as $5/3$, its "rise over run." A vector orthogonal to $\vec{u}$ will have slope $-3/5$. There are many such choices, though all parallel:

   $$\langle -5, 3\rangle \quad \text{or} \quad \langle 5, -3\rangle \quad \text{or} \quad \langle -10, 6\rangle \quad \text{or} \quad \langle 15, -9\rangle, \text{etc.}$$

2. There are infinitely many directions in space orthogonal to any given direction, so there are an infinite number of non–parallel vectors orthogonal to $\vec{v}$. Since there are so many, we have great leeway in finding some.

   One way is to arbitrarily pick values for the first two components, leaving the third unknown. For instance, let $\vec{v}_1 = \langle 2, 7, z\rangle$. If $\vec{v}_1$ is to be orthogonal to $\vec{v}$, then $\vec{v}_1 \cdot \vec{v} = 0$, so

   $$2 + 14 + 3z = 0 \quad \Rightarrow z = \frac{-16}{3}.$$

   So $\vec{v}_1 = \langle 2, 7, -16/3\rangle$ is orthogonal to $\vec{v}$. We can apply a similar technique by leaving the first or second component unknown.

   Another method of finding a vector orthogonal to $\vec{v}$ mirrors what we did in part 1. Let $\vec{v}_2 = \langle -2, 1, 0\rangle$. Here we switched the first two components of $\vec{v}$, changing the sign of one of them (similar to the "opposite reciprocal" concept before). Letting the third component be 0 effectively ignores the third component of $\vec{v}$, and it is easy to see that

   $$\vec{v}_2 \cdot \vec{v} = \langle -2, 1, 0\rangle \cdot \langle 1, 2, 3\rangle = 0.$$

   Clearly $\vec{v}_1$ and $\vec{v}_2$ are not parallel.

An important construction is illustrated in Figure 10.3.5, where vectors $\vec{u}$ and $\vec{v}$ are sketched. In part (a), a dotted line is drawn from the tip of $\vec{u}$ to the line containing $\vec{v}$, where the dotted line is orthogonal to $\vec{v}$. In part (b), the dotted line is replaced with the vector $\vec{z}$ and $\vec{w}$ is formed, parallel to $\vec{v}$. It is clear by the diagram that $\vec{u} = \vec{w} + \vec{z}$. What is important about this construction is this: $\vec{u}$ is *decomposed* as the sum of two vectors, one of which is parallel to $\vec{v}$ and one that is perpendicular to $\vec{v}$. It is hard to overstate the importance of this construction (as we'll see in upcoming examples).

The vectors $\vec{w}$, $\vec{z}$ and $\vec{u}$ as shown in Figure 10.3.5 (b) form a right triangle, where the angle between $\vec{v}$ and $\vec{u}$ is labeled $\theta$. We can find $\vec{w}$ in terms of $\vec{v}$ and $\vec{u}$.

Using trigonometry, we can state that

$$||\vec{w}|| = ||\vec{u}||\cos\theta. \tag{10.2}$$

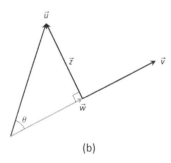

Figure 10.3.5: Developing the construction of the *orthogonal projection*.

Notes:

We also know that $\vec{w}$ is parallel to to $\vec{v}$; that is, the direction of $\vec{w}$ is the direction of $\vec{v}$, described by the unit vector $\vec{v}/\|\vec{v}\|$. The vector $\vec{w}$ is the vector in the direction $\vec{v}/\|\vec{v}\|$ with magnitude $\|\vec{u}\|\cos\theta$:

$$\vec{w} = \left(\|\vec{u}\|\cos\theta\right)\frac{1}{\|\vec{v}\|}\vec{v}.$$

Replace $\cos\theta$ using Theorem 10.3.2:

$$= \left(\|\vec{u}\|\frac{\vec{u}\cdot\vec{v}}{\|\vec{u}\|\|\vec{v}\|}\right)\frac{1}{\|\vec{v}\|}\vec{v}$$

$$= \frac{\vec{u}\cdot\vec{v}}{\|\vec{v}\|^2}\vec{v}.$$

Now apply Theorem 10.3.1.

$$= \frac{\vec{u}\cdot\vec{v}}{\vec{v}\cdot\vec{v}}\vec{v}.$$

Since this construction is so important, it is given a special name.

---

**Definition 10.3.3**     **Orthogonal Projection**

Let nonzero vectors $\vec{u}$ and $\vec{v}$ be given. The **orthogonal projection of $\vec{u}$ onto $\vec{v}$**, denoted $\operatorname{proj}_{\vec{v}}\vec{u}$, is

$$\operatorname{proj}_{\vec{v}}\vec{u} = \frac{\vec{u}\cdot\vec{v}}{\vec{v}\cdot\vec{v}}\vec{v}.$$

---

**Example 10.3.5**     **Computing the orthogonal projection**

1. Let $\vec{u} = \langle -2, 1\rangle$ and $\vec{v} = \langle 3, 1\rangle$. Find $\operatorname{proj}_{\vec{v}}\vec{u}$, and sketch all three vectors with initial points at the origin.

2. Let $\vec{w} = \langle 2, 1, 3\rangle$ and $\vec{x} = \langle 1, 1, 1\rangle$. Find $\operatorname{proj}_{\vec{x}}\vec{w}$, and sketch all three vectors with initial points at the origin.

**Solution**

Notes:

1. Applying Definition 10.3.3, we have

$$\text{proj}_{\vec{v}}\,\vec{u} = \frac{\vec{u}\cdot\vec{v}}{\vec{v}\cdot\vec{v}}\vec{v}$$
$$= \frac{-5}{10}\langle 3, 1\rangle$$
$$= \left\langle -\frac{3}{2}, -\frac{1}{2}\right\rangle.$$

Vectors $\vec{u}$, $\vec{v}$ and $\text{proj}_{\vec{v}}\,\vec{u}$ are sketched in Figure 10.3.6(a). Note how the projection is parallel to $\vec{v}$; that is, it lies on the same line through the origin as $\vec{v}$, although it points in the opposite direction. That is because the angle between $\vec{u}$ and $\vec{v}$ is obtuse (i.e., greater than 90°).

2. Apply the definition:

$$\text{proj}_{\vec{x}}\,\vec{w} = \frac{\vec{w}\cdot\vec{x}}{\vec{x}\cdot\vec{x}}\vec{x}$$
$$= \frac{6}{3}\langle 1, 1, 1\rangle$$
$$= \langle 2, 2, 2\rangle.$$

These vectors are sketched in Figure 10.3.6(b), and again in part (c) from a different perspective. Because of the nature of graphing these vectors, the sketch in part (b) makes it difficult to recognize that the drawn projection has the geometric properties it should. The graph shown in part (c) illustrates these properties better.

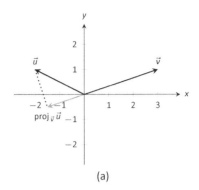

(a)

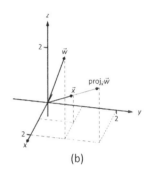

(b)

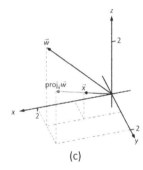

(c)

Figure 10.3.6: Graphing the vectors used in Example 10.3.5.

We can use the properties of the dot product found in Theorem 10.3.1 to rearrange the formula found in Definition 10.3.3:

$$\text{proj}_{\vec{v}}\,\vec{u} = \frac{\vec{u}\cdot\vec{v}}{\vec{v}\cdot\vec{v}}\vec{v}$$
$$= \frac{\vec{u}\cdot\vec{v}}{\|\vec{v}\|^2}\vec{v}$$
$$= \left(\vec{u}\cdot\frac{\vec{v}}{\|\vec{v}\|}\right)\frac{\vec{v}}{\|\vec{v}\|}.$$

The above formula shows that the orthogonal projection of $\vec{u}$ onto $\vec{v}$ is only concerned with the *direction* of $\vec{v}$, as both instances of $\vec{v}$ in the formula come in the form $\vec{v}/\|\vec{v}\|$, the unit vector in the direction of $\vec{v}$.

A special case of orthogonal projection occurs when $\vec{v}$ is a unit vector. In this situation, the formula for the orthogonal projection of a vector $\vec{u}$ onto $\vec{v}$ reduces to just $\text{proj}_{\vec{v}}\,\vec{u} = (\vec{u}\cdot\vec{v})\vec{v}$, as $\vec{v}\cdot\vec{v} = 1$.

Notes:

This gives us a new understanding of the dot product. When $\vec{v}$ is a unit vector, essentially providing only direction information, the dot product of $\vec{u}$ and $\vec{v}$ gives "how much of $\vec{u}$ is in the direction of $\vec{v}$." This use of the dot product will be very useful in future sections.

Now consider Figure 10.3.7 where the concept of the orthogonal projection is again illustrated. It is clear that

$$\vec{u} = \text{proj}_{\vec{v}}\, \vec{u} + \vec{z}. \tag{10.3}$$

As we know what $\vec{u}$ and $\text{proj}_{\vec{v}}\, \vec{u}$ are, we can solve for $\vec{z}$ and state that

$$\vec{z} = \vec{u} - \text{proj}_{\vec{v}}\, \vec{u}.$$

This leads us to rewrite Equation (10.3) in a seemingly silly way:

$$\vec{u} = \text{proj}_{\vec{v}}\, \vec{u} + (\vec{u} - \text{proj}_{\vec{v}}\, \vec{u}).$$

This is not nonsense, as pointed out in the following Key Idea. (Notation note: the expression "$\parallel \vec{y}$" means "is parallel to $\vec{y}$." We can use this notation to state "$\vec{x} \parallel \vec{y}$" which means "$\vec{x}$ is parallel to $\vec{y}$." The expression "$\perp \vec{y}$" means "is orthogonal to $\vec{y}$," and is used similarly.)

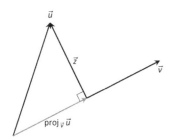

Figure 10.3.7: Illustrating the orthogonal projection.

---

**Key Idea 10.3.1**     **Orthogonal Decomposition of Vectors**

Let nonzero vectors $\vec{u}$ and $\vec{v}$ be given. Then $\vec{u}$ can be written as the sum of two vectors, one of which is parallel to $\vec{v}$, and one of which is orthogonal to $\vec{v}$:

$$\vec{u} = \underbrace{\text{proj}_{\vec{v}}\, \vec{u}}_{\parallel\, \vec{v}} + \underbrace{(\vec{u} - \text{proj}_{\vec{v}}\, \vec{u})}_{\perp\, \vec{v}}.$$

---

We illustrate the use of this equality in the following example.

**Example 10.3.6**     **Orthogonal decomposition of vectors**

1. Let $\vec{u} = \langle -2, 1\rangle$ and $\vec{v} = \langle 3, 1\rangle$ as in Example 10.3.5. Decompose $\vec{u}$ as the sum of a vector parallel to $\vec{v}$ and a vector orthogonal to $\vec{v}$.

2. Let $\vec{w} = \langle 2, 1, 3\rangle$ and $\vec{x} = \langle 1, 1, 1\rangle$ as in Example 10.3.5. Decompose $\vec{w}$ as the sum of a vector parallel to $\vec{x}$ and a vector orthogonal to $\vec{x}$.

**Solution**

---

Notes:

1. In Example 10.3.5, we found that $\text{proj}_{\vec{v}}\,\vec{u} = \langle -1.5, -0.5\rangle$. Let
$$\vec{z} = \vec{u} - \text{proj}_{\vec{v}}\,\vec{u} = \langle -2, 1\rangle - \langle -1.5, -0.5\rangle = \langle -0.5, 1.5\rangle.$$

Is $\vec{z}$ orthogonal to $\vec{v}$? (I.e, is $\vec{z} \perp \vec{v}$?) We check for orthogonality with the dot product:
$$\vec{z} \cdot \vec{v} = \langle -0.5, 1.5\rangle \cdot \langle 3, 1\rangle = 0.$$

Since the dot product is 0, we know $\vec{z} \perp \vec{v}$. Thus:
$$\vec{u} = \text{proj}_{\vec{v}}\,\vec{u} + (\vec{u} - \text{proj}_{\vec{v}}\,\vec{u})$$
$$\langle -2, 1\rangle = \underbrace{\langle -1.5, -0.5\rangle}_{\|\,\vec{v}} + \underbrace{\langle -0.5, 1.5\rangle}_{\perp\,\vec{v}}.$$

2. We found in Example 10.3.5 that $\text{proj}_{\vec{x}}\,\vec{w} = \langle 2, 2, 2\rangle$. Applying the Key Idea, we have:
$$\vec{z} = \vec{w} - \text{proj}_{\vec{x}}\,\vec{w} = \langle 2, 1, 3\rangle - \langle 2, 2, 2\rangle = \langle 0, -1, 1\rangle.$$

We check to see if $\vec{z} \perp \vec{x}$:
$$\vec{z} \cdot \vec{x} = \langle 0, -1, 1\rangle \cdot \langle 1, 1, 1\rangle = 0.$$

Since the dot product is 0, we know the two vectors are orthogonal. We now write $\vec{w}$ as the sum of two vectors, one parallel and one orthogonal to $\vec{x}$:
$$\vec{w} = \text{proj}_{\vec{x}}\,\vec{w} + (\vec{w} - \text{proj}_{\vec{x}}\,\vec{w})$$
$$\langle 2, 1, 3\rangle = \underbrace{\langle 2, 2, 2\rangle}_{\|\,\vec{x}} + \underbrace{\langle 0, -1, 1\rangle}_{\perp\,\vec{x}}.$$

We give an example of where this decomposition is useful.

**Example 10.3.7  Orthogonally decomposing a force vector**
Consider Figure 10.3.8(a), showing a box weighing 50lb on a ramp that rises 5ft over a span of 20ft. Find the components of force, and their magnitudes, acting on the box (as sketched in part (b) of the figure):

1. in the direction of the ramp, and
2. orthogonal to the ramp.

**SOLUTION**   As the ramp rises 5ft over a horizontal distance of 20ft, we can represent the direction of the ramp with the vector $\vec{r} = \langle 20, 5\rangle$. Gravity pulls down with a force of 50lb, which we represent with $\vec{g} = \langle 0, -50\rangle$.

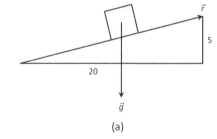

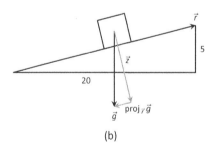

Figure 10.3.8: Sketching the ramp and box in Example 10.3.7. Note: *The vectors are not drawn to scale.*

1. To find the force of gravity in the direction of the ramp, we compute $\text{proj}_{\vec{r}}\,\vec{g}$:

$$\text{proj}_{\vec{r}}\,\vec{g} = \frac{\vec{g}\cdot\vec{r}}{\vec{r}\cdot\vec{r}}\vec{r}$$
$$= \frac{-250}{425}\langle 20, 5\rangle$$
$$= \left\langle -\frac{200}{17}, -\frac{50}{17}\right\rangle \approx \langle -11.76, -2.94\rangle.$$

The magnitude of $\text{proj}_{\vec{r}}\,\vec{g}$ is $\|\text{proj}_{\vec{r}}\,\vec{g}\| = 50/\sqrt{17} \approx 12.13$lb. Though the box weighs 50lb, a force of about 12lb is enough to keep the box from sliding down the ramp.

2. To find the component $\vec{z}$ of gravity orthogonal to the ramp, we use Key Idea 10.3.1.

$$\vec{z} = \vec{g} - \text{proj}_{\vec{r}}\,\vec{g}$$
$$= \left\langle \frac{200}{17}, -\frac{800}{17}\right\rangle \approx \langle 11.76, -47.06\rangle.$$

The magnitude of this force is $\|\vec{z}\| \approx 48.51$lb. In physics and engineering, knowing this force is important when computing things like static frictional force. (For instance, we could easily compute if the static frictional force alone was enough to keep the box from sliding down the ramp.)

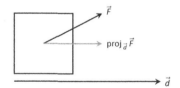

Figure 10.3.9: Finding work when the force and direction of travel are given as vectors.

## Application to Work

In physics, the application of a force $F$ to move an object in a straight line a distance $d$ produces *work*; the amount of work $W$ is $W = Fd$, (where $F$ is in the direction of travel). The orthogonal projection allows us to compute work when the force is not in the direction of travel.

Consider Figure 10.3.9, where a force $\vec{F}$ is being applied to an object moving in the direction of $\vec{d}$. (The distance the object travels is the magnitude of $\vec{d}$.) The

Notes:

work done is the amount of force in the direction of $\vec{d}$, $\|\operatorname{proj}_{\vec{d}}\vec{F}\|$, times $\|\vec{d}\|$:

$$\|\operatorname{proj}_{\vec{d}}\vec{F}\| \cdot \|\vec{d}\| = \left\|\frac{\vec{F}\cdot\vec{d}}{\vec{d}\cdot\vec{d}}\vec{d}\right\| \cdot \|\vec{d}\|$$

$$= \left|\frac{\vec{F}\cdot\vec{d}}{\|\vec{d}\|^2}\right| \cdot \|\vec{d}\| \cdot \|\vec{d}\|$$

$$= \frac{|\vec{F}\cdot\vec{d}|}{\|\vec{d}\|^2}\|\vec{d}\|^2$$

$$= |\vec{F}\cdot\vec{d}|.$$

The expression $\vec{F}\cdot\vec{d}$ will be positive if the angle between $\vec{F}$ and $\vec{d}$ is acute; when the angle is obtuse (hence $\vec{F}\cdot\vec{d}$ is negative), the force is causing motion in the opposite direction of $\vec{d}$, resulting in "negative work." We want to capture this sign, so we drop the absolute value and find that $W = \vec{F}\cdot\vec{d}$.

---

**Definition 10.3.4    Work**

Let $\vec{F}$ be a constant force that moves an object in a straight line from point $P$ to point $Q$. Let $\vec{d} = \overrightarrow{PQ}$. The **work** $W$ done by $\vec{F}$ along $\vec{d}$ is $W = \vec{F}\cdot\vec{d}$.

---

**Example 10.3.8    Computing work**

A man slides a box along a ramp that rises 3ft over a distance of 15ft by applying 50lb of force as shown in Figure 10.3.10. Compute the work done.

**Solution**    The figure indicates that the force applied makes a 30° angle with the horizontal, so $\vec{F} = 50\langle\cos 30°, \sin 30°\rangle \approx \langle 43.3, 25\rangle$. The ramp is represented by $\vec{d} = \langle 15, 3\rangle$. The work done is simply

$$\vec{F}\cdot\vec{d} = 50\langle\cos 30°, \sin 30°\rangle \cdot \langle 15, 3\rangle \approx 724.5\text{ft–lb}.$$

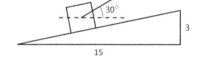

Figure 10.3.10: Computing work when sliding a box up a ramp in Example 10.3.8.

Note how we did not actually compute the distance the object traveled, nor the magnitude of the force in the direction of travel; this is all inherently computed by the dot product!

The dot product is a powerful way of evaluating computations that depend on angles without actually using angles. The next section explores another "product" on vectors, the *cross product*. Once again, angles play an important role, though in a much different way.

---

Notes:

# Exercises 10.3

## Terms and Concepts

1. The dot product of two vectors is a _____, not a vector.

2. How are the concepts of the dot product and vector magnitude related?

3. How can one quickly tell if the angle between two vectors is acute or obtuse?

4. Give a synonym for "orthogonal."

## Problems

**In Exercises 5 – 10, find the dot product of the given vectors.**

5. $\vec{u} = \langle 2, -4 \rangle$, $\vec{v} = \langle 3, 7 \rangle$

6. $\vec{u} = \langle 5, 3 \rangle$, $\vec{v} = \langle 6, 1 \rangle$

7. $\vec{u} = \langle 1, -1, 2 \rangle$, $\vec{v} = \langle 2, 5, 3 \rangle$

8. $\vec{u} = \langle 3, 5, -1 \rangle$, $\vec{v} = \langle 4, -1, 7 \rangle$

9. $\vec{u} = \langle 1, 1 \rangle$, $\vec{v} = \langle 1, 2, 3 \rangle$

10. $\vec{u} = \langle 1, 2, 3 \rangle$, $\vec{v} = \langle 0, 0, 0 \rangle$

11. Create your own vectors $\vec{u}$, $\vec{v}$ and $\vec{w}$ in $\mathbb{R}^2$ and show that $\vec{u} \cdot (\vec{v} + \vec{w}) = \vec{u} \cdot \vec{v} + \vec{u} \cdot \vec{w}$.

12. Create your own vectors $\vec{u}$ and $\vec{v}$ in $\mathbb{R}^3$ and scalar $c$ and show that $c(\vec{u} \cdot \vec{v}) = \vec{u} \cdot (c\vec{v})$.

**In Exercises 13 – 16, find the measure of the angle between the two vectors in both radians and degrees.**

13. $\vec{u} = \langle 1, 1 \rangle$, $\vec{v} = \langle 1, 2 \rangle$

14. $\vec{u} = \langle -2, 1 \rangle$, $\vec{v} = \langle 3, 5 \rangle$

15. $\vec{u} = \langle 8, 1, -4 \rangle$, $\vec{v} = \langle 2, 2, 0 \rangle$

16. $\vec{u} = \langle 1, 7, 2 \rangle$, $\vec{v} = \langle 4, -2, 5 \rangle$

**In Exercises 17 – 20, a vector $\vec{v}$ is given. Give two vectors that are orthogonal to $\vec{v}$.**

17. $\vec{v} = \langle 4, 7 \rangle$

18. $\vec{v} = \langle -3, 5 \rangle$

19. $\vec{v} = \langle 1, 1, 1 \rangle$

20. $\vec{v} = \langle 1, -2, 3 \rangle$

**In Exercises 21 – 26, vectors $\vec{u}$ and $\vec{v}$ are given. Find $\text{proj}_{\vec{v}}\,\vec{u}$, the orthogonal projection of $\vec{u}$ onto $\vec{v}$, and sketch all three vectors with the same initial point.**

21. $\vec{u} = \langle 1, 2 \rangle$, $\vec{v} = \langle -1, 3 \rangle$

22. $\vec{u} = \langle 5, 5 \rangle$, $\vec{v} = \langle 1, 3 \rangle$

23. $\vec{u} = \langle -3, 2 \rangle$, $\vec{v} = \langle 1, 1 \rangle$

24. $\vec{u} = \langle -3, 2 \rangle$, $\vec{v} = \langle 2, 3 \rangle$

25. $\vec{u} = \langle 1, 5, 1 \rangle$, $\vec{v} = \langle 1, 2, 3 \rangle$

26. $\vec{u} = \langle 3, -1, 2 \rangle$, $\vec{v} = \langle 2, 2, 1 \rangle$

**In Exercises 27 – 32, vectors $\vec{u}$ and $\vec{v}$ are given. Write $\vec{u}$ as the sum of two vectors, one of which is parallel to $\vec{v}$ and one of which is perpendicular to $\vec{v}$. Note: these are the same pairs of vectors as found in Exercises 21 – 26.**

27. $\vec{u} = \langle 1, 2 \rangle$, $\vec{v} = \langle -1, 3 \rangle$

28. $\vec{u} = \langle 5, 5 \rangle$, $\vec{v} = \langle 1, 3 \rangle$

29. $\vec{u} = \langle -3, 2 \rangle$, $\vec{v} = \langle 1, 1 \rangle$

30. $\vec{u} = \langle -3, 2 \rangle$, $\vec{v} = \langle 2, 3 \rangle$

31. $\vec{u} = \langle 1, 5, 1 \rangle$, $\vec{v} = \langle 1, 2, 3 \rangle$

32. $\vec{u} = \langle 3, -1, 2 \rangle$, $\vec{v} = \langle 2, 2, 1 \rangle$

33. A 10lb box sits on a ramp that rises 4ft over a distance of 20ft. How much force is required to keep the box from sliding down the ramp?

34. A 10lb box sits on a 15ft ramp that makes a 30° angle with the horizontal. How much force is required to keep the box from sliding down the ramp?

35. How much work is performed in moving a box horizontally 10ft with a force of 20lb applied at an angle of 45° to the horizontal?

36. How much work is performed in moving a box horizontally 10ft with a force of 20lb applied at an angle of 10° to the horizontal?

37. How much work is performed in moving a box up the length of a ramp that rises 2ft over a distance of 10ft, with a force of 50lb applied horizontally?

38. How much work is performed in moving a box up the length of a ramp that rises 2ft over a distance of 10ft, with a force of 50lb applied at an angle of 45° to the horizontal?

39. How much work is performed in moving a box up the length of a 10ft ramp that makes a 5° angle with the horizontal, with 50lb of force applied in the direction of the ramp?

## 10.4 The Cross Product

"Orthogonality" is immensely important. A quick scan of your current environment will undoubtedly reveal numerous surfaces and edges that are perpendicular to each other (including the edges of this page). The dot product provides a quick test for orthogonality: vectors $\vec{u}$ and $\vec{v}$ are perpendicular if, and only if, $\vec{u} \cdot \vec{v} = 0$.

Given two non–parallel, nonzero vectors $\vec{u}$ and $\vec{v}$ in space, it is very useful to find a vector $\vec{w}$ that is perpendicular to both $\vec{u}$ and $\vec{v}$. There is a operation, called the **cross product**, that creates such a vector. This section defines the cross product, then explores its properties and applications.

> **Definition 10.4.1    Cross Product**
>
> Let $\vec{u} = \langle u_1, u_2, u_3 \rangle$ and $\vec{v} = \langle v_1, v_2, v_3 \rangle$ be vectors in $\mathbb{R}^3$. The **cross product of $\vec{u}$ and $\vec{v}$**, denoted $\vec{u} \times \vec{v}$, is the vector
>
> $$\vec{u} \times \vec{v} = \langle u_2 v_3 - u_3 v_2, -(u_1 v_3 - u_3 v_1), u_1 v_2 - u_2 v_1 \rangle.$$

This definition can be a bit cumbersome to remember. After an example we will give a convenient method for computing the cross product. For now, careful examination of the products and differences given in the definition should reveal a pattern that is not too difficult to remember. (For instance, in the first component only 2 and 3 appear as subscripts; in the second component, only 1 and 3 appear as subscripts. Further study reveals the order in which they appear.)

Let's practice using this definition by computing a cross product.

**Example 10.4.1    Computing a cross product**
Let $\vec{u} = \langle 2, -1, 4 \rangle$ and $\vec{v} = \langle 3, 2, 5 \rangle$. Find $\vec{u} \times \vec{v}$, and verify that it is orthogonal to both $\vec{u}$ and $\vec{v}$.

**SOLUTION**    Using Definition 10.4.1, we have

$$\vec{u} \times \vec{v} = \langle (-1)5 - (4)2, -((2)5 - (4)3), (2)2 - (-1)3 \rangle = \langle -13, 2, 7 \rangle.$$

(We encourage the reader to compute this product on their own, then verify their result.)

We test whether or not $\vec{u} \times \vec{v}$ is orthogonal to $\vec{u}$ and $\vec{v}$ using the dot product:

$$(\vec{u} \times \vec{v}) \cdot \vec{u} = \langle -13, 2, 7 \rangle \cdot \langle 2, -1, 4 \rangle = 0,$$

$$(\vec{u} \times \vec{v}) \cdot \vec{v} = \langle -13, 2, 7 \rangle \cdot \langle 3, 2, 5 \rangle = 0.$$

Since both dot products are zero, $\vec{u} \times \vec{v}$ is indeed orthogonal to both $\vec{u}$ and $\vec{v}$.

---

Notes:

A convenient method of computing the cross product starts with forming a particular 3 × 3 *matrix*, or rectangular array. The first row comprises the standard unit vectors $\vec{i}, \vec{j},$ and $\vec{k}$. The second and third rows are the vectors $\vec{u}$ and $\vec{v}$, respectively. Using $\vec{u}$ and $\vec{v}$ from Example 10.4.1, we begin with:

$$\begin{array}{ccc} \vec{i} & \vec{j} & \vec{k} \\ 2 & -1 & 4 \\ 3 & 2 & 5 \end{array}$$

Now repeat the first two columns after the original three:

$$\begin{array}{ccccc} \vec{i} & \vec{j} & \vec{k} & \vec{i} & \vec{j} \\ 2 & -1 & 4 & 2 & -1 \\ 3 & 2 & 5 & 3 & 2 \end{array}$$

This gives three full "upper left to lower right" diagonals, and three full "upper right to lower left" diagonals, as shown. Compute the products along each diagonal, then add the products on the right and subtract the products on the left:

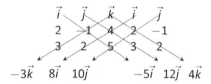

$$\vec{u} \times \vec{v} = \left(-5\vec{i} + 12\vec{j} + 4\vec{k}\right) - \left(-3\vec{k} + 8\vec{i} + 10\vec{j}\right) = -13\vec{i} + 2\vec{j} + 7\vec{k} = \langle -13, 2, 7\rangle.$$

We practice using this method.

**Example 10.4.2   Computing a cross product**
Let $\vec{u} = \langle 1, 3, 6 \rangle$ and $\vec{v} = \langle -1, 2, 1 \rangle$. Compute both $\vec{u} \times \vec{v}$ and $\vec{v} \times \vec{u}$.

**Solution**   To compute $\vec{u} \times \vec{v}$, we form the matrix as prescribed above, complete with repeated first columns:

$$\begin{array}{ccccc} \vec{i} & \vec{j} & \vec{k} & \vec{i} & \vec{j} \\ 1 & 3 & 6 & 1 & 3 \\ -1 & 2 & 1 & -1 & 2 \end{array}$$

We let the reader compute the products of the diagonals; we give the result:

$$\vec{u} \times \vec{v} = \left(3\vec{i} - 6\vec{j} + 2\vec{k}\right) - \left(-3\vec{k} + 12\vec{i} + \vec{j}\right) = \langle -9, -7, 5\rangle.$$

Notes:

To compute $\vec{v} \times \vec{u}$, we switch the second and third rows of the above matrix, then multiply along diagonals and subtract:

$$\begin{array}{ccccc} \vec{i} & \vec{j} & \vec{k} & \vec{i} & \vec{j} \\ -1 & 2 & 1 & -1 & 2 \\ 1 & 3 & 6 & 1 & 3 \end{array}$$

Note how with the rows being switched, the products that once appeared on the right now appear on the left, and vice–versa. Thus the result is:

$$\vec{v} \times \vec{u} = \left(12\vec{i} + \vec{j} - 3\vec{k}\right) - \left(2\vec{k} + 3\vec{i} - 6\vec{j}\right) = \langle 9, 7, -5 \rangle,$$

which is the opposite of $\vec{u} \times \vec{v}$. We leave it to the reader to verify that each of these vectors is orthogonal to $\vec{u}$ and $\vec{v}$.

## Properties of the Cross Product

It is not coincidence that $\vec{v} \times \vec{u} = -(\vec{u} \times \vec{v})$ in the preceding example; one can show using Definition 10.4.1 that this will always be the case. The following theorem states several useful properties of the cross product, each of which can be verified by referring to the definition.

---

**Theorem 10.4.1**      **Properties of the Cross Product**

Let $\vec{u}$, $\vec{v}$ and $\vec{w}$ be vectors in $\mathbb{R}^3$ and let $c$ be a scalar. The following identities hold:

1. $\vec{u} \times \vec{v} = -(\vec{v} \times \vec{u})$      Anticommutative Property

2. (a) $(\vec{u} + \vec{v}) \times \vec{w} = \vec{u} \times \vec{w} + \vec{v} \times \vec{w}$      Distributive Properties
   (b) $\vec{u} \times (\vec{v} + \vec{w}) = \vec{u} \times \vec{v} + \vec{u} \times \vec{w}$

3. $c(\vec{u} \times \vec{v}) = (c\vec{u}) \times \vec{v} = \vec{u} \times (c\vec{v})$

4. (a) $(\vec{u} \times \vec{v}) \cdot \vec{u} = 0$      Orthogonality Properties
   (b) $(\vec{u} \times \vec{v}) \cdot \vec{v} = 0$

5. $\vec{u} \times \vec{u} = \vec{0}$

6. $\vec{u} \times \vec{0} = \vec{0}$

7. $\vec{u} \cdot (\vec{v} \times \vec{w}) = (\vec{u} \times \vec{v}) \cdot \vec{w}$      Triple Scalar Product

---

Notes:

We introduced the cross product as a way to find a vector orthogonal to two given vectors, but we did not give a proof that the construction given in Definition 10.4.1 satisfies this property. Theorem 10.4.1 asserts this property holds; we leave it as a problem in the Exercise section to verify this.

Property 5 from the theorem is also left to the reader to prove in the Exercise section, but it reveals something more interesting than "the cross product of a vector with itself is $\vec{0}$." Let $\vec{u}$ and $\vec{v}$ be parallel vectors; that is, let there be a scalar $c$ such that $\vec{v} = c\vec{u}$. Consider their cross product:

$$\begin{aligned}\vec{u} \times \vec{v} &= \vec{u} \times (c\vec{u}) \\ &= c(\vec{u} \times \vec{u}) \quad \text{(by Property 3 of Theorem 10.4.1)} \\ &= \vec{0}. \quad \text{(by Property 5 of Theorem 10.4.1)}\end{aligned}$$

We have just shown that the cross product of parallel vectors is $\vec{0}$. This hints at something deeper. Theorem 10.3.2 related the angle between two vectors and their dot product; there is a similar relationship relating the cross product of two vectors and the angle between them, given by the following theorem.

**Note:** We could rewrite Definition 10.3.2 and Theorem 10.4.2 to include $\vec{0}$, then define that $\vec{u}$ and $\vec{v}$ are parallel if $\vec{u} \times \vec{v} = \vec{0}$. Since $\vec{0} \cdot \vec{v} = 0$ and $\vec{0} \times \vec{v} = \vec{0}$, this would mean that $\vec{0}$ is both parallel *and* orthogonal to all vectors. Apparent paradoxes such as this are not uncommon in mathematics and can be very useful. (See also the marginal note on page 582.)

---

**Theorem 10.4.2**      **The Cross Product and Angles**

Let $\vec{u}$ and $\vec{v}$ be nonzero vectors in $\mathbb{R}^3$. Then

$$\|\vec{u} \times \vec{v}\| = \|\vec{u}\|\,\|\vec{v}\| \sin\theta,$$

where $\theta$, $0 \leq \theta \leq \pi$, is the angle between $\vec{u}$ and $\vec{v}$.

---

Note that this theorem makes a statement about the *magnitude* of the cross product. When the angle between $\vec{u}$ and $\vec{v}$ is 0 or $\pi$ (i.e., the vectors are parallel), the magnitude of the cross product is 0. The only vector with a magnitude of 0 is $\vec{0}$ (see Property 9 of Theorem 10.2.1), hence the cross product of parallel vectors is $\vec{0}$.

We demonstrate the truth of this theorem in the following example.

**Example 10.4.3**      **The cross product and angles**
Let $\vec{u} = \langle 1, 3, 6 \rangle$ and $\vec{v} = \langle -1, 2, 1 \rangle$ as in Example 10.4.2. Verify Theorem 10.4.2 by finding $\theta$, the angle between $\vec{u}$ and $\vec{v}$, and the magnitude of $\vec{u} \times \vec{v}$.

---

Notes:

**SOLUTION** We use Theorem 10.3.2 to find the angle between $\vec{u}$ and $\vec{v}$.

$$\theta = \cos^{-1}\left(\frac{\vec{u}\cdot\vec{v}}{\|\vec{u}\|\|\vec{v}\|}\right)$$
$$= \cos^{-1}\left(\frac{11}{\sqrt{46}\sqrt{6}}\right)$$
$$\approx 0.8471 = 48.54°.$$

Our work in Example 10.4.2 showed that $\vec{u}\times\vec{v} = \langle -9,-7,5\rangle$, hence $\|\vec{u}\times\vec{v}\| = \sqrt{155}$. Is $\|\vec{u}\times\vec{v}\| = \|\vec{u}\|\|\vec{v}\|\sin\theta$? Using numerical approximations, we find:

$$\|\vec{u}\times\vec{v}\| = \sqrt{155} \qquad \|\vec{u}\|\|\vec{v}\|\sin\theta = \sqrt{46}\sqrt{6}\sin 0.8471$$
$$\approx 12.45. \qquad\qquad\qquad \approx 12.45.$$

Numerically, they seem equal. Using a right triangle, one can show that

$$\sin\left(\cos^{-1}\left(\frac{11}{\sqrt{46}\sqrt{6}}\right)\right) = \frac{\sqrt{155}}{\sqrt{46}\sqrt{6}},$$

which allows us to verify the theorem exactly.

### Right Hand Rule

The anticommutative property of the cross product demonstrates that $\vec{u}\times\vec{v}$ and $\vec{v}\times\vec{u}$ differ only by a sign – these vectors have the same magnitude but point in the opposite direction. When seeking a vector perpendicular to $\vec{u}$ and $\vec{v}$, we essentially have two directions to choose from, one in the direction of $\vec{u}\times\vec{v}$ and one in the direction of $\vec{v}\times\vec{u}$. Does it matter which we choose? How can we tell which one we will get without graphing, etc.?

Another wonderful property of the cross product, as defined, is that it follows the **right hand rule.** Given $\vec{u}$ and $\vec{v}$ in $\mathbb{R}^3$ with the same initial point, point the index finger of your right hand in the direction of $\vec{u}$ and let your middle finger point in the direction of $\vec{v}$ (much as we did when establishing the right hand rule for the 3-dimensional coordinate system). Your thumb will naturally extend in the direction of $\vec{u}\times\vec{v}$. One can "practice" this using Figure 10.4.1. If you switch, and point the index finder in the direction of $\vec{v}$ and the middle finger in the direction of $\vec{u}$, your thumb will now point in the opposite direction, allowing you to "visualize" the anticommutative property of the cross product.

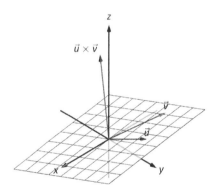

Figure 10.4.1: Illustrating the Right Hand Rule of the cross product.

## Applications of the Cross Product

There are a number of ways in which the cross product is useful in mathematics, physics and other areas of science beyond "just" finding a vector perpendicular to two others. We highlight a few here.

Notes:

# Chapter 10 Vectors

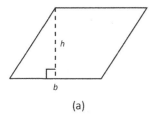

(a)

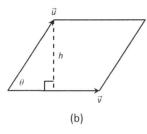

(b)

Figure 10.4.2: Using the cross product to find the area of a parallelogram.

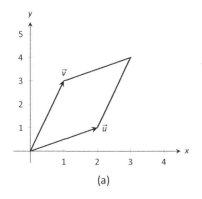

(a)

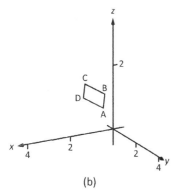

(b)

Figure 10.4.3: Sketching the parallelograms in Example 10.4.4.

**Area of a Parallelogram**

It is a standard geometry fact that the area of a parallelogram is $A = bh$, where $b$ is the length of the base and $h$ is the height of the parallelogram, as illustrated in Figure 10.4.2(a). As shown when defining the Parallelogram Law of vector addition, two vectors $\vec{u}$ and $\vec{v}$ define a parallelogram when drawn from the same initial point, as illustrated in Figure 10.4.2(b). Trigonometry tells us that $h = \|\vec{u}\|\sin\theta$, hence the area of the parallelogram is

$$A = \|\vec{u}\|\|\vec{v}\|\sin\theta = \|\vec{u}\times\vec{v}\|, \tag{10.4}$$

where the second equality comes from Theorem 10.4.2. We illustrate using Equation (10.4) in the following example.

**Example 10.4.4    Finding the area of a parallelogram**

1. Find the area of the parallelogram defined by the vectors $\vec{u} = \langle 2, 1\rangle$ and $\vec{v} = \langle 1, 3\rangle$.

2. Verify that the points $A = (1, 1, 1)$, $B = (2, 3, 2)$, $C = (4, 5, 3)$ and $D = (3, 3, 2)$ are the vertices of a parallelogram. Find the area of the parallelogram.

**SOLUTION**

1. Figure 10.4.3(a) sketches the parallelogram defined by the vectors $\vec{u}$ and $\vec{v}$. We have a slight problem in that our vectors exist in $\mathbb{R}^2$, not $\mathbb{R}^3$, and the cross product is only defined on vectors in $\mathbb{R}^3$. We skirt this issue by viewing $\vec{u}$ and $\vec{v}$ as vectors in the $x-y$ plane of $\mathbb{R}^3$, and rewrite them as $\vec{u} = \langle 2, 1, 0\rangle$ and $\vec{v} = \langle 1, 3, 0\rangle$. We can now compute the cross product. It is easy to show that $\vec{u}\times\vec{v} = \langle 0, 0, 5\rangle$; therefore the area of the parallelogram is $A = \|\vec{u}\times\vec{v}\| = 5$.

2. To show that the quadrilateral $ABCD$ is a parallelogram (shown in Figure 10.4.3(b)), we need to show that the opposite sides are parallel. We can quickly show that $\overrightarrow{AB} = \overrightarrow{DC} = \langle 1, 2, 1\rangle$ and $\overrightarrow{BC} = \overrightarrow{AD} = \langle 2, 2, 1\rangle$. We find the area by computing the magnitude of the cross product of $\overrightarrow{AB}$ and $\overrightarrow{BC}$:

$$\overrightarrow{AB}\times\overrightarrow{BC} = \langle 0, 1, -2\rangle \quad\Rightarrow\quad \|\overrightarrow{AB}\times\overrightarrow{BC}\| = \sqrt{5} \approx 2.236.$$

This application is perhaps more useful in finding the area of a triangle (in short, triangles are used more often than parallelograms). We illustrate this in the following example.

Notes:

### Example 10.4.5 Area of a triangle

Find the area of the triangle with vertices $A = (1, 2)$, $B = (2, 3)$ and $C = (3, 1)$, as pictured in Figure 10.4.4.

**SOLUTION** We found the area of this triangle in Example 7.1.4 to be 1.5 using integration. There we discussed the fact that finding the area of a triangle can be inconvenient using the "$\frac{1}{2}bh$" formula as one has to compute the height, which generally involves finding angles, etc. Using a cross product is much more direct.

We can choose any two sides of the triangle to use to form vectors; we choose $\vec{AB} = \langle 1, 1 \rangle$ and $\vec{AC} = \langle 2, -1 \rangle$. As in the previous example, we will rewrite these vectors with a third component of 0 so that we can apply the cross product. The area of the triangle is

$$\frac{1}{2} \| \vec{AB} \times \vec{AC} \| = \frac{1}{2} \| \langle 1, 1, 0 \rangle \times \langle 2, -1, 0 \rangle \| = \frac{1}{2} \| \langle 0, 0, -3 \rangle \| = \frac{3}{2}.$$

We arrive at the same answer as before with less work.

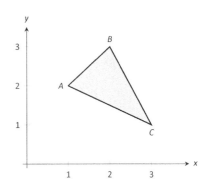

Figure 10.4.4: Finding the area of a triangle in Example 10.4.5.

**Volume of a Parallelepiped**

The three dimensional analogue to the parallelogram is the **parallelepiped**. Each face is parallel to the opposite face, as illustrated in Figure 10.4.5. By crossing $\vec{v}$ and $\vec{w}$, one gets a vector whose magnitude is the area of the base. Dotting this vector with $\vec{u}$ computes the volume of parallelepiped! (Up to a sign; take the absolute value.)

Thus the volume of a parallelepiped defined by vectors $\vec{u}$, $\vec{v}$ and $\vec{w}$ is

$$V = |\vec{u} \cdot (\vec{v} \times \vec{w})|. \qquad (10.5)$$

Note how this is the Triple Scalar Product, first seen in Theorem 10.4.1. Applying the identities given in the theorem shows that we can apply the Triple Scalar Product in any "order" we choose to find the volume. That is,

$$V = |\vec{u} \cdot (\vec{v} \times \vec{w})| = |\vec{u} \cdot (\vec{w} \times \vec{v})| = |(\vec{u} \times \vec{v}) \cdot \vec{w}|, \quad \text{etc.}$$

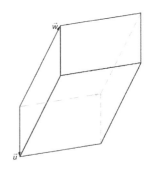

Figure 10.4.5: A parallelepiped is the three dimensional analogue to the parallelogram.

**Note:** The word "parallelepiped" is pronounced "parallel–eh–pipe–ed."

### Example 10.4.6 Finding the volume of parallelepiped

Find the volume of the parallelepiped defined by the vectors $\vec{u} = \langle 1, 1, 0 \rangle$, $\vec{v} = \langle -1, 1, 0 \rangle$ and $\vec{w} = \langle 0, 1, 1 \rangle$.

**SOLUTION** We apply Equation (10.5). We first find $\vec{v} \times \vec{w} = \langle 1, 1, -1 \rangle$. Then

$$|\vec{u} \cdot (\vec{v} \times \vec{w})| = |\langle 1, 1, 0 \rangle \cdot \langle 1, 1, -1 \rangle| = 2.$$

So the volume of the parallelepiped is 2 cubic units.

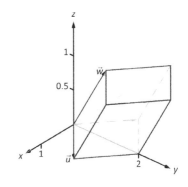

Figure 10.4.6: A parallelepiped in Example 10.4.6.

While this application of the Triple Scalar Product is interesting, it is not used all that often: parallelepipeds are not a common shape in physics and engineering. The last application of the cross product is very applicable in engineering.

**Torque**

**Torque** is a measure of the turning force applied to an object. A classic scenario involving torque is the application of a wrench to a bolt. When a force is applied to the wrench, the bolt turns. When we represent the force and wrench with vectors $\vec{F}$ and $\vec{\ell}$, we see that the bolt moves (because of the threads) in a direction orthogonal to $\vec{F}$ and $\vec{\ell}$. Torque is usually represented by the Greek letter $\tau$, or tau, and has units of N·m, a Newton–meter, or ft·lb, a foot–pound.

While a full understanding of torque is beyond the purposes of this book, when a force $\vec{F}$ is applied to a lever arm $\vec{\ell}$, the resulting torque is

$$\vec{\tau} = \vec{\ell} \times \vec{F}. \tag{10.6}$$

**Example 10.4.7  Computing torque**

A lever of length 2ft makes an angle with the horizontal of 45°. Find the resulting torque when a force of 10lb is applied to the end of the level where:

1. the force is perpendicular to the lever, and
2. the force makes an angle of 60° with the lever, as shown in Figure 10.4.7.

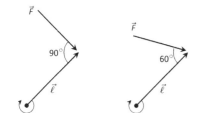

Figure 10.4.7: Showing a force being applied to a lever in Example 10.4.7.

**Solution**

1. We start by determining vectors for the force and lever arm. Since the lever arm makes a 45° angle with the horizontal and is 2ft long, we can state that $\vec{\ell} = 2\langle \cos 45°, \sin 45° \rangle = \langle \sqrt{2}, \sqrt{2} \rangle$.

   Since the force vector is perpendicular to the lever arm (as seen in the left hand side of Figure 10.4.7), we can conclude it is making an angle of $-45°$ with the horizontal. As it has a magnitude of 10lb, we can state $\vec{F} = 10\langle \cos(-45°), \sin(-45°) \rangle = \langle 5\sqrt{2}, -5\sqrt{2} \rangle$.

   Using Equation (10.6) to find the torque requires a cross product. We again let the third component of each vector be 0 and compute the cross product:

$$\vec{\tau} = \vec{\ell} \times \vec{F}$$
$$= \langle \sqrt{2}, \sqrt{2}, 0 \rangle \times \langle 5\sqrt{2}, -5\sqrt{2}, 0 \rangle$$
$$= \langle 0, 0, -20 \rangle$$

Notes:

This clearly has a magnitude of 20 ft-lb.

We can view the force and lever arm vectors as lying "on the page"; our computation of $\vec{\tau}$ shows that the torque goes "into the page." This follows the Right Hand Rule of the cross product, and it also matches well with the example of the wrench turning the bolt. Turning a bolt clockwise moves it in.

2. Our lever arm can still be represented by $\vec{\ell} = \langle \sqrt{2}, \sqrt{2} \rangle$. As our force vector makes a 60° angle with $\vec{\ell}$, we can see (referencing the right hand side of the figure) that $\vec{F}$ makes a $-15°$ angle with the horizontal. Thus

$$\vec{F} = 10 \langle \cos -15°, \sin -15° \rangle = \left\langle \frac{5(1+\sqrt{3})}{\sqrt{2}}, \frac{5(-1+\sqrt{3})}{\sqrt{2}} \right\rangle$$
$$\approx \langle 9.659, -2.588 \rangle.$$

We again make the third component 0 and take the cross product to find the torque:

$$\vec{\tau} = \vec{\ell} \times \vec{F}$$
$$= \langle \sqrt{2}, \sqrt{2}, 0 \rangle \times \left\langle \frac{5(1+\sqrt{3})}{\sqrt{2}}, \frac{5(-1+\sqrt{3})}{\sqrt{2}}, 0 \right\rangle$$
$$= \langle 0, 0, -10\sqrt{3} \rangle$$
$$\approx \langle 0, 0, -17.321 \rangle.$$

As one might expect, when the force and lever arm vectors *are* orthogonal, the magnitude of force is greater than when the vectors *are not* orthogonal.

While the cross product has a variety of applications (as noted in this chapter), its fundamental use is finding a vector perpendicular to two others. Knowing a vector is orthogonal to two others is of incredible importance, as it allows us to find the equations of lines and planes in a variety of contexts. The importance of the cross product, in some sense, relies on the importance of lines and planes, which see widespread use throughout engineering, physics and mathematics. We study lines and planes in the next two sections.

Notes:

# Exercises 10.4

## Terms and Concepts

1. The cross product of two vectors is a _____, not a scalar.

2. One can visualize the direction of $\vec{u} \times \vec{v}$ using the _____ _____ _____.

3. Give a synonym for "orthogonal."

4. T/F: A fundamental principle of the cross product is that $\vec{u} \times \vec{v}$ is orthogonal to $\vec{u}$ and $\vec{v}$.

5. _____ is a measure of the turning force applied to an object.

6. T/F: If $\vec{u}$ and $\vec{v}$ are parallel, then $\vec{u} \times \vec{v} = \vec{0}$.

## Problems

In Exercises 7 – 16, vectors $\vec{u}$ and $\vec{v}$ are given. Compute $\vec{u} \times \vec{v}$ and show this is orthogonal to both $\vec{u}$ and $\vec{v}$.

7. $\vec{u} = \langle 3, 2, -2 \rangle$, $\vec{v} = \langle 0, 1, 5 \rangle$

8. $\vec{u} = \langle 5, -4, 3 \rangle$, $\vec{v} = \langle 2, -5, 1 \rangle$

9. $\vec{u} = \langle 4, -5, -5 \rangle$, $\vec{v} = \langle 3, 3, 4 \rangle$

10. $\vec{u} = \langle -4, 7, -10 \rangle$, $\vec{v} = \langle 4, 4, 1 \rangle$

11. $\vec{u} = \langle 1, 0, 1 \rangle$, $\vec{v} = \langle 5, 0, 7 \rangle$

12. $\vec{u} = \langle 1, 5, -4 \rangle$, $\vec{v} = \langle -2, -10, 8 \rangle$

13. $\vec{u} = \langle a, b, 0 \rangle$, $\vec{v} = \langle c, d, 0 \rangle$

14. $\vec{u} = \vec{i}$, $\vec{v} = \vec{j}$

15. $\vec{u} = \vec{i}$, $\vec{v} = \vec{k}$

16. $\vec{u} = \vec{j}$, $\vec{v} = \vec{k}$

17. Pick any vectors $\vec{u}$, $\vec{v}$ and $\vec{w}$ in $\mathbb{R}^3$ and show that $\vec{u} \times (\vec{v} + \vec{w}) = \vec{u} \times \vec{v} + \vec{u} \times \vec{w}$.

18. Pick any vectors $\vec{u}$, $\vec{v}$ and $\vec{w}$ in $\mathbb{R}^3$ and show that $\vec{u} \cdot (\vec{v} \times \vec{w}) = (\vec{u} \times \vec{v}) \cdot \vec{w}$.

In Exercises 19 – 22, the magnitudes of vectors $\vec{u}$ and $\vec{v}$ in $\mathbb{R}^3$ are given, along with the angle $\theta$ between them. Use this information to find the magnitude of $\vec{u} \times \vec{v}$.

19. $\|\vec{u}\| = 2$, $\|\vec{v}\| = 5$, $\theta = 30°$

20. $\|\vec{u}\| = 3$, $\|\vec{v}\| = 7$, $\theta = \pi/2$

21. $\|\vec{u}\| = 3$, $\|\vec{v}\| = 4$, $\theta = \pi$

22. $\|\vec{u}\| = 2$, $\|\vec{v}\| = 5$, $\theta = 5\pi/6$

In Exercises 23 – 26, find the area of the parallelogram defined by the given vectors.

23. $\vec{u} = \langle 1, 1, 2 \rangle$, $\vec{v} = \langle 2, 0, 3 \rangle$

24. $\vec{u} = \langle -2, 1, 5 \rangle$, $\vec{v} = \langle -1, 3, 1 \rangle$

25. $\vec{u} = \langle 1, 2 \rangle$, $\vec{v} = \langle 2, 1 \rangle$

26. $\vec{u} = \langle 2, 0 \rangle$, $\vec{v} = \langle 0, 3 \rangle$

In Exercises 27 – 30, find the area of the triangle with the given vertices.

27. Vertices: $(0, 0, 0)$, $(1, 3, -1)$ and $(2, 1, 1)$.

28. Vertices: $(5, 2, -1)$, $(3, 6, 2)$ and $(1, 0, 4)$.

29. Vertices: $(1, 1)$, $(1, 3)$ and $(2, 2)$.

30. Vertices: $(3, 1)$, $(1, 2)$ and $(4, 3)$.

In Exercises 31 – 32, find the area of the quadrilateral with the given vertices. (Hint: break the quadrilateral into 2 triangles.)

31. Vertices: $(0, 0)$, $(1, 2)$, $(3, 0)$ and $(4, 3)$.

32. Vertices: $(0, 0, 0)$, $(2, 1, 1)$, $(-1, 2, -8)$ and $(1, -1, 5)$.

In Exercises 33 – 34, find the volume of the parallelepiped defined by the given vectors.

33. $\vec{u} = \langle 1, 1, 1 \rangle$, $\vec{v} = \langle 1, 2, 3 \rangle$, $\vec{w} = \langle 1, 0, 1 \rangle$

34. $\vec{u} = \langle -1, 2, 1 \rangle$, $\vec{v} = \langle 2, 2, 1 \rangle$, $\vec{w} = \langle 3, 1, 3 \rangle$

In Exercises 35 – 38, find a unit vector orthogonal to both $\vec{u}$ and $\vec{v}$.

35. $\vec{u} = \langle 1, 1, 1 \rangle$, $\vec{v} = \langle 2, 0, 1 \rangle$

36. $\vec{u} = \langle 1, -2, 1 \rangle$, $\vec{v} = \langle 3, 2, 1 \rangle$

37. $\vec{u} = \langle 5, 0, 2 \rangle$, $\vec{v} = \langle -3, 0, 7 \rangle$

38. $\vec{u} = \langle 1, -2, 1 \rangle$, $\vec{v} = \langle -2, 4, -2 \rangle$

39. A bicycle rider applies 150lb of force, straight down, onto a pedal that extends 7in horizontally from the crankshaft. Find the magnitude of the torque applied to the crankshaft.

40. A bicycle rider applies 150lb of force, straight down, onto a pedal that extends 7in from the crankshaft, making a 30° angle with the horizontal. Find the magnitude of the torque applied to the crankshaft.

41. To turn a stubborn bolt, 80lb of force is applied to a 10in wrench. What is the maximum amount of torque that can be applied to the bolt?

42. To turn a stubborn bolt, 80lb of force is applied to a 10in wrench in a confined space, where the direction of applied force makes a 10° angle with the wrench. How much torque is subsequently applied to the wrench?

43. Show, using the definition of the Cross Product, that $\vec{u} \cdot (\vec{u} \times \vec{v}) = 0$; that is, that $\vec{u}$ is orthogonal to the cross product of $\vec{u}$ and $\vec{v}$.

44. Show, using the definition of the Cross Product, that $\vec{u} \times \vec{u} = \vec{0}$.

## 10.5 Lines

To find the equation of a line in the x-y plane, we need two pieces of information: a point and the slope. The slope conveys *direction* information. As vertical lines have an undefined slope, the following statement is more accurate:

> To define a line, one needs a point on the line and the direction of the line.

This holds true for lines in space.

Let $P$ be a point in space, let $\vec{p}$ be the vector with initial point at the origin and terminal point at $P$ (i.e., $\vec{p}$ "points" to $P$), and let $\vec{d}$ be a vector. Consider the points on the line through $P$ in the direction of $\vec{d}$.

Clearly one point on the line is $P$; we can say that the *vector* $\vec{p}$ lies at this point on the line. To find another point on the line, we can start at $\vec{p}$ and move in a direction parallel to $\vec{d}$. For instance, starting at $\vec{p}$ and traveling one length of $\vec{d}$ places one at another point on the line. Consider Figure 10.5.2 where certain points along the line are indicated.

The figure illustrates how every point on the line can be obtained by starting with $\vec{p}$ and moving a certain distance in the direction of $\vec{d}$. That is, we can define the line as a function of $t$:

$$\vec{\ell}(t) = \vec{p} + t\,\vec{d}. \tag{10.7}$$

In many ways, this is *not* a new concept. Compare Equation (10.7) to the familiar "$y = mx + b$" equation of a line:

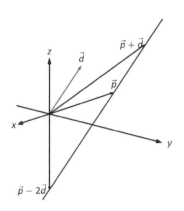

Figure 10.5.2: Defining a line in space.

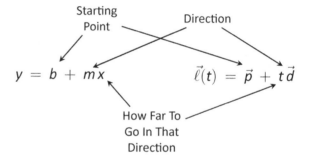

Figure 10.5.1: Understanding the vector equation of a line.

The equations exhibit the same structure: they give a starting point, define a direction, and state how far in that direction to travel.

Equation (10.7) is an example of a **vector–valued function**; the input of the function is a real number and the output is a vector. We will cover vector–valued functions extensively in the next chapter.

Notes:

There are other ways to represent a line. Let $\vec{p} = \langle x_0, y_0, z_0 \rangle$ and let $\vec{d} = \langle a, b, c \rangle$. Then the equation of the line through $\vec{p}$ in the direction of $\vec{d}$ is:

$$\vec{\ell}(t) = \vec{p} + t\vec{d}$$
$$= \langle x_0, y_0, z_0 \rangle + t \langle a, b, c \rangle$$
$$= \langle x_0 + at, y_0 + bt, z_0 + ct \rangle.$$

The last line states that the $x$ values of the line are given by $x = x_0 + at$, the $y$ values are given by $y = y_0 + bt$, and the $z$ values are given by $z = z_0 + ct$. These three equations, taken together, are the **parametric equations of the line** through $\vec{p}$ in the direction of $\vec{d}$.

Finally, each of the equations for $x$, $y$ and $z$ above contain the variable $t$. We can solve for $t$ in each equation:

$$x = x_0 + at \quad \Rightarrow \quad t = \frac{x - x_0}{a},$$
$$y = y_0 + bt \quad \Rightarrow \quad t = \frac{y - y_0}{b},$$
$$z = z_0 + ct \quad \Rightarrow \quad t = \frac{z - z_0}{c},$$

assuming $a, b, c \neq 0$. Since $t$ is equal to each expression on the right, we can set these equal to each other, forming the **symmetric equations of the line** through $\vec{p}$ in the direction of $\vec{d}$:

$$\frac{x - x_0}{a} = \frac{y - y_0}{b} = \frac{z - z_0}{c}.$$

Each representation has its own advantages, depending on the context. We summarize these three forms in the following definition, then give examples of their use.

Notes:

# Chapter 10  Vectors

> **Definition 10.5.1  Equations of Lines in Space**
>
> Consider the line in space that passes through $\vec{p} = \langle x_0, y_0, z_0 \rangle$ in the direction of $\vec{d} = \langle a, b, c \rangle$.
>
> 1. The **vector equation** of the line is
> $$\vec{\ell}(t) = \vec{p} + t\vec{d}.$$
>
> 2. The **parametric equations** of the line are
> $$x = x_0 + at, \quad y = y_0 + bt, \quad z = z_0 + ct.$$
>
> 3. The **symmetric equations** of the line are
> $$\frac{x - x_0}{a} = \frac{y - y_0}{b} = \frac{z - z_0}{c}.$$

**Example 10.5.1  Finding the equation of a line**

Give all three equations, as given in Definition 10.5.1, of the line through $P = (2, 3, 1)$ in the direction of $\vec{d} = \langle -1, 1, 2 \rangle$. Does the point $Q = (-1, 6, 6)$ lie on this line?

**Solution**  We identify the point $P = (2, 3, 1)$ with the vector $\vec{p} = \langle 2, 3, 1 \rangle$. Following the definition, we have

- the vector equation of the line is $\vec{\ell}(t) = \langle 2, 3, 1 \rangle + t\langle -1, 1, 2 \rangle$;

- the parametric equations of the line are
$$x = 2 - t, \quad y = 3 + t, \quad z = 1 + 2t; \text{ and}$$

- the symmetric equations of the line are
$$\frac{x - 2}{-1} = \frac{y - 3}{1} = \frac{z - 1}{2}.$$

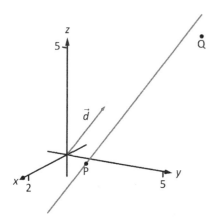

Figure 10.5.3: Graphing a line in Example 10.5.1.

The first two equations of the line are useful when a $t$ value is given: one can immediately find the corresponding point on the line. These forms are good when calculating with a computer; most software programs easily handle equations in these formats. (For instance, the graphics program that made Figure 10.5.3 can be given the input "(2-t,3+t,1+2*t)" for $-1 \leq t \leq 3$.).

Does the point $Q = (-1, 6, 6)$ lie on the line? The graph in Figure 10.5.3 makes it clear that it does not. We can answer this question without the graph

Notes:

using any of the three equation forms. Of the three, the symmetric equations are probably best suited for this task. Simply plug in the values of x, y and z and see if equality is maintained:

$$\frac{-1-2}{-1} \stackrel{?}{=} \frac{6-3}{1} \stackrel{?}{=} \frac{6-1}{2} \quad \Rightarrow \quad 3 = 3 \neq 2.5.$$

We see that Q does not lie on the line as it did not satisfy the symmetric equations.

**Example 10.5.2   Finding the equation of a line through two points**
Find the parametric equations of the line through the points $P = (2, -1, 2)$ and $Q = (1, 3, -1)$.

**SOLUTION**   Recall the statement made at the beginning of this section: to find the equation of a line, we need a point and a direction. We have *two* points; either one will suffice. The direction of the line can be found by the vector with initial point P and terminal point Q: $\vec{PQ} = \langle -1, 4, -3 \rangle$.

The parametric equations of the line $\ell$ through P in the direction of $\vec{PQ}$ are:

$$\ell: \quad x = 2 - t \quad y = -1 + 4t \quad z = 2 - 3t.$$

A graph of the points and line are given in Figure 10.5.4. Note how in the given parametrization of the line, $t = 0$ corresponds to the point P, and $t = 1$ corresponds to the point Q. This relates to the understanding of the vector equation of a line described in Figure 10.5.1. The parametric equations "start" at the point P, and t determines how far in the direction of $\vec{PQ}$ to travel. When $t = 0$, we travel 0 lengths of $\vec{PQ}$; when $t = 1$, we travel one length of $\vec{PQ}$, resulting in the point Q.

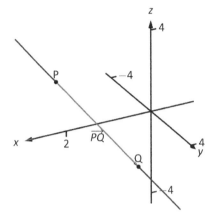

Figure 10.5.4: A graph of the line in Example 10.5.2.

## Parallel, Intersecting and Skew Lines

In the plane, two *distinct* lines can either be parallel or they will intersect at exactly one point. In space, given equations of two lines, it can sometimes be difficult to tell whether the lines are distinct or not (i.e., the same line can be represented in different ways). Given lines $\vec{\ell_1}(t) = \vec{p_1} + t\vec{d_1}$ and $\vec{\ell_2}(t) = \vec{p_2} + t\vec{d_2}$, we have four possibilities: $\vec{\ell_1}$ and $\vec{\ell_2}$ are

| | |
|---|---|
| the same line | they share all points; |
| intersecting lines | share only 1 point; |
| parallel lines | $\vec{d_1} \parallel \vec{d_2}$, no points in common; or |
| skew lines | $\vec{d_1} \nparallel \vec{d_2}$, no points in common. |

Notes:

The next two examples investigate these possibilities.

**Example 10.5.3  Comparing lines**

Consider lines $\ell_1$ and $\ell_2$, given in parametric equation form:

$$\ell_1: \begin{array}{rcl} x &=& 1+3t \\ y &=& 2-t \\ z &=& t \end{array} \qquad \ell_2: \begin{array}{rcl} x &=& -2+4s \\ y &=& 3+s \\ z &=& 5+2s. \end{array}$$

Determine whether $\ell_1$ and $\ell_2$ are the same line, intersect, are parallel, or skew.

**SOLUTION**  We start by looking at the directions of each line. Line $\ell_1$ has the direction given by $\vec{d_1} = \langle 3, -1, 1 \rangle$ and line $\ell_2$ has the direction given by $\vec{d_2} = \langle 4, 1, 2 \rangle$. It should be clear that $\vec{d_1}$ and $\vec{d_2}$ are not parallel, hence $\ell_1$ and $\ell_2$ are not the same line, nor are they parallel. Figure 10.5.5 verifies this fact (where the points and directions indicated by the equations of each line are identified).

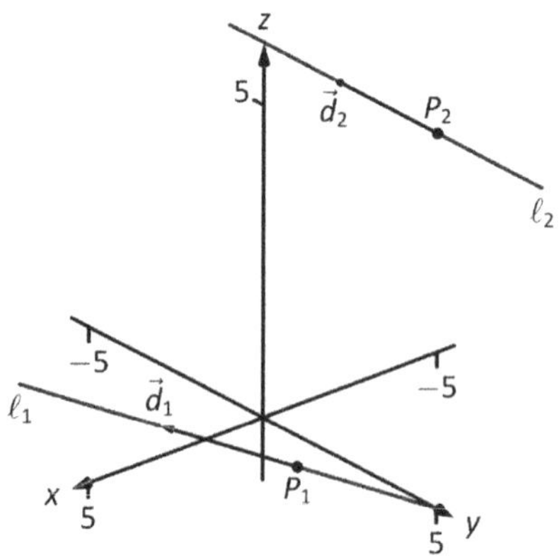

We next check to see if they intersect (if they do not, they are skew lines). To find if they intersect, we look for $t$ and $s$ values such that the respective $x$, $y$ and $z$ values are the same. That is, we want $s$ and $t$ such that:

$$\begin{array}{rcl} 1+3t &=& -2+4s \\ 2-t &=& 3+s \\ t &=& 5+2s. \end{array}$$

This is a relatively simple system of linear equations. Since the last equation is already solved for $t$, substitute that value of $t$ into the equation above it:

$$2-(5+2s) = 3+s \quad \Rightarrow \quad s = -2, \, t = 1.$$

A key to remember is that we have *three* equations; we need to check if $s = -2$, $t = 1$ satisfies the first equation as well:

$$1 + 3(1) \neq -2 + 4(-2).$$

It does not. Therefore, we conclude that the lines $\ell_1$ and $\ell_2$ are skew.

**Example 10.5.4  Comparing lines**

Consider lines $\ell_1$ and $\ell_2$, given in parametric equation form:

$$\ell_1: \begin{array}{rcl} x &=& -0.7+1.6t \\ y &=& 4.2+2.72t \\ z &=& 2.3-3.36t \end{array} \qquad \ell_2: \begin{array}{rcl} x &=& 2.8-2.9s \\ y &=& 10.15-4.93s \\ z &=& -5.05+6.09s. \end{array}$$

Determine whether $\ell_1$ and $\ell_2$ are the same line, intersect, are parallel, or skew.

Figure 10.5.5: Sketching the lines from Example 10.5.3.

Notes:

**SOLUTION**    It is obviously very difficult to simply look at these equations and discern anything. This is done intentionally. In the "real world," most equations that are used do not have nice, integer coefficients. Rather, there are lots of digits after the decimal and the equations can look "messy."

We again start by deciding whether or not each line has the same direction. The direction of $\ell_1$ is given by $\vec{d}_1 = \langle 1.6, 2.72, -3.36 \rangle$ and the direction of $\ell_2$ is given by $\vec{d}_2 = \langle -2.9, -4.93, 6.09 \rangle$. When it is not clear through observation whether two vectors are parallel or not, the standard way of determining this is by comparing their respective unit vectors. Using a calculator, we find:

$$\vec{u}_1 = \frac{\vec{d}_1}{\|\vec{d}_1\|} = \langle 0.3471, 0.5901, -0.7289 \rangle$$

$$\vec{u}_2 = \frac{\vec{d}_2}{\|\vec{d}_2\|} = \langle -0.3471, -0.5901, 0.7289 \rangle.$$

The two vectors seem to be parallel (at least, their components are equal to 4 decimal places). In most situations, it would suffice to conclude that the lines are at least parallel, if not the same. One way to be sure is to rewrite $\vec{d}_1$ and $\vec{d}_2$ in terms of fractions, not decimals. We have

$$\vec{d}_1 = \left\langle \frac{16}{10}, \frac{272}{100}, -\frac{336}{100} \right\rangle \qquad \vec{d}_2 = \left\langle -\frac{29}{10}, -\frac{493}{100}, \frac{609}{100} \right\rangle.$$

One can then find the magnitudes of each vector in terms of fractions, then compute the unit vectors likewise. After a lot of manual arithmetic (or after briefly using a computer algebra system), one finds that

$$\vec{u}_1 = \left\langle \sqrt{\frac{10}{83}}, \frac{17}{\sqrt{830}}, -\frac{21}{\sqrt{830}} \right\rangle \qquad \vec{u}_2 = \left\langle -\sqrt{\frac{10}{83}}, -\frac{17}{\sqrt{830}}, \frac{21}{\sqrt{830}} \right\rangle.$$

We can now say without equivocation that these lines are parallel.

Are they the same line? The parametric equations for a line describe one point that lies on the line, so we know that the point $P_1 = (-0.7, 4.2, 2.3)$ lies on $\ell_1$. To determine if this point also lies on $\ell_2$, plug in the $x$, $y$ and $z$ values of $P_1$ into the symmetric equations for $\ell_2$:

$$\frac{(-0.7) - 2.8}{-2.9} \stackrel{?}{=} \frac{(4.2) - 10.15}{-4.93} \stackrel{?}{=} \frac{(2.3) - (-5.05)}{6.09} \quad \Rightarrow \quad 1.2069 = 1.2069 = 1.2069.$$

The point $P_1$ lies on both lines, so we conclude they are the same line, just parametrized differently. Figure 10.5.6 graphs this line along with the points and vectors described by the parametric equations. Note how $\vec{d}_1$ and $\vec{d}_2$ are parallel, though point in opposite directions (as indicated by their unit vectors above).

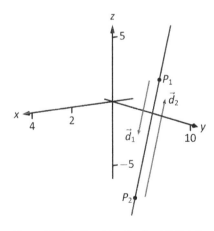

Figure 10.5.6: Graphing the lines in Example 10.5.4.

## Distances

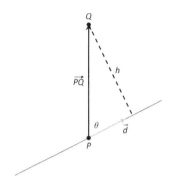

Figure 10.5.7: Establishing the distance from a point to a line.

Given a point $Q$ and a line $\vec{\ell}(t) = \vec{p} + t\vec{d}$ in space, it is often useful to know the distance from the point to the line. (Here we use the standard definition of "distance," i.e., the length of the shortest line segment from the point to the line.) Identifying $\vec{p}$ with the point $P$, Figure 10.5.7 will help establish a general method of computing this distance $h$.

From trigonometry, we know $h = ||\overrightarrow{PQ}|| \sin\theta$. We have a similar identity involving the cross product: $||\overrightarrow{PQ} \times \vec{d}|| = ||\overrightarrow{PQ}|| \, ||\vec{d}|| \sin\theta$. Divide both sides of this latter equation by $||\vec{d}||$ to obtain $h$:

$$h = \frac{||\overrightarrow{PQ} \times \vec{d}||}{||\vec{d}||}. \tag{10.8}$$

It is also useful to determine the distance between lines, which we define as the length of the shortest line segment that connects the two lines (an argument from geometry shows that this line segments is perpendicular to both lines). Let lines $\vec{\ell}_1(t) = \vec{p}_1 + t\vec{d}_1$ and $\vec{\ell}_2(t) = \vec{p}_2 + t\vec{d}_2$ be given, as shown in Figure 10.5.8. To find the direction orthogonal to both $\vec{d}_1$ and $\vec{d}_2$, we take the cross product: $\vec{c} = \vec{d}_1 \times \vec{d}_2$. The magnitude of the orthogonal projection of $\overrightarrow{P_1P_2}$ onto $\vec{c}$ is the distance $h$ we seek:

$$\begin{aligned}
h &= ||\operatorname{proj}_{\vec{c}} \overrightarrow{P_1P_2}|| \\
&= \left|\left|\frac{\overrightarrow{P_1P_2} \cdot \vec{c}}{\vec{c} \cdot \vec{c}} \vec{c}\right|\right| \\
&= \frac{|\overrightarrow{P_1P_2} \cdot \vec{c}|}{||\vec{c}||^2} ||\vec{c}|| \\
&= \frac{|\overrightarrow{P_1P_2} \cdot \vec{c}|}{||\vec{c}||}.
\end{aligned}$$

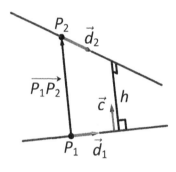

Figure 10.5.8: Establishing the distance between lines.

A problem in the Exercise section is to show that this distance is 0 when the lines intersect. Note the use of the Triple Scalar Product: $\overrightarrow{P_1P_2} \cdot \vec{c} = \overrightarrow{P_1P_2} \cdot (\vec{d}_1 \times \vec{d}_2)$.

The following Key Idea restates these two distance formulas.

Notes:

## 10.5 Lines

> **Key Idea 10.5.1  Distances to Lines**
>
> 1. Let $P$ be a point on a line $\ell$ that is parallel to $\vec{d}$. The distance $h$ from a point $Q$ to the line $\ell$ is:
>
> $$h = \frac{\|\overrightarrow{PQ} \times \vec{d}\|}{\|\vec{d}\|}.$$
>
> 2. Let $P_1$ be a point on line $\ell_1$ that is parallel to $\vec{d}_1$, and let $P_2$ be a point on line $\ell_2$ parallel to $\vec{d}_2$, and let $\vec{c} = \vec{d}_1 \times \vec{d}_2$, where lines $\ell_1$ and $\ell_2$ are not parallel. The distance $h$ between the two lines is:
>
> $$h = \frac{|\overrightarrow{P_1P_2} \cdot \vec{c}|}{\|\vec{c}\|}.$$

**Example 10.5.5  Finding the distance from a point to a line**
Find the distance from the point $Q = (1, 1, 3)$ to the line $\vec{\ell}(t) = \langle 1, -1, 1 \rangle + t\langle 2, 3, 1 \rangle$.

**SOLUTION**  The equation of the line gives us the point $P = (1, -1, 1)$ that lies on the line, hence $\overrightarrow{PQ} = \langle 0, 2, 2 \rangle$. The equation also gives $\vec{d} = \langle 2, 3, 1 \rangle$. Following Key Idea 10.5.1, we have the distance as

$$\begin{aligned} h &= \frac{\|\overrightarrow{PQ} \times \vec{d}\|}{\|\vec{d}\|} \\ &= \frac{\|\langle -4, 4, -4 \rangle\|}{\sqrt{14}} \\ &= \frac{4\sqrt{3}}{\sqrt{14}} \approx 1.852. \end{aligned}$$

The point $Q$ is approximately 1.852 units from the line $\vec{\ell}(t)$.

**Example 10.5.6  Finding the distance between lines**
Find the distance between the lines

$$\ell_1 : \begin{array}{rcl} x &=& 1 + 3t \\ y &=& 2 - t \\ z &=& t \end{array} \qquad \ell_2 : \begin{array}{rcl} x &=& -2 + 4s \\ y &=& 3 + s \\ z &=& 5 + 2s. \end{array}$$

**SOLUTION**  These are the sames lines as given in Example 10.5.3, where

Notes:

we showed them to be skew. The equations allow us to identify the following points and vectors:

$$P_1 = (1, 2, 0) \quad P_2 = (-2, 3, 5) \quad \Rightarrow \quad \overrightarrow{P_1P_2} = \langle -3, 1, 5 \rangle.$$

$$\vec{d_1} = \langle 3, -1, 1 \rangle \quad \vec{d_2} = \langle 4, 1, 2 \rangle \quad \Rightarrow \quad \vec{c} = \vec{d_1} \times \vec{d_2} = \langle -3, -2, 7 \rangle.$$

From Key Idea 10.5.1 we have the distance $h$ between the two lines is

$$h = \frac{|\overrightarrow{P_1P_2} \cdot \vec{c}|}{\|\vec{c}\|}$$
$$= \frac{42}{\sqrt{62}} \approx 5.334.$$

The lines are approximately 5.334 units apart.

One of the key points to understand from this section is this: to describe a line, we need a point and a direction. Whenever a problem is posed concerning a line, one needs to take whatever information is offered and glean point and direction information. Many questions can be asked (and *are* asked in the Exercise section) whose answer immediately follows from this understanding.

Lines are one of two fundamental objects of study in space. The other fundamental object is the *plane*, which we study in detail in the next section. Many complex three dimensional objects are studied by approximating their surfaces with lines and planes.

Notes:

# Exercises 10.5

## Terms and Concepts

1. To find an equation of a line, what two pieces of information are needed?

2. Two distinct lines in the plane can intersect or be _____.

3. Two distinct lines in space can intersect, be _____ or be _____.

4. Use your own words to describe what it means for two lines in space to be skew.

## Problems

**In Exercises 5 – 14, write the vector, parametric and symmetric equations of the lines described.**

5. Passes through $P = (2, -4, 1)$, parallel to $\vec{d} = \langle 9, 2, 5 \rangle$.

6. Passes through $P = (6, 1, 7)$, parallel to $\vec{d} = \langle -3, 2, 5 \rangle$.

7. Passes through $P = (2, 1, 5)$ and $Q = (7, -2, 4)$.

8. Passes through $P = (1, -2, 3)$ and $Q = (5, 5, 5)$.

9. Passes through $P = (0, 1, 2)$ and orthogonal to both $\vec{d_1} = \langle 2, -1, 7 \rangle$ and $\vec{d_2} = \langle 7, 1, 3 \rangle$.

10. Passes through $P = (5, 1, 9)$ and orthogonal to both $\vec{d_1} = \langle 1, 0, 1 \rangle$ and $\vec{d_2} = \langle 2, 0, 3 \rangle$.

11. Passes through the point of intersection of $\vec{\ell_1}(t)$ and $\vec{\ell_2}(t)$ and orthogonal to both lines, where
$\vec{\ell_1}(t) = \langle 2, 1, 1 \rangle + t\langle 5, 1, -2 \rangle$ and
$\vec{\ell_2}(t) = \langle -2, -1, 2 \rangle + t\langle 3, 1, -1 \rangle$.

12. Passes through the point of intersection of $\ell_1(t)$ and $\ell_2(t)$ and orthogonal to both lines, where
$\ell_1 = \begin{cases} x = t \\ y = -2 + 2t \\ z = 1 + t \end{cases}$ and $\ell_2 = \begin{cases} x = 2 + t \\ y = 2 - t \\ z = 3 + 2t \end{cases}$.

13. Passes through $P = (1, 1)$, parallel to $\vec{d} = \langle 2, 3 \rangle$.

14. Passes through $P = (-2, 5)$, parallel to $\vec{d} = \langle 0, 1 \rangle$.

**In Exercises 15 – 22, determine if the described lines are the same line, parallel lines, intersecting or skew lines. If intersecting, give the point of intersection.**

15. $\vec{\ell_1}(t) = \langle 1, 2, 1 \rangle + t\langle 2, -1, 1 \rangle$,
$\vec{\ell_2}(t) = \langle 3, 3, 3 \rangle + t\langle -4, 2, -2 \rangle$.

16. $\vec{\ell_1}(t) = \langle 2, 1, 1 \rangle + t\langle 5, 1, 3 \rangle$,
$\vec{\ell_2}(t) = \langle 14, 5, 9 \rangle + t\langle 1, 1, 1 \rangle$.

17. $\vec{\ell_1}(t) = \langle 3, 4, 1 \rangle + t\langle 2, -3, 4 \rangle$,
$\vec{\ell_2}(t) = \langle -3, 3, -3 \rangle + t\langle 3, -2, 4 \rangle$.

18. $\vec{\ell_1}(t) = \langle 1, 1, 1 \rangle + t\langle 3, 1, 3 \rangle$,
$\vec{\ell_2}(t) = \langle 7, 3, 7 \rangle + t\langle 6, 2, 6 \rangle$.

19. $\ell_1 = \begin{cases} x = 1 + 2t \\ y = 3 - 2t \\ z = t \end{cases}$ and $\ell_2 = \begin{cases} x = 3 - t \\ y = 3 + 5t \\ z = 2 + 7t \end{cases}$

20. $\ell_1 = \begin{cases} x = 1.1 + 0.6t \\ y = 3.77 + 0.9t \\ z = -2.3 + 1.5t \end{cases}$ and $\ell_2 = \begin{cases} x = 3.11 + 3.4t \\ y = 2 + 5.1t \\ z = 2.5 + 8.5t \end{cases}$

21. $\ell_1 = \begin{cases} x = 0.2 + 0.6t \\ y = 1.33 - 0.45t \\ z = -4.2 + 1.05t \end{cases}$ and $\ell_2 = \begin{cases} x = 0.86 + 9.2t \\ y = 0.835 - 6.9t \\ z = -3.045 + 16.1t \end{cases}$

22. $\ell_1 = \begin{cases} x = 0.1 + 1.1t \\ y = 2.9 - 1.5t \\ z = 3.2 + 1.6t \end{cases}$ and $\ell_2 = \begin{cases} x = 4 - 2.1t \\ y = 1.8 + 7.2t \\ z = 3.1 + 1.1t \end{cases}$

**In Exercises 23 – 26, find the distance from the point to the line.**

23. $Q = (1, 1, 1)$,  $\vec{\ell}(t) = \langle 2, 1, 3 \rangle + t\langle 2, 1, -2 \rangle$

24. $Q = (2, 5, 6)$,  $\vec{\ell}(t) = \langle -1, 1, 1 \rangle + t\langle 1, 0, 1 \rangle$

25. $Q = (0, 3)$,  $\vec{\ell}(t) = \langle 2, 0 \rangle + t\langle 1, 1 \rangle$

26. $Q = (1, 1)$,  $\vec{\ell}(t) = \langle 4, 5 \rangle + t\langle -4, 3 \rangle$

**In Exercises 27 – 28, find the distance between the two lines.**

27. $\vec{\ell_1}(t) = \langle 1, 2, 1 \rangle + t\langle 2, -1, 1 \rangle$,
$\vec{\ell_2}(t) = \langle 3, 3, 3 \rangle + t\langle 4, 2, -2 \rangle$.

28. $\vec{\ell_1}(t) = \langle 0, 0, 1 \rangle + t\langle 1, 0, 0 \rangle$,
$\vec{\ell_2}(t) = \langle 0, 0, 3 \rangle + t\langle 0, 1, 0 \rangle$.

**Exercises 29 – 31 explore special cases of the distance formulas found in Key Idea 10.5.1.**

29. Let $Q$ be a point on the line $\vec{\ell}(t)$. Show why the distance formula correctly gives the distance from the point to the line as 0.

30. Let lines $\vec{\ell_1}(t)$ and $\vec{\ell_2}(t)$ be intersecting lines. Show why the distance formula correctly gives the distance between these lines as 0.

31. Let lines $\vec{\ell}_1(t)$ and $\vec{\ell}_2(t)$ be parallel.

    (a) Show why the distance formula for distance between lines cannot be used as stated to find the distance between the lines.

    (b) Show why letting $\vec{c} = (\overrightarrow{P_1P_2} \times \vec{d}_2) \times \vec{d}_2$ allows one to use the formula.

    (c) Show how one can use the formula for the distance between a point and a line to find the distance between parallel lines.

## 10.6 Planes

Any flat surface, such as a wall, table top or stiff piece of cardboard can be thought of as representing part of a plane. Consider a piece of cardboard with a point P marked on it. One can take a nail and stick it into the cardboard at P such that the nail is perpendicular to the cardboard; see Figure 10.6.1.

This nail provides a "handle" for the cardboard. Moving the cardboard around moves P to different locations in space. Tilting the nail (but keeping P fixed) tilts the cardboard. Both moving and tilting the cardboard defines a different plane in space. In fact, we can define a plane by: 1) the location of P in space, and 2) the direction of the nail.

The previous section showed that one can define a line given a point on the line and the direction of the line (usually given by a vector). One can make a similar statement about planes: we can define a plane in space given a point on the plane and the direction the plane "faces" (using the description above, the direction of the nail). Once again, the direction information will be supplied by a vector, called a **normal vector**, that is orthogonal to the plane.

What exactly does "orthogonal to the plane" mean? Choose any two points P and Q in the plane, and consider the vector $\vec{PQ}$. We say a vector $\vec{n}$ is orthogonal to the plane if $\vec{n}$ is perpendicular to $\vec{PQ}$ for all choices of P and Q; that is, if $\vec{n} \cdot \vec{PQ} = 0$ for all P and Q.

This gives us way of writing an equation describing the plane. Let $P = (x_0, y_0, z_0)$ be a point in the plane and let $\vec{n} = \langle a, b, c \rangle$ be a normal vector to the plane. A point $Q = (x, y, z)$ lies in the plane defined by P and $\vec{n}$ if, and only if, $\vec{PQ}$ is orthogonal to $\vec{n}$. Knowing $\vec{PQ} = \langle x - x_0, y - y_0, z - z_0 \rangle$, consider:

$$\vec{PQ} \cdot \vec{n} = 0$$
$$\langle x - x_0, y - y_0, z - z_0 \rangle \cdot \langle a, b, c \rangle = 0$$
$$a(x - x_0) + b(y - y_0) + c(z - z_0) = 0 \qquad (10.9)$$

Equation (10.9) defines an *implicit* function describing the plane. More algebra produces:

$$ax + by + cz = ax_0 + by_0 + cz_0.$$

The right hand side is just a number, so we replace it with $d$:

$$ax + by + cz = d. \qquad (10.10)$$

As long as $c \neq 0$, we can solve for $z$:

$$z = \frac{1}{c}(d - ax - by). \qquad (10.11)$$

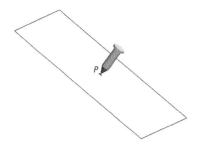

Figure 10.6.1: Illustrating defining a plane with a sheet of cardboard and a nail.

Equation (10.11) is especially useful as many computer programs can graph functions in this form. Equations (10.9) and (10.10) have specific names, given next.

> **Definition 10.6.1**     **Equations of a Plane in Standard and General Forms**
>
> The plane passing through the point $P = (x_0, y_0, z_0)$ with normal vector $\vec{n} = \langle a, b, c \rangle$ can be described by an equation with **standard form**
>
> $$a(x - x_0) + b(y - y_0) + c(z - z_0) = 0;$$
>
> the equation's **general form** is
>
> $$ax + by + cz = d.$$

A key to remember throughout this section is this: to find the equation of a plane, we need a point and a normal vector. We will give several examples of finding the equation of a plane, and in each one different types of information are given. In each case, we need to use the given information to find a point on the plane and a normal vector.

**Example 10.6.1**     **Finding the equation of a plane.**
Write the equation of the plane that passes through the points $P = (1, 1, 0)$, $Q = (1, 2, -1)$ and $R = (0, 1, 2)$ in standard form.

**SOLUTION**     We need a vector $\vec{n}$ that is orthogonal to the plane. Since $P$, $Q$ and $R$ are in the plane, so are the vectors $\vec{PQ}$ and $\vec{PR}$; $\vec{PQ} \times \vec{PR}$ is orthogonal to $\vec{PQ}$ and $\vec{PR}$ and hence the plane itself.

It is straightforward to compute $\vec{n} = \vec{PQ} \times \vec{PR} = \langle 2, 1, 1 \rangle$. We can use any point we wish in the plane (any of $P$, $Q$ or $R$ will do) and we arbitrarily choose $P$. Following Definition 10.6.1, the equation of the plane in standard form is

$$2(x - 1) + (y - 1) + z = 0.$$

The plane is sketched in Figure 10.6.2.

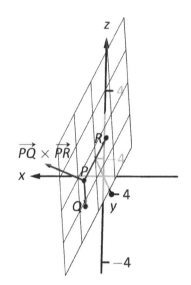

Figure 10.6.2: Sketching the plane in Example 10.6.1.

We have just demonstrated the fact that any three non-collinear points define a plane. (This is why a three-legged stool does not "rock;" it's three feet always lie in a plane. A four-legged stool will rock unless all four feet lie in the same plane.)

**Example 10.6.2**     **Finding the equation of a plane.**
Verify that lines $\ell_1$ and $\ell_2$, whose parametric equations are given below, inter-

Notes:

sect, then give the equation of the plane that contains these two lines in general form.

$$\ell_1: \begin{array}{rcl} x & = & -5 + 2s \\ y & = & 1 + s \\ z & = & -4 + 2s \end{array} \qquad \ell_2: \begin{array}{rcl} x & = & 2 + 3t \\ y & = & 1 - 2t \\ z & = & 1 + t \end{array}$$

**SOLUTION** The lines clearly are not parallel. If they do not intersect, they are skew, meaning there is not a plane that contains them both. If they do intersect, there is such a plane.

To find their point of intersection, we set the $x$, $y$ and $z$ equations equal to each other and solve for $s$ and $t$:

$$\begin{array}{rcl} -5 + 2s & = & 2 + 3t \\ 1 + s & = & 1 - 2t \\ -4 + 2s & = & 1 + t \end{array} \Rightarrow s = 2, \quad t = -1.$$

When $s = 2$ and $t = -1$, the lines intersect at the point $P = (-1, 3, 0)$.

Let $\vec{d}_1 = \langle 2, 1, 2 \rangle$ and $\vec{d}_2 = \langle 3, -2, 1 \rangle$ be the directions of lines $\ell_1$ and $\ell_2$, respectively. A normal vector to the plane containing these the two lines will also be orthogonal to $\vec{d}_1$ and $\vec{d}_2$. Thus we find a normal vector $\vec{n}$ by computing $\vec{n} = \vec{d}_1 \times \vec{d}_2 = \langle 5, 4 - 7 \rangle$.

We can pick any point in the plane with which to write our equation; each line gives us infinite choices of points. We choose $P$, the point of intersection. We follow Definition 10.6.1 to write the plane's equation in general form:

$$5(x + 1) + 4(y - 3) - 7z = 0$$
$$5x + 5 + 4y - 12 - 7z = 0$$
$$5x + 4y - 7z = 7.$$

The plane's equation in general form is $5x + 4y - 7z = 7$; it is sketched in Figure 10.6.3.

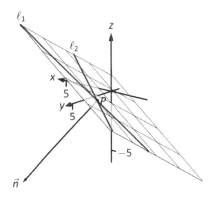

Figure 10.6.3: Sketching the plane in Example 10.6.2.

### Example 10.6.3 Finding the equation of a plane

Give the equation, in standard form, of the plane that passes through the point $P = (-1, 0, 1)$ and is orthogonal to the line with vector equation $\vec{\ell}(t) = \langle -1, 0, 1 \rangle + t \langle 1, 2, 2 \rangle$.

**SOLUTION** As the plane is to be orthogonal to the line, the plane must be orthogonal to the direction of the line given by $\vec{d} = \langle 1, 2, 2 \rangle$. We use this as our normal vector. Thus the plane's equation, in standard form, is

$$(x + 1) + 2y + 2(z - 1) = 0.$$

The line and plane are sketched in Figure 10.6.4.

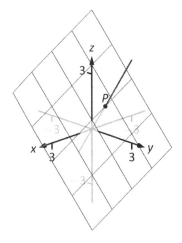

Figure 10.6.4: The line and plane in Example 10.6.3.

## Example 10.6.4  Finding the intersection of two planes

Give the parametric equations of the line that is the intersection of the planes $p_1$ and $p_2$, where:

$$p_1 : x - (y - 2) + (z - 1) = 0$$
$$p_2 : -2(x - 2) + (y + 1) + (z - 3) = 0$$

**Solution**  To find an equation of a line, we need a point on the line and the direction of the line.

We can find a point on the line by solving each equation of the planes for $z$:

$$p_1 : z = -x + y - 1$$
$$p_2 : z = 2x - y - 2$$

We can now set these two equations equal to each other (i.e., we are finding values of $x$ and $y$ where the planes have the same $z$ value):

$$-x + y - 1 = 2x - y - 2$$
$$2y = 3x - 1$$
$$y = \frac{1}{2}(3x - 1)$$

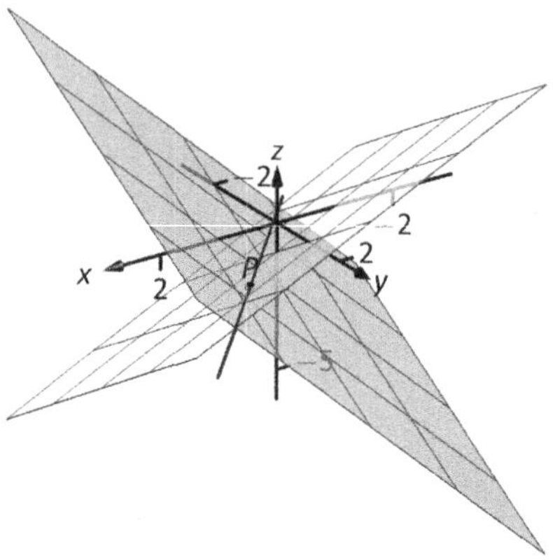

We can choose any value for $x$; we choose $x = 1$. This determines that $y = 1$. We can now use the equations of either plane to find $z$: when $x = 1$ and $y = 1$, $z = -1$ on both planes. We have found a point $P$ on the line: $P = (1, 1, -1)$.

We now need the direction of the line. Since the line lies in each plane, its direction is orthogonal to a normal vector for each plane. Considering the equations for $p_1$ and $p_2$, we can quickly determine their normal vectors. For $p_1$, $\vec{n}_1 = \langle 1, -1, 1 \rangle$ and for $p_2$, $\vec{n}_2 = \langle -2, 1, 1 \rangle$. A direction orthogonal to both of these directions is their cross product: $\vec{d} = \vec{n}_1 \times \vec{n}_2 = \langle -2, -3, -1 \rangle$.

The parametric equations of the line through $P = (1, 1, -1)$ in the direction of $d = \langle -2, -3, -1 \rangle$ is:

$$\ell : \quad x = -2t + 1 \quad y = -3t + 1 \quad z = -t - 1.$$

The planes and line are graphed in Figure 10.6.5.

Figure 10.6.5: Graphing the planes and their line of intersection in Example 10.6.4.

## Example 10.6.5  Finding the intersection of a plane and a line

Find the point of intersection, if any, of the line $\ell(t) = \langle 3, -3, -1 \rangle + t \langle -1, 2, 1 \rangle$ and the plane with equation in general form $2x + y + z = 4$.

**Solution**  The equation of the plane shows that the vector $\vec{n} = \langle 2, 1, 1 \rangle$ is a normal vector to the plane, and the equation of the line shows that the line

Notes:

moves parallel to $\vec{d} = \langle -1, 2, 1 \rangle$. Since these are not orthogonal, we know there is a point of intersection. (If there were orthogonal, it would mean that the plane and line were parallel to each other, either never intersecting or the line was in the plane itself.)

To find the point of intersection, we need to find a $t$ value such that $\ell(t)$ satisfies the equation of the plane. Rewriting the equation of the line with parametric equations will help:

$$\ell(t) = \begin{cases} x = 3 - t \\ y = -3 + 2t \\ z = -1 + t \end{cases}.$$

Replacing $x$, $y$ and $z$ in the equation of the plane with the expressions containing $t$ found in the equation of the line allows us to determine a $t$ value that indicates the point of intersection:

$$2x + y + z = 4$$
$$2(3 - t) + (-3 + 2t) + (-1 + t) = 4$$
$$t = 2.$$

When $t = 2$, the point on the line satisfies the equation of the plane; that point is $\ell(2) = \langle 1, 1, 1 \rangle$. Thus the point $(1, 1, 1)$ is the point of intersection between the plane and the line, illustrated in Figure 10.6.6.

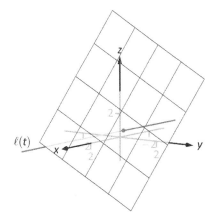

Figure 10.6.6: Illustrating the intersection of a line and a plane in Example 10.6.5.

## Distances

Just as it was useful to find distances between points and lines in the previous section, it is also often necessary to find the distance from a point to a plane.

Consider Figure 10.6.7, where a plane with normal vector $\vec{n}$ is sketched containing a point $P$ and a point $Q$, not on the plane, is given. We measure the distance from $Q$ to the plane by measuring the length of the projection of $\overrightarrow{PQ}$ onto $\vec{n}$. That is, we want:

$$\left\| \text{proj}_{\vec{n}} \overrightarrow{PQ} \right\| = \left\| \frac{\vec{n} \cdot \overrightarrow{PQ}}{\|\vec{n}\|^2} \vec{n} \right\| = \frac{|\vec{n} \cdot \overrightarrow{PQ}|}{\|\vec{n}\|} \quad (10.12)$$

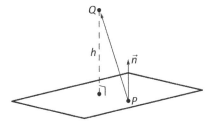

Figure 10.6.7: Illustrating finding the distance from a point to a plane.

Equation (10.12) is important as it does more than just give the distance between a point and a plane. We will see how it allows us to find several other distances as well: the distance between parallel planes and the distance from a line and a plane. Because Equation (10.12) is important, we restate it as a Key Idea.

Notes:

> **Key Idea 10.6.1**     **Distance from a Point to a Plane**
>
> Let a plane with normal vector $\vec{n}$ be given, and let $Q$ be a point. The distance $h$ from $Q$ to the plane is
>
> $$h = \frac{|\vec{n} \cdot \overrightarrow{PQ}|}{\|\vec{n}\|},$$
>
> where $P$ is any point in the plane.

**Example 10.6.6**     **Distance between a point and a plane**
Find the distance between the point $Q = (2, 1, 4)$ and the plane with equation $2x - 5y + 6z = 9$.

**SOLUTION**     Using the equation of the plane, we find the normal vector $\vec{n} = \langle 2, -5, 6 \rangle$. To find a point on the plane, we can let $x$ and $y$ be anything we choose, then let $z$ be whatever satisfies the equation. Letting $x$ and $y$ be 0 seems simple; this makes $z = 1.5$. Thus we let $P = \langle 0, 0, 1.5 \rangle$, and $\overrightarrow{PQ} = \langle 2, 1, 2.5 \rangle$.

The distance $h$ from $Q$ to the plane is given by Key Idea 10.6.1:

$$\begin{aligned} h &= \frac{|\vec{n} \cdot \overrightarrow{PQ}|}{\|\vec{n}\|} \\ &= \frac{|\langle 2, -5, 6 \rangle \cdot \langle 2, 1, 2.5 \rangle|}{\|\langle 2, -5, 6 \rangle\|} \\ &= \frac{|14|}{\sqrt{65}} \\ &\approx 1.74. \end{aligned}$$

We can use Key Idea 10.6.1 to find other distances. Given two parallel planes, we can find the distance between these planes by letting $P$ be a point on one plane and $Q$ a point on the other. If $\ell$ is a line parallel to a plane, we can use the Key Idea to find the distance between them as well: again, let $P$ be a point in the plane and let $Q$ be any point on the line. (One can also use Key Idea 10.5.1.) The Exercise section contains problems of these types.

These past two sections have not explored lines and planes in space as an exercise of mathematical curiosity. However, there are many, many applications of these fundamental concepts. Complex shapes can be modeled (or, *approximated*) using planes. For instance, part of the exterior of an aircraft may have a complex, yet smooth, shape, and engineers will want to know how air flows across this piece as well as how heat might build up due to air friction. Many equations that help determine air flow and heat dissipation are difficult to apply to arbitrary surfaces, but simple to apply to planes. By approximating a surface with millions of small planes one can more readily model the needed behavior.

Notes:

# Exercises 10.6

## Terms and Concepts

1. In order to find the equation of a plane, what two pieces of information must one have?

2. What is the relationship between a plane and one of its normal vectors?

## Problems

**In Exercises 3 – 6, give any two points in the given plane.**

3. $2x - 4y + 7z = 2$

4. $3(x + 2) + 5(y - 9) - 4z = 0$

5. $x = 2$

6. $4(y + 2) - (z - 6) = 0$

**In Exercises 7 – 20, give the equation of the described plane in standard and general forms.**

7. Passes through $(2, 3, 4)$ and has normal vector $\vec{n} = \langle 3, -1, 7 \rangle$.

8. Passes through $(1, 3, 5)$ and has normal vector $\vec{n} = \langle 0, 2, 4 \rangle$.

9. Passes through the points $(1, 2, 3)$, $(3, -1, 4)$ and $(1, 0, 1)$.

10. Passes through the points $(5, 3, 8)$, $(6, 4, 9)$ and $(3, 3, 3)$.

11. Contains the intersecting lines
$\vec{\ell_1}(t) = \langle 2, 1, 2 \rangle + t \langle 1, 2, 3 \rangle$ and
$\vec{\ell_2}(t) = \langle 2, 1, 2 \rangle + t \langle 2, 5, 4 \rangle$.

12. Contains the intersecting lines
$\vec{\ell_1}(t) = \langle 5, 0, 3 \rangle + t \langle -1, 1, 1 \rangle$ and
$\vec{\ell_2}(t) = \langle 1, 4, 7 \rangle + t \langle 3, 0, -3 \rangle$.

13. Contains the parallel lines
$\vec{\ell_1}(t) = \langle 1, 1, 1 \rangle + t \langle 1, 2, 3 \rangle$ and
$\vec{\ell_2}(t) = \langle 1, 1, 2 \rangle + t \langle 1, 2, 3 \rangle$.

14. Contains the parallel lines
$\vec{\ell_1}(t) = \langle 1, 1, 1 \rangle + t \langle 4, 1, 3 \rangle$ and
$\vec{\ell_2}(t) = \langle 4, 4, 4 \rangle + t \langle 4, 1, 3 \rangle$.

15. Contains the point $(2, -6, 1)$ and the line
$\ell(t) = \begin{cases} x = 2 + 5t \\ y = 2 + 2t \\ z = -1 + 2t \end{cases}$

16. Contains the point $(5, 7, 3)$ and the line
$\ell(t) = \begin{cases} x = t \\ y = t \\ z = t \end{cases}$

17. Contains the point $(5, 7, 3)$ and is orthogonal to the line
$\vec{\ell}(t) = \langle 4, 5, 6 \rangle + t \langle 1, 1, 1 \rangle$.

18. Contains the point $(4, 1, 1)$ and is orthogonal to the line
$\ell(t) = \begin{cases} x = 4 + 4t \\ y = 1 + 1t \\ z = 1 + 1t \end{cases}$

19. Contains the point $(-4, 7, 2)$ and is parallel to the plane $3(x - 2) + 8(y + 1) - 10z = 0$.

20. Contains the point $(1, 2, 3)$ and is parallel to the plane $x = 5$.

**In Exercises 21 – 22, give the equation of the line that is the intersection of the given planes.**

21. $p1: 3(x - 2) + (y - 1) + 4z = 0$, and
$p2: 2(x - 1) - 2(y + 3) + 6(z - 1) = 0$.

22. $p1: 5(x - 5) + 2(y + 2) + 4(z - 1) = 0$, and
$p2: 3x - 4(y - 1) + 2(z - 1) = 0$.

**In Exercises 23 – 26, find the point of intersection between the line and the plane.**

23. line: $\langle 5, 1, -1 \rangle + t \langle 2, 2, 1 \rangle$,
plane: $5x - y - z = -3$

24. line: $\langle 4, 1, 0 \rangle + t \langle 1, 0, -1 \rangle$,
plane: $3x + y - 2z = 8$

25. line: $\langle 1, 2, 3 \rangle + t \langle 3, 5, -1 \rangle$,
plane: $3x - 2y - z = 4$

26. line: $\langle 1, 2, 3 \rangle + t \langle 3, 5, -1 \rangle$,
plane: $3x - 2y - z = -4$

**In Exercises 27 – 30, find the given distances.**

27. The distance from the point $(1, 2, 3)$ to the plane $3(x - 1) + (y - 2) + 5(z - 2) = 0$.

28. The distance from the point $(2, 6, 2)$ to the plane $2(x - 1) - y + 4(z + 1) = 0$.

29. The distance between the parallel planes
$x + y + z = 0$ and
$(x - 2) + (y - 3) + (z + 4) = 0$

30. The distance between the parallel planes
$2(x-1) + 2(y+1) + (z-2) = 0$ and
$2(x-3) + 2(y-1) + (z-3) = 0$

31. Show why if the point $Q$ lies in a plane, then the distance formula correctly gives the distance from the point to the plane as 0.

32. How is Exercise 30 in Section 10.5 easier to answer once we have an understanding of planes?

# 11: Vector Valued Functions

In the previous chapter, we learned about vectors and were introduced to the power of vectors within mathematics. In this chapter, we'll build on this foundation to define functions whose input is a real number and whose output is a vector. We'll see how to graph these functions and apply calculus techniques to analyze their behavior. Most importantly, we'll see *why* we are interested in doing this: we'll see beautiful applications to the study of moving objects.

## 11.1 Vector–Valued Functions

We are very familiar with **real valued functions**, that is, functions whose output is a real number. This section introduces **vector–valued functions** – functions whose output is a vector.

---

**Definition 11.1.1**     **Vector–Valued Functions**

A **vector–valued function** is a function of the form

$$\vec{r}(t) = \langle f(t), g(t) \rangle \quad \text{or} \quad \vec{r}(t) = \langle f(t), g(t), h(t) \rangle,$$

where $f$, $g$ and $h$ are real valued functions.

The **domain** of $\vec{r}$ is the set of all values of $t$ for which $\vec{r}(t)$ is defined. The **range** of $\vec{r}$ is the set of all possible output vectors $\vec{r}(t)$.

---

**Evaluating and Graphing Vector–Valued Functions**

Evaluating a vector–valued function at a specific value of $t$ is straightforward; simply evaluate each component function at that value of $t$. For instance, if $\vec{r}(t) = \langle t^2, t^2 + t - 1 \rangle$, then $\vec{r}(-2) = \langle 4, 1 \rangle$. We can sketch this vector, as is done in Figure 11.1.1(a). Plotting lots of vectors is cumbersome, though, so generally we do not sketch the whole vector but just the terminal point. The **graph** of a vector–valued function is the set of all terminal points of $\vec{r}(t)$, where the initial point of each vector is always the origin. In Figure 11.1.1(b) we sketch the graph of $\vec{r}$; we can indicate individual points on the graph with their respective vector, as shown.

Vector–valued functions are closely related to parametric equations of graphs. While in both methods we plot points $(x(t), y(t))$ or $(x(t), y(t), z(t))$ to produce a graph, in the context of vector–valued functions each such point represents a vector. The implications of this will be more fully realized in the next section as we apply calculus ideas to these functions.

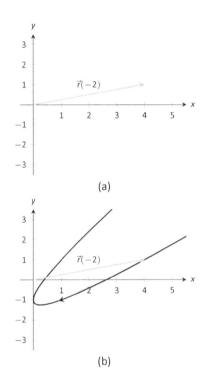

Figure 11.1.1: Sketching the graph of a vector–valued function.

## Chapter 11  Vector Valued Functions

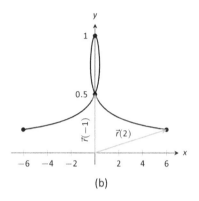

| $t$ | $t^3 - t$ | $\dfrac{1}{t^2+1}$ |
|---|---|---|
| $-2$ | $-6$ | $1/5$ |
| $-1$ | $0$ | $1/2$ |
| $0$ | $0$ | $1$ |
| $1$ | $0$ | $1/2$ |
| $2$ | $6$ | $1/5$ |

(a)

(b)

Figure 11.1.2: Sketching the vector–valued function of Example 11.1.1.

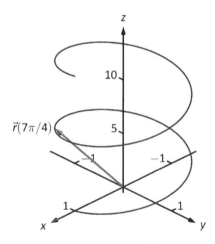

Figure 11.1.3: The graph of $\vec{r}(t)$ in Example 11.1.2.

**Example 11.1.1  Graphing vector–valued functions**

Graph $\vec{r}(t) = \left\langle t^3 - t, \dfrac{1}{t^2+1} \right\rangle$, for $-2 \leq t \leq 2$. Sketch $\vec{r}(-1)$ and $\vec{r}(2)$.

**SOLUTION**    We start by making a table of $t$, $x$ and $y$ values as shown in Figure 11.1.2(a). Plotting these points gives an indication of what the graph looks like. In Figure 11.1.2(b), we indicate these points and sketch the full graph. We also highlight $\vec{r}(-1)$ and $\vec{r}(2)$ on the graph.

**Example 11.1.2  Graphing vector–valued functions.**

Graph $\vec{r}(t) = \langle \cos t, \sin t, t \rangle$ for $0 \leq t \leq 4\pi$.

**SOLUTION**    We can again plot points, but careful consideration of this function is very revealing. Momentarily ignoring the third component, we see the $x$ and $y$ components trace out a circle of radius 1 centered at the origin. Noticing that the $z$ component is $t$, we see that as the graph winds around the $z$-axis, it is also increasing at a constant rate in the positive $z$ direction, forming a spiral. This is graphed in Figure 11.1.3. In the graph $\vec{r}(7\pi/4) \approx (0.707, -0.707, 5.498)$ is highlighted to help us understand the graph.

### Algebra of Vector–Valued Functions

---

**Definition 11.1.2    Operations on Vector–Valued Functions**

Let $\vec{r}_1(t) = \langle f_1(t), g_1(t) \rangle$ and $\vec{r}_2(t) = \langle f_2(t), g_2(t) \rangle$ be vector–valued functions in $\mathbb{R}^2$ and let $c$ be a scalar. Then:

1. $\vec{r}_1(t) \pm \vec{r}_2(t) = \langle f_1(t) \pm f_2(t), g_1(t) \pm g_2(t) \rangle$.

2. $c\vec{r}_1(t) = \langle cf_1(t), cg_1(t) \rangle$.

A similar definition holds for vector–valued functions in $\mathbb{R}^3$.

---

This definition states that we add, subtract and scale vector-valued functions component–wise. Combining vector–valued functions in this way can be very useful (as well as create interesting graphs).

**Example 11.1.3  Adding and scaling vector–valued functions.**

Let $\vec{r}_1(t) = \langle 0.2t, 0.3t \rangle$, $\vec{r}_2(t) = \langle \cos t, \sin t \rangle$ and $\vec{r}(t) = \vec{r}_1(t) + \vec{r}_2(t)$. Graph $\vec{r}_1(t)$, $\vec{r}_2(t)$, $\vec{r}(t)$ and $5\vec{r}(t)$ on $-10 \leq t \leq 10$.

---

Notes:

**SOLUTION** We can graph $\vec{r}_1$ and $\vec{r}_2$ easily by plotting points (or just using technology). Let's think about each for a moment to better understand how vector–valued functions work.

We can rewrite $\vec{r}_1(t) = \langle 0.2t, 0.3t \rangle$ as $\vec{r}_1(t) = t\langle 0.2, 0.3 \rangle$. That is, the function $\vec{r}_1$ scales the vector $\langle 0.2, 0.3 \rangle$ by $t$. This scaling of a vector produces a line in the direction of $\langle 0.2, 0.3 \rangle$.

We are familiar with $\vec{r}_2(t) = \langle \cos t, \sin t \rangle$; it traces out a circle, centered at the origin, of radius 1. Figure 11.1.4(a) graphs $\vec{r}_1(t)$ and $\vec{r}_2(t)$.

Adding $\vec{r}_1(t)$ to $\vec{r}_2(t)$ produces $\vec{r}(t) = \langle \cos t + 0.2t, \sin t + 0.3t \rangle$, graphed in Figure 11.1.4(b). The linear movement of the line combines with the circle to create loops that move in the direction of $\langle 0.2, 0.3 \rangle$. (We encourage the reader to experiment by changing $\vec{r}_1(t)$ to $\langle 2t, 3t \rangle$, etc., and observe the effects on the loops.)

Multiplying $\vec{r}(t)$ by 5 scales the function by 5, producing $5\vec{r}(t) = \langle 5\cos t + 1, 5\sin t + 1.5 \rangle$, which is graphed in Figure 11.1.4(c) along with $\vec{r}(t)$. The new function is "5 times bigger" than $\vec{r}(t)$. Note how the graph of $5\vec{r}(t)$ in (c) looks identical to the graph of $\vec{r}(t)$ in (b). This is due to the fact that the $x$ and $y$ bounds of the plot in (c) are exactly 5 times larger than the bounds in (b).

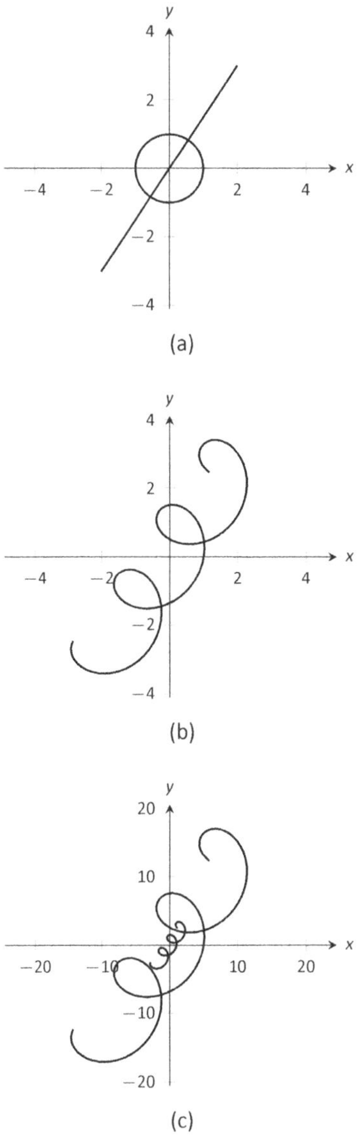

**Example 11.1.4  Adding and scaling vector–valued functions.**
A **cycloid** is a graph traced by a point $p$ on a rolling circle, as shown in Figure 11.1.5. Find an equation describing the cycloid, where the circle has radius 1.

Figure 11.1.5: Tracing a cycloid.

**SOLUTION** This problem is not very difficult if we approach it in a clever way. We start by letting $\vec{p}(t)$ describe the position of the point $p$ on the circle, where the circle is centered at the origin and only rotates clockwise (i.e., it does not roll). This is relatively simple given our previous experiences with parametric equations; $\vec{p}(t) = \langle \cos t, -\sin t \rangle$.

We now want the circle to roll. We represent this by letting $\vec{c}(t)$ represent the location of the center of the circle. It should be clear that the $y$ component of $\vec{c}(t)$ should be 1; the center of the circle is always going to be 1 if it rolls on a horizontal surface.

The $x$ component of $\vec{c}(t)$ is a linear function of $t$: $f(t) = mt$ for some scalar $m$. When $t = 0$, $f(t) = 0$ (the circle starts centered on the $y$-axis). When $t = 2\pi$, the circle has made one complete revolution, traveling a distance equal to its

Figure 11.1.4: Graphing the functions in Example 11.1.3.

Notes:

## Chapter 11 Vector Valued Functions

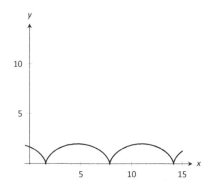

Figure 11.1.6: The cycloid in Example 11.1.4.

circumference, which is also $2\pi$. This gives us a point on our line $f(t) = mt$, the point $(2\pi, 2\pi)$. It should be clear that $m = 1$ and $f(t) = t$. So $\vec{c}(t) = \langle t, 1 \rangle$.

We now combine $\vec{p}$ and $\vec{c}$ together to form the equation of the cycloid: $\vec{r}(t) = \vec{p}(t) + \vec{c}(t) = \langle \cos t + t, -\sin t + 1 \rangle$, which is graphed in Figure 11.1.6.

### Displacement

A vector–valued function $\vec{r}(t)$ is often used to describe the position of a moving object at time $t$. At $t = t_0$, the object is at $\vec{r}(t_0)$; at $t = t_1$, the object is at $\vec{r}(t_1)$. Knowing the locations $\vec{r}(t_0)$ and $\vec{r}(t_1)$ give no indication of the path taken between them, but often we only care about the difference of the locations, $\vec{r}(t_1) - \vec{r}(t_0)$, the **displacement**.

---

**Definition 11.1.3   Displacement**

Let $\vec{r}(t)$ be a vector–valued function and let $t_0 < t_1$ be values in the domain. The **displacement** $\vec{d}$ of $\vec{r}$, from $t = t_0$ to $t = t_1$, is

$$\vec{d} = \vec{r}(t_1) - \vec{r}(t_0).$$

---

When the displacement vector is drawn with initial point at $\vec{r}(t_0)$, its terminal point is $\vec{r}(t_1)$. We think of it as the vector which points from a starting position to an ending position.

**Example 11.1.5   Finding and graphing displacement vectors**
Let $\vec{r}(t) = \langle \cos(\frac{\pi}{2}t), \sin(\frac{\pi}{2}t) \rangle$. Graph $\vec{r}(t)$ on $-1 \leq t \leq 1$, and find the displacement of $\vec{r}(t)$ on this interval.

**Solution**   The function $\vec{r}(t)$ traces out the unit circle, though at a different rate than the "usual" $\langle \cos t, \sin t \rangle$ parametrization. At $t_0 = -1$, we have $\vec{r}(t_0) = \langle 0, -1 \rangle$; at $t_1 = 1$, we have $\vec{r}(t_1) = \langle 0, 1 \rangle$. The displacement of $\vec{r}(t)$ on $[-1, 1]$ is thus $\vec{d} = \langle 0, 1 \rangle - \langle 0, -1 \rangle = \langle 0, 2 \rangle$.

A graph of $\vec{r}(t)$ on $[-1, 1]$ is given in Figure 11.1.7, along with the displacement vector $\vec{d}$ on this interval.

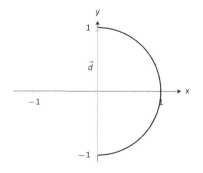

Figure 11.1.7: Graphing the displacement of a position function in Example 11.1.5.

Measuring displacement makes us contemplate related, yet very different, concepts. Considering the semi–circular path the object in Example 11.1.5 took, we can quickly verify that the object ended up a distance of 2 units from its initial location. That is, we can compute $||\vec{d}|| = 2$. However, measuring *distance from the starting point* is different from measuring *distance traveled*. Being a semi–

---

Notes:

circle, we can measure the distance traveled by this object as $\pi \approx 3.14$ units. Knowing *distance from the starting point* allows us to compute **average rate of change.**

> **Definition 11.1.4**     **Average Rate of Change**
>
> Let $\vec{r}(t)$ be a vector–valued function, where each of its component functions is continuous on its domain, and let $t_0 < t_1$. The **average rate of change** of $\vec{r}(t)$ on $[t_0, t_1]$ is
> $$\text{average rate of change} = \frac{\vec{r}(t_1) - \vec{r}(t_0)}{t_1 - t_0}.$$

**Example 11.1.6**     **Average rate of change**
Let $\vec{r}(t) = \langle \cos(\frac{\pi}{2}t), \sin(\frac{\pi}{2}t) \rangle$ as in Example 11.1.5. Find the average rate of change of $\vec{r}(t)$ on $[-1, 1]$ and on $[-1, 5]$.

**SOLUTION**     We computed in Example 11.1.5 that the displacement of $\vec{r}(t)$ on $[-1, 1]$ was $\vec{d} = \langle 0, 2 \rangle$. Thus the average rate of change of $\vec{r}(t)$ on $[-1, 1]$ is:
$$\frac{\vec{r}(1) - \vec{r}(-1)}{1 - (-1)} = \frac{\langle 0, 2 \rangle}{2} = \langle 0, 1 \rangle.$$

We interpret this as follows: the object followed a semi–circular path, meaning it moved towards the right then moved back to the left, while climbing slowly, then quickly, then slowly again. *On average*, however, it progressed straight up at a constant rate of $\langle 0, 1 \rangle$ per unit of time.

We can quickly see that the displacement on $[-1, 5]$ is the same as on $[-1, 1]$, so $\vec{d} = \langle 0, 2 \rangle$. The average rate of change is different, though:
$$\frac{\vec{r}(5) - \vec{r}(-1)}{5 - (-1)} = \frac{\langle 0, 2 \rangle}{6} = \langle 0, 1/3 \rangle.$$

As it took "3 times as long" to arrive at the same place, this average rate of change on $[-1, 5]$ is $1/3$ the average rate of change on $[-1, 1]$.

We considered average rates of change in Sections 1.1 and 2.1 as we studied limits and derivatives. The same is true here; in the following section we apply calculus concepts to vector–valued functions as we find limits, derivatives, and integrals. Understanding the average rate of change will give us an understanding of the derivative; displacement gives us one application of integration.

Notes:

# Exercises 11.1

## Terms and Concepts

1. Vector–valued functions are closely related to _____ _____ of graphs.

2. When sketching vector–valued functions, technically one isn't graphing points, but rather _____.

3. It can be useful to think of _____ as a vector that points from a starting position to an ending position.

4. In the context of vector–valued functions, average rate of change is _____ divided by time.

## Problems

**In Exercises 5 – 12, sketch the vector–valued function on the given interval.**

5. $\vec{r}(t) = \langle t^2, t^2 - 1\rangle$, for $-2 \leq t \leq 2$.

6. $\vec{r}(t) = \langle t^2, t^3\rangle$, for $-2 \leq t \leq 2$.

7. $\vec{r}(t) = \langle 1/t, 1/t^2\rangle$, for $-2 \leq t \leq 2$.

8. $\vec{r}(t) = \langle \frac{1}{10}t^2, \sin t\rangle$, for $-2\pi \leq t \leq 2\pi$.

9. $\vec{r}(t) = \langle \frac{1}{10}t^2, \sin t\rangle$, for $-2\pi \leq t \leq 2\pi$.

10. $\vec{r}(t) = \langle 3\sin(\pi t), 2\cos(\pi t)\rangle$, on $[0, 2]$.

11. $\vec{r}(t) = \langle 3\cos t, 2\sin(2t)\rangle$, on $[0, 2\pi]$.

12. $\vec{r}(t) = \langle 2\sec t, \tan t\rangle$, on $[-\pi, \pi]$.

**In Exercises 13 – 16, sketch the vector–valued function on the given interval in $\mathbb{R}^3$. Technology may be useful in creating the sketch.**

13. $\vec{r}(t) = \langle 2\cos t, t, 2\sin t\rangle$, on $[0, 2\pi]$.

14. $\vec{r}(t) = \langle 3\cos t, \sin t, t/\pi\rangle$ on $[0, 2\pi]$.

15. $\vec{r}(t) = \langle \cos t, \sin t, \sin t\rangle$ on $[0, 2\pi]$.

16. $\vec{r}(t) = \langle \cos t, \sin t, \sin(2t)\rangle$ on $[0, 2\pi]$.

**In Exercises 17 – 20, find $\|\vec{r}(t)\|$.**

17. $\vec{r}(t) = \langle t, t^2\rangle$.

18. $\vec{r}(t) = \langle 5\cos t, 3\sin t\rangle$.

19. $\vec{r}(t) = \langle 2\cos t, 2\sin t, t\rangle$.

20. $\vec{r}(t) = \langle \cos t, t, t^2\rangle$.

**In Exercises 21 – 30, create a vector–valued function whose graph matches the given description.**

21. A circle of radius 2, centered at $(1, 2)$, traced counter-clockwise once on $[0, 2\pi]$.

22. A circle of radius 3, centered at $(5, 5)$, traced clockwise once on $[0, 2\pi]$.

23. An ellipse, centered at $(0, 0)$ with vertical major axis of length 10 and minor axis of length 3, traced once counter-clockwise on $[0, 2\pi]$.

24. An ellipse, centered at $(3, -2)$ with horizontal major axis of length 6 and minor axis of length 4, traced once clockwise on $[0, 2\pi]$.

25. A line through $(2, 3)$ with a slope of 5.

26. A line through $(1, 5)$ with a slope of $-1/2$.

27. The line through points $(1, 2, 3)$ and $(4, 5, 6)$, where $\vec{r}(0) = \langle 1, 2, 3\rangle$ and $\vec{r}(1) = \langle 4, 5, 6\rangle$.

28. The line through points $(1, 2)$ and $(4, 4)$, where $\vec{r}(0) = \langle 1, 2\rangle$ and $\vec{r}(1) = \langle 4, 4\rangle$.

29. A vertically oriented helix with radius of 2 that starts at $(2, 0, 0)$ and ends at $(2, 0, 4\pi)$ after 1 revolution on $[0, 2\pi]$.

30. A vertically oriented helix with radius of 3 that starts at $(3, 0, 0)$ and ends at $(3, 0, 3)$ after 2 revolutions on $[0, 1]$.

**In Exercises 31 – 34, find the average rate of change of $\vec{r}(t)$ on the given interval.**

31. $\vec{r}(t) = \langle t, t^2\rangle$ on $[-2, 2]$.

32. $\vec{r}(t) = \langle t, t + \sin t\rangle$ on $[0, 2\pi]$.

33. $\vec{r}(t) = \langle 3\cos t, 2\sin t, t\rangle$ on $[0, 2\pi]$.

34. $\vec{r}(t) = \langle t, t^2, t^3\rangle$ on $[-1, 3]$.

## 11.2 Calculus and Vector–Valued Functions

The previous section introduced us to a new mathematical object, the vector–valued function. We now apply calculus concepts to these functions. We start with the limit, then work our way through derivatives to integrals.

### Limits of Vector–Valued Functions

The initial definition of the limit of a vector–valued function is a bit intimidating, as was the definition of the limit in Definition 1.2.1. The theorem following the definition shows that in practice, taking limits of vector–valued functions is no more difficult than taking limits of real–valued functions.

---

**Definition 11.2.1**  **Limits of Vector–Valued Functions**

Let $I$ be an open interval containing $c$, and let $\vec{r}(t)$ be a vector–valued function defined on $I$, except possibly at $c$. The **limit of $\vec{r}(t)$, as $t$ approaches $c$, is $\vec{L}$**, expressed as

$$\lim_{t \to c} \vec{r}(t) = \vec{L},$$

means that given any $\varepsilon > 0$, there exists a $\delta > 0$ such that for all $t \neq c$, if $|t - c| < \delta$, we have $|| \vec{r}(t) - \vec{L} || < \varepsilon$.

---

**Note:** we can define one-sided limits in a manner very similar to Definition 11.2.1.

Note how the measurement of distance between real numbers is the absolute value of their difference; the measure of distance between vectors is the vector norm, or magnitude, of their difference.

Theorem 11.2.1 states that we can compute limits of vector–valued functions component–wise.

---

**Theorem 11.2.1**  **Limits of Vector–Valued Functions**

1. Let $\vec{r}(t) = \langle f(t), g(t) \rangle$ be a vector–valued function in $\mathbb{R}^2$ defined on an open interval $I$ containing $c$, except possibly at $c$. Then

$$\lim_{t \to c} \vec{r}(t) = \left\langle \lim_{t \to c} f(t), \lim_{t \to c} g(t) \right\rangle.$$

2. Let $\vec{r}(t) = \langle f(t), g(t), h(t) \rangle$ be a vector–valued function in $\mathbb{R}^3$ defined on an open interval $I$ containing $c$, except possibly at $c$. Then

$$\lim_{t \to c} \vec{r}(t) = \left\langle \lim_{t \to c} f(t), \lim_{t \to c} g(t), \lim_{t \to c} h(t) \right\rangle$$

---

Notes:

**Example 11.2.1**  **Finding limits of vector–valued functions**

Let $\vec{r}(t) = \left\langle \dfrac{\sin t}{t},\, t^2 - 3t + 3,\, \cos t \right\rangle$. Find $\lim\limits_{t \to 0} \vec{r}(t)$.

**SOLUTION**   We apply the theorem and compute limits component–wise.

$$\lim_{t \to 0} \vec{r}(t) = \left\langle \lim_{t \to 0} \frac{\sin t}{t},\, \lim_{t \to 0} t^2 - 3t + 3,\, \lim_{t \to 0} \cos t \right\rangle$$
$$= \langle 1, 3, 1 \rangle.$$

## Continuity

---

**Definition 11.2.2**   **Continuity of Vector–Valued Functions**

Let $\vec{r}(t)$ be a vector–valued function defined on an open interval $I$ containing $c$.

1. $\vec{r}(t)$ is **continuous at** $c$ if $\lim\limits_{t \to c} \vec{r}(t) = \vec{r}(c)$.

2. If $\vec{r}(t)$ is continuous at all $c$ in $I$, then $\vec{r}(t)$ is **continuous on** $I$.

---

**Note:** Using one-sided limits, we can also define continuity on closed intervals as done before.

We again have a theorem that lets us evaluate continuity component–wise.

---

**Theorem 11.2.2**   **Continuity of Vector–Valued Functions**

Let $\vec{r}(t)$ be a vector–valued function defined on an open interval $I$ containing $c$. Then $\vec{r}(t)$ is continuous at $c$ if, and only if, each of its component functions is continuous at $c$.

---

**Example 11.2.2**   **Evaluating continuity of vector–valued functions**

Let $\vec{r}(t) = \left\langle \dfrac{\sin t}{t},\, t^2 - 3t + 3,\, \cos t \right\rangle$. Determine whether $\vec{r}$ is continuous at $t = 0$ and $t = 1$.

**SOLUTION**   While the second and third components of $\vec{r}(t)$ are defined at $t = 0$, the first component, $(\sin t)/t$, is not. Since the first component is not even defined at $t = 0$, $\vec{r}(t)$ is not defined at $t = 0$, and hence it is not continuous at $t = 0$.

At $t = 1$ each of the component functions is continuous. Therefore $\vec{r}(t)$ is continuous at $t = 1$.

---

Notes:

## 11.2 Calculus and Vector–Valued Functions

## Derivatives

Consider a vector–valued function $\vec{r}$ defined on an open interval $I$ containing $t_0$ and $t_1$. We can compute the displacement of $\vec{r}$ on $[t_0, t_1]$, as shown in Figure 11.2.1(a). Recall that dividing the displacement vector by $t_1 - t_0$ gives the average rate of change on $[t_0, t_1]$, as shown in (b).

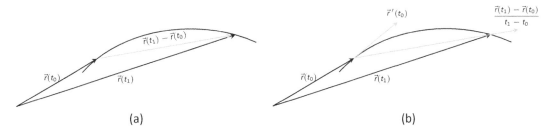

Figure 11.2.1: Illustrating displacement, leading to an understanding of the derivative of vector–valued functions.

The **derivative** of a vector–valued function is a measure of the *instantaneous* rate of change, measured by taking the limit as the length of $[t_0, t_1]$ goes to 0. Instead of thinking of an interval as $[t_0, t_1]$, we think of it as $[c, c + h]$ for some value of $h$ (hence the interval has length $h$). The *average* rate of change is

$$\frac{\vec{r}(c+h) - \vec{r}(c)}{h}$$

for any value of $h \neq 0$. We take the limit as $h \to 0$ to measure the instantaneous rate of change; this is the derivative of $\vec{r}$.

---

**Definition 11.2.3**     **Derivative of a Vector–Valued Function**

Let $\vec{r}(t)$ be continuous on an open interval $I$ containing $c$.

1. The **derivative of $\vec{r}$ at $t = c$** is

$$\vec{r}\,'(c) = \lim_{h \to 0} \frac{\vec{r}(c+h) - \vec{r}(c)}{h}.$$

2. The **derivative of $\vec{r}$** is

$$\vec{r}\,'(t) = \lim_{h \to 0} \frac{\vec{r}(t+h) - \vec{r}(t)}{h}.$$

---

Alternate notations for the derivative of $\vec{r}$ include:

$$\vec{r}\,'(t) = \frac{d}{dt}\big(\vec{r}(t)\big) = \frac{d\vec{r}}{dt}.$$

Notes:

Chapter 11  Vector Valued Functions

**Note:** again, using one-sided limits, we can define differentiability on closed intervals. We'll make use of this a few times in this chapter.

If a vector–valued function has a derivative for all $c$ in an open interval $I$, we say that $\vec{r}(t)$ is **differentiable** on $I$.

Once again we might view this definition as intimidating, but recall that we can evaluate limits component–wise. The following theorem verifies that this means we can compute derivatives component–wise as well, making the task not too difficult.

---

**Theorem 11.2.3**  **Derivatives of Vector–Valued Functions**

1. Let $\vec{r}(t) = \langle f(t), g(t) \rangle$. Then
$$\vec{r}\,'(t) = \langle f'(t), g'(t) \rangle.$$

2. Let $\vec{r}(t) = \langle f(t), g(t), h(t) \rangle$. Then
$$\vec{r}\,'(t) = \langle f'(t), g'(t), h'(t) \rangle.$$

---

**Example 11.2.3**  **Derivatives of vector–valued functions**
Let $\vec{r}(t) = \langle t^2, t \rangle$.

1. Sketch $\vec{r}(t)$ and $\vec{r}\,'(t)$ on the same axes.

2. Compute $\vec{r}\,'(1)$ and sketch this vector with its initial point at the origin and at $\vec{r}(1)$.

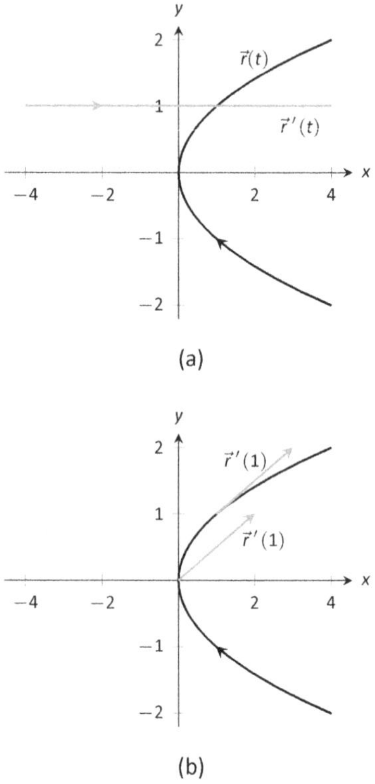

**Solution**

1. Theorem 11.2.3 allows us to compute derivatives component–wise, so
$$\vec{r}\,'(t) = \langle 2t, 1 \rangle.$$

   $\vec{r}(t)$ and $\vec{r}\,'(t)$ are graphed together in Figure 11.2.2(a). Note how plotting the two of these together, in this way, is not very illuminating. When dealing with real–valued functions, plotting $f(x)$ with $f'(x)$ gave us useful information as we were able to compare $f$ and $f'$ at the same $x$-values. When dealing with vector–valued functions, it is hard to tell which points on the graph of $\vec{r}\,'$ correspond to which points on the graph of $\vec{r}$.

2. We easily compute $\vec{r}\,'(1) = \langle 2, 1 \rangle$, which is drawn in Figure 11.2.2 with its initial point at the origin, as well as at $\vec{r}(1) = \langle 1, 1 \rangle$. These are sketched in Figure 11.2.2(b).

Figure 11.2.2: Graphing the derivative of a vector–valued function in Example 11.2.3.

Notes:

## 11.2 Calculus and Vector–Valued Functions

**Example 11.2.4  Derivatives of vector–valued functions**
Let $\vec{r}(t) = \langle \cos t, \sin t, t \rangle$. Compute $\vec{r}\,'(t)$ and $\vec{r}\,'(\pi/2)$. Sketch $\vec{r}\,'(\pi/2)$ with its initial point at the origin and at $\vec{r}(\pi/2)$.

**Solution** We compute $\vec{r}\,'$ as $\vec{r}\,'(t) = \langle -\sin t, \cos t, 1 \rangle$. At $t = \pi/2$, we have $\vec{r}\,'(\pi/2) = \langle -1, 0, 1 \rangle$. Figure 11.2.3 shows two graphs of $\vec{r}(t)$, from different perspectives, with $\vec{r}\,'(\pi/2)$ plotted with its initial point at the origin and at $\vec{r}(\pi/2)$.

In Examples 11.2.3 and 11.2.4, sketching a particular derivative with its initial point at the origin did not seem to reveal anything significant. However, when we sketched the vector with its initial point on the corresponding point on the graph, we did see something significant: the vector appeared to be *tangent* to the graph. We have not yet defined what "tangent" means in terms of curves in space; in fact, we use the derivative to define this term.

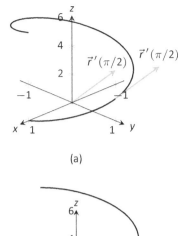

(a)

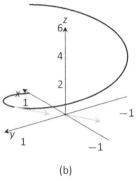

(b)

Figure 11.2.3: Viewing a vector–valued function, and its derivative at one point, from two different perspectives.

---

**Definition 11.2.4  Tangent Vector, Tangent Line**

Let $\vec{r}(t)$ be a differentiable vector–valued function on an open interval $I$ containing $c$, where $\vec{r}\,'(c) \neq \vec{0}$.

1. A vector $\vec{v}$ is **tangent to the graph of** $\vec{r}(t)$ **at** $t = c$ if $\vec{v}$ is parallel to $\vec{r}\,'(c)$.

2. The **tangent line** to the graph of $\vec{r}(t)$ at $t = c$ is the line through $\vec{r}(c)$ with direction parallel to $\vec{r}\,'(c)$. An equation of the tangent line is
$$\vec{\ell}(t) = \vec{r}(c) + t\,\vec{r}\,'(c).$$

---

**Example 11.2.5  Finding tangent lines to curves in space**
Let $\vec{r}(t) = \langle t, t^2, t^3 \rangle$ on $[-1.5, 1.5]$. Find the vector equation of the line tangent to the graph of $\vec{r}$ at $t = -1$.

**Solution** To find the equation of a line, we need a point on the line and the line's direction. The point is given by $\vec{r}(-1) = \langle -1, 1, -1 \rangle$. (To be clear, $\langle -1, 1, -1 \rangle$ is a *vector*, not a point, but we use the point "pointed to" by this vector.)

The direction comes from $\vec{r}\,'(-1)$. We compute, component-wise, $\vec{r}\,'(t) = \langle 1, 2t, 3t^2 \rangle$. Thus $\vec{r}\,'(-1) = \langle 1, -2, 3 \rangle$.

The vector equation of the line is $\vec{\ell}(t) = \langle -1, 1, -1 \rangle + t \langle 1, -2, 3 \rangle$. This line and $\vec{r}(t)$ are sketched, from two perspectives, in Figure 11.2.4 (a) and (b).

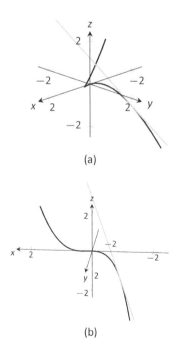

Figure 11.2.4: Graphing a curve in space with its tangent line.

Notes:

### Example 11.2.6 Finding tangent lines to curves

Find the equations of the lines tangent to $\vec{r}(t) = \langle t^3, t^2 \rangle$ at $t = -1$ and $t = 0$.

**SOLUTION** We find that $\vec{r}\,'(t) = \langle 3t^2, 2t \rangle$. At $t = -1$, we have

$$\vec{r}(-1) = \langle -1, 1 \rangle \quad \text{and} \quad \vec{r}\,'(-1) = \langle 3, -2 \rangle,$$

so the equation of the line tangent to the graph of $\vec{r}(t)$ at $t = -1$ is

$$\ell(t) = \langle -1, 1 \rangle + t \langle 3, -2 \rangle.$$

This line is graphed with $\vec{r}(t)$ in Figure 11.2.5.

At $t = 0$, we have $\vec{r}\,'(0) = \langle 0, 0 \rangle = \vec{0}$! This implies that the tangent line "has no direction." We cannot apply Definition 11.2.4, hence cannot find the equation of the tangent line.

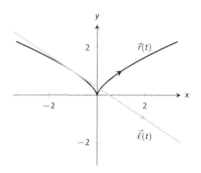

Figure 11.2.5: Graphing $\vec{r}(t)$ and its tangent line in Example 11.2.6.

We were unable to compute the equation of the tangent line to $\vec{r}(t) = \langle t^3, t^2 \rangle$ at $t = 0$ because $\vec{r}\,'(0) = \vec{0}$. The graph in Figure 11.2.5 shows that there is a cusp at this point. This leads us to another definition of **smooth**, previously defined by Definition 9.2.2 in Section 9.2.

---

**Definition 11.2.5  Smooth Vector–Valued Functions**

Let $\vec{r}(t)$ be a differentiable vector–valued function on an open interval $I$ where $\vec{r}\,'(t)$ is continuous on $I$. $\vec{r}(t)$ is **smooth** on $I$ if $\vec{r}\,'(t) \neq \vec{0}$ on $I$.

---

Having established derivatives of vector–valued functions, we now explore the relationships between the derivative and other vector operations. The following theorem states how the derivative interacts with vector addition and the various vector products.

Notes:

## 11.2 Calculus and Vector–Valued Functions

**Theorem 11.2.4  Properties of Derivatives of Vector–Valued Functions**

Let $\vec{r}$ and $\vec{s}$ be differentiable vector–valued functions, let $f$ be a differentiable real–valued function, and let $c$ be a real number.

1. $\dfrac{d}{dt}\Big(\vec{r}(t) \pm \vec{s}(t)\Big) = \vec{r}\,'(t) \pm \vec{s}\,'(t)$

2. $\dfrac{d}{dt}\Big(c\vec{r}(t)\Big) = c\vec{r}\,'(t)$

3. $\dfrac{d}{dt}\Big(f(t)\vec{r}(t)\Big) = f'(t)\vec{r}(t) + f(t)\vec{r}\,'(t)$      **Product Rule**

4. $\dfrac{d}{dt}\Big(\vec{r}(t) \cdot \vec{s}(t)\Big) = \vec{r}\,'(t) \cdot \vec{s}(t) + \vec{r}(t) \cdot \vec{s}\,'(t)$      **Product Rule**

5. $\dfrac{d}{dt}\Big(\vec{r}(t) \times \vec{s}(t)\Big) = \vec{r}\,'(t) \times \vec{s}(t) + \vec{r}(t) \times \vec{s}\,'(t)$      **Product Rule**

6. $\dfrac{d}{dt}\Big(\vec{r}(f(t))\Big) = \vec{r}\,'(f(t))f'(t)$      **Chain Rule**

**Example 11.2.7  Using derivative properties of vector–valued functions**
Let $\vec{r}(t) = \langle t, t^2 - 1 \rangle$ and let $\vec{u}(t)$ be the unit vector that points in the direction of $\vec{r}(t)$.

1. Graph $\vec{r}(t)$ and $\vec{u}(t)$ on the same axes, on $[-2, 2]$.

2. Find $\vec{u}\,'(t)$ and sketch $\vec{u}\,'(-2)$, $\vec{u}\,'(-1)$ and $\vec{u}\,'(0)$. Sketch each with initial point the corresponding point on the graph of $\vec{u}$.

**Solution**

1. To form the unit vector that points in the direction of $\vec{r}$, we need to divide $\vec{r}(t)$ by its magnitude.

$$\|\vec{r}(t)\| = \sqrt{t^2 + (t^2 - 1)^2} \quad \Rightarrow \quad \vec{u}(t) = \dfrac{1}{\sqrt{t^2 + (t^2 - 1)^2}} \langle t, t^2 - 1 \rangle.$$

$\vec{r}(t)$ and $\vec{u}(t)$ are graphed in Figure 11.2.6. Note how the graph of $\vec{u}(t)$ forms part of a circle; this must be the case, as the length of $\vec{u}(t)$ is 1 for all $t$.

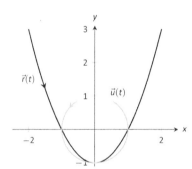

Figure 11.2.6: Graphing $\vec{r}(t)$ and $\vec{u}(t)$ in Example 11.2.7.

Notes:

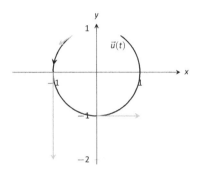

Figure 11.2.7: Graphing some of the derivatives of $\vec{u}(t)$ in Example 11.2.7.

2. To compute $\vec{u}\,'(t)$, we use Theorem 11.2.4, writing

$$\vec{u}(t) = f(t)\vec{r}(t), \quad \text{where} \quad f(t) = \frac{1}{\sqrt{t^2 + (t^2-1)^2}} = \left(t^2+(t^2-1)^2\right)^{-1/2}.$$

(We *could* write

$$\vec{u}(t) = \left\langle \frac{t}{\sqrt{t^2+(t^2-1)^2}}, \frac{t^2-1}{\sqrt{t^2+(t^2-1)^2}} \right\rangle$$

and then take the derivative. It is a matter of preference; this latter method requires two applications of the Quotient Rule where our method uses the Product and Chain Rules.)

We find $f'(t)$ using the Chain Rule:

$$f'(t) = -\frac{1}{2}\left(t^2+(t^2-1)^2\right)^{-3/2}\left(2t + 2(t^2-1)(2t)\right)$$

$$= -\frac{2t(2t^2-1)}{2\left(\sqrt{t^2+(t^2-1)^2}\right)^3}$$

We now find $\vec{u}\,'(t)$ using part 3 of Theorem 11.2.4:

$$\vec{u}\,'(t) = f'(t)\vec{u}(t) + f(t)\vec{u}\,'(t)$$

$$= -\frac{2t(2t^2-1)}{2\left(\sqrt{t^2+(t^2-1)^2}\right)^3}\langle t, t^2-1\rangle + \frac{1}{\sqrt{t^2+(t^2-1)^2}}\langle 1, 2t\rangle.$$

This is admittedly very "messy;" such is usually the case when we deal with unit vectors. We can use this formula to compute $\vec{u}\,'(-2)$, $\vec{u}\,'(-1)$ and $\vec{u}\,'(0)$:

$$\vec{u}\,'(-2) = \left\langle -\frac{15}{13\sqrt{13}}, -\frac{10}{13\sqrt{13}}\right\rangle \approx \langle -0.320, -0.213\rangle$$

$$\vec{u}\,'(-1) = \langle 0, -2\rangle$$

$$\vec{u}\,'(0) = \langle 1, 0\rangle$$

Each of these is sketched in Figure 11.2.7. Note how the length of the vector gives an indication of how quickly the circle is being traced at that point. When $t = -2$, the circle is being drawn relatively slow; when $t = -1$, the circle is being traced much more quickly.

It is a basic geometric fact that a line tangent to a circle at a point $P$ is perpendicular to the line passing through the center of the circle and $P$. This is

Notes:

illustrated in Figure 11.2.7; each tangent vector is perpendicular to the line that passes through its initial point and the center of the circle. Since the center of the circle is the origin, we can state this another way: $\vec{u}\,'(t)$ is orthogonal to $\vec{u}(t)$.

Recall that the dot product serves as a test for orthogonality: if $\vec{u} \cdot \vec{v} = 0$, then $\vec{u}$ is orthogonal to $\vec{v}$. Thus in the above example, $\vec{u}(t) \cdot \vec{u}\,'(t) = 0$.

This is true of any vector–valued function that has a constant length, that is, that traces out part of a circle. It has important implications later on, so we state it as a theorem (and leave its formal proof as an Exercise.)

---

**Theorem 11.2.5**      **Vector–Valued Functions of Constant Length**

Let $\vec{r}(t)$ be a vector–valued function of constant length that is differentiable on an open interval $I$. That is, $\|\vec{r}(t)\| = c$ for all $t$ in $I$ (equivalently, $\vec{r}(t) \cdot \vec{r}(t) = c^2$ for all $t$ in $I$). Then $\vec{r}(t) \cdot \vec{r}\,'(t) = 0$ for all $t$ in $I$.

---

## Integration

Before formally defining integrals of vector–valued functions, consider the following equation that our calculus experience tells us *should* be true:

$$\int_a^b \vec{r}\,'(t)\, dt = \vec{r}(b) - \vec{r}(a).$$

That is, the integral of a rate of change function should give total change. In the context of vector–valued functions, this total change is displacement. The above equation *is* true; we now develop the theory to show why.

We can define antiderivatives and the indefinite integral of vector–valued functions in the same manner we defined indefinite integrals in Definition 5.1.1. However, we cannot define the definite integral of a vector–valued function as we did in Definition 5.2.1. That definition was based on the signed area between a function $y = f(x)$ and the $x$-axis. An area–based definition will not be useful in the context of vector–valued functions. Instead, we define the definite integral of a vector–valued function in a manner similar to that of Theorem 5.3.2, utilizing Riemann sums.

---

Notes:

## Chapter 11 Vector Valued Functions

> **Definition 11.2.6**    **Antiderivatives, Indefinite and Definite Integrals of Vector–Valued Functions**
>
> Let $\vec{r}(t)$ be a continuous vector–valued function on $[a,b]$. An **antiderivative** of $\vec{r}(t)$ is a function $\vec{R}(t)$ such that $\vec{R}\,'(t) = \vec{r}(t)$.
>
> The set of all antiderivatives of $\vec{r}(t)$ is the **indefinite integral of $\vec{r}(t)$**, denoted by
> $$\int \vec{r}(t)\,dt.$$
>
> The definite integral of $\vec{r}(t)$ on $[a,b]$ is
> $$\int_a^b \vec{r}(t)\,dt = \lim_{||\Delta t|| \to 0} \sum_{i=1}^n \vec{r}(c_i)\Delta t_i,$$
>
> where $\Delta t_i$ is the length of the $i^{\text{th}}$ subinterval of a partition of $[a,b]$, $||\Delta t||$ is the length of the largest subinterval in the partition, and $c_i$ is any value in the $i^{\text{th}}$ subinterval of the partition.

It is probably difficult to infer meaning from the definition of the definite integral. The important thing to realize from the definition is that it is built upon limits, which we can evaluate component–wise.

The following theorem simplifies the computation of definite integrals; the rest of this section and the following section will give meaning and application to these integrals.

> **Theorem 11.2.6**    **Indefinite and Definite Integrals of Vector–Valued Functions**
>
> Let $\vec{r}(t) = \langle f(t), g(t) \rangle$ be a vector–valued function in $\mathbb{R}^2$ that are continuous on $[a,b]$.
>
> 1. $\displaystyle \int \vec{r}(t)\,dt = \left\langle \int f(t)\,dt, \int g(t)\,dt \right\rangle$
>
> 2. $\displaystyle \int_a^b \vec{r}(t)\,dt = \left\langle \int_a^b f(t)\,dt, \int_a^b g(t)\,dt \right\rangle$
>
> A similar statement holds for vector–valued functions in $\mathbb{R}^3$.

Notes:

**Example 11.2.8  Evaluating a definite integral of a vector–valued function**

Let $\vec{r}(t) = \langle e^{2t}, \sin t \rangle$. Evaluate $\displaystyle\int_0^1 \vec{r}(t)\,dt$.

**SOLUTION**  We follow Theorem 11.2.6.

$$\int_0^1 \vec{r}(t)\,dt = \int_0^1 \langle e^{2t}, \sin t \rangle\,dt$$

$$= \left\langle \int_0^1 e^{2t}\,dt,\, \int_0^1 \sin t\,dt \right\rangle$$

$$= \left\langle \left.\frac{1}{2}e^{2t}\right|_0^1,\, -\cos t\Big|_0^1 \right\rangle$$

$$= \left\langle \frac{1}{2}(e^2 - 1),\, -\cos(1) + 1 \right\rangle$$

$$\approx \langle 3.19, 0.460 \rangle.$$

**Example 11.2.9  Solving an initial value problem**

Let $\vec{r}\,''(t) = \langle 2, \cos t, 12t \rangle$. Find $\vec{r}(t)$, where $\vec{r}(0) = \langle -7, -1, 2 \rangle$ and $\vec{r}\,'(0) = \langle 5, 3, 0 \rangle$.

**SOLUTION**  Knowing $\vec{r}\,''(t) = \langle 2, \cos t, 12t \rangle$, we find $\vec{r}\,'(t)$ by evaluating the indefinite integral.

$$\int \vec{r}\,''(t)\,dt = \left\langle \int 2\,dt,\, \int \cos t\,dt,\, \int 12t\,dt \right\rangle$$

$$= \langle 2t + C_1, \sin t + C_2, 6t^2 + C_3 \rangle$$

$$= \langle 2t, \sin t, 6t^2 \rangle + \langle C_1, C_2, C_3 \rangle$$

$$= \langle 2t, \sin t, 6t^2 \rangle + \vec{C}.$$

Note how each indefinite integral creates its own constant which we collect as one constant vector $\vec{C}$. Knowing $\vec{r}\,'(0) = \langle 5, 3, 0 \rangle$ allows us to solve for $\vec{C}$:

$$\vec{r}\,'(t) = \langle 2t, \sin t, 6t^2 \rangle + \vec{C}$$

$$\vec{r}\,'(0) = \langle 0, 0, 0 \rangle + \vec{C}$$

$$\langle 5, 3, 0 \rangle = \vec{C}.$$

So $\vec{r}\,'(t) = \langle 2t, \sin t, 6t^2 \rangle + \langle 5, 3, 0 \rangle = \langle 2t + 5, \sin t + 3, 6t^2 \rangle$. To find $\vec{r}(t)$, we integrate once more.

$$\int \vec{r}\,'(t)\,dt = \left\langle \int 2t + 5\,dt,\, \int \sin t + 3\,dt,\, \int 6t^2\,dt \right\rangle$$

$$= \langle t^2 + 5t, -\cos t + 3t, 2t^3 \rangle + \vec{C}.$$

Notes:

With $\vec{r}(0) = \langle -7, -1, 2 \rangle$, we solve for $\vec{C}$:

$$\vec{r}(t) = \langle t^2 + 5t, -\cos t + 3t, 2t^3 \rangle + \vec{C}$$
$$\vec{r}(0) = \langle 0, -1, 0 \rangle + \vec{C}$$
$$\langle -7, -1, 2 \rangle = \langle 0, -1, 0 \rangle + \vec{C}$$
$$\langle -7, 0, 2 \rangle = \vec{C}.$$

So $\vec{r}(t) = \langle t^2 + 5t, -\cos t + 3t, 2t^3 \rangle + \langle -7, 0, 2 \rangle = \langle t^2 + 5t - 7, -\cos t + 3t, 2t^3 + 2 \rangle$.

What does the integration of a vector–valued function *mean*? There are many applications, but none as direct as "the area under the curve" that we used in understanding the integral of a real–valued function.

A key understanding for us comes from considering the integral of a derivative:

$$\int_a^b \vec{r}\,'(t)\,dt = \vec{r}(t)\Big|_a^b = \vec{r}(b) - \vec{r}(a).$$

Integrating a *rate of change* function gives *displacement*.

Noting that vector–valued functions are closely related to parametric equations, we can describe the arc length of the graph of a vector–valued function as an integral. Given parametric equations $x = f(t)$, $y = g(t)$, the arc length on $[a, b]$ of the graph is

$$\text{Arc Length} = \int_a^b \sqrt{f'(t)^2 + g'(t)^2}\,dt,$$

as stated in Theorem 9.3.1. If $\vec{r}(t) = \langle f(t), g(t) \rangle$, note that $\sqrt{f'(t)^2 + g'(t)^2} = \|\vec{r}\,'(t)\|$. Therefore we can express the arc length of the graph of a vector-valued function as an integral of the magnitude of its derivative.

---

**Theorem 11.2.7    Arc Length of a Vector–Valued Function**

Let $\vec{r}(t)$ be a vector–valued function where $\vec{r}\,'(t)$ is continuous on $[a, b]$. The arc length $L$ of the graph of $\vec{r}(t)$ is

$$L = \int_a^b \|\vec{r}\,'(t)\|\,dt.$$

---

Note that we are actually integrating a scalar–function here, not a vector–valued function.

The next section takes what we have established thus far and applies it to objects in motion. We will let $\vec{r}(t)$ describe the path of an object in the plane or in space and will discover the information provided by $\vec{r}\,'(t)$ and $\vec{r}\,''(t)$.

Notes:

# Exercises 11.2

## Terms and Concepts

1. Limits, derivatives and integrals of vector–valued functions are all evaluated _____-wise.

2. The definite integral of a rate of change function gives _____.

3. Why is it generally not useful to graph both $\vec{r}(t)$ and $\vec{r}\,'(t)$ on the same axes?

4. Theorem 11.2.4 contains three product rules. What are the three different types of products used in these rules?

## Problems

In Exercises 5 – 8, evaluate the given limit.

5. $\lim\limits_{t\to 5} \langle 2t+1, 3t^2-1, \sin t\rangle$

6. $\lim\limits_{t\to 3} \left\langle e^t, \dfrac{t^2-9}{t+3}\right\rangle$

7. $\lim\limits_{t\to 0} \left\langle \dfrac{t}{\sin t}, (1+t)^{\frac{1}{t}}\right\rangle$

8. $\lim\limits_{h\to 0} \dfrac{\vec{r}(t+h)-\vec{r}(t)}{h}$, where $\vec{r}(t) = \langle t^2, t, 1\rangle$.

In Exercises 9 – 10, identify the interval(s) on which $\vec{r}(t)$ is continuous.

9. $\vec{r}(t) = \langle t^2, 1/t\rangle$

10. $\vec{r}(t) = \langle \cos t, e^t, \ln t\rangle$

In Exercises 11 – 16, find the derivative of the given function.

11. $\vec{r}(t) = \langle \cos t, e^t, \ln t\rangle$

12. $\vec{r}(t) = \left\langle \dfrac{1}{t}, \dfrac{2t-1}{3t+1}, \tan t\right\rangle$

13. $\vec{r}(t) = (t^2)\langle \sin t, 2t+5\rangle$

14. $r(t) = \langle t^2+1, t-1\rangle \cdot \langle \sin t, 2t+5\rangle$

15. $\vec{r}(t) = \langle t^2+1, t-1, 1\rangle \times \langle \sin t, 2t+5, 1\rangle$

16. $\vec{r}(t) = \langle \cosh t, \sinh t\rangle$

In Exercises 17 – 20, find $\vec{r}\,'(t)$. Sketch $\vec{r}(t)$ and $\vec{r}\,'(1)$, with the initial point of $\vec{r}\,'(1)$ at $\vec{r}(1)$.

17. $\vec{r}(t) = \langle t^2+t, t^2-t\rangle$

18. $\vec{r}(t) = \langle t^2-2t+2, t^3-3t^2+2t\rangle$

19. $\vec{r}(t) = \langle t^2+1, t^3-t\rangle$

20. $\vec{r}(t) = \langle t^2-4t+5, t^3-6t^2+11t-6\rangle$

In Exercises 21 – 24, give the equation of the line tangent to the graph of $\vec{r}(t)$ at the given $t$ value.

21. $\vec{r}(t) = \langle t^2+t, t^2-t\rangle$ at $t=1$.

22. $\vec{r}(t) = \langle 3\cos t, \sin t\rangle$ at $t=\pi/4$.

23. $\vec{r}(t) = \langle 3\cos t, 3\sin t, t\rangle$ at $t=\pi$.

24. $\vec{r}(t) = \langle e^t, \tan t, t\rangle$ at $t=0$.

In Exercises 25 – 28, find the value(s) of $t$ for which $\vec{r}(t)$ is not smooth.

25. $\vec{r}(t) = \langle \cos t, \sin t - t\rangle$

26. $\vec{r}(t) = \langle t^2-2t+1, t^3+t^2-5t+3\rangle$

27. $\vec{r}(t) = \langle \cos t - \sin t, \sin t - \cos t, \cos(4t)\rangle$

28. $\vec{r}(t) = \langle t^3-3t+2, -\cos(\pi t), \sin^2(\pi t)\rangle$

Exercises 29 – 32 ask you to verify parts of Theorem 11.2.4. In each let $f(t) = t^3$, $\vec{r}(t) = \langle t^2, t-1, 1\rangle$ and $\vec{s}(t) = \langle \sin t, e^t, t\rangle$. Compute the various derivatives as indicated.

29. Simplify $f(t)\vec{r}(t)$, then find its derivative; show this is the same as $f'(t)\vec{r}(t) + f(t)\vec{r}\,'(t)$.

30. Simplify $\vec{r}(t)\cdot \vec{s}(t)$, then find its derivative; show this is the same as $\vec{r}\,'(t)\cdot \vec{s}(t) + \vec{r}(t)\cdot \vec{s}\,'(t)$.

31. Simplify $\vec{r}(t)\times \vec{s}(t)$, then find its derivative; show this is the same as $\vec{r}\,'(t)\times \vec{s}(t) + \vec{r}(t)\times \vec{s}\,'(t)$.

32. Simplify $\vec{r}(f(t))$, then find its derivative; show this is the same as $\vec{r}\,'(f(t))f'(t)$.

In Exercises 33 – 36, evaluate the given definite or indefinite integral.

33. $\displaystyle\int \langle t^3, \cos t, te^t\rangle\, dt$

34. $\displaystyle\int \left\langle \dfrac{1}{1+t^2}, \sec^2 t\right\rangle dt$

35. $\displaystyle\int_0^\pi \langle -\sin t, \cos t\rangle\, dt$

36. $\displaystyle\int_{-2}^2 \langle 2t+1, 2t-1\rangle\, dt$

**In Exercises 37 – 40, solve the given initial value problems.**

37. Find $\vec{r}(t)$, given that $\vec{r}\,'(t) = \langle t, \sin t \rangle$ and $\vec{r}(0) = \langle 2, 2 \rangle$.

38. Find $\vec{r}(t)$, given that $\vec{r}\,'(t) = \langle 1/(t+1), \tan t \rangle$ and $\vec{r}(0) = \langle 1, 2 \rangle$.

39. Find $\vec{r}(t)$, given that $\vec{r}\,''(t) = \langle t^2, t, 1 \rangle$, $\vec{r}\,'(0) = \langle 1, 2, 3 \rangle$ and $\vec{r}(0) = \langle 4, 5, 6 \rangle$.

40. Find $\vec{r}(t)$, given that $\vec{r}\,''(t) = \langle \cos t, \sin t, e^t \rangle$, $\vec{r}\,'(0) = \langle 0, 0, 0 \rangle$ and $\vec{r}(0) = \langle 0, 0, 0 \rangle$.

**In Exercises 41 – 44, find the arc length of $\vec{r}(t)$ on the indicated interval.**

41. $\vec{r}(t) = \langle 2\cos t, 2\sin t, 3t \rangle$ on $[0, 2\pi]$.

42. $\vec{r}(t) = \langle 5\cos t, 3\sin t, 4\sin t \rangle$ on $[0, 2\pi]$.

43. $\vec{r}(t) = \langle t^3, t^2, t^3 \rangle$ on $[0, 1]$.

44. $\vec{r}(t) = \langle e^{-t}\cos t, e^{-t}\sin t \rangle$ on $[0, 1]$.

45. Prove Theorem 11.2.5; that is, show if $\vec{r}(t)$ has constant length and is differentiable, then $\vec{r}(t) \cdot \vec{r}\,'(t) = 0$. (Hint: use the Product Rule to compute $\frac{d}{dt}\big(\vec{r}(t) \cdot \vec{r}(t)\big)$.)

## 11.3 The Calculus of Motion

A common use of vector–valued functions is to describe the motion of an object in the plane or in space. A **position function** $\vec{r}(t)$ gives the position of an object at **time** $t$. This section explores how derivatives and integrals are used to study the motion described by such a function.

---

**Definition 11.3.1**     **Velocity, Speed and Acceleration**

Let $\vec{r}(t)$ be a position function in $\mathbb{R}^2$ or $\mathbb{R}^3$.

1. **Velocity**, denoted $\vec{v}(t)$, is the instantaneous rate of position change; that is, $\vec{v}(t) = \vec{r}\,'(t)$.

2. **Speed** is the magnitude of velocity, $\|\vec{v}(t)\|$.

3. **Acceleration**, denoted $\vec{a}(t)$, is the instantaneous rate of velocity change; that is, $\vec{a}(t) = \vec{v}\,'(t) = \vec{r}\,''(t)$.

---

**Example 11.3.1**     **Finding velocity and acceleration**

An object is moving with position function $\vec{r}(t) = \langle t^2 - t, t^2 + t \rangle$, $-3 \leq t \leq 3$, where distances are measured in feet and time is measured in seconds.

1. Find $\vec{v}(t)$ and $\vec{a}(t)$.

2. Sketch $\vec{r}(t)$; plot $\vec{v}(-1)$, $\vec{a}(-1)$, $\vec{v}(1)$ and $\vec{a}(1)$, each with their initial point at their corresponding point on the graph of $\vec{r}(t)$.

3. When is the object's speed minimized?

**Solution**

1. Taking derivatives, we find

$$\vec{v}(t) = \vec{r}\,'(t) = \langle 2t - 1, 2t + 1 \rangle \quad \text{and} \quad \vec{a}(t) = \vec{r}\,''(t) = \langle 2, 2 \rangle.$$

Note that acceleration is constant.

2. $\vec{v}(-1) = \langle -3, -1 \rangle$, $\vec{a}(-1) = \langle 2, 2 \rangle$; $\vec{v}(1) = \langle 1, 3 \rangle$, $\vec{a}(1) = \langle 2, 2 \rangle$. These are plotted with $\vec{r}(t)$ in Figure 11.3.1(a).

We can think of acceleration as "pulling" the velocity vector in a certain direction. At $t = -1$, the velocity vector points down and to the left; at $t = 1$, the velocity vector has been pulled in the $\langle 2, 2 \rangle$ direction and is

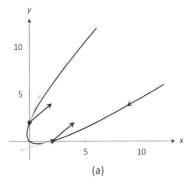

(a)

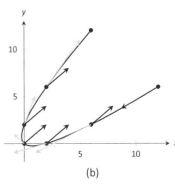

(b)

Figure 11.3.1: Graphing the position, velocity and acceleration of an object in Example 11.3.1.

Notes:

now pointing up and to the right. In Figure 11.3.1(b) we plot more velocity/acceleration vectors, making more clear the effect acceleration has on velocity.

Since $\vec{a}(t)$ is constant in this example, as $t$ grows large $\vec{v}(t)$ becomes almost parallel to $\vec{a}(t)$. For instance, when $t = 10$, $\vec{v}(10) = \langle 19, 21 \rangle$, which is nearly parallel to $\langle 2, 2 \rangle$.

3. The object's speed is given by
$$|| \vec{v}(t) || = \sqrt{(2t-1)^2 + (2t+1)^2} = \sqrt{8t^2 + 2}.$$

To find the minimal speed, we could apply calculus techniques (such as set the derivative equal to 0 and solve for $t$, etc.) but we can find it by inspection. Inside the square root we have a quadratic which is minimized when $t = 0$. Thus the speed is minimized at $t = 0$, with a speed of $\sqrt{2}$ ft/s.

The graph in Figure 11.3.1(b) also implies speed is minimized here. The filled dots on the graph are located at integer values of $t$ between $-3$ and 3. Dots that are far apart imply the object traveled a far distance in 1 second, indicating high speed; dots that are close together imply the object did not travel far in 1 second, indicating a low speed. The dots are closest together near $t = 0$, implying the speed is minimized near that value.

**Example 11.3.2   Analyzing Motion**
Two objects follow an identical path at different rates on $[-1, 1]$. The position function for Object 1 is $\vec{r}_1(t) = \langle t, t^2 \rangle$; the position function for Object 2 is $\vec{r}_2(t) = \langle t^3, t^6 \rangle$, where distances are measured in feet and time is measured in seconds. Compare the velocity, speed and acceleration of the two objects on the path.

**SOLUTION**   We begin by computing the velocity and acceleration function for each object:

$$\vec{v}_1(t) = \langle 1, 2t \rangle \qquad \vec{v}_2(t) = \langle 3t^2, 6t^5 \rangle$$
$$\vec{a}_1(t) = \langle 0, 2 \rangle \qquad \vec{a}_2(t) = \langle 6t, 30t^4 \rangle$$

We immediately see that Object 1 has constant acceleration, whereas Object 2 does not.

At $t = -1$, we have $\vec{v}_1(-1) = \langle 1, -2 \rangle$ and $\vec{v}_2(-1) = \langle 3, -6 \rangle$; the velocity of Object 2 is three times that of Object 1 and so it follows that the speed of Object 2 is three times that of Object 1 ($3\sqrt{5}$ ft/s compared to $\sqrt{5}$ ft/s.)

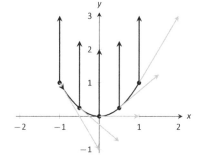

Figure 11.3.2: Plotting velocity and acceleration vectors for Object 1 in Example 11.3.2.

Notes:

At $t = 0$, the velocity of Object 1 is $\vec{v}(1) = \langle 1, 0 \rangle$ and the velocity of Object 2 is $\vec{0}$! This tells us that Object 2 comes to a complete stop at $t = 0$.

In Figure 11.3.2, we see the velocity and acceleration vectors for Object 1 plotted for $t = -1, -1/2, 0, 1/2$ and $t = 1$. Note again how the constant acceleration vector seems to "pull" the velocity vector from pointing down, right to up, right. We could plot the analogous picture for Object 2, but the velocity and acceleration vectors are rather large ($\vec{a}_2(-1) = \langle -6, 30 \rangle$!)

Instead, we simply plot the locations of Object 1 and 2 on intervals of $1/5^{\text{th}}$ of a second, shown in Figure 11.3.3(a) and (b). Note how the x-values of Object 1 increase at a steady rate. This is because the x-component of $\vec{a}(t)$ is 0; there is no acceleration in the x-component. The dots are not evenly spaced; the object is moving faster near $t = -1$ and $t = 1$ than near $t = 0$.

In part (b) of the Figure, we see the points plotted for Object 2. Note the large change in position from $t = -1$ to $t = -0.8$; the object starts moving very quickly. However, it slows considerably at it approaches the origin, and comes to a complete stop at $t = 0$. While it looks like there are 3 points near the origin, there are in reality 5 points there.

Since the objects begin and end at the same location, they have the same displacement. Since they begin and end at the same time, with the same displacement, they have the same average rate of change (i.e, they have the same average velocity). Since they follow the same path, they have the same distance traveled. Even though these three measurements are the same, the objects obviously travel the path in very different ways.

**Example 11.3.3   Analyzing the motion of a whirling ball on a string**
A young boy whirls a ball, attached to a string, above his head in a counterclockwise circle. The ball follows a circular path and makes 2 revolutions per second. The string has length 2ft.

1. Find the position function $\vec{r}(t)$ that describes this situation.

2. Find the acceleration of the ball and give a physical interpretation of it.

3. A tree stands 10ft in front of the boy. At what $t$-values should the boy release the string so that the ball hits the tree?

**Solution**

1. The ball whirls in a circle. Since the string is 2ft long, the radius of the circle is 2. The position function $\vec{r}(t) = \langle 2\cos t, 2\sin t \rangle$ describes a circle with radius 2, centered at the origin, but makes a full revolution every $2\pi$ seconds, not two revolutions per second. We modify the period of the

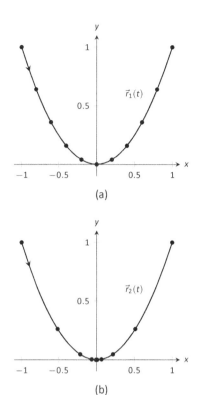

Figure 11.3.3: Comparing the positions of Objects 1 and 2 in Example 11.3.2.

trigonometric functions to be 1/2 by multiplying $t$ by $4\pi$. The final position function is thus
$$\vec{r}(t) = \langle 2\cos(4\pi t), 2\sin(4\pi t) \rangle.$$

(Plot this for $0 \leq t \leq 1/2$ to verify that one revolution is made in 1/2 a second.)

2. To find $\vec{a}(t)$, we take the derivative of $\vec{r}(t)$ twice.
$$\vec{v}(t) = \vec{r}\,'(t) = \langle -8\pi\sin(4\pi t), 8\pi\cos(4\pi t) \rangle$$
$$\vec{a}(t) = \vec{r}\,''(t) = \langle -32\pi^2\cos(4\pi t), -32\pi^2\sin(4\pi t) \rangle$$
$$= -32\pi^2 \langle \cos(4\pi t), \sin(4\pi t) \rangle.$$

Note how $\vec{a}(t)$ is parallel to $\vec{r}(t)$, but has a different magnitude and points in the opposite direction. Why is this?

Recall the classic physics equation, "Force = mass × acceleration." A force acting on a mass induces acceleration (i.e., the mass moves); acceleration acting on a mass induces a force (gravity gives our mass a *weight*). Thus force and acceleration are closely related. A moving ball "wants" to travel in a straight line. Why does the ball in our example move in a circle? It is attached to the boy's hand by a string. The string applies a force to the ball, affecting it's motion: the string *accelerates* the ball. This is not acceleration in the sense of "it travels faster;" rather, this acceleration is changing the velocity of the ball. In what direction is this force/acceleration being applied? In the direction of the string, towards the boy's hand.

The magnitude of the acceleration is related to the speed at which the ball is traveling. A ball whirling quickly is rapidly changing direction/velocity. When velocity is changing rapidly, the acceleration must be "large."

3. When the boy releases the string, the string no longer applies a force to the ball, meaning acceleration is $\vec{0}$ and the ball can now move in a straight line in the direction of $\vec{v}(t)$.

Let $t = t_0$ be the time when the boy lets go of the string. The ball will be at $\vec{r}(t_0)$, traveling in the direction of $\vec{v}(t_0)$. We want to find $t_0$ so that this line contains the point $(0, 10)$ (since the tree is 10ft directly in front of the boy).

There are many ways to find this time value. We choose one that is relatively simple computationally. As shown in Figure 11.3.4, the vector from the release point to the tree is $\langle 0, 10 \rangle - \vec{r}(t_0)$. This line segment is tangent to the circle, which means it is also perpendicular to $\vec{r}(t_0)$ itself, so their dot product is 0.

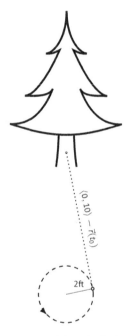

Figure 11.3.4: Modeling the flight of a ball in Example 11.3.3.

$$\vec{r}(t_0) \cdot \big(\langle 0, 10\rangle - \vec{r}(t_0)\big) = 0$$
$$\langle 2\cos(4\pi t_0), 2\sin(4\pi t_0)\rangle \cdot \langle -2\cos(4\pi t_0), 10 - 2\sin(4\pi t_0)\rangle = 0$$
$$-4\cos^2(4\pi t_0) + 20\sin(4\pi t_0) - 4\sin^2(4\pi t_0) = 0$$
$$20\sin(4\pi t_0) - 4 = 0$$
$$\sin(4\pi t_0) = 1/5$$
$$4\pi t_0 = \sin^{-1}(1/5)$$
$$4\pi t_0 \approx 0.2 + 2\pi n,$$

where $n$ is an integer. Solving for $t_0$ we have:

$$t_0 \approx 0.016 + n/2$$

This is a wonderful formula. Every 1/2 second after $t = 0.016$s the boy can release the string (since the ball makes 2 revolutions per second, he has two chances each second to release the ball).

### Example 11.3.4  Analyzing motion in space

An object moves in a spiral with position function $\vec{r}(t) = \langle \cos t, \sin t, t\rangle$, where distances are measured in meters and time is in minutes. Describe the object's speed and acceleration at time $t$.

**SOLUTION**  With $\vec{r}(t) = \langle \cos t, \sin t, t\rangle$, we have:

$$\vec{v}(t) = \langle -\sin t, \cos t, 1\rangle \quad \text{and}$$
$$\vec{a}(t) = \langle -\cos t, -\sin t, 0\rangle.$$

The speed of the object is $\|\vec{v}(t)\| = \sqrt{(-\sin t)^2 + \cos^2 t + 1} = \sqrt{2}$m/min; it moves at a constant speed. Note that the object does not accelerate in the $z$-direction, but rather moves up at a constant rate of 1m/min.

The objects in Examples 11.3.3 and 11.3.4 traveled at a constant speed. That is, $\|\vec{v}(t)\| = c$ for some constant $c$. Recall Theorem 11.2.5, which states that if a vector–valued function $\vec{r}(t)$ has constant length, then $\vec{r}(t)$ is perpendicular to its derivative: $\vec{r}(t) \cdot \vec{r}\,'(t) = 0$. In these examples, the velocity function has constant length, therefore we can conclude that the velocity is perpendicular to the acceleration: $\vec{v}(t) \cdot \vec{a}(t) = 0$. A quick check verifies this.

There is an intuitive understanding of this. If acceleration is parallel to velocity, then it is only affecting the object's speed; it does not change the direction of travel. (For example, consider a dropped stone. Acceleration and velocity are

Notes:

## Chapter 11 Vector Valued Functions

parallel – straight down – and the direction of velocity never changes, though speed does increase.) If acceleration is not perpendicular to velocity, then there is some acceleration in the direction of travel, influencing the speed. If speed is constant, then acceleration must be orthogonal to velocity, as it then only affects direction, and not speed.

> **Key Idea 11.3.1    Objects With Constant Speed**
>
> If an object moves with constant speed, then its velocity and acceleration vectors are orthogonal. That is, $\vec{v}(t) \cdot \vec{a}(t) = 0$.

**Projectile Motion**

An important application of vector–valued position functions is *projectile motion*: the motion of objects under only the influence of gravity. We will measure time in seconds, and distances will either be in meters or feet. We will show that we can completely describe the path of such an object knowing its initial position and initial velocity (i.e., where it *is* and where it *is going*.)

Suppose an object has initial position $\vec{r}(0) = \langle x_0, y_0 \rangle$ and initial velocity $\vec{v}(0) = \langle v_x, v_y \rangle$. It is customary to rewrite $\vec{v}(0)$ in terms of its speed $v_0$ and direction $\vec{u}$, where $\vec{u}$ is a unit vector. Recall all unit vectors in $\mathbb{R}^2$ can be written as $\langle \cos\theta, \sin\theta \rangle$, where $\theta$ is an angle measure counter–clockwise from the *x*-axis. (We refer to $\theta$ as the **angle of elevation**.) Thus $\vec{v}(0) = v_0 \langle \cos\theta, \sin\theta \rangle$.

Since the acceleration of the object is known, namely $\vec{a}(t) = \langle 0, -g \rangle$, where $g$ is the gravitational constant, we can find $\vec{r}(t)$ knowing our two initial conditions. We first find $\vec{v}(t)$:

**Note:** This text uses $g = 32\text{ft/s}^2$ when using Imperial units, and $g = 9.8\text{m/s}^2$ when using SI units.

$$\vec{v}(t) = \int \vec{a}(t)\, dt$$
$$\vec{v}(t) = \int \langle 0, -g \rangle\, dt$$
$$\vec{v}(t) = \langle 0, -gt \rangle + \vec{C}.$$

Knowing $\vec{v}(0) = v_0 \langle \cos\theta, \sin\theta \rangle$, we have $\vec{C} = v_0 \langle \cos\theta, \sin\theta \rangle$ and so

$$\vec{v}(t) = \langle v_0 \cos\theta,\, -gt + v_0 \sin\theta \rangle.$$

Notes:

## 11.3 The Calculus of Motion

We integrate once more to find $\vec{r}(t)$:

$$\vec{r}(t) = \int \vec{v}(t)\, dt$$

$$\vec{r}(t) = \int \langle v_0 \cos\theta, -gt + v_0 \sin\theta \rangle\, dt$$

$$\vec{r}(t) = \left\langle (v_0 \cos\theta)t, -\frac{1}{2}gt^2 + (v_0 \sin\theta)t \right\rangle + \vec{C}.$$

Knowing $\vec{r}(0) = \langle x_0, y_0 \rangle$, we conclude $\vec{C} = \langle x_0, y_0 \rangle$ and

$$\vec{r}(t) = \left\langle (v_0 \cos\theta)t + x_0,\; -\frac{1}{2}gt^2 + (v_0 \sin\theta)t + y_0 \right\rangle.$$

---

**Key Idea 11.3.2    Projectile Motion**

The position function of a projectile propelled from an initial position of $\vec{r}_0 = \langle x_0, y_0 \rangle$, with initial speed $v_0$, with angle of elevation $\theta$ and neglecting all accelerations but gravity is

$$\vec{r}(t) = \left\langle (v_0 \cos\theta)t + x_0,\; -\frac{1}{2}gt^2 + (v_0 \sin\theta)t + y_0 \right\rangle.$$

Letting $\vec{v}_0 = v_0 \langle \cos\theta, \sin\theta \rangle$, $\vec{r}(t)$ can be written as

$$\vec{r}(t) = \left\langle 0, -\frac{1}{2}gt^2 \right\rangle + \vec{v}_0 t + \vec{r}_0.$$

---

We demonstrate how to use this position function in the next two examples.

**Example 11.3.5    Projectile Motion**

Sydney shoots her Red Ryder® bb gun across level ground from an elevation of 4ft, where the barrel of the gun makes a 5° angle with the horizontal. Find how far the bb travels before landing, assuming the bb is fired at the advertised rate of 350ft/s and ignoring air resistance.

**SOLUTION**    A direct application of Key Idea 11.3.2 gives

$$\vec{r}(t) = \langle (350 \cos 5°)t, -16t^2 + (350 \sin 5°)t + 4 \rangle$$
$$\approx \langle 346.67t, -16t^2 + 30.50t + 4 \rangle,$$

Notes:

where we set her initial position to be $\langle 0, 4\rangle$. We need to find *when* the bb lands, then we can find *where*. We accomplish this by setting the *y*-component equal to 0 and solving for *t*:

$$-16t^2 + 30.50t + 4 = 0$$
$$t = \frac{-30.50 \pm \sqrt{30.50^2 - 4(-16)(4)}}{-32}$$
$$t \approx 2.03s.$$

(We discarded a negative solution that resulted from our quadratic equation.)

We have found that the bb lands 2.03s after firing; with $t = 2.03$, we find the *x*-component of our position function is $346.67(2.03) = 703.74$ft. The bb lands about 704 feet away.

**Example 11.3.6    Projectile Motion**

Alex holds his sister's bb gun at a height of 3ft and wants to shoot a target that is 6ft above the ground, 25ft away. At what angle should he hold the gun to hit his target? (We still assume the muzzle velocity is 350ft/s.)

**Solution**    The position function for the path of Alex's bb is

$$\vec{r}(t) = \langle (350\cos\theta)t, -16t^2 + (350\sin\theta)t + 3\rangle.$$

We need to find $\theta$ so that $\vec{r}(t) = \langle 25, 6\rangle$ for some value of *t*. That is, we want to find $\theta$ and *t* such that

$$(350\cos\theta)t = 25 \quad\text{and}\quad -16t^2 + (350\sin\theta)t + 3 = 6.$$

This is not trivial (though not "hard"). We start by solving each equation for $\cos\theta$ and $\sin\theta$, respectively.

$$\cos\theta = \frac{25}{350t} \quad\text{and}\quad \sin\theta = \frac{3 + 16t^2}{350t}.$$

Using the Pythagorean Identity $\cos^2\theta + \sin^2\theta = 1$, we have

$$\left(\frac{25}{350t}\right)^2 + \left(\frac{3 + 16t^2}{350t}\right)^2 = 1$$

Multiply both sides by $(350t)^2$:

$$25^2 + (3 + 16t^2)^2 = 350^2 t^2$$
$$256t^4 - 122{,}404t^2 + 634 = 0.$$

Notes:

This is a quadratic *in* $t^2$. That is, we can apply the quadratic formula to find $t^2$, then solve for *t* itself.

$$t^2 = \frac{122,404 \pm \sqrt{122,404^2 - 4(256)(634)}}{512}$$

$$t^2 = 0.0052, \, 478.135$$

$$t = \pm 0.072, \, \pm 21.866$$

Clearly the negative *t* values do not fit our context, so we have $t = 0.072$ and $t = 21.866$. Using $\cos\theta = 25/(350t)$, we can solve for $\theta$:

$$\theta = \cos^{-1}\left(\frac{25}{350 \cdot 0.072}\right) \quad \text{and} \quad \cos^{-1}\left(\frac{25}{350 \cdot 21.866}\right)$$

$$\theta = 7.03° \quad \text{and} \quad 89.8°.$$

Alex has two choices of angle. He can hold the rifle at an angle of about 7° with the horizontal and hit his target 0.07s after firing, or he can hold his rifle almost straight up, with an angle of 89.8°, where he'll hit his target about 22s later. The first option is clearly the option he should choose.

## Distance Traveled

Consider a driver who sets her cruise–control to 60mph, and travels at this speed for an hour. We can ask:

1. How far did the driver travel?

2. How far from her starting position is the driver?

The first is easy to answer: she traveled 60 miles. The second is impossible to answer with the given information. We do not know if she traveled in a straight line, on an oval racetrack, or along a slowly–winding highway.

This highlights an important fact: to compute distance traveled, we need only to know the speed, given by $\|\vec{v}(t)\|$.

---

**Theorem 11.3.1**    **Distance Traveled**

Let $\vec{v}(t)$ be a velocity function for a moving object. The distance traveled by the object on $[a, b]$ is:

$$\text{distance traveled} = \int_a^b \|\vec{v}(t)\| \, dt.$$

---

Note that this is just a restatement of Theorem 11.2.7: arc length is the same as distance traveled, just viewed in a different context.

Notes:

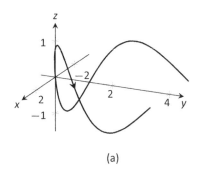

(a)

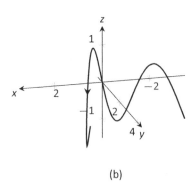

(b)

Figure 11.3.5: The path of the particle, from two perspectives, in Example 11.3.7.

**Example 11.3.7**    **Distance Traveled, Displacement, and Average Speed**

A particle moves in space with position function $\vec{r}(t) = \langle t, t^2, \sin(\pi t) \rangle$ on $[-2, 2]$, where $t$ is measured in seconds and distances are in meters. Find:

1. The distance traveled by the particle on $[-2, 2]$.
2. The displacement of the particle on $[-2, 2]$.
3. The particle's average speed.

**Solution**

1. We use Theorem 11.3.1 to establish the integral:

$$\text{distance traveled} = \int_{-2}^{2} \| \vec{v}(t) \| \, dt$$

$$= \int_{-2}^{2} \sqrt{1 + (2t)^2 + \pi^2 \cos^2(\pi t)} \, dt.$$

This cannot be solved in terms of elementary functions so we turn to numerical integration, finding the distance to be 12.88m.

2. The displacement is the vector

$$\vec{r}(2) - \vec{r}(-2) = \langle 2, 4, 0 \rangle - \langle -2, 4, 0 \rangle = \langle 4, 0, 0 \rangle.$$

That is, the particle ends with an $x$-value increased by 4 and with $y$- and $z$-values the same (see Figure 11.3.5).

3. We found above that the particle traveled 12.88m over 4 seconds. We can compute average speed by dividing: 12.88/4 = 3.22m/s.

We should also consider Definition 5.4.1 of Section 5.4, which says that the average value of a function $f$ on $[a, b]$ is $\frac{1}{b-a} \int_a^b f(x) \, dx$. In our context, the average value of the speed is

$$\text{average speed} = \frac{1}{2 - (-2)} \int_{-2}^{2} \| \vec{v}(t) \| \, dt \approx \frac{1}{4} 12.88 = 3.22 \text{m/s}.$$

Note how the physical context of a particle traveling gives meaning to a more abstract concept learned earlier.

In Definition 5.4.1 of Chapter 5 we defined the average value of a function $f(x)$ on $[a, b]$ to be

$$\frac{1}{b-a} \int_a^b f(x) \, dx.$$

Notes:

Note how in Example 11.3.7 we computed the average speed as

$$\frac{\text{distance traveled}}{\text{travel time}} = \frac{1}{2-(-2)}\int_{-2}^{2} \|\vec{v}(t)\|\, dt;$$

that is, we just found the average value of $\|\vec{v}(t)\|$ on $[-2, 2]$.

Likewise, given position function $\vec{r}(t)$, the average velocity on $[a, b]$ is

$$\frac{\text{displacement}}{\text{travel time}} = \frac{1}{b-a}\int_{a}^{b} \vec{r}\,'(t)\, dt = \frac{\vec{r}(b) - \vec{r}(a)}{b-a};$$

that is, it is the average value of $\vec{r}\,'(t)$, or $\vec{v}(t)$, on $[a, b]$.

---

**Key Idea 11.3.3     Average Speed, Average Velocity**

Let $\vec{r}(t)$ be a differentiable position function on $[a, b]$.

The **average speed** is:

$$\frac{\text{distance traveled}}{\text{travel time}} = \frac{\int_a^b \|\vec{v}(t)\|\, dt}{b-a} = \frac{1}{b-a}\int_a^b \|\vec{v}(t)\|\, dt.$$

The **average velocity** is:

$$\frac{\text{displacement}}{\text{travel time}} = \frac{\int_a^b \vec{r}\,'(t)\, dt}{b-a} = \frac{1}{b-a}\int_a^b \vec{r}\,'(t)\, dt.$$

---

The next two sections investigate more properties of the graphs of vector-valued functions and we'll apply these new ideas to what we just learned about motion.

Notes:

# Exercises 11.3

## Terms and Concepts

1. How is *velocity* different from *speed*?

2. What is the difference between *displacement* and *distance traveled*?

3. What is the difference between *average velocity* and *average speed*?

4. *Distance traveled* is the same as _____ _____, just viewed in a different context.

5. Describe a scenario where an object's average speed is a large number, but the magnitude of the average velocity is not a large number.

6. Explain why it is not possible to have an average velocity with a large magnitude but a small average speed.

## Problems

In Exercises 7 – 10, a position function $\vec{r}(t)$ is given. Find $\vec{v}(t)$ and $\vec{a}(t)$.

7. $\vec{r}(t) = \langle 2t+1, 5t-2, 7 \rangle$

8. $\vec{r}(t) = \langle 3t^2 - 2t + 1, -t^2 + t + 14 \rangle$

9. $\vec{r}(t) = \langle \cos t, \sin t \rangle$

10. $\vec{r}(t) = \langle t/10, -\cos t, \sin t \rangle$

In Exercises 11 – 14, a position function $\vec{r}(t)$ is given. Sketch $\vec{r}(t)$ on the indicated interval. Find $\vec{v}(t)$ and $\vec{a}(t)$, then add $\vec{v}(t_0)$ and $\vec{a}(t_0)$ to your sketch, with their initial points at $\vec{r}(t_0)$, for the given value of $t_0$.

11. $\vec{r}(t) = \langle t, \sin t \rangle$ on $[0, \pi/2]$; $t_0 = \pi/4$

12. $\vec{r}(t) = \langle t^2, \sin t^2 \rangle$ on $[0, \pi/2]$; $t_0 = \sqrt{\pi/4}$

13. $\vec{r}(t) = \langle t^2 + t, -t^2 + 2t \rangle$ on $[-2, 2]$; $t_0 = 1$

14. $\vec{r}(t) = \left\langle \dfrac{2t+3}{t^2+1}, t^2 \right\rangle$ on $[-1, 1]$; $t_0 = 0$

In Exercises 15 – 24, a position function $\vec{r}(t)$ of an object is given. Find the speed of the object in terms of $t$, and find where the speed is minimized/maximized on the indicated interval.

15. $\vec{r}(t) = \langle t^2, t \rangle$ on $[-1, 1]$

16. $\vec{r}(t) = \langle t^2, t^2 - t^3 \rangle$ on $[-1, 1]$

17. $\vec{r}(t) = \langle 5\cos t, 5\sin t \rangle$ on $[0, 2\pi]$

18. $\vec{r}(t) = \langle 2\cos t, 5\sin t \rangle$ on $[0, 2\pi]$

19. $\vec{r}(t) = \langle \sec t, \tan t \rangle$ on $[0, \pi/4]$

20. $\vec{r}(t) = \langle t + \cos t, 1 - \sin t \rangle$ on $[0, 2\pi]$

21. $\vec{r}(t) = \langle 12t, 5\cos t, 5\sin t \rangle$ on $[0, 4\pi]$

22. $\vec{r}(t) = \langle t^2 - t, t^2 + t, t \rangle$ on $[0, 1]$

23. $\vec{r}(t) = \left\langle t, t^2, \sqrt{1-t^2} \right\rangle$ on $[-1, 1]$

24. **Projectile Motion:** $\vec{r}(t) = \left\langle (v_0 \cos\theta)t, -\dfrac{1}{2}gt^2 + (v_0 \sin\theta)t \right\rangle$ on $\left[0, \dfrac{2v_0 \sin\theta}{g}\right]$

In Exercises 25 – 28, position functions $\vec{r}_1(t)$ and $\vec{r}_2(s)$ for two objects are given that follow the same path on the respective intervals.

(a) Show that the positions are the same at the indicated $t_0$ and $s_0$ values; i.e., show $\vec{r}_1(t_0) = \vec{r}_2(s_0)$.

(b) Find the velocity, speed and acceleration of the two objects at $t_0$ and $s_0$, respectively.

25. $\vec{r}_1(t) = \langle t, t^2 \rangle$ on $[0, 1]$; $t_0 = 1$
    $\vec{r}_2(s) = \langle s^2, s^4 \rangle$ on $[0, 1]$; $s_0 = 1$

26. $\vec{r}_1(t) = \langle 3\cos t, 3\sin t \rangle$ on $[0, 2\pi]$; $t_0 = \pi/2$
    $\vec{r}_2(s) = \langle 3\cos(4s), 3\sin(4s) \rangle$ on $[0, \pi/2]$; $s_0 = \pi/8$

27. $\vec{r}_1(t) = \langle 3t, 2t \rangle$ on $[0, 2]$; $t_0 = 2$
    $\vec{r}_2(s) = \langle 6s - 6, 4s - 4 \rangle$ on $[1, 2]$; $s_0 = 2$

28. $\vec{r}_1(t) = \langle t, \sqrt{t} \rangle$ on $[0, 1]$; $t_0 = 1$
    $\vec{r}_2(s) = \langle \sin t, \sqrt{\sin t} \rangle$ on $[0, \pi/2]$; $s_0 = \pi/2$

In Exercises 29 – 32, find the position function of an object given its acceleration and initial velocity and position.

29. $\vec{a}(t) = \langle 2, 3 \rangle$; $\vec{v}(0) = \langle 1, 2 \rangle$, $\vec{r}(0) = \langle 5, -2 \rangle$

30. $\vec{a}(t) = \langle 2, 3 \rangle$; $\vec{v}(1) = \langle 1, 2 \rangle$, $\vec{r}(1) = \langle 5, -2 \rangle$

31. $\vec{a}(t) = \langle \cos t, -\sin t \rangle$; $\vec{v}(0) = \langle 0, 1 \rangle$, $\vec{r}(0) = \langle 0, 0 \rangle$

32. $\vec{a}(t) = \langle 0, -32 \rangle$; $\vec{v}(0) = \langle 10, 50 \rangle$, $\vec{r}(0) = \langle 0, 0 \rangle$

In Exercises 33 – 36, find the displacement, distance traveled, average velocity and average speed of the described object on the given interval.

33. An object with position function $\vec{r}(t) = \langle 2\cos t, 2\sin t, 3t \rangle$, where distances are measured in feet and time is in seconds, on $[0, 2\pi]$.

34. An object with position function $\vec{r}(t) = \langle 5\cos t, -5\sin t\rangle$, where distances are measured in feet and time is in seconds, on $[0, \pi]$.

35. An object with velocity function $\vec{v}(t) = \langle \cos t, \sin t\rangle$, where distances are measured in feet and time is in seconds, on $[0, 2\pi]$.

36. An object with velocity function $\vec{v}(t) = \langle 1, 2, -1\rangle$, where distances are measured in feet and time is in seconds, on $[0, 10]$.

**Exercises 37 – 42 ask you to solve a variety of problems based on the principles of projectile motion.**

37. A boy whirls a ball, attached to a 3ft string, above his head in a counter–clockwise circle. The ball makes 2 revolutions per second.
    At what $t$-values should the boy release the string so that the ball heads directly for a tree standing 10ft in front of him?

38. David faces Goliath with only a stone in a 3ft sling, which he whirls above his head at 4 revolutions per second. They stand 20ft apart.
    (a) At what $t$-values must David release the stone in his sling in order to hit Goliath?
    (b) What is the speed at which the stone is traveling when released?
    (c) Assume David releases the stone from a height of 6ft and Goliath's forehead is 9ft above the ground. What angle of elevation must David apply to the stone to hit Goliath's head?

39. A hunter aims at a deer which is 40 yards away. Her crossbow is at a height of 5ft, and she aims for a spot on the deer 4ft above the ground. The crossbow fires her arrows at 300ft/s.
    (a) At what angle of elevation should she hold the crossbow to hit her target?
    (b) If the deer is moving perpendicularly to her line of sight at a rate of 20mph, by approximately how much should she lead the deer in order to hit it in the desired location?

40. A baseball player hits a ball at 100mph, with an initial height of 3ft and an angle of elevation of $20°$, at Boston's Fenway Park. The ball flies towards the famed "Green Monster," a wall 37ft high located 310ft from home plate.
    (a) Show that as hit, the ball hits the wall.
    (b) Show that if the angle of elevation is $21°$, the ball clears the Green Monster.

41. A Cessna flies at 1000ft at 150mph and drops a box of supplies to the professor (and his wife) on an island. Ignoring wind resistance, how far horizontally will the supplies travel before they land?

42. A football quarterback throws a pass from a height of 6ft, intending to hit his receiver 20yds away at a height of 5ft.
    (a) If the ball is thrown at a rate of 50mph, what angle of elevation is needed to hit his intended target?
    (b) If the ball is thrown at with an angle of elevation of $8°$, what initial ball speed is needed to hit his target?

## 11.4 Unit Tangent and Normal Vectors

### Unit Tangent Vector

Given a smooth vector–valued function $\vec{r}(t)$, we defined in Definition 11.2.4 that any vector parallel to $\vec{r}\,'(t_0)$ is *tangent* to the graph of $\vec{r}(t)$ at $t = t_0$. It is often useful to consider just the *direction* of $\vec{r}\,'(t)$ and not its magnitude. Therefore we are interested in the unit vector in the direction of $\vec{r}\,'(t)$. This leads to a definition.

> **Definition 11.4.1**     **Unit Tangent Vector**
>
> Let $\vec{r}(t)$ be a smooth function on an open interval $I$. The unit tangent vector $\vec{T}(t)$ is
> $$\vec{T}(t) = \frac{1}{\|\vec{r}\,'(t)\|} \vec{r}\,'(t).$$

**Example 11.4.1**     **Computing the unit tangent vector**
Let $\vec{r}(t) = \langle 3\cos t, 3\sin t, 4t \rangle$. Find $\vec{T}(t)$ and compute $\vec{T}(0)$ and $\vec{T}(1)$.

**SOLUTION**     We apply Definition 11.4.1 to find $\vec{T}(t)$.

$$\vec{T}(t) = \frac{1}{\|\vec{r}\,'(t)\|} \vec{r}\,'(t)$$

$$= \frac{1}{\sqrt{(-3\sin t)^2 + (3\cos t)^2 + 4^2}} \langle -3\sin t, 3\cos t, 4 \rangle$$

$$= \left\langle -\frac{3}{5}\sin t, \frac{3}{5}\cos t, \frac{4}{5} \right\rangle.$$

We can now easily compute $\vec{T}(0)$ and $\vec{T}(1)$:

$$\vec{T}(0) = \left\langle 0, \frac{3}{5}, \frac{4}{5} \right\rangle; \quad \vec{T}(1) = \left\langle -\frac{3}{5}\sin 1, \frac{3}{5}\cos 1, \frac{4}{5} \right\rangle \approx \langle -0.505, 0.324, 0.8 \rangle.$$

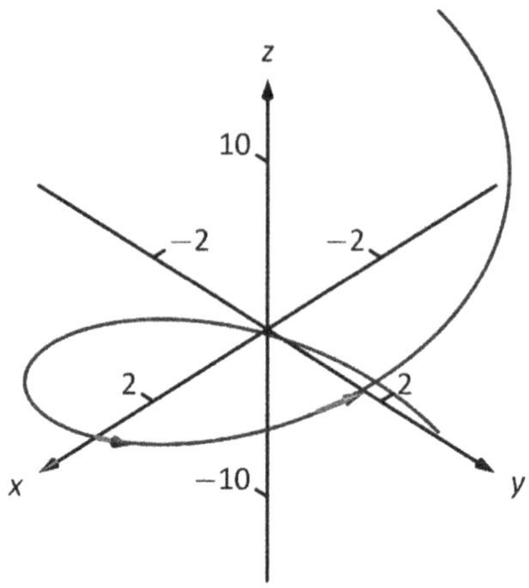

These are plotted in Figure 11.4.1 with their initial points at $\vec{r}(0)$ and $\vec{r}(1)$, respectively. (They look rather "short" since they are only length 1.)

The unit tangent vector $\vec{T}(t)$ always has a magnitude of 1, though it is sometimes easy to doubt that is true. We can help solidify this thought in our minds by computing $\|\vec{T}(1)\|$:

$$\|\vec{T}(1)\| \approx \sqrt{(-0.505)^2 + 0.324^2 + 0.8^2} = 1.000001.$$

Figure 11.4.1: Plotting unit tangent vectors in Example 11.4.1.

Notes:

We have rounded in our computation of $\vec{T}(1)$, so we don't get 1 exactly. We leave it to the reader to use the exact representation of $\vec{T}(1)$ to verify it has length 1.

In many ways, the previous example was "too nice." It turned out that $\vec{r}'(t)$ was always of length 5. In the next example the length of $\vec{r}'(t)$ is variable, leaving us with a formula that is not as clean.

**Example 11.4.2  Computing the unit tangent vector**
Let $\vec{r}(t) = \langle t^2 - t, t^2 + t \rangle$. Find $\vec{T}(t)$ and compute $\vec{T}(0)$ and $\vec{T}(1)$.

**SOLUTION**  We find $\vec{r}'(t) = \langle 2t - 1, 2t + 1 \rangle$, and
$$\|\vec{r}'(t)\| = \sqrt{(2t-1)^2 + (2t+1)^2} = \sqrt{8t^2 + 2}.$$
Therefore
$$\vec{T}(t) = \frac{1}{\sqrt{8t^2 + 2}} \langle 2t - 1, 2t + 1 \rangle = \left\langle \frac{2t-1}{\sqrt{8t^2+2}}, \frac{2t+1}{\sqrt{8t^2+2}} \right\rangle.$$
When $t = 0$, we have $\vec{T}(0) = \langle -1/\sqrt{2}, 1/\sqrt{2} \rangle$; when $t = 1$, we have $\vec{T}(1) = \langle 1/\sqrt{10}, 3/\sqrt{10} \rangle$. We leave it to the reader to verify each of these is a unit vector. They are plotted in Figure 11.4.2

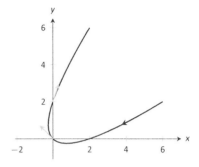

Figure 11.4.2: Plotting unit tangent vectors in Example 11.4.2.

## Unit Normal Vector

Just as knowing the direction tangent to a path is important, knowing a direction orthogonal to a path is important. When dealing with real-valued functions, we defined the normal line at a point to the be the line through the point that was perpendicular to the tangent line at that point. We can do a similar thing with vector–valued functions. Given $\vec{r}(t)$ in $\mathbb{R}^2$, we have 2 directions perpendicular to the tangent vector, as shown in Figure 11.4.3. It is good to wonder "Is one of these two directions preferable over the other?"

Given $\vec{r}(t)$ in $\mathbb{R}^3$, there are infinitely many vectors orthogonal to the tangent vector at a given point. Again, we might wonder "Is one of these infinite choices preferable over the others? Is one of these the 'right' choice?"

The answer in both $\mathbb{R}^2$ and $\mathbb{R}^3$ is "Yes, there is one vector that is not only preferable, it is the 'right' one to choose." Recall Theorem 11.2.5, which states that if $\vec{r}(t)$ has constant length, then $\vec{r}(t)$ is orthogonal to $\vec{r}'(t)$ for all $t$. We know $\vec{T}(t)$, the unit tangent vector, has constant length. Therefore $\vec{T}(t)$ is orthogonal to $\vec{T}'(t)$.

We'll see that $\vec{T}'(t)$ is more than just a convenient choice of vector that is orthogonal to $\vec{r}'(t)$; rather, it is the "right" choice. Since all we care about is the direction, we define this newly found vector to be a unit vector.

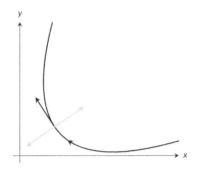

Figure 11.4.3: Given a direction in the plane, there are always two directions orthogonal to it.

**Note:** $\vec{T}(t)$ is a unit vector, by definition. This *does not* imply that $\vec{T}'(t)$ is also a unit vector.

Notes:

Chapter 11  Vector Valued Functions

> **Definition 11.4.2**  **Unit Normal Vector**
>
> Let $\vec{r}(t)$ be a vector–valued function where the unit tangent vector, $\vec{T}(t)$, is smooth on an open interval $I$. The **unit normal vector** $\vec{N}(t)$ is
> $$\vec{N}(t) = \frac{1}{||\vec{T}'(t)||}\vec{T}'(t).$$

**Example 11.4.3**  **Computing the unit normal vector**

Let $\vec{r}(t) = \langle 3\cos t, 3\sin t, 4t\rangle$ as in Example 11.4.1. Sketch both $\vec{T}(\pi/2)$ and $\vec{N}(\pi/2)$ with initial points at $\vec{r}(\pi/2)$.

**Solution**   In Example 11.4.1, we found $\vec{T}(t) = \langle (-3/5)\sin t, (3/5)\cos t, 4/5\rangle$. Therefore

$$\vec{T}'(t) = \left\langle -\frac{3}{5}\cos t, -\frac{3}{5}\sin t, 0\right\rangle \quad \text{and} \quad ||\vec{T}'(t)|| = \frac{3}{5}.$$

Thus

$$\vec{N}(t) = \frac{\vec{T}'(t)}{3/5} = \langle -\cos t, -\sin t, 0\rangle.$$

We compute $\vec{T}(\pi/2) = \langle -3/5, 0, 4/5\rangle$ and $\vec{N}(\pi/2) = \langle 0, -1, 0\rangle$. These are sketched in Figure 11.4.4.

The previous example was once again "too nice." In general, the expression for $\vec{T}(t)$ contains fractions of square–roots, hence the expression of $\vec{T}'(t)$ is very messy. We demonstrate this in the next example.

**Example 11.4.4**  **Computing the unit normal vector**

Let $\vec{r}(t) = \langle t^2 - t, t^2 + t\rangle$ as in Example 11.4.2. Find $\vec{N}(t)$ and sketch $\vec{r}(t)$ with the unit tangent and normal vectors at $t = -1, 0$ and $1$.

**Solution**   In Example 11.4.2, we found

$$\vec{T}(t) = \left\langle \frac{2t-1}{\sqrt{8t^2+2}}, \frac{2t+1}{\sqrt{8t^2+2}}\right\rangle.$$

Finding $\vec{T}'(t)$ requires two applications of the Quotient Rule:

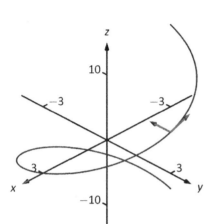

Figure 11.4.4: Plotting unit tangent and normal vectors in Example 11.4.4.

Notes:

$$T'(t) = \left\langle \frac{\sqrt{8t^2+2}(2) - (2t-1)\left(\frac{1}{2}(8t^2+2)^{-1/2}(16t)\right)}{8t^2+2}, \right.$$
$$\left. \frac{\sqrt{8t^2+2}(2) - (2t+1)\left(\frac{1}{2}(8t^2+2)^{-1/2}(16t)\right)}{8t^2+2} \right\rangle$$
$$= \left\langle \frac{4(2t+1)}{(8t^2+2)^{3/2}}, \frac{4(1-2t)}{(8t^2+2)^{3/2}} \right\rangle$$

This is not a unit vector; to find $\vec{N}(t)$, we need to divide $\vec{T}'(t)$ by it's magnitude.

$$\|\vec{T}'(t)\| = \sqrt{\frac{16(2t+1)^2}{(8t^2+2)^3} + \frac{16(1-2t)^2}{(8t^2+2)^3}}$$
$$= \sqrt{\frac{16(8t^2+2)}{(8t^2+2)^3}}$$
$$= \frac{4}{8t^2+2}.$$

Finally,

$$\vec{N}(t) = \frac{1}{4/(8t^2+2)} \left\langle \frac{4(2t+1)}{(8t^2+2)^{3/2}}, \frac{4(1-2t)}{(8t^2+2)^{3/2}} \right\rangle$$
$$= \left\langle \frac{2t+1}{\sqrt{8t^2+2}}, -\frac{2t-1}{\sqrt{8t^2+2}} \right\rangle.$$

Using this formula for $\vec{N}(t)$, we compute the unit tangent and normal vectors for $t = -1, 0$ and $1$ and sketch them in Figure 11.4.5.

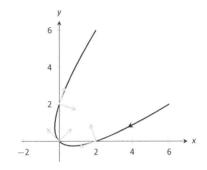

Figure 11.4.5: Plotting unit tangent and normal vectors in Example 11.4.4.

The final result for $\vec{N}(t)$ in Example 11.4.4 is suspiciously similar to $\vec{T}(t)$. There is a clear reason for this. If $\vec{u} = \langle u_1, u_2 \rangle$ is a unit vector in $\mathbb{R}^2$, then the *only* unit vectors orthogonal to $\vec{u}$ are $\langle -u_2, u_1 \rangle$ and $\langle u_2, -u_1 \rangle$. Given $\vec{T}(t)$, we can quickly determine $\vec{N}(t)$ if we know which term to multiply by $(-1)$.

Consider again Figure 11.4.5, where we have plotted some unit tangent and normal vectors. Note how $\vec{N}(t)$ always points "inside" the curve, or to the concave side of the curve. This is not a coincidence; this is true in general. Knowing the direction that $\vec{r}(t)$ "turns" allows us to quickly find $\vec{N}(t)$.

Notes:

## Chapter 11 Vector Valued Functions

> **Theorem 11.4.1**     **Unit Normal Vectors in $\mathbb{R}^2$**
>
> Let $\vec{r}(t)$ be a vector–valued function in $\mathbb{R}^2$ where $\vec{T}'(t)$ is smooth on an open interval $I$. Let $t_0$ be in $I$ and $\vec{T}(t_0) = \langle t_1, t_2 \rangle$ Then $\vec{N}(t_0)$ is either
>
> $$\vec{N}(t_0) = \langle -t_2, t_1 \rangle \quad \text{or} \quad \vec{N}(t_0) = \langle t_2, -t_1 \rangle,$$
>
> whichever is the vector that points to the concave side of the graph of $\vec{r}$.

### Application to Acceleration

Let $\vec{r}(t)$ be a position function. It is a fact (stated later in Theorem 11.4.2) that acceleration, $\vec{a}(t)$, lies in the plane defined by $\vec{T}$ and $\vec{N}$. That is, there are scalar functions $a_T(t)$ and $a_N(t)$ such that

$$\vec{a}(t) = a_T(t)\vec{T}(t) + a_N(t)\vec{N}(t).$$

We generally drop the "of $t$" part of the notation and just write $a_T$ and $a_N$.

The scalar $a_T$ measures "how much" acceleration is in the direction of travel, that is, it measures the component of acceleration that affects the speed. The scalar $a_N$ measures "how much" acceleration is perpendicular to the direction of travel, that is, it measures the component of acceleration that affects the direction of travel.

We can find $a_T$ using the orthogonal projection of $\vec{a}(t)$ onto $\vec{T}(t)$ (review Definition 10.3.3 in Section 10.3 if needed). Recalling that since $\vec{T}(t)$ is a unit vector, $\vec{T}(t) \cdot \vec{T}(t) = 1$, so we have

$$\operatorname{proj}_{\vec{T}(t)} \vec{a}(t) = \frac{\vec{a}(t) \cdot \vec{T}(t)}{\vec{T}(t) \cdot \vec{T}(t)} \vec{T}(t) = \underbrace{\left(\vec{a}(t) \cdot \vec{T}(t)\right)}_{a_T} \vec{T}(t).$$

Thus the amount of $\vec{a}(t)$ in the direction of $\vec{T}(t)$ is $a_T = \vec{a}(t) \cdot \vec{T}(t)$. The same logic gives $a_N = \vec{a}(t) \cdot \vec{N}(t)$.

While this is a fine way of computing $a_T$, there are simpler ways of finding $a_N$ (as finding $\vec{N}$ itself can be complicated). The following theorem gives alternate formulas for $a_T$ and $a_N$.

**Note:** Keep in mind that both $a_T$ and $a_N$ are functions of $t$; that is, the scalar changes depending on $t$. It is convention to drop the "$(t)$" notation from $a_T(t)$ and simply write $a_T$.

Notes:

## 11.4 Unit Tangent and Normal Vectors

> **Theorem 11.4.2**     **Acceleration in the Plane Defined by $\vec{T}$ and $\vec{N}$**
>
> Let $\vec{r}(t)$ be a position function with acceleration $\vec{a}(t)$ and unit tangent and normal vectors $\vec{T}(t)$ and $\vec{N}(t)$. Then $\vec{a}(t)$ lies in the plane defined by $\vec{T}(t)$ and $\vec{N}(t)$; that is, there exists scalars $a_T$ and $a_N$ such that
>
> $$\vec{a}(t) = a_T \vec{T}(t) + a_N \vec{N}(t).$$
>
> Moreover,
>
> $$a_T = \vec{a}(t) \cdot \vec{T}(t) = \frac{d}{dt}\left(\|\vec{v}(t)\|\right)$$
>
> $$a_N = \vec{a}(t) \cdot \vec{N}(t) = \sqrt{\|\vec{a}(t)\|^2 - a_T^2} = \frac{\|\vec{a}(t) \times \vec{v}(t)\|}{\|\vec{v}(t)\|} = \|\vec{v}(t)\|\,\|\vec{T}'(t)\|$$

Note the second formula for $a_T$: $\frac{d}{dt}\left(\|\vec{v}(t)\|\right)$. This measures the rate of change of speed, which again is the amount of acceleration in the direction of travel.

**Example 11.4.5**    **Computing $a_T$ and $a_N$**
Let $\vec{r}(t) = \langle 3\cos t, 3\sin t, 4t \rangle$ as in Examples 11.4.1 and 11.4.3. Find $a_T$ and $a_N$.

    **SOLUTION**      The previous examples give $\vec{a}(t) = \langle -3\cos t, -3\sin t, 0 \rangle$ and

$$\vec{T}(t) = \left\langle -\frac{3}{5}\sin t, \frac{3}{5}\cos t, \frac{4}{5} \right\rangle \quad \text{and} \quad \vec{N}(t) = \langle -\cos t, -\sin t, 0 \rangle.$$

We can find $a_T$ and $a_N$ directly with dot products:

$$a_T = \vec{a}(t) \cdot \vec{T}(t) = \frac{9}{5}\cos t \sin t - \frac{9}{5}\cos t \sin t + 0 = 0.$$

$$a_N = \vec{a}(t) \cdot \vec{N}(t) = 3\cos^2 t + 3\sin^2 t + 0 = 3.$$

Thus $\vec{a}(t) = 0\vec{T}(t) + 3\vec{N}(t) = 3\vec{N}(t)$, which is clearly the case.

What is the practical interpretation of these numbers? $a_T = 0$ means the object is moving at a constant speed, and hence all acceleration comes in the form of direction change.

**Example 11.4.6**    **Computing $a_T$ and $a_N$**
Let $\vec{r}(t) = \langle t^2 - t, t^2 + t \rangle$ as in Examples 11.4.2 and 11.4.4. Find $a_T$ and $a_N$.

---

Notes:

# Chapter 11  Vector Valued Functions

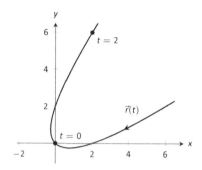

Figure 11.4.6: Graphing $\vec{r}(t)$ in Example 11.4.6.

**SOLUTION**   The previous examples give $\vec{a}(t) = \langle 2, 2 \rangle$ and

$$\vec{T}(t) = \left\langle \frac{2t-1}{\sqrt{8t^2+2}}, \frac{2t+1}{\sqrt{8t^2+2}} \right\rangle \quad \text{and} \quad \vec{N}(t) = \left\langle \frac{2t+1}{\sqrt{8t^2+2}}, -\frac{2t-1}{\sqrt{8t^2+2}} \right\rangle.$$

While we can compute $a_N$ using $\vec{N}(t)$, we instead demonstrate using another formula from Theorem 11.4.2.

$$a_T = \vec{a}(t) \cdot \vec{T}(t) = \frac{4t-2}{\sqrt{8t^2+2}} + \frac{4t+2}{\sqrt{8t^2+2}} = \frac{8t}{\sqrt{8t^2+2}}.$$

$$a_N = \sqrt{\|\vec{a}(t)\|^2 - a_T^2} = \sqrt{8 - \left(\frac{8t}{\sqrt{8t^2+2}}\right)^2} = \frac{4}{\sqrt{8t^2+2}}.$$

When $t=2$, $a_T = \frac{16}{\sqrt{34}} \approx 2.74$ and $a_N = \frac{4}{\sqrt{34}} \approx 0.69$. We interpret this to mean that at $t=2$, the particle is accelerating mostly by increasing speed, not by changing direction. As the path near $t=2$ is relatively straight, this should make intuitive sense. Figure 11.4.6 gives a graph of the path for reference.

Contrast this with $t=0$, where $a_T = 0$ and $a_N = 4/\sqrt{2} \approx 2.82$. Here the particle's speed is not changing and all acceleration is in the form of direction change.

### Example 11.4.7    Analyzing projectile motion
A ball is thrown from a height of 240ft with an initial speed of 64ft/s and an angle of elevation of 30°. Find the position function $\vec{r}(t)$ of the ball and analyze $a_T$ and $a_N$.

**SOLUTION**   Using Key Idea 11.3.2 of Section 11.3 we form the position function of the ball:

$$\vec{r}(t) = \left\langle (64\cos 30°)t, -16t^2 + (64\sin 30°)t + 240 \right\rangle,$$

which we plot in Figure 11.4.7.

From this we find $\vec{v}(t) = \langle 64\cos 30°, -32t + 64\sin 30° \rangle$ and $\vec{a}(t) = \langle 0, -32 \rangle$. Computing $\vec{T}(t)$ is not difficult, and with some simplification we find

$$\vec{T}(t) = \left\langle \frac{\sqrt{3}}{\sqrt{t^2-2t+4}}, \frac{1-t}{\sqrt{t^2-2t+4}} \right\rangle.$$

With $\vec{a}(t)$ as simple as it is, finding $a_T$ is also simple:

$$a_T = \vec{a}(t) \cdot \vec{T}(t) = \frac{32t-32}{\sqrt{t^2-2t+4}}.$$

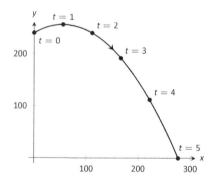

Figure 11.4.7: Plotting the position of a thrown ball, with 1s increments shown.

Notes:

## 11.4 Unit Tangent and Normal Vectors

We choose to not find $\vec{N}(t)$ and find $a_N$ through the formula $a_N = \sqrt{||\vec{a}(t)||^2 - a_T^2}$ :

$$a_N = \sqrt{32^2 - \left(\frac{32t - 32}{\sqrt{t^2 - 2t + 4}}\right)^2} = \frac{32\sqrt{3}}{\sqrt{t^2 - 2t + 4}}.$$

Figure 11.4.8 gives a table of values of $a_T$ and $a_N$. When $t = 0$, we see the ball's speed is decreasing; when $t = 1$ the speed of the ball is unchanged. This corresponds to the fact that at $t = 1$ the ball reaches its highest point.

After $t = 1$ we see that $a_N$ is decreasing in value. This is because as the ball falls, it's path becomes straighter and most of the acceleration is in the form of speeding up the ball, and not in changing its direction.

| $t$ | $a_T$ | $a_N$ |
|---|---|---|
| 0 | $-16$ | 27.7 |
| 1 | 0 | 32 |
| 2 | 16 | 27.7 |
| 3 | 24.2 | 20.9 |
| 4 | 27.7 | 16 |
| 5 | 29.4 | 12.7 |

Figure 11.4.8: A table of values of $a_T$ and $a_N$ in Example 11.4.7.

Our understanding of the unit tangent and normal vectors is aiding our understanding of motion. The work in Example 11.4.7 gave quantitative analysis of what we intuitively knew.

The next section provides two more important steps towards this analysis. We currently describe position only in terms of time. In everyday life, though, we often describe position in terms of distance ("The gas station is about 2 miles ahead, on the left."). The *arc length parameter* allows us to reference position in terms of distance traveled.

We also intuitively know that some paths are straighter than others – and some are curvier than others, but we lack a measurement of "curviness." The arc length parameter provides a way for us to compute *curvature*, a quantitative measurement of how curvy a curve is.

Notes:

# Exercises 11.4

## Terms and Concepts

1. If $\vec{T}(t)$ is a unit tangent vector, what is $\|\vec{T}(t)\|$?

2. If $\vec{N}(t)$ is a unit normal vector, what is $\vec{N}(t) \cdot \vec{r}\,'(t)$?

3. The acceleration vector $\vec{a}(t)$ lies in the plane defined by what two vectors?

4. $a_T$ measures how much the acceleration is affecting the _____ of an object.

## Problems

In Exercises 5 – 8, given $\vec{r}(t)$, find $\vec{T}(t)$ and evaluate it at the indicated value of $t$.

5. $\vec{r}(t) = \langle 2t^2, t^2 - t \rangle, \quad t = 1$

6. $\vec{r}(t) = \langle t, \cos t \rangle, \quad t = \pi/4$

7. $\vec{r}(t) = \langle \cos^3 t, \sin^3 t \rangle, \quad t = \pi/4$

8. $\vec{r}(t) = \langle \cos t, \sin t \rangle, \quad t = \pi$

In Exercises 9 – 12, find the equation of the line tangent to the curve at the indicated $t$-value using the unit tangent vector. Note: these are the same problems as in Exercises 5 – 8.

9. $\vec{r}(t) = \langle 2t^2, t^2 - t \rangle, \quad t = 1$

10. $\vec{r}(t) = \langle t, \cos t \rangle, \quad t = \pi/4$

11. $\vec{r}(t) = \langle \cos^3 t, \sin^3 t \rangle, \quad t = \pi/4$

12. $\vec{r}(t) = \langle \cos t, \sin t \rangle, \quad t = \pi$

In Exercises 13 – 16, find $\vec{N}(t)$ using Definition 11.4.2. Confirm the result using Theorem 11.4.1.

13. $\vec{r}(t) = \langle 3\cos t, 3\sin t \rangle$

14. $\vec{r}(t) = \langle t, t^2 \rangle$

15. $\vec{r}(t) = \langle \cos t, 2\sin t \rangle$

16. $\vec{r}(t) = \langle e^t, e^{-t} \rangle$

In Exercises 17 – 20, a position function $\vec{r}(t)$ is given along with its unit tangent vector $\vec{T}(t)$ evaluated at $t = a$, for some value of $a$.

(a) Confirm that $\vec{T}(a)$ is as stated.

(b) Using a graph of $\vec{r}(t)$ and Theorem 11.4.1, find $\vec{N}(a)$.

17. $\vec{r}(t) = \langle 3\cos t, 5\sin t \rangle; \quad \vec{T}(\pi/4) = \left\langle -\dfrac{3}{\sqrt{34}}, \dfrac{5}{\sqrt{34}} \right\rangle$.

18. $\vec{r}(t) = \left\langle t, \dfrac{1}{t^2+1} \right\rangle; \quad \vec{T}(1) = \left\langle \dfrac{2}{\sqrt{5}}, -\dfrac{1}{\sqrt{5}} \right\rangle$.

19. $\vec{r}(t) = (1 + 2\sin t)\langle \cos t, \sin t \rangle; \quad \vec{T}(0) = \left\langle \dfrac{2}{\sqrt{5}}, \dfrac{1}{\sqrt{5}} \right\rangle$.

20. $\vec{r}(t) = \langle \cos^3 t, \sin^3 t \rangle; \quad \vec{T}(\pi/4) = \left\langle -\dfrac{1}{\sqrt{2}}, \dfrac{1}{\sqrt{2}} \right\rangle$.

In Exercises 21 – 24, find $\vec{N}(t)$.

21. $\vec{r}(t) = \langle 4t, 2\sin t, 2\cos t \rangle$

22. $\vec{r}(t) = \langle 5\cos t, 3\sin t, 4\sin t \rangle$

23. $\vec{r}(t) = \langle a\cos t, a\sin t, bt \rangle; \quad a > 0$

24. $\vec{r}(t) = \langle \cos(at), \sin(at), t \rangle$

In Exercises 25 – 30, find $a_T$ and $a_N$ given $\vec{r}(t)$. Sketch $\vec{r}(t)$ on the indicated interval, and comment on the relative sizes of $a_T$ and $a_N$ at the indicated $t$ values.

25. $\vec{r}(t) = \langle t, t^2 \rangle$ on $[-1, 1]$; consider $t = 0$ and $t = 1$.

26. $\vec{r}(t) = \langle t, 1/t \rangle$ on $(0, 4]$; consider $t = 1$ and $t = 2$.

27. $\vec{r}(t) = \langle 2\cos t, 2\sin t \rangle$ on $[0, 2\pi]$; consider $t = 0$ and $t = \pi/2$.

28. $\vec{r}(t) = \langle \cos(t^2), \sin(t^2) \rangle$ on $(0, 2\pi]$; consider $t = \sqrt{\pi/2}$ and $t = \sqrt{\pi}$.

29. $\vec{r}(t) = \langle a\cos t, a\sin t, bt \rangle$ on $[0, 2\pi]$, where $a, b > 0$; consider $t = 0$ and $t = \pi/2$.

30. $\vec{r}(t) = \langle 5\cos t, 4\sin t, 3\sin t \rangle$ on $[0, 2\pi]$; consider $t = 0$ and $t = \pi/2$.

## 11.5 The Arc Length Parameter and Curvature

In normal conversation we describe position in terms of both *time* and *distance*. For instance, imagine driving to visit a friend. If she calls and asks where you are, you might answer "I am 20 minutes from your house," or you might say "I am 10 miles from your house." Both answers provide your friend with a general idea of where you are.

Currently, our vector–valued functions have defined points with a parameter $t$, which we often take to represent time. Consider Figure 11.5.1(a), where $\vec{r}(t) = \langle t^2 - t, t^2 + t \rangle$ is graphed and the points corresponding to $t = 0$, 1 and 2 are shown. Note how the arc length between $t = 0$ and $t = 1$ is smaller than the arc length between $t = 1$ and $t = 2$; if the parameter $t$ is time and $\vec{r}$ is position, we can say that the particle traveled faster on $[1, 2]$ than on $[0, 1]$.

Now consider Figure 11.5.1(b), where the same graph is parametrized by a different variable $s$. Points corresponding to $s = 0$ through $s = 6$ are plotted. The arc length of the graph between each adjacent pair of points is 1. We can view this parameter $s$ as distance; that is, the arc length of the graph from $s = 0$ to $s = 3$ is 3, the arc length from $s = 2$ to $s = 6$ is 4, etc. If one wants to find the point 2.5 units from an initial location (i.e., $s = 0$), one would compute $\vec{r}(2.5)$. This parameter $s$ is very useful, and is called the **arc length parameter**.

How do we find the arc length parameter?

Start with any parametrization of $\vec{r}$. We can compute the arc length of the graph of $\vec{r}$ on the interval $[0, t]$ with

$$\text{arc length} = \int_0^t \| \vec{r}'(u) \| \, du.$$

We can turn this into a function: as $t$ varies, we find the arc length $s$ from 0 to $t$. This function is

$$s(t) = \int_0^t \| \vec{r}'(u) \| \, du. \qquad (11.1)$$

This establishes a relationship between $s$ and $t$. Knowing this relationship explicitly, we can rewrite $\vec{r}(t)$ as a function of $s$: $\vec{r}(s)$. We demonstrate this in an example.

**Example 11.5.1  Finding the arc length parameter**
Let $\vec{r}(t) = \langle 3t - 1, 4t + 2 \rangle$. Parametrize $\vec{r}$ with the arc length parameter $s$.

**SOLUTION**  Using Equation (11.1), we write

$$s(t) = \int_0^t \| \vec{r}'(u) \| \, du.$$

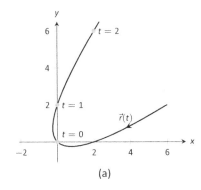

(a)

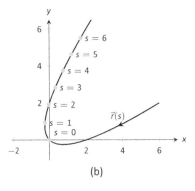

(b)

Figure 11.5.1: Introducing the arc length parameter.

Notes:

We can integrate this, explicitly finding a relationship between $s$ and $t$:

$$s(t) = \int_0^t \|\vec{r}\,'(u)\| \, du$$
$$= \int_0^t \sqrt{3^2 + 4^2} \, du$$
$$= \int_0^t 5 \, du$$
$$= 5t.$$

Since $s = 5t$, we can write $t = s/5$ and replace $t$ in $\vec{r}(t)$ with $s/5$:

$$\vec{r}(s) = \langle 3(s/5) - 1, 4(s/5) + 2 \rangle = \left\langle \frac{3}{5}s - 1, \frac{4}{5}s + 2 \right\rangle.$$

Clearly, as shown in Figure 11.5.2, the graph of $\vec{r}$ is a line, where $t = 0$ corresponds to the point $(-1, 2)$. What point on the line is 2 units away from this initial point? We find it with $\vec{r}(2) = \langle 1/5, 18/5 \rangle$.

Is the point $(1/5, 18/5)$ really 2 units away from $(-1, 2)$? We use the Distance Formula to check:

$$d = \sqrt{\left(\frac{1}{5} - (-1)\right)^2 + \left(\frac{18}{5} - 2\right)^2} = \sqrt{\frac{36}{25} + \frac{64}{25}} = \sqrt{4} = 2.$$

Yes, $\vec{r}(2)$ is indeed 2 units away, in the direction of travel, from the initial point.

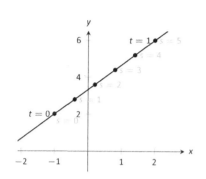

Figure 11.5.2: Graphing $\vec{r}$ in Example 11.5.1 with parameters $t$ and $s$.

Things worked out very nicely in Example 11.5.1; we were able to establish directly that $s = 5t$. Usually, the arc length parameter is much more difficult to describe in terms of $t$, a result of integrating a square–root. There are a number of things that we can learn about the arc length parameter from Equation (11.1), though, that are incredibly useful.

First, take the derivative of $s$ with respect to $t$. The Fundamental Theorem of Calculus (see Theorem 5.4.1) states that

$$\frac{ds}{dt} = s'(t) = \|\vec{r}\,'(t)\|. \qquad (11.2)$$

Letting $t$ represent time and $\vec{r}(t)$ represent position, we see that the rate of change of $s$ with respect to $t$ is speed; that is, the rate of change of "distance traveled" is speed, which should match our intuition.

The Chain Rule states that

$$\frac{d\vec{r}}{dt} = \frac{d\vec{r}}{ds} \cdot \frac{ds}{dt}$$
$$\vec{r}\,'(t) = \vec{r}\,'(s) \cdot \|\vec{r}\,'(t)\|.$$

Notes:

## 11.5 The Arc Length Parameter and Curvature

Solving for $\vec{r}'(s)$, we have

$$\vec{r}'(s) = \frac{\vec{r}'(t)}{||\vec{r}'(t)||} = \vec{T}(t), \qquad (11.3)$$

where $\vec{T}(t)$ is the unit tangent vector. Equation 11.3 is often misinterpreted, as one is tempted to think it states $\vec{r}'(t) = \vec{T}(t)$, but there is a big difference between $\vec{r}'(s)$ and $\vec{r}'(t)$. The key to take from it is that $\vec{r}'(s)$ is a unit vector. In fact, the following theorem states that this characterizes the arc length parameter.

---

**Theorem 11.5.1    Arc Length Parameter**

Let $\vec{r}(s)$ be a vector–valued function. The parameter $s$ is the arc length parameter if, and only if, $||\vec{r}'(s)|| = 1$.

---

## Curvature

Consider points $A$ and $B$ on the curve graphed in Figure 11.5.3(a). One can readily argue that the curve curves more sharply at $A$ than at $B$. It is useful to use a number to describe how sharply the curve bends; that number is the **curvature** of the curve.

We derive this number in the following way. Consider Figure 11.5.3(b), where unit tangent vectors are graphed around points $A$ and $B$. Notice how the direction of the unit tangent vector changes quite a bit near $A$, whereas it does not change as much around $B$. This leads to an important concept: measuring the rate of change of the unit tangent vector with respect to arc length gives us a measurement of curvature.

---

**Definition 11.5.1    Curvature**

Let $\vec{r}(s)$ be a vector–valued function where $s$ is the arc length parameter. The curvature $\kappa$ of the graph of $\vec{r}(s)$ is

$$\kappa = \left|\left|\frac{d\vec{T}}{ds}\right|\right| = ||\vec{T}'(s)||.$$

---

If $\vec{r}(s)$ is parametrized by the arc length parameter, then

$$\vec{T}(s) = \frac{\vec{r}'(s)}{||\vec{r}'(s)||} \quad \text{and} \quad \vec{N}(s) = \frac{\vec{T}'(s)}{||\vec{T}'(s)||}.$$

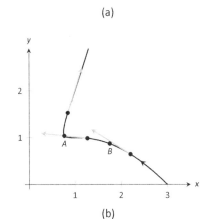

Figure 11.5.3: Establishing the concept of curvature.

Having defined $\|\vec{T}'(s)\| = \kappa$, we can rewrite the second equation as

$$\vec{T}'(s) = \kappa \vec{N}(s). \qquad (11.4)$$

We already knew that $\vec{T}'(s)$ is in the same direction as $\vec{N}(s)$; that is, we can think of $\vec{T}(s)$ as being "pulled" in the direction of $\vec{N}(s)$. How "hard" is it being pulled? By a factor of $\kappa$. When the curvature is large, $\vec{T}(s)$ is being "pulled hard" and the direction of $\vec{T}(s)$ changes rapidly. When $\kappa$ is small, $T(s)$ is not being pulled hard and hence its direction is not changing rapidly.

We use Definition 11.5.1 to find the curvature of the line in Example 11.5.1.

**Example 11.5.2  Finding the curvature of a line**
Use Definition 11.5.1 to find the curvature of $\vec{r}(t) = \langle 3t - 1, 4t + 2 \rangle$.

**Solution**  In Example 11.5.1, we found that the arc length parameter was defined by $s = 5t$, so $\vec{r}(s) = \langle 3s/5 - 1, 4s/5 + 2 \rangle$ parametrized $\vec{r}$ with the arc length parameter. To find $\kappa$, we need to find $\vec{T}'(s)$.

$$\vec{T}(s) = \vec{r}'(s) \quad \text{(recall this is a unit vector)}$$
$$= \langle 3/5, 4/5 \rangle.$$

Therefore

$$\vec{T}'(s) = \langle 0, 0 \rangle$$

and

$$\kappa = \|\vec{T}'(s)\| = 0.$$

It probably comes as no surprise that the curvature of a line is 0. (How "curvy" is a line? It is not curvy at all.)

While the definition of curvature is a beautiful mathematical concept, it is nearly impossible to use most of the time; writing $\vec{r}$ in terms of the arc length parameter is generally very hard. Fortunately, there are other methods of calculating this value that are much easier. There is a tradeoff: the definition is "easy" to understand though hard to compute, whereas these other formulas are easy to compute though it may be hard to understand why they work.

Notes:

## 11.5 The Arc Length Parameter and Curvature

**Theorem 11.5.2**     **Formulas for Curvature**

Let $C$ be a smooth curve in the plane or in space.

1. If $C$ is defined by $y = f(x)$, then
$$\kappa = \frac{|f''(x)|}{\left(1 + (f'(x))^2\right)^{3/2}}.$$

2. If $C$ is defined as a vector–valued function in the plane, $\vec{r}(t) = \langle x(t), y(t) \rangle$, then
$$\kappa = \frac{|x'y'' - x''y'|}{\left((x')^2 + (y')^2\right)^{3/2}}.$$

3. If $C$ is defined in space by a vector–valued function $\vec{r}(t)$, then
$$\kappa = \frac{||\vec{T}'(t)||}{||\vec{r}'(t)||} = \frac{||\vec{r}'(t) \times \vec{r}''(t)||}{||\vec{r}'(t)||^3} = \frac{\vec{a}(t) \cdot \vec{N}(t)}{||\vec{v}(t)||^2}.$$

We practice using these formulas.

**Example 11.5.3**     **Finding the curvature of a circle**

Find the curvature of a circle with radius $r$, defined by $\vec{c}(t) = \langle r\cos t, r\sin t \rangle$.

**SOLUTION**     Before we start, we should expect the curvature of a circle to be constant, and not dependent on $t$. (Why?)

We compute $\kappa$ using the second part of Theorem 11.5.2.

$$\kappa = \frac{|(-r\sin t)(-r\sin t) - (-r\cos t)(r\cos t)|}{\left((-r\sin t)^2 + (r\cos t)^2\right)^{3/2}}$$

$$= \frac{r^2(\sin^2 t + \cos^2 t)}{\left(r^2(\sin^2 t + \cos^2 t)\right)^{3/2}}$$

$$= \frac{r^2}{r^3} = \frac{1}{r}.$$

We have found that a circle with radius $r$ has curvature $\kappa = 1/r$.

Example 11.5.3 gives a great result. Before this example, if we were told

---

Notes:

"The curve has a curvature of 5 at point A," we would have no idea what this really meant. Is 5 "big" – does is correspond to a really sharp turn, or a not-so-sharp turn? Now we can think of 5 in terms of a circle with radius 1/5. Knowing the units (inches vs. miles, for instance) allows us to determine how sharply the curve is curving.

Let a point $P$ on a smooth curve $C$ be given, and let $\kappa$ be the curvature of the curve at $P$. A circle that:

- passes through $P$,
- lies on the concave side of $C$,
- has a common tangent line as $C$ at $P$ and
- has radius $r = 1/\kappa$ (hence has curvature $\kappa$)

is the **osculating circle**, or **circle of curvature**, to $C$ at $P$, and $r$ is the **radius of curvature**. Figure 11.5.4 shows the graph of the curve seen earlier in Figure 11.5.3 and its osculating circles at $A$ and $B$. A sharp turn corresponds to a circle with a small radius; a gradual turn corresponds to a circle with a large radius. Being able to think of curvature in terms of the radius of a circle is very useful. (The word "osculating" comes from a Latin word related to kissing; an osculating circle "kisses" the graph at a particular point. Many beautiful ideas in mathematics have come from studying the osculating circles to a curve.)

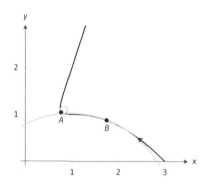

Figure 11.5.4: Illustrating the osculating circles for the curve seen in Figure 11.5.3.

**Example 11.5.4    Finding curvature**
Find the curvature of the parabola defined by $y = x^2$ at the vertex and at $x = 1$.

**Solution**    We use the first formula found in Theorem 11.5.2.

$$\kappa(x) = \frac{|2|}{\left(1 + (2x)^2\right)^{3/2}}$$
$$= \frac{2}{\left(1 + 4x^2\right)^{3/2}}.$$

At the vertex ($x = 0$), the curvature is $\kappa = 2$. At $x = 1$, the curvature is $\kappa = 2/(5)^{3/2} \approx 0.179$. So at $x = 0$, the curvature of $y = x^2$ is that of a circle of radius $1/2$; at $x = 1$, the curvature is that of a circle with radius $\approx 1/0.179 \approx 5.59$. This is illustrated in Figure 11.5.5. At $x = 3$, the curvature is $0.009$; the graph is nearly straight as the curvature is very close to 0.

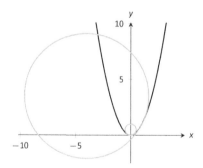

Figure 11.5.5: Examining the curvature of $y = x^2$.

**Example 11.5.5    Finding curvature**
Find where the curvature of $\vec{r}(t) = \langle t, t^2, 2t^3 \rangle$ is maximized.

Notes:

## 11.5 The Arc Length Parameter and Curvature

**SOLUTION** We use the third formula in Theorem 11.5.2 as $\vec{r}(t)$ is defined in space. We leave it to the reader to verify that

$$\vec{r}\,'(t) = \langle 1, 2t, 6t^2 \rangle, \quad \vec{r}\,''(t) = \langle 0, 2, 12t \rangle, \quad \text{and} \quad \vec{r}\,'(t) \times \vec{r}\,''(t) = \langle 12t^2, -12t, 2 \rangle.$$

Thus

$$\kappa(t) = \frac{\|\vec{r}\,'(t) \times \vec{r}\,''(t)\|}{\|\vec{r}\,'(t)\|^3}$$

$$= \frac{\|\langle 12t^2, -12t, 2 \rangle\|}{\|\langle 1, 2t, 6t^2 \rangle\|^3}$$

$$= \frac{\sqrt{144t^4 + 144t^2 + 4}}{\left(\sqrt{1 + 4t^2 + 36t^4}\right)^3}$$

While this is not a particularly "nice" formula, it does explicitly tell us what the curvature is at a given $t$ value. To maximize $\kappa(t)$, we should solve $\kappa'(t) = 0$ for $t$. This is doable, but *very* time consuming. Instead, consider the graph of $\kappa(t)$ as given in Figure 11.5.6(a). We see that $\kappa$ is maximized at two $t$ values; using a numerical solver, we find these values are $t \approx \pm 0.189$. In part (b) of the figure we graph $\vec{r}(t)$ and indicate the points where curvature is maximized.

### Curvature and Motion

Let $\vec{r}(t)$ be a position function of an object, with velocity $\vec{v}(t) = \vec{r}\,'(t)$ and acceleration $\vec{a}(t) = \vec{r}\,''(t)$. In Section 11.4 we established that acceleration is in the plane formed by $\vec{T}(t)$ and $\vec{N}(t)$, and that we can find scalars $a_T$ and $a_N$ such that

$$\vec{a}(t) = a_T \vec{T}(t) + a_N \vec{N}(t).$$

Theorem 11.4.2 gives formulas for $a_T$ and $a_N$:

$$a_T = \frac{d}{dt}\big(\|\vec{v}(t)\|\big) \quad \text{and} \quad a_N = \frac{\|\vec{v}(t) \times \vec{a}(t)\|}{\|\vec{v}(t)\|}.$$

We understood that the amount of acceleration in the direction of $\vec{T}$ relates only to how the speed of the object is changing, and that the amount of acceleration in the direction of $\vec{N}$ relates to how the direction of travel of the object is changing. (That is, if the object travels at constant speed, $a_T = 0$; if the object travels in a constant direction, $a_N = 0$.)

In Equation (11.2) at the beginning of this section, we found $s'(t) = \|\vec{v}(t)\|$. We can combine this fact with the above formula for $a_T$ to write

$$a_T = \frac{d}{dt}\big(\|\vec{v}(t)\|\big) = \frac{d}{dt}\big(s'(t)\big) = s''(t).$$

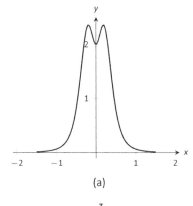

(a)

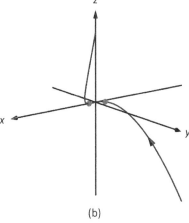

(b)

Figure 11.5.6: Understanding the curvature of a curve in space.

Notes:

Since $s'(t)$ is speed, $s''(t)$ is the rate at which speed is changing with respect to time. We see once more that the component of acceleration in the direction of travel relates only to speed, not to a change in direction.

Now compare the formula for $a_N$ above to the formula for curvature in Theorem 11.5.2:

$$a_N = \frac{||\vec{v}(t) \times \vec{a}(t)||}{||\vec{v}(t)||} \quad \text{and} \quad \kappa = \frac{||\vec{r}'(t) \times \vec{r}''(t)||}{||\vec{r}'(t)||^3} = \frac{||\vec{v}(t) \times \vec{a}(t)||}{||\vec{v}(t)||^3}.$$

Thus

$$\begin{aligned} a_N &= \kappa ||\vec{v}(t)||^2 \\ &= \kappa \left(s'(t)\right)^2 \end{aligned} \qquad (11.5)$$

This last equation shows that the component of acceleration that changes the object's direction is dependent on two things: the curvature of the path and the speed of the object.

Imagine driving a car in a clockwise circle. You will naturally feel a force pushing you towards the door (more accurately, the door is pushing you as the car is turning and you want to travel in a straight line). If you keep the radius of the circle constant but speed up (i.e., increasing $s'(t)$), the door pushes harder against you ($a_N$ has increased). If you keep your speed constant but tighten the turn (i.e., increase $\kappa$), once again the door will push harder against you.

Putting our new formulas for $a_T$ and $a_N$ together, we have

$$\vec{a}(t) = s''(t)\vec{T}(t) + \kappa ||\vec{v}(t)||^2 \vec{N}(t).$$

This is not a particularly practical way of finding $a_T$ and $a_N$, but it reveals some great concepts about how acceleration interacts with speed and the shape of a curve.

**Example 11.5.6  Curvature and road design**
The minimum radius of the curve in a highway cloverleaf is determined by the operating speed, as given in the table in Figure 11.5.7. For each curve and speed, compute $a_N$.

**Solution**  Using Equation (11.5), we can compute the acceleration normal to the curve in each case. We start by converting each speed from "miles per hour" to "feet per second" by multiplying by 5280/3600.

| Operating Speed (mph) | Minimum Radius (ft) |
|---|---|
| 35 | 310 |
| 40 | 430 |
| 45 | 540 |

Figure 11.5.7: Operating speed and minimum radius in highway cloverleaf design.

Notes:

$$35\text{mph}, 310\text{ft} \Rightarrow 51.33\text{ft/s}, \quad \kappa = 1/310$$
$$a_N = \kappa \| \vec{v}(t) \|^2$$
$$= \frac{1}{310}(51.33)^2$$
$$= 8.50\text{ft/s}^2.$$

$$40\text{mph}, 430\text{ft} \Rightarrow 58.67\text{ft/s}, \quad \kappa = 1/430$$
$$a_N = \frac{1}{430}(58.67)^2$$
$$= 8.00\text{ft/s}^2.$$

$$45\text{mph}, 540\text{ft} \Rightarrow 66\text{ft/s}, \quad \kappa = 1/540$$
$$a_N = \frac{1}{540}(66)^2$$
$$= 8.07\text{ft/s}^2.$$

Note that each acceleration is similar; this is by design. Considering the classic "Force = mass × acceleration" formula, this acceleration must be kept small in order for the tires of a vehicle to keep a "grip" on the road. If one travels on a turn of radius 310ft at a rate of 50mph, the acceleration is double, at 17.35ft/s$^2$. If the acceleration is too high, the frictional force created by the tires may not be enough to keep the car from sliding. Civil engineers routinely compute a "safe" design speed, then subtract 5-10mph to create the posted speed limit for additional safety.

We end this chapter with a reflection on what we've covered. We started with vector–valued functions, which may have seemed at the time to be just another way of writing parametric equations. However, we have seen that the vector perspective has given us great insight into the behavior of functions and the study of motion. Vector–valued position functions convey displacement, distance traveled, speed, velocity, acceleration and curvature information, each of which has great importance in science and engineering.

Notes:

# Exercises 11.5

## Terms and Concepts

1. It is common to describe position in terms of both _____ and/or _____.

2. A measure of the "curviness" of a curve is _____.

3. Give two shapes with constant curvature.

4. Describe in your own words what an "osculating circle" is.

5. Complete the identity: $\vec{T}'(s) =$ _____ $\vec{N}(s)$.

6. Given a position function $\vec{r}(t)$, how are $a_T$ and $a_N$ affected by the curvature?

## Problems

In Exercises 7 – 10, a position function $\vec{r}(t)$ is given, where $t = 0$ corresponds to the initial position. Find the arc length parameter $s$, and rewrite $\vec{r}(t)$ in terms of $s$; that is, find $\vec{r}(s)$.

7. $\vec{r}(t) = \langle 2t, t, -2t \rangle$

8. $\vec{r}(t) = \langle 7\cos t, 7\sin t \rangle$

9. $\vec{r}(t) = \langle 3\cos t, 3\sin t, 2t \rangle$

10. $\vec{r}(t) = \langle 5\cos t, 13\sin t, 12\cos t \rangle$

In Exercises 11 – 22, a curve $C$ is described along with 2 points on $C$.

(a) Using a sketch, determine at which of these points the curvature is greater.

(b) Find the curvature $\kappa$ of $C$, and evaluate $\kappa$ at each of the 2 given points.

11. $C$ is defined by $y = x^3 - x$; points given at $x = 0$ and $x = 1/2$.

12. $C$ is defined by $y = \dfrac{1}{x^2 + 1}$; points given at $x = 0$ and $x = 2$.

13. $C$ is defined by $y = \cos x$; points given at $x = 0$ and $x = \pi/2$.

14. $C$ is defined by $y = \sqrt{1 - x^2}$ on $(-1, 1)$; points given at $x = 0$ and $x = 1/2$.

15. $C$ is defined by $\vec{r}(t) = \langle \cos t, \sin(2t) \rangle$; points given at $t = 0$ and $t = \pi/4$.

16. $C$ is defined by $\vec{r}(t) = \langle \cos^2 t, \sin t \cos t \rangle$; points given at $t = 0$ and $t = \pi/3$.

17. $C$ is defined by $\vec{r}(t) = \langle t^2 - 1, t^3 - t \rangle$; points given at $t = 0$ and $t = 5$.

18. $C$ is defined by $\vec{r}(t) = \langle \tan t, \sec t \rangle$; points given at $t = 0$ and $t = \pi/6$.

19. $C$ is defined by $\vec{r}(t) = \langle 4t + 2, 3t - 1, 2t + 5 \rangle$; points given at $t = 0$ and $t = 1$.

20. $C$ is defined by $\vec{r}(t) = \langle t^3 - t, t^3 - 4, t^2 - 1 \rangle$; points given at $t = 0$ and $t = 1$.

21. $C$ is defined by $\vec{r}(t) = \langle 3\cos t, 3\sin t, 2t \rangle$; points given at $t = 0$ and $t = \pi/2$.

22. $C$ is defined by $\vec{r}(t) = \langle 5\cos t, 13\sin t, 12\cos t \rangle$; points given at $t = 0$ and $t = \pi/2$.

In Exercises 23 – 26, find the value of $x$ or $t$ where curvature is maximized.

23. $y = \dfrac{1}{6}x^3$

24. $y = \sin x$

25. $\vec{r}(t) = \langle t^2 + 2t, 3t - t^2 \rangle$

26. $\vec{r}(t) = \langle t, 4/t, 3/t \rangle$

In Exercises 27 – 30, find the radius of curvature at the indicated value.

27. $y = \tan x$, at $x = \pi/4$

28. $y = x^2 + x - 3$, at $x = \pi/4$

29. $\vec{r}(t) = \langle \cos t, \sin(3t) \rangle$, at $t = 0$

30. $\vec{r}(t) = \langle 5\cos(3t), t \rangle$, at $t = 0$

In Exercises 31 – 34, find the equation of the osculating circle to the curve at the indicated $t$-value.

31. $\vec{r}(t) = \langle t, t^2 \rangle$, at $t = 0$

32. $\vec{r}(t) = \langle 3\cos t, \sin t \rangle$, at $t = 0$

33. $\vec{r}(t) = \langle 3\cos t, \sin t \rangle$, at $t = \pi/2$

34. $\vec{r}(t) = \langle t^2 - t, t^2 + t \rangle$, at $t = 0$

# 12: FUNCTIONS OF SEVERAL VARIABLES

A function of the form $y = f(x)$ is a function of a single variable; given a value of $x$, we can find a value $y$. Even the vector–valued functions of Chapter 11 are single–variable functions; the input is a single variable though the output is a vector.

There are many situations where a desired quantity is a function of two or more variables. For instance, wind chill is measured by knowing the temperature and wind speed; the volume of a gas can be computed knowing the pressure and temperature of the gas; to compute a baseball player's batting average, one needs to know the number of hits and the number of at–bats.

This chapter studies **multivariable** functions, that is, functions with more than one input.

## 12.1 Introduction to Multivariable Functions

---

**Definition 12.1.1**  **Function of Two Variables**

Let $D$ be a subset of $\mathbb{R}^2$. A **function $f$ of two variables** is a rule that assigns each pair $(x, y)$ in $D$ a value $z = f(x, y)$ in $\mathbb{R}$. $D$ is the **domain** of $f$; the set of all outputs of $f$ is the **range**.

---

**Example 12.1.1**  **Understanding a function of two variables**
Let $z = f(x, y) = x^2 - y$. Evaluate $f(1, 2)$, $f(2, 1)$, and $f(-2, 4)$; find the domain and range of $f$.

**SOLUTION**  Using the definition $f(x, y) = x^2 - y$, we have:
$$f(1, 2) = 1^2 - 2 = -1$$
$$f(2, 1) = 2^2 - 1 = 3$$
$$f(-2, 4) = (-2)^2 - 4 = 0$$

The domain is not specified, so we take it to be all possible pairs in $\mathbb{R}^2$ for which $f$ is defined. In this example, $f$ is defined for *all* pairs $(x, y)$, so the domain $D$ of $f$ is $\mathbb{R}^2$.

The output of $f$ can be made as large or small as possible; any real number $r$ can be the output. (In fact, given any real number $r$, $f(0, -r) = r$.) So the range $R$ of $f$ is $\mathbb{R}$.

# Chapter 12 Functions of Several Variables

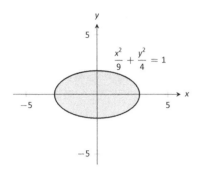

Figure 12.1.1: Illustrating the domain of $f(x,y)$ in Example 12.1.2.

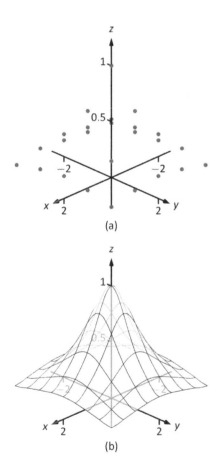

Figure 12.1.2: Graphing a function of two variables.

**Example 12.1.2**     Understanding a function of two variables

Let $f(x,y) = \sqrt{1 - \dfrac{x^2}{9} - \dfrac{y^2}{4}}$. Find the domain and range of $f$.

**Solution**     The domain is all pairs $(x,y)$ allowable as input in $f$. Because of the square–root, we need $(x,y)$ such that $0 \leq 1 - \frac{x^2}{9} - \frac{y^2}{4}$:

$$0 \leq 1 - \frac{x^2}{9} - \frac{y^2}{4}$$

$$\frac{x^2}{9} + \frac{y^2}{4} \leq 1$$

The above equation describes an ellipse and its interior as shown in Figure 12.1.1. We can represent the domain $D$ graphically with the figure; in set notation, we can write $D = \{(x,y) | \frac{x^2}{9} + \frac{y^2}{4} \leq 1\}$.

The range is the set of all possible output values. The square–root ensures that all output is $\geq 0$. Since the $x$ and $y$ terms are squared, then subtracted, inside the square–root, the largest output value comes at $x = 0, y = 0$: $f(0,0) = 1$. Thus the range $R$ is the interval $[0, 1]$.

## Graphing Functions of Two Variables

The **graph** of a function $f$ of two variables is the set of all points $(x, y, f(x, y))$ where $(x, y)$ is in the domain of $f$. This creates a **surface** in space.

One can begin sketching a graph by plotting points, but this has limitations. Consider Figure 12.1.2(a) where 25 points have been plotted of $f(x,y) = \dfrac{1}{x^2 + y^2 + 1}$. More points have been plotted than one would reasonably want to do by hand, yet it is not clear at all what the graph of the function looks like. Technology allows us to plot lots of points, connect adjacent points with lines and add shading to create a graph like Figure 12.1.2b which does a far better job of illustrating the behavior of $f$.

While technology is readily available to help us graph functions of two variables, there is still a paper–and–pencil approach that is useful to understand and master as it, combined with high–quality graphics, gives one great insight into the behavior of a function. This technique is known as sketching **level curves**.

## Level Curves

It may be surprising to find that the problem of representing a three dimensional surface on paper is familiar to most people (they just don't realize it). Topographical maps, like the one shown in Figure 12.1.3, represent the surface of Earth by indicating points with the same elevation with **contour lines**. The

Notes:

elevations marked are equally spaced; in this example, each thin line indicates an elevation change in 50ft increments and each thick line indicates a change of 200ft. When lines are drawn close together, elevation changes rapidly (as one does not have to travel far to rise 50ft). When lines are far apart, such as near "Aspen Campground," elevation changes more gradually as one has to walk farther to rise 50ft.

Given a function $z = f(x, y)$, we can draw a "topographical map" of $f$ by drawing **level curves** (or, contour lines). A level curve at $z = c$ is a curve in the x-y plane such that for all points $(x, y)$ on the curve, $f(x, y) = c$.

When drawing level curves, it is important that the $c$ values are spaced equally apart as that gives the best insight to how quickly the "elevation" is changing. Examples will help one understand this concept.

**Example 12.1.3  Drawing Level Curves**
Let $f(x, y) = \sqrt{1 - \frac{x^2}{9} - \frac{y^2}{4}}$. Find the level curves of $f$ for $c = 0, 0.2, 0.4, 0.6, 0.8$ and $1$.

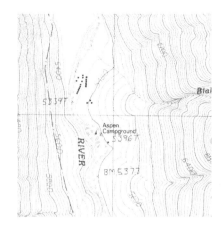

Figure 12.1.3: A topographical map displays elevation by drawing contour lines, along with the elevation is constant.
Sample taken from the public domain USGS Digital Raster Graphics, http://topmaps.usgs.gove/drg/.

**SOLUTION** Consider first $c = 0$. The level curve for $c = 0$ is the set of all points $(x, y)$ such that $0 = \sqrt{1 - \frac{x^2}{9} - \frac{y^2}{4}}$. Squaring both sides gives us

$$\frac{x^2}{9} + \frac{y^2}{4} = 1,$$

an ellipse centered at $(0, 0)$ with horizontal major axis of length 6 and minor axis of length 4. Thus for any point $(x, y)$ on this curve, $f(x, y) = 0$.

Now consider the level curve for $c = 0.2$

$$0.2 = \sqrt{1 - \frac{x^2}{9} - \frac{y^2}{4}}$$
$$0.04 = 1 - \frac{x^2}{9} - \frac{y^2}{4}$$
$$\frac{x^2}{9} + \frac{y^2}{4} = 0.96$$
$$\frac{x^2}{8.64} + \frac{y^2}{3.84} = 1.$$

This is also an ellipse, where $a = \sqrt{8.64} \approx 2.94$ and $b = \sqrt{3.84} \approx 1.96$.

Notes:

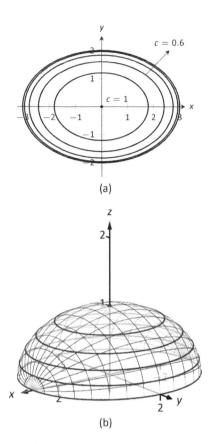

(a)

(b)

Figure 12.1.4: Graphing the level curves in Example 12.1.3.

In general, for $z = c$, the level curve is:

$$c = \sqrt{1 - \frac{x^2}{9} - \frac{y^2}{4}}$$

$$c^2 = 1 - \frac{x^2}{9} - \frac{y^2}{4}$$

$$\frac{x^2}{9} + \frac{y^2}{4} = 1 - c^2$$

$$\frac{x^2}{9(1-c^2)} + \frac{y^2}{4(1-c^2)} = 1,$$

ellipses that are decreasing in size as $c$ increases. A special case is when $c = 1$; there the ellipse is just the point $(0,0)$.

The level curves are shown in Figure 12.1.4(a). Note how the level curves for $c = 0$ and $c = 0.2$ are very, very close together: this indicates that $f$ is growing rapidly along those curves.

In Figure 12.1.4(b), the curves are drawn on a graph of $f$ in space. Note how the elevations are evenly spaced. Near the level curves of $c = 0$ and $c = 0.2$ we can see that $f$ indeed is growing quickly.

**Example 12.1.4   Analyzing Level Curves**

Let $f(x,y) = \dfrac{x+y}{x^2+y^2+1}$. Find the level curves for $z = c$.

**Solution**   We begin by setting $f(x,y) = c$ for an arbitrary $c$ and seeing if algebraic manipulation of the equation reveals anything significant.

$$\frac{x+y}{x^2+y^2+1} = c$$

$$x+y = c(x^2+y^2+1).$$

We recognize this as a circle, though the center and radius are not yet clear. By completing the square, we can obtain:

$$\left(x - \frac{1}{2c}\right)^2 + \left(y - \frac{1}{2c}\right)^2 = \frac{1}{2c^2} - 1,$$

a circle centered at $(1/(2c), 1/(2c))$ with radius $\sqrt{1/(2c^2) - 1}$, where $|c| < 1/\sqrt{2}$. The level curves for $c = \pm 0.2$, $\pm 0.4$ and $\pm 0.6$ are sketched in Figure 12.1.5(a). To help illustrate "elevation," we use thicker lines for $c$ values near 0, and dashed lines indicate where $c < 0$.

There is one special level curve, when $c = 0$. The level curve in this situation is $x + y = 0$, the line $y = -x$.

Notes:

In Figure 12.1.5(b) we see a graph of the surface. Note how the y-axis is pointing away from the viewer to more closely resemble the orientation of the level curves in (a).

Seeing the level curves helps us understand the graph. For instance, the graph does not make it clear that one can "walk" along the line $y = -x$ without elevation change, though the level curve does.

## Functions of Three Variables

We extend our study of multivariable functions to functions of three variables. (One can make a function of as many variables as one likes; we limit our study to three variables.)

> **Definition 12.1.2   Function of Three Variables**
>
> Let $D$ be a subset of $\mathbb{R}^3$. A **function $f$ of three variables** is a rule that assigns each triple $(x, y, z)$ in $D$ a value $w = f(x, y, z)$ in $\mathbb{R}$. $D$ is the **domain** of $f$; the set of all outputs of $f$ is the **range**.

Note how this definition closely resembles that of Definition 12.1.1.

**Example 12.1.5   Understanding a function of three variables**

Let $f(x, y, z) = \dfrac{x^2 + z + 3\sin y}{x + 2y - z}$. Evaluate $f$ at the point $(3, 0, 2)$ and find the domain and range of $f$.

**SOLUTION**    $f(3, 0, 2) = \dfrac{3^2 + 2 + 3\sin 0}{3 + 2(0) - 2} = 11.$

As the domain of $f$ is not specified, we take it to be the set of all triples $(x, y, z)$ for which $f(x, y, z)$ is defined. As we cannot divide by 0, we find the domain $D$ is

$$D = \{(x, y, z) \mid x + 2y - z \neq 0\}.$$

We recognize that the set of all points in $\mathbb{R}^3$ that *are not* in $D$ form a plane in space that passes through the origin (with normal vector $\langle 1, 2, -1 \rangle$).

We determine the range $R$ is $\mathbb{R}$; that is, all real numbers are possible outputs of $f$. There is no set way of establishing this. Rather, to get numbers near 0 we can let $y = 0$ and choose $z \approx -x^2$. To get numbers of arbitrarily large magnitude, we can let $z \approx x + 2y$.

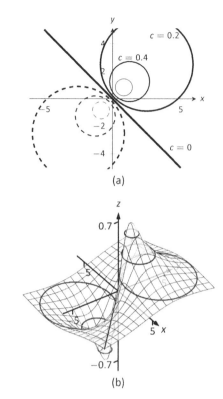

Figure 12.1.5: Graphing the level curves in Example 12.1.4.

Notes:

## Level Surfaces

It is very difficult to produce a meaningful graph of a function of three variables. A function of *one* variable is a *curve* drawn in *2* dimensions; a function of *two* variables is a *surface* drawn in *3* dimensions; a function of *three* variables is a *hypersurface* drawn in *4* dimensions.

There are a few techniques one can employ to try to "picture" a graph of three variables. One is an analogue of level curves: **level surfaces**. Given $w = f(x, y, z)$, the level surface at $w = c$ is the surface in space formed by all points $(x, y, z)$ where $f(x, y, z) = c$.

### Example 12.1.6   Finding level surfaces
If a point source $S$ is radiating energy, the intensity $I$ at a given point $P$ in space is inversely proportional to the square of the distance between $S$ and $P$. That is, when $S = (0, 0, 0)$, $I(x, y, z) = \dfrac{k}{x^2 + y^2 + z^2}$ for some constant $k$.

Let $k = 1$; find the level surfaces of $I$.

**Solution**     We can (mostly) answer this question using "common sense." If energy (say, in the form of light) is emanating from the origin, its intensity will be the same at all points equidistant from the origin. That is, at any point on the surface of a sphere centered at the origin, the intensity should be the same. Therefore, the level surfaces are spheres.

We now find this mathematically. The level surface at $I = c$ is defined by

$$c = \frac{1}{x^2 + y^2 + z^2}.$$

A small amount of algebra reveals

$$x^2 + y^2 + z^2 = \frac{1}{c}.$$

Given an intensity $c$, the level surface $I = c$ is a sphere of radius $1/\sqrt{c}$, centered at the origin.

Figure 12.1.6 gives a table of the radii of the spheres for given $c$ values. Normally one would use equally spaced $c$ values, but these values have been chosen purposefully. At a distance of 0.25 from the point source, the intensity is 16; to move to a point of half that intensity, one just moves out 0.1 to 0.35 — not much at all. To again halve the intensity, one moves 0.15, a little more than before.

Note how each time the intensity if halved, the distance required to move away grows. We conclude that the closer one is to the source, the more rapidly the intensity changes.

In the next section we apply the concepts of limits to functions of two or more variables.

| c | r |
|---|---|
| 16. | 0.25 |
| 8. | 0.35 |
| 4. | 0.5 |
| 2. | 0.71 |
| 1. | 1. |
| 0.5 | 1.41 |
| 0.25 | 2. |
| 0.125 | 2.83 |
| 0.0625 | 4. |

Figure 12.1.6: A table of $c$ values and the corresponding radius $r$ of the spheres of constant value in Example 12.1.6.

Notes:

# Exercises 12.1

## Terms and Concepts

1. Give two examples (other than those given in the text) of "real world" functions that require more than one input.

2. The graph of a function of two variables is a _____.

3. Most people are familiar with the concept of level curves in the context of _____ maps.

4. T/F: Along a level curve, the output of a function does not change.

5. The analogue of a level curve for functions of three variables is a level _____.

6. What does it mean when level curves are close together? Far apart?

## Problems

**In Exercises 7 – 14, give the domain and range of the multivariable function.**

7. $f(x, y) = x^2 + y^2 + 2$

8. $f(x, y) = x + 2y$

9. $f(x, y) = x - 2y$

10. $f(x, y) = \dfrac{1}{x + 2y}$

11. $f(x, y) = \dfrac{1}{x^2 + y^2 + 1}$

12. $f(x, y) = \sin x \cos y$

13. $f(x, y) = \sqrt{9 - x^2 - y^2}$

14. $f(x, y) = \dfrac{1}{\sqrt{x^2 + y^2 - 9}}$

**In Exercises 15 – 22, describe in words and sketch the level curves for the function and given $c$ values.**

15. $f(x, y) = 3x - 2y;\ c = -2, 0, 2$

16. $f(x, y) = x^2 - y^2;\ c = -1, 0, 1$

17. $f(x, y) = x - y^2;\ c = -2, 0, 2$

18. $f(x, y) = \dfrac{1 - x^2 - y^2}{2y - 2x};\ c = -2, 0, 2$

19. $f(x, y) = \dfrac{2x - 2y}{x^2 + y^2 + 1};\ c = -1, 0, 1$

20. $f(x, y) = \dfrac{y - x^3 - 1}{x};\ c = -3, -1, 0, 1, 3$

21. $f(x, y) = \sqrt{x^2 + 4y^2};\ c = 1, 2, 3, 4$

22. $f(x, y) = x^2 + 4y^2;\ c = 1, 2, 3, 4$

**In Exercises 23 – 26, give the domain and range of the functions of three variables.**

23. $f(x, y, z) = \dfrac{x}{x + 2y - 4z}$

24. $f(x, y, z) = \dfrac{1}{1 - x^2 - y^2 - z^2}$

25. $f(x, y, z) = \sqrt{z - x^2 + y^2}$

26. $f(x, y, z) = z^2 \sin x \cos y$

**In Exercises 27 – 30, describe the level surfaces of the given functions of three variables.**

27. $f(x, y, z) = x^2 + y^2 + z^2$

28. $f(x, y, z) = z - x^2 + y^2$

29. $f(x, y, z) = \dfrac{x^2 + y^2}{z}$

30. $f(x, y, z) = \dfrac{z}{x - y}$

31. Compare the level curves of Exercises 21 and 22. How are they similar, and how are they different? Each surface is a quadric surface; describe how the level curves are consistent with what we know about each surface.

# Chapter 12 Functions of Several Variables

## 12.2 Limits and Continuity of Multivariable Functions

We continue with the pattern we have established in this text: after defining a new kind of function, we apply calculus ideas to it. The previous section defined functions of two and three variables; this section investigates what it means for these functions to be "continuous."

We begin with a series of definitions. We are used to "open intervals" such as $(1, 3)$, which represents the set of all $x$ such that $1 < x < 3$, and "closed intervals" such as $[1, 3]$, which represents the set of all $x$ such that $1 \leq x \leq 3$. We need analogous definitions for open and closed sets in the $x$-$y$ plane.

> **Definition 12.2.1**     Open Disk, Boundary and Interior Points, Open and Closed Sets, Bounded Sets
>
> An **open disk** $B$ in $\mathbb{R}^2$ centered at $(x_0, y_0)$ with radius $r$ is the set of all points $(x, y)$ such that $\sqrt{(x-x_0)^2 + (y-y_0)^2} < r$.
>
> Let $S$ be a set of points in $\mathbb{R}^2$. A point $P$ in $\mathbb{R}^2$ is a **boundary point** of $S$ if all open disks centered at $P$ contain both points in $S$ and points not in $S$.
>
> A point $P$ in $S$ is an **interior point** of $S$ if there is an open disk centered at $P$ that contains only points in $S$.
>
> A set $S$ is **open** if every point in $S$ is an interior point.
>
> A set $S$ is **closed** if it contains all of its boundary points.
>
> A set $S$ is **bounded** if there is an $M > 0$ such that the open disk, centered at the origin with radius $M$, contains $S$. A set that is not bounded is **unbounded**.

Figure 12.2.1 shows several sets in the $x$-$y$ plane. In each set, point $P_1$ lies on the boundary of the set as all open disks centered there contain both points in, and not in, the set. In contrast, point $P_2$ is an interior point for there is an open disk centered there that lies entirely within the set.

The set depicted in Figure 12.2.1(a) is a closed set as it contains all of its boundary points. The set in (b) is open, for all of its points are interior points (or, equivalently, it does not contain any of its boundary points). The set in (c) is neither open nor closed as it contains some of its boundary points.

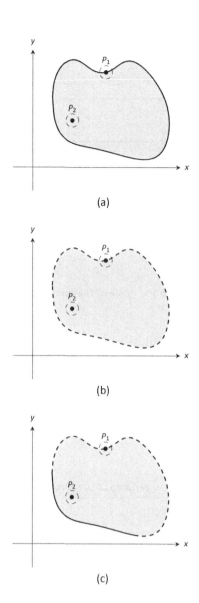

Figure 12.2.1: Illustrating open and closed sets in the $x$-$y$ plane.

Notes:

## 12.2 Limits and Continuity of Multivariable Functions

**Example 12.2.1    Determining open/closed, bounded/unbounded**
Determine if the domain of the function $f(x, y) = \sqrt{1 - x^2/9 - y^2/4}$ is open, closed, or neither, and if it is bounded.

**SOLUTION**    This domain of this function was found in Example 12.1.2 to be $D = \{(x, y) \mid \frac{x^2}{9} + \frac{y^2}{4} \leq 1\}$, the region *bounded* by the ellipse $\frac{x^2}{9} + \frac{y^2}{4} = 1$. Since the region includes the boundary (indicated by the use of "$\leq$"), the set contains all of its boundary points and hence is closed. The region is bounded as a disk of radius 4, centered at the origin, contains $D$.

**Example 12.2.2    Determining open/closed, bounded/unbounded**
Determine if the domain of $f(x, y) = \frac{1}{x-y}$ is open, closed, or neither.

**SOLUTION**    As we cannot divide by 0, we find the domain to be $D = \{(x, y) \mid x - y \neq 0\}$. In other words, the domain is the set of all points $(x, y)$ *not* on the line $y = x$.

The domain is sketched in Figure 12.2.2. Note how we can draw an open disk around any point in the domain that lies entirely inside the domain, and also note how the only boundary points of the domain are the points on the line $y = x$. We conclude the domain is an open set. The set is unbounded.

## Limits

Recall a pseudo–definition of the limit of a function of one variable: "$\lim\limits_{x \to c} f(x) = L$" means that if $x$ is "really close" to $c$, then $f(x)$ is "really close" to $L$. A similar pseudo–definition holds for functions of two variables. We'll say that

$$\lim_{(x,y) \to (x_0, y_0)} f(x, y) = L"$$

means "if the point $(x, y)$ is really close to the point $(x_0, y_0)$, then $f(x, y)$ is really close to $L$." The formal definition is given below.

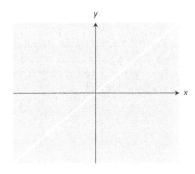

Figure 12.2.2: Sketching the domain of the function in Example 12.2.2.

---

**Definition 12.2.2    Limit of a Function of Two Variables**

Let $S$ be a set containing $P = (x_0, y_0)$ where every open disk centered at $P$ contains points in $S$ other than $P$, let $f$ be a function of two variables defined on $S$, except possibly at $P$, and let $L$ be a real number. The **limit of $f(x, y)$ as $(x, y)$ approaches $(x_0, y_0)$ is $L$**, denoted

$$\lim_{(x,y) \to (x_0, y_0)} f(x, y) = L,$$

means that given any $\varepsilon > 0$, there exists $\delta > 0$ such that for all $(x, y)$ in $S$, where $(x, y) \neq (x_0, y_0)$, if $(x, y)$ is in the open disk centered at $(x_0, y_0)$ with radius $\delta$, then $|f(x, y) - L| < \varepsilon$.

---

**Note:** While our first limit definition was defined over an open interval, we now define limits over a set $S$ in the plane (where $S$ does not have to be open). As planar sets can be far more complicated than intervals, our definition adds the restriction "... where every open disk centered at $P$ contains points in $S$ other than $P$." In this text, all sets we'll consider will satisfy this condition and we won't bother to check; it is included in the definition for completeness.

Notes:

# Chapter 12  Functions of Several Variables

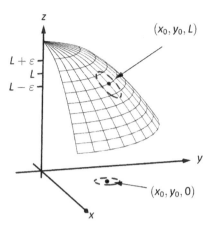

Figure 12.2.3: **Illustrating the definition of a limit.** The open disk in the x-y plane has radius $\delta$. Let $(x, y)$ be any point in this disk; $f(x, y)$ is within $\varepsilon$ of $L$.

The concept behind Definition 12.2.2 is sketched in Figure 12.2.3. Given $\varepsilon > 0$, find $\delta > 0$ such that if $(x, y)$ is any point in the open disk centered at $(x_0, y_0)$ in the x-y plane with radius $\delta$, then $f(x, y)$ should be within $\varepsilon$ of $L$.

Computing limits using this definition is rather cumbersome. The following theorem allows us to evaluate limits much more easily.

---

**Theorem 12.2.1**  **Basic Limit Properties of Functions of Two Variables**

Let $b$, $x_0$, $y_0$, $L$ and $K$ be real numbers, let $n$ be a positive integer, and let $f$ and $g$ be functions with the following limits:

$$\lim_{(x,y)\to(x_0,y_0)} f(x,y) = L \quad \text{and} \quad \lim_{(x,y)\to(x_0,y_0)} g(x,y) = K.$$

The following limits hold.

1. Constants: $\displaystyle\lim_{(x,y)\to(x_0,y_0)} b = b$

2. Identity: $\displaystyle\lim_{(x,y)\to(x_0,y_0)} x = x_0;\quad \lim_{(x,y)\to(x_0,y_0)} y = y_0$

3. Sums/Differences: $\displaystyle\lim_{(x,y)\to(x_0,y_0)} (f(x,y) \pm g(x,y)) = L \pm K$

4. Scalar Multiples: $\displaystyle\lim_{(x,y)\to(x_0,y_0)} b \cdot f(x,y) = bL$

5. Products: $\displaystyle\lim_{(x,y)\to(x_0,y_0)} f(x,y) \cdot g(x,y) = LK$

6. Quotients: $\displaystyle\lim_{(x,y)\to(x_0,y_0)} f(x,y)/g(x,y) = L/K,\ (K \neq 0)$

7. Powers: $\displaystyle\lim_{(x,y)\to(x_0,y_0)} f(x,y)^n = L^n$

---

This theorem, combined with Theorems 1.3.2 and 1.3.3 of Section 1.3, allows us to evaluate many limits.

**Example 12.2.3  Evaluating a limit**
Evaluate the following limits:

1. $\displaystyle\lim_{(x,y)\to(1,\pi)} \left(\frac{y}{x} + \cos(xy)\right)$
2. $\displaystyle\lim_{(x,y)\to(0,0)} \frac{3xy}{x^2+y^2}$

---

Notes:

**SOLUTION**

1. The aforementioned theorems allow us to simply evaluate $y/x + \cos(xy)$ when $x = 1$ and $y = \pi$. If an indeterminate form is returned, we must do more work to evaluate the limit; otherwise, the result is the limit. Therefore

$$\lim_{(x,y)\to(1,\pi)} \left(\frac{y}{x} + \cos(xy)\right) = \frac{\pi}{1} + \cos \pi$$
$$= \pi - 1.$$

2. We attempt to evaluate the limit by substituting 0 in for $x$ and $y$, but the result is the indeterminate form "0/0." To evaluate this limit, we must "do more work," but we have not yet learned what "kind" of work to do. Therefore we cannot yet evaluate this limit.

When dealing with functions of a single variable we also considered one–sided limits and stated

$$\lim_{x \to c} f(x) = L \quad \text{if, and only if,} \quad \lim_{x \to c^+} f(x) = L \quad \text{and} \quad \lim_{x \to c^-} f(x) = L.$$

That is, the limit is $L$ if and only if $f(x)$ approaches $L$ when $x$ approaches $c$ from **either** direction, the left or the right.

In the plane, there are infinitely many directions from which $(x, y)$ might approach $(x_0, y_0)$. In fact, we do not have to restrict ourselves to approaching $(x_0, y_0)$ from a particular direction, but rather we can approach that point along a path that is not a straight line. It is possible to arrive at different limiting values by approaching $(x_0, y_0)$ along different paths. If this happens, we say that $\lim_{(x,y)\to(x_0,y_0)} f(x, y)$ does not exist (this is analogous to the left and right hand limits of single variable functions not being equal).

Our theorems tell us that we can evaluate most limits quite simply, without worrying about paths. When indeterminate forms arise, the limit may or may not exist. If it does exist, it can be difficult to prove this as we need to show the same limiting value is obtained regardless of the path chosen. The case where the limit does not exist is often easier to deal with, for we can often pick two paths along which the limit is different.

**Example 12.2.4  Showing limits do not exist**

1. Show $\lim_{(x,y)\to(0,0)} \dfrac{3xy}{x^2 + y^2}$ does not exist by finding the limits along the lines $y = mx$.

Notes:

2. Show $\lim\limits_{(x,y)\to(0,0)} \dfrac{\sin(xy)}{x+y}$ does not exist by finding the limit along the path $y = -\sin x$.

**SOLUTION**

1. Evaluating $\lim\limits_{(x,y)\to(0,0)} \dfrac{3xy}{x^2+y^2}$ along the lines $y = mx$ means replace all $y$'s with $mx$ and evaluating the resulting limit:

$$\lim_{(x,mx)\to(0,0)} \frac{3x(mx)}{x^2+(mx)^2} = \lim_{x\to 0} \frac{3mx^2}{x^2(m^2+1)}$$
$$= \lim_{x\to 0} \frac{3m}{m^2+1}$$
$$= \frac{3m}{m^2+1}.$$

While the limit exists for each choice of $m$, we get a *different* limit for each choice of $m$. That is, along different lines we get differing limiting values, meaning *the* limit does not exist.

2. Let $f(x,y) = \frac{\sin(xy)}{x+y}$. We are to show that $\lim\limits_{(x,y)\to(0,0)} f(x,y)$ does not exist by finding the limit along the path $y = -\sin x$. First, however, consider the limits found along the lines $y = mx$ as done above.

$$\lim_{(x,mx)\to(0,0)} \frac{\sin(x(mx))}{x+mx} = \lim_{x\to 0} \frac{\sin(mx^2)}{x(m+1)}$$
$$= \lim_{x\to 0} \frac{\sin(mx^2)}{x} \cdot \frac{1}{m+1}.$$

By applying L'Hôpital's Rule, we can show this limit is 0 *except* when $m = -1$, that is, along the line $y = -x$. This line is not in the domain of $f$, so we have found the following fact: along every line $y = mx$ in the domain of $f$, $\lim\limits_{(x,y)\to(0,0)} f(x,y) = 0$.

Now consider the limit along the path $y = -\sin x$:

$$\lim_{(x,-\sin x)\to(0,0)} \frac{\sin(-x\sin x)}{x-\sin x} = \lim_{x\to 0} \frac{\sin(-x\sin x)}{x-\sin x}$$

Now apply L'Hôpital's Rule twice:

$$= \lim_{x\to 0} \frac{\cos(-x\sin x)(-\sin x - x\cos x)}{1-\cos x} \quad (\text{``}=0/0\text{''})$$
$$= \lim_{x\to 0} \frac{-\sin(-x\sin x)(-\sin x - x\cos x)^2 + \cos(-x\sin x)(-2\cos x + x\sin x)}{\sin x}$$
$$= \text{``}-2/0\text{''} \Rightarrow \text{ the limit does not exist.}$$

---

Notes:

Step back and consider what we have just discovered. Along any line $y = mx$ in the domain of the $f(x, y)$, the limit is 0. However, along the path $y = -\sin x$, which lies in the domain of $f(x, y)$ for all $x \neq 0$, the limit does not exist. Since the limit is not the same along every path to $(0, 0)$, we say

$$\lim_{(x,y) \to (0,0)} \frac{\sin(xy)}{x + y} \text{ does not exist.}$$

**Example 12.2.5  Finding a limit**

Let $f(x, y) = \dfrac{5x^2 y^2}{x^2 + y^2}$. Find $\lim\limits_{(x,y) \to (0,0)} f(x, y)$.

**Solution**  It is relatively easy to show that along any line $y = mx$, the limit is 0. This is not enough to prove that the limit exists, as demonstrated in the previous example, but it tells us that if the limit does exist then it must be 0.

To prove the limit is 0, we apply Definition 12.2.2. Let $\varepsilon > 0$ be given. We want to find $\delta > 0$ such that if $\sqrt{(x-0)^2 + (y-0)^2} < \delta$, then $|f(x, y) - 0| < \varepsilon$.

Set $\delta < \sqrt{\varepsilon/5}$. Note that $\left|\dfrac{5y^2}{x^2 + y^2}\right| < 5$ for all $(x, y) \neq (0, 0)$, and that if $\sqrt{x^2 + y^2} < \delta$, then $x^2 < \delta^2$.

Let $\sqrt{(x-0)^2 + (y-0)^2} = \sqrt{x^2 + y^2} < \delta$. Consider $|f(x, y) - 0|$:

$$\begin{aligned}
|f(x, y) - 0| &= \left|\frac{5x^2 y^2}{x^2 + y^2} - 0\right| \\
&= \left|x^2 \cdot \frac{5y^2}{x^2 + y^2}\right| \\
&< \delta^2 \cdot 5 \\
&< \frac{\varepsilon}{5} \cdot 5 \\
&= \varepsilon.
\end{aligned}$$

Thus if $\sqrt{(x-0)^2 + (y-0)^2} < \delta$ then $|f(x, y) - 0| < \varepsilon$, which is what we wanted to show. Thus $\lim\limits_{(x,y) \to (0,0)} \dfrac{5x^2 y^2}{x^2 + y^2} = 0$.

## Continuity

Definition 1.5.1 defines what it means for a function of one variable to be continuous. In brief, it meant that the graph of the function did not have breaks, holes, jumps, etc. We define continuity for functions of two variables in a similar way as we did for functions of one variable.

Notes:

## Chapter 12 Functions of Several Variables

> **Definition 12.2.3**    **Continuous**
>
> Let a function $f(x,y)$ be defined on a set $S$ containing the point $(x_0, y_0)$.
>
> 1. $f$ is **continuous** at $(x_0, y_0)$ if $\lim\limits_{(x,y) \to (x_0,y_0)} f(x,y) = f(x_0, y_0)$.
>
> 2. $f$ is **continuous on** $S$ if $f$ is continuous at all points in $S$. If $f$ is continuous at all points in $\mathbb{R}^2$, we say that $f$ is **continuous everywhere**.

**Example 12.2.6**    **Continuity of a function of two variables**

Let $f(x,y) = \begin{cases} \frac{\cos y \sin x}{x} & x \neq 0 \\ \cos y & x = 0 \end{cases}$. Is $f$ continuous at $(0,0)$? Is $f$ continuous everywhere?

**Solution**    To determine if $f$ is continuous at $(0,0)$, we need to compare $\lim\limits_{(x,y) \to (0,0)} f(x,y)$ to $f(0,0)$.

Applying the definition of $f$, we see that $f(0,0) = \cos 0 = 1$.

We now consider the limit $\lim\limits_{(x,y) \to (0,0)} f(x,y)$. Substituting 0 for $x$ and $y$ in $(\cos y \sin x)/x$ returns the indeterminate form "0/0", so we need to do more work to evaluate this limit.

Consider two related limits: $\lim\limits_{(x,y) \to (0,0)} \cos y$ and $\lim\limits_{(x,y) \to (0,0)} \frac{\sin x}{x}$. The first limit does not contain $x$, and since $\cos y$ is continuous,

$$\lim_{(x,y) \to (0,0)} \cos y = \lim_{y \to 0} \cos y = \cos 0 = 1.$$

The second limit does not contain $y$. By Theorem 1.3.5 we can say

$$\lim_{(x,y) \to (0,0)} \frac{\sin x}{x} = \lim_{x \to 0} \frac{\sin x}{x} = 1.$$

Finally, Theorem 12.2.1 of this section states that we can combine these two limits as follows:

$$\lim_{(x,y) \to (0,0)} \frac{\cos y \sin x}{x} = \lim_{(x,y) \to (0,0)} (\cos y)\left(\frac{\sin x}{x}\right)$$

$$= \left(\lim_{(x,y) \to (0,0)} \cos y\right)\left(\lim_{(x,y) \to (0,0)} \frac{\sin x}{x}\right)$$

$$= (1)(1)$$

$$= 1.$$

Notes:

## 12.2 Limits and Continuity of Multivariable Functions

We have found that $\lim\limits_{(x,y)\to(0,0)} \dfrac{\cos y \sin x}{x} = f(0,0)$, so $f$ is continuous at $(0,0)$.

A similar analysis shows that $f$ is continuous at all points in $\mathbb{R}^2$. As long as $x \neq 0$, we can evaluate the limit directly; when $x = 0$, a similar analysis shows that the limit is $\cos y$. Thus we can say that $f$ is continuous everywhere. A graph of $f$ is given in Figure 12.2.4. Notice how it has no breaks, jumps, etc.

The following theorem is very similar to Theorem 1.5.1, giving us ways to combine continuous functions to create other continuous functions.

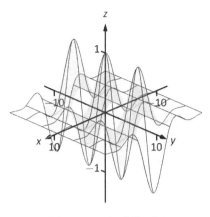

Figure 12.2.4: A graph of $f(x,y)$ in Example 12.2.6.

---

**Theorem 12.2.2**     **Properties of Continuous Functions**

Let $f$ and $g$ be continuous on a set $S$, let $c$ be a real number, and let $n$ be a positive integer. The following functions are continuous on $S$.

1. Sums/Differences:    $f \pm g$
2. Constant Multiples:    $c \cdot f$
3. Products:           $f \cdot g$
4. Quotients:           $f/g$    (as longs as $g \neq 0$ on $S$)
5. Powers:            $f^n$
6. Roots:             $\sqrt[n]{f}$    (if $n$ is even then $f \geq 0$ on $S$; if $n$ is odd, then true for all values of $f$ on $S$.)
7. Compositions:    Adjust the definitions of $f$ and $g$ to: Let $f$ be continuous on $S$, where the range of $f$ on $S$ is $J$, and let $g$ be a single variable function that is continuous on $J$. Then $g \circ f$, i.e., $g(f(x,y))$, is continuous on $S$.

---

**Example 12.2.7**    **Establishing continuity of a function**

Let $f(x,y) = \sin(x^2 \cos y)$. Show $f$ is continuous everywhere.

**Solution**    We will apply both Theorems 1.5.1 and 12.2.2. Let $f_1(x,y) = x^2$. Since $y$ is not actually used in the function, and polynomials are continuous (by Theorem 1.5.1), we conclude $f_1$ is continuous everywhere. A similar statement can be made about $f_2(x,y) = \cos y$. Part 3 of Theorem 12.2.2 states that $f_3 = f_1 \cdot f_2$ is continuous everywhere, and Part 7 of the theorem states the composition of sine with $f_3$ is continuous: that is, $\sin(f_3) = \sin(x^2 \cos y)$ is continuous everywhere.

Notes:

## Functions of Three Variables

The definitions and theorems given in this section can be extended in a natural way to definitions and theorems about functions of three (or more) variables. We cover the key concepts here; some terms from Definitions 12.2.1 and 12.2.3 are not redefined but their analogous meanings should be clear to the reader.

---

**Definition 12.2.4**     **Open Balls, Limit, Continuous**

1. An **open ball** in $\mathbb{R}^3$ centered at $(x_0, y_0, z_0)$ with radius $r$ is the set of all points $(x, y, z)$ such that $\sqrt{(x-x_0)^2 + (y-y_0)^2 + (z-z_0)^2} = r$.

2. Let $D$ be a set in $\mathbb{R}^3$ containing $(x_0, y_0, z_0)$ where every open ball centered at $(x_0, y_0, z_0)$ contains points of $D$ other than $(x_0, y_0, z_0)$, and let $f(x, y, z)$ be a function of three variables defined on $D$, except possibly at $(x_0, y_0, z_0)$. The **limit** of $f(x, y, z)$ as $(x, y, z)$ approaches $(x_0, y_0, z_0)$ is $L$, denoted
$$\lim_{(x,y,z) \to (x_0,y_0,z_0)} f(x, y, z) = L,$$
means that given any $\varepsilon > 0$, there is a $\delta > 0$ such that for all $(x, y, z)$ in $D$, $(x, y, z) \neq (x_0, y_0, z_0)$, if $(x, y, z)$ is in the open ball centered at $(x_0, y_0, z_0)$ with radius $\delta$, then $|f(x, y, z) - L| < \varepsilon$.

3. Let $f(x, y, z)$ be defined on a set $D$ containing $(x_0, y_0, z_0)$. $f$ is **continuous** at $(x_0, y_0, z_0)$ if $\lim_{(x,y,z) \to (x_0,y_0,z_0)} f(x, y, z) = f(x_0, y_0, z_0)$; if $f$ is continuous at all points in $D$, we say $f$ is **continuous on $D$**.

---

These definitions can also be extended naturally to apply to functions of four or more variables. Theorem 12.2.2 also applies to function of three or more variables, allowing us to say that the function
$$f(x, y, z) = \frac{e^{x^2+y}\sqrt{y^2 + z^2 + 3}}{\sin(xyz) + 5}$$
is continuous everywhere.

When considering single variable functions, we studied limits, then continuity, then the derivative. In our current study of multivariable functions, we have studied limits and continuity. In the next section we study derivation, which takes on a slight twist as we are in a multivarible context.

---

Notes:

# Exercises 12.2

## Terms and Concepts

1. Describe in your own words the difference between boundary and interior points of a set.

2. Use your own words to describe (informally) what $\lim_{(x,y)\to(1,2)} f(x,y) = 17$ means.

3. Give an example of a closed, bounded set.

4. Give an example of a closed, unbounded set.

5. Give an example of a open, bounded set.

6. Give an example of a open, unbounded set.

## Problems

In Exercises 7 – 10, a set $S$ is given.

(a) Give one boundary point and one interior point, when possible, of $S$.

(b) State whether $S$ is open, closed, or neither.

(c) State whether $S$ is bounded or unbounded.

7. $S = \left\{ (x,y) \,\bigg|\, \dfrac{(x-1)^2}{4} + \dfrac{(y-3)^2}{9} \leq 1 \right\}$

8. $S = \{(x,y) \mid y \neq x^2\}$

9. $S = \{(x,y) \mid x^2 + y^2 = 1\}$

10. $S = \{(x,y) \mid y > \sin x\}$

In Exercises 11 – 14:

(a) Find the domain $D$ of the given function.

(b) State whether $D$ is an open or closed set.

(c) State whether $D$ is bounded or unbounded.

11. $f(x,y) = \sqrt{9 - x^2 - y^2}$

12. $f(x,y) = \sqrt{y - x^2}$

13. $f(x,y) = \dfrac{1}{\sqrt{y - x^2}}$

14. $f(x,y) = \dfrac{x^2 - y^2}{x^2 + y^2}$

In Exercises 15 – 20, a limit is given. Evaluate the limit along the paths given, then state why these results show the given limit does not exist.

15. $\lim\limits_{(x,y)\to(0,0)} \dfrac{x^2 - y^2}{x^2 + y^2}$

   (a) Along the path $y = 0$.

   (b) Along the path $x = 0$.

16. $\lim\limits_{(x,y)\to(0,0)} \dfrac{x + y}{x - y}$

   (a) Along the path $y = mx$.

17. $\lim\limits_{(x,y)\to(0,0)} \dfrac{xy - y^2}{y^2 + x}$

   (a) Along the path $y = mx$.

   (b) Along the path $x = 0$.

18. $\lim\limits_{(x,y)\to(0,0)} \dfrac{\sin(x^2)}{y}$

   (a) Along the path $y = mx$.

   (b) Along the path $y = x^2$.

19. $\lim\limits_{(x,y)\to(1,2)} \dfrac{x + y - 3}{x^2 - 1}$

   (a) Along the path $y = 2$.

   (b) Along the path $y = x + 1$.

20. $\lim\limits_{(x,y)\to(\pi,\pi/2)} \dfrac{\sin x}{\cos y}$

   (a) Along the path $x = \pi$.

   (b) Along the path $y = x - \pi/2$.

## 12.3 Partial Derivatives

Let $y$ be a function of $x$. We have studied in great detail the derivative of $y$ with respect to $x$, that is, $\frac{dy}{dx}$, which measures the rate at which $y$ changes with respect to $x$. Consider now $z = f(x, y)$. It makes sense to want to know how $z$ changes with respect to $x$ and/or $y$. This section begins our investigation into these rates of change.

Consider the function $z = f(x, y) = x^2 + 2y^2$, as graphed in Figure 12.3.1(a). By fixing $y = 2$, we focus our attention to all points on the surface where the $y$-value is 2, shown in both parts (a) and (b) of the figure. These points form a curve in space: $z = f(x, 2) = x^2 + 8$ which is a function of just one variable. We can take the derivative of $z$ with respect to $x$ along this curve and find equations of tangent lines, etc.

The key notion to extract from this example is: by treating $y$ as constant (it does not vary) we can consider how $z$ changes with respect to $x$. In a similar fashion, we can hold $x$ constant and consider how $z$ changes with respect to $y$. This is the underlying principle of **partial derivatives**. We state the formal, limit–based definition first, then show how to compute these partial derivatives without directly taking limits.

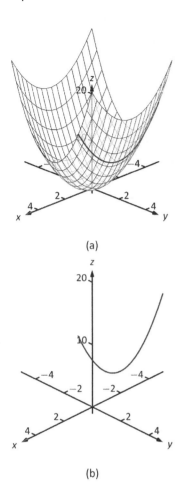

(a)

(b)

Figure 12.3.1: By fixing $y = 2$, the surface $f(x, y) = x^2 + 2y^2$ is a curve in space.

Alternate notations for $f_x(x, y)$ include:

$$\frac{\partial}{\partial x} f(x, y), \quad \frac{\partial f}{\partial x}, \quad \frac{\partial z}{\partial x}, \quad \text{and } z_x,$$

with similar notations for $f_y(x, y)$. For ease of notation, $f_x(x, y)$ is often abbreviated $f_x$.

---

**Definition 12.3.1  Partial Derivative**

Let $z = f(x, y)$ be a continuous function on a set $S$ in $\mathbb{R}^2$.

1. The **partial derivative of $f$ with respect to $x$** is:

$$f_x(x, y) = \lim_{h \to 0} \frac{f(x + h, y) - f(x, y)}{h}.$$

2. The **partial derivative of $f$ with respect to $y$** is:

$$f_y(x, y) = \lim_{h \to 0} \frac{f(x, y + h) - f(x, y)}{h}.$$

---

**Example 12.3.1  Computing partial derivatives with the limit definition**
Let $f(x, y) = x^2 y + 2x + y^3$. Find $f_x(x, y)$ using the limit definition.

Notes:

**SOLUTION** Using Definition 12.3.1, we have:

$$f_x(x,y) = \lim_{h \to 0} \frac{f(x+h,y) - f(x,y)}{h}$$

$$= \lim_{h \to 0} \frac{(x+h)^2 y + 2(x+h) + y^3 - (x^2 y + 2x + y^3)}{h}$$

$$= \lim_{h \to 0} \frac{x^2 y + 2xhy + h^2 y + 2x + 2h + y^3 - (x^2 y + 2x + y^3)}{h}$$

$$= \lim_{h \to 0} \frac{2xhy + h^2 y + 2h}{h}$$

$$= \lim_{h \to 0} 2xy + hy + 2$$

$$= 2xy + 2.$$

We have found $f_x(x,y) = 2xy + 2$.

Example 12.3.1 found a partial derivative using the formal, limit–based definition. Using limits is not necessary, though, as we can rely on our previous knowledge of derivatives to compute partial derivatives easily. When computing $f_x(x,y)$, we hold $y$ fixed – it does not vary. Therefore we can compute the derivative with respect to $x$ by treating $y$ as a constant or coefficient.

Just as $\frac{d}{dx}(5x^2) = 10x$, we compute $\frac{\partial}{\partial x}(x^2 y) = 2xy$. Here we are treating $y$ as a coefficient.

Just as $\frac{d}{dx}(5^3) = 0$, we compute $\frac{\partial}{\partial x}(y^3) = 0$. Here we are treating $y$ as a constant. More examples will help make this clear.

**Example 12.3.2    Finding partial derivatives**
Find $f_x(x,y)$ and $f_y(x,y)$ in each of the following.

1. $f(x,y) = x^3 y^2 + 5y^2 - x + 7$

2. $f(x,y) = \cos(xy^2) + \sin x$

3. $f(x,y) = e^{x^2 y^3}\sqrt{x^2 + 1}$

**SOLUTION**

1. We have $f(x,y) = x^3 y^2 + 5y^2 - x + 7$.
   Begin with $f_x(x,y)$. Keep $y$ fixed, treating it as a constant or coefficient, as appropriate:
   $$f_x(x,y) = 3x^2 y^2 - 1.$$
   Note how the $5y^2$ and 7 terms go to zero.

Notes:

To compute $f_y(x,y)$, we hold $x$ fixed:
$$f_y(x,y) = 2x^3y + 10y.$$
Note how the $-x$ and 7 terms go to zero.

2. We have $f(x,y) = \cos(xy^2) + \sin x$.
   Begin with $f_x(x,y)$. We need to apply the Chain Rule with the cosine term; $y^2$ is the coefficient of the $x$-term inside the cosine function.
   $$f_x(x,y) = -\sin(xy^2)(y^2) + \cos x = -y^2\sin(xy^2) + \cos x.$$
   To find $f_y(x,y)$, note that $x$ is the coefficient of the $y^2$ term inside of the cosine term; also note that since $x$ is fixed, $\sin x$ is also fixed, and we treat it as a constant.
   $$f_y(x,y) = -\sin(xy^2)(2xy) = -2xy\sin(xy^2).$$

3. We have $f(x,y) = e^{x^2y^3}\sqrt{x^2+1}$.
   Beginning with $f_x(x,y)$, note how we need to apply the Product Rule.
   $$f_x(x,y) = e^{x^2y^3}(2xy^3)\sqrt{x^2+1} + e^{x^2y^3}\frac{1}{2}(x^2+1)^{-1/2}(2x)$$
   $$= 2xy^3 e^{x^2y^3}\sqrt{x^2+1} + \frac{xe^{x^2y^3}}{\sqrt{x^2+1}}.$$
   Note that when finding $f_y(x,y)$ we do not have to apply the Product Rule; since $\sqrt{x^2+1}$ does not contain $y$, we treat it as fixed and hence becomes a coefficient of the $e^{x^2y^3}$ term.
   $$f_y(x,y) = e^{x^2y^3}(3x^2y^2)\sqrt{x^2+1} = 3x^2y^2 e^{x^2y^3}\sqrt{x^2+1}.$$

We have shown *how* to compute a partial derivative, but it may still not be clear what a partial derivative *means*. Given $z = f(x,y)$, $f_x(x,y)$ measures the rate at which $z$ changes as only $x$ varies: $y$ is held constant.

Imagine standing in a rolling meadow, then beginning to walk due east. Depending on your location, you might walk up, sharply down, or perhaps not change elevation at all. This is similar to measuring $z_x$: you are moving only east (in the "$x$"-direction) and not north/south at all. Going back to your original location, imagine now walking due north (in the "$y$"-direction). Perhaps walking due north does not change your elevation at all. This is analogous to $z_y = 0$: $z$ does not change with respect to $y$. We can see that $z_x$ and $z_y$ do not have to be the same, or even similar, as it is easy to imagine circumstances where walking east means you walk downhill, though walking north makes you walk uphill.

Notes:

## 12.3 Partial Derivatives

The following example helps us visualize this more.

**Example 12.3.3  Evaluating partial derivatives**
Let $z = f(x, y) = -x^2 - \frac{1}{2}y^2 + xy + 10$. Find $f_x(2, 1)$ and $f_y(2, 1)$ and interpret their meaning.

**SOLUTION**  We begin by computing $f_x(x, y) = -2x + y$ and $f_y(x, y) = -y + x$. Thus

$$f_x(2, 1) = -3 \quad \text{and} \quad f_y(2, 1) = 1.$$

It is also useful to note that $f(2, 1) = 7.5$. What does each of these numbers mean?

Consider $f_x(2, 1) = -3$, along with Figure 12.3.2(a). If one "stands" on the surface at the point $(2, 1, 7.5)$ and moves parallel to the $x$-axis (i.e., only the $x$-value changes, not the $y$-value), then the instantaneous rate of change is $-3$. Increasing the $x$-value will decrease the $z$-value; decreasing the $x$-value will increase the $z$-value.

Now consider $f_y(2, 1) = 1$, illustrated in Figure 12.3.2(b). Moving along the curve drawn on the surface, i.e., parallel to the $y$-axis and not changing the $x$-values, increases the $z$-value instantaneously at a rate of 1. Increasing the $y$-value by 1 would increase the $z$-value by approximately 1.

Since the magnitude of $f_x$ is greater than the magnitude of $f_y$ at $(2, 1)$, it is "steeper" in the $x$-direction than in the $y$-direction.

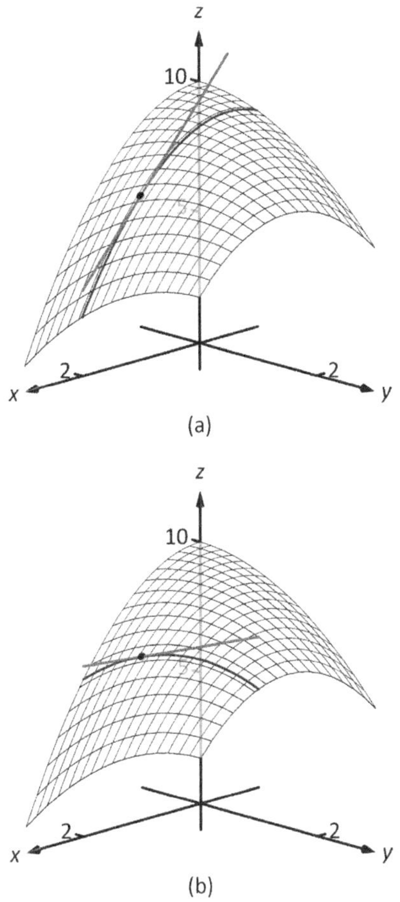

Figure 12.3.2: Illustrating the meaning of partial derivatives.

### Second Partial Derivatives

Let $z = f(x, y)$. We have learned to find the partial derivatives $f_x(x, y)$ and $f_y(x, y)$, which are each functions of $x$ and $y$. Therefore we can take partial derivatives of them, each with respect to $x$ and $y$. We define these "second partials" along with the notation, give examples, then discuss their meaning.

Notes:

Chapter 12  Functions of Several Variables

> **Definition 12.3.2**  **Second Partial Derivative, Mixed Partial Derivative**
>
> Let $z = f(x, y)$ be continuous on a set $S$.
>
> 1. The **second partial derivative of $f$ with respect to $x$ then $x$** is
>
> $$\frac{\partial}{\partial x}\left(\frac{\partial f}{\partial x}\right) = \frac{\partial^2 f}{\partial x^2} = (f_x)_x = f_{xx}$$
>
> 2. The **second partial derivative of $f$ with respect to $x$ then $y$** is
>
> $$\frac{\partial}{\partial y}\left(\frac{\partial f}{\partial x}\right) = \frac{\partial^2 f}{\partial y \partial x} = (f_x)_y = f_{xy}$$
>
> Similar definitions hold for $\dfrac{\partial^2 f}{\partial y^2} = f_{yy}$ and $\dfrac{\partial^2 f}{\partial x \partial y} = f_{yx}$.
>
> The second partial derivatives $f_{xy}$ and $f_{yx}$ are **mixed partial derivatives**.

**Note:** The terms in Definition 12.3.2 all depend on limits, so each definition comes with the caveat "where the limit exists."

The notation of second partial derivatives gives some insight into the notation of the second derivative of a function of a single variable. If $y = f(x)$, then $f''(x) = \dfrac{d^2 y}{dx^2}$. The "$d^2 y$" portion means "take the derivative of $y$ twice," while "$dx^2$" means "with respect to $x$ both times." When we only know of functions of a single variable, this latter phrase seems silly: there is only one variable to take the derivative with respect to. Now that we understand functions of multiple variables, we see the importance of specifying which variables we are referring to.

**Example 12.3.4**  **Second partial derivatives**
For each of the following, find all six first and second partial derivatives. That is, find

$$f_x, \quad f_y, \quad f_{xx}, \quad f_{yy}, \quad f_{xy} \quad \text{and} \quad f_{yx}.$$

1. $f(x, y) = x^3 y^2 + 2xy^3 + \cos x$
2. $f(x, y) = \dfrac{x^3}{y^2}$
3. $f(x, y) = e^x \sin(x^2 y)$

**Solution**  In each, we give $f_x$ and $f_y$ immediately and then spend time deriving the second partial derivatives.

1. $f(x,y) = x^3y^2 + 2xy^3 + \cos x$

   $f_x(x,y) = 3x^2y^2 + 2y^3 - \sin x$

   $f_y(x,y) = 2x^3y + 6xy^2$

   $f_{xx}(x,y) = \dfrac{\partial}{\partial x}(f_x) = \dfrac{\partial}{\partial x}(3x^2y^2 + 2y^3 - \sin x) = 6xy^2 - \cos x$

   $f_{yy}(x,y) = \dfrac{\partial}{\partial y}(f_y) = \dfrac{\partial}{\partial y}(2x^3y + 6xy^2) = 2x^3 + 12xy$

   $f_{xy}(x,y) = \dfrac{\partial}{\partial y}(f_x) = \dfrac{\partial}{\partial y}(3x^2y^2 + 2y^3 - \sin x) = 6x^2y + 6y^2$

   $f_{yx}(x,y) = \dfrac{\partial}{\partial x}(f_y) = \dfrac{\partial}{\partial x}(2x^3y + 6xy^2) = 6x^2y + 6y^2$

2. $f(x,y) = \dfrac{x^3}{y^2} = x^3y^{-2}$

   $f_x(x,y) = \dfrac{3x^2}{y^2}$

   $f_y(x,y) = -\dfrac{2x^3}{y^3}$

   $f_{xx}(x,y) = \dfrac{\partial}{\partial x}(f_x) = \dfrac{\partial}{\partial x}\left(\dfrac{3x^2}{y^2}\right) = \dfrac{6x}{y^2}$

   $f_{yy}(x,y) = \dfrac{\partial}{\partial y}(f_y) = \dfrac{\partial}{\partial y}\left(-\dfrac{2x^3}{y^3}\right) = \dfrac{6x^3}{y^4}$

   $f_{xy}(x,y) = \dfrac{\partial}{\partial y}(f_x) = \dfrac{\partial}{\partial y}\left(\dfrac{3x^2}{y^2}\right) = -\dfrac{6x^2}{y^3}$

   $f_{yx}(x,y) = \dfrac{\partial}{\partial x}(f_y) = \dfrac{\partial}{\partial x}\left(-\dfrac{2x^3}{y^3}\right) = -\dfrac{6x^2}{y^3}$

3. $f(x,y) = e^x \sin(x^2 y)$

   Because the following partial derivatives get rather long, we omit the extra notation and just give the results. In several cases, multiple applications of the Product and Chain Rules will be necessary, followed by some basic combination of like terms.

   $f_x(x,y) = e^x \sin(x^2y) + 2xye^x \cos(x^2y)$

   $f_y(x,y) = x^2 e^x \cos(x^2y)$

   $f_{xx}(x,y) = e^x \sin(x^2y) + 4xye^x \cos(x^2y) + 2ye^x \cos(x^2y) - 4x^2y^2 e^x \sin(x^2y)$

   $f_{yy}(x,y) = -x^4 e^x \sin(x^2y)$

   $f_{xy}(x,y) = x^2 e^x \cos(x^2y) + 2xe^x \cos(x^2y) - 2x^3 ye^x \sin(x^2y)$

   $f_{yx}(x,y) = x^2 e^x \cos(x^2y) + 2xe^x \cos(x^2y) - 2x^3 ye^x \sin(x^2y)$

Notes:

Notice how in each of the three functions in Example 12.3.4, $f_{xy} = f_{yx}$. Due to the complexity of the examples, this likely is not a coincidence. The following theorem states that it is not.

---

**Theorem 12.3.1    Mixed Partial Derivatives**

Let $f$ be defined such that $f_{xy}$ and $f_{yx}$ are continuous on a set $S$. Then for each point $(x, y)$ in $S$, $f_{xy}(x, y) = f_{yx}(x, y)$.

---

Finding $f_{xy}$ and $f_{yx}$ independently and comparing the results provides a convenient way of checking our work.

### Understanding Second Partial Derivatives

Now that we know *how* to find second partials, we investigate *what* they tell us.

Again we refer back to a function $y = f(x)$ of a single variable. The second derivative of $f$ is "the derivative of the derivative," or "the rate of change of the rate of change." The second derivative measures how much the derivative is changing. If $f''(x) < 0$, then the derivative is getting smaller (so the graph of $f$ is concave down); if $f''(x) > 0$, then the derivative is growing, making the graph of $f$ concave up.

Now consider $z = f(x, y)$. Similar statements can be made about $f_{xx}$ and $f_{yy}$ as could be made about $f''(x)$ above. When taking derivatives with respect to $x$ twice, we measure how much $f_x$ changes with respect to $x$. If $f_{xx}(x, y) < 0$, it means that as $x$ increases, $f_x$ decreases, and the graph of $f$ will be concave down *in the x-direction*. Using the analogy of standing in the rolling meadow used earlier in this section, $f_{xx}$ measures whether one's path is concave up/down when walking due east.

Similarly, $f_{yy}$ measures the concavity in the $y$-direction. If $f_{yy}(x, y) > 0$, then $f_y$ is increasing with respect to $y$ and the graph of $f$ will be concave up in the $y$-direction. Appealing to the rolling meadow analogy again, $f_{yy}$ measures whether one's path is concave up/down when walking due north.

We now consider the mixed partials $f_{xy}$ and $f_{yx}$. The mixed partial $f_{xy}$ measures how much $f_x$ changes with respect to $y$. Once again using the rolling meadow analogy, $f_x$ measures the slope if one walks due east. Looking east, begin walking *north* (side–stepping). Is the path towards the east getting steeper? If so, $f_{xy} > 0$. Is the path towards the east not changing in steepness? If so, then $f_{xy} = 0$. A similar thing can be said about $f_{yx}$: consider the steepness of paths heading north while side–stepping to the east.

The following example examines these ideas with concrete numbers and

---

Notes:

graphs.

**Example 12.3.5  Understanding second partial derivatives**
Let $z = x^2 - y^2 + xy$. Evaluate the 6 first and second partial derivatives at $(-1/2, 1/2)$ and interpret what each of these numbers mean.

**SOLUTION**    We find that:
$f_x(x, y) = 2x + y$,  $f_y(x, y) = -2y + x$,  $f_{xx}(x, y) = 2$,  $f_{yy}(x, y) = -2$ and $f_{xy}(x, y) = f_{yx}(x, y) = 1$. Thus at $(-1/2, 1/2)$ we have

$$f_x(-1/2, 1/2) = -1/2, \qquad f_y(-1/2, 1/2) = -3/2.$$

The slope of the tangent line at $(-1/2, 1/2, -1/4)$ in the direction of $x$ is $-1/2$: if one moves from that point parallel to the $x$-axis, the instantaneous rate of change will be $-1/2$. The slope of the tangent line at this point in the direction of $y$ is $-3/2$: if one moves from this point parallel to the $y$-axis, the instantaneous rate of change will be $-3/2$. These tangents lines are graphed in Figure 12.3.3(a) and (b), respectively, where the tangent lines are drawn in a solid line.

Now consider only Figure 12.3.3(a). Three directed tangent lines are drawn (two are dashed), each in the direction of $x$; that is, each has a slope determined by $f_x$. Note how as $y$ increases, the slope of these lines get closer to 0. Since the slopes are all negative, getting closer to 0 means the *slopes are increasing*. The slopes given by $f_x$ are increasing as $y$ increases, meaning $f_{xy}$ must be positive.

Since $f_{xy} = f_{yx}$, we also expect $f_y$ to increase as $x$ increases. Consider Figure 12.3.3(b) where again three directed tangent lines are drawn, this time each in the direction of $y$ with slopes determined by $f_y$. As $x$ increases, the slopes become less steep (closer to 0). Since these are negative slopes, this means the slopes are increasing.

Thus far we have a visual understanding of $f_x$, $f_y$, and $f_{xy} = f_{yx}$. We now interpret $f_{xx}$ and $f_{yy}$. In Figure 12.3.3(a), we see a curve drawn where $x$ is held constant at $x = -1/2$: only $y$ varies. This curve is clearly concave down, corresponding to the fact that $f_{yy} < 0$. In part (b) of the figure, we see a similar curve where $y$ is constant and only $x$ varies. This curve is concave up, corresponding to the fact that $f_{xx} > 0$.

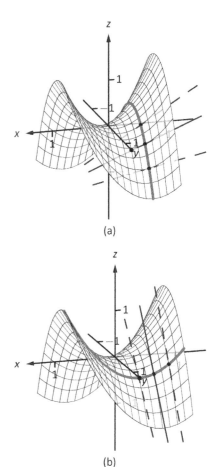

Figure 12.3.3: Understanding the second partial derivatives in Example 12.3.5.

## Partial Derivatives and Functions of Three Variables

The concepts underlying partial derivatives can be easily extend to more than two variables. We give some definitions and examples in the case of three variables and trust the reader can extend these definitions to more variables if needed.

Notes:

> **Definition 12.3.3** **Partial Derivatives with Three Variables**
>
> Let $w = f(x, y, z)$ be a continuous function on a set $D$ in $\mathbb{R}^3$.
> The **partial derivative of $f$ with respect to $x$** is:
>
> $$f_x(x, y, z) = \lim_{h \to 0} \frac{f(x+h, y, z) - f(x, y, z)}{h}.$$
>
> Similar definitions hold for $f_y(x, y, z)$ and $f_z(x, y, z)$.

By taking partial derivatives of partial derivatives, we can find second partial derivatives of $f$ with respect to $z$ then $y$, for instance, just as before.

**Example 12.3.6** **Partial derivatives of functions of three variables**
For each of the following, find $f_x$, $f_y$, $f_z$, $f_{xz}$, $f_{yz}$, and $f_{zz}$.

1. $f(x, y, z) = x^2 y^3 z^4 + x^2 y^2 + x^3 z^3 + y^4 z^4$
2. $f(x, y, z) = x \sin(yz)$

**Solution**

1. $f_x = 2xy^3 z^4 + 2xy^2 + 3x^2 z^3$;  $f_y = 3x^2 y^2 z^4 + 2x^2 y + 4y^3 z^4$;
   $f_z = 4x^2 y^3 z^3 + 3x^3 z^2 + 4y^4 z^3$;  $f_{xz} = 8xy^3 z^3 + 9x^2 z^2$;
   $f_{yz} = 12x^2 y^2 z^3 + 16y^3 z^3$;  $f_{zz} = 12x^2 y^3 z^2 + 6x^3 z + 12y^4 z^2$

2. $f_x = \sin(yz)$;  $f_y = xz\cos(yz)$;  $f_z = xy\cos(yz)$;
   $f_{xz} = y\cos(yz)$;  $f_{yz} = x\cos(yz) - xyz\sin(yz)$;  $f_{zz} = -xy^2 \sin(xy)$

## Higher Order Partial Derivatives

We can continue taking partial derivatives of partial derivatives of partial derivatives of ...; we do not have to stop with second partial derivatives. These higher order partial derivatives do not have a tidy graphical interpretation; nevertheless they are not hard to compute and worthy of some practice.

We do not formally define each higher order derivative, but rather give just a few examples of the notation.

$$f_{xyx}(x, y) = \frac{\partial}{\partial x}\left(\frac{\partial}{\partial y}\left(\frac{\partial f}{\partial x}\right)\right) \quad \text{and}$$

$$f_{xyz}(x, y, z) = \frac{\partial}{\partial z}\left(\frac{\partial}{\partial y}\left(\frac{\partial f}{\partial x}\right)\right).$$

Notes:

## Example 12.3.7   Higher order partial derivatives

1. Let $f(x, y) = x^2y^2 + \sin(xy)$. Find $f_{xxy}$ and $f_{yxx}$.

2. Let $f(x, y, z) = x^3 e^{xy} + \cos(z)$. Find $f_{xyz}$.

**SOLUTION**

1. To find $f_{xxy}$, we first find $f_x$, then $f_{xx}$, then $f_{xxy}$:

$$f_x = 2xy^2 + y\cos(xy) \qquad f_{xx} = 2y^2 - y^2\sin(xy)$$
$$f_{xxy} = 4y - 2y\sin(xy) - xy^2\cos(xy).$$

To find $f_{yxx}$, we first find $f_y$, then $f_{yx}$, then $f_{yxx}$:

$$f_y = 2x^2y + x\cos(xy) \qquad f_{yx} = 4xy + \cos(xy) - xy\sin(xy)$$
$$f_{yxx} = 4y - y\sin(xy) - \big(y\sin(xy) + xy^2\cos(xy)\big)$$
$$= 4y - 2y\sin(xy) - xy^2\cos(xy).$$

Note how $f_{xxy} = f_{yxx}$.

2. To find $f_{xyz}$, we find $f_x$, then $f_{xy}$, then $f_{xyz}$:

$$f_x = 3x^2 e^{xy} + x^3 y e^{xy} \qquad f_{xy} = 3x^3 e^{xy} + x^3 e^{xy} + x^4 y e^{xy} = 4x^3 e^{xy} + x^4 y e^{xy}$$
$$f_{xyz} = 0.$$

In the previous example we saw that $f_{xxy} = f_{yxx}$; this is not a coincidence. While we do not state this as a formal theorem, as long as each partial derivative is continuous, it does not matter in the order in which the partial derivatives are taken. For instance, $f_{xxy} = f_{xyx} = f_{yxx}$.

This can be useful at times. Had we known this, the second part of Example 12.3.7 would have been much simpler to compute. Instead of computing $f_{xyz}$ in the $x$, $y$ then $z$ orders, we could have applied the $z$, then $x$ then $y$ order (as $f_{xyz} = f_{zxy}$). It is easy to see that $f_z = -\sin z$; then $f_{zx}$ and $f_{zxy}$ are clearly 0 as $f_z$ does not contain an $x$ or $y$.

Notes:

A brief review of this section: partial derivatives measure the instantaneous rate of change of a multivariable function with respect to one variable. With $z = f(x,y)$, the partial derivatives $f_x$ and $f_y$ measure the instantaneous rate of change of $z$ when moving parallel to the $x$- and $y$-axes, respectively. How do we measure the rate of change at a point when we do not move parallel to one of these axes? What if we move in the direction given by the vector $\langle 2, 1 \rangle$? Can we measure that rate of change? The answer is, of course, yes, we can. This is the topic of Section 12.6. First, we need to define what it means for a function of two variables to be *differentiable*.

Notes:

# Exercises 12.3

## Terms and Concepts

1. What is the difference between a constant and a coefficient?

2. Given a function $z = f(x, y)$, explain in your own words how to compute $f_x$.

3. In the mixed partial fraction $f_{xy}$, which is computed first, $f_x$ or $f_y$?

4. In the mixed partial fraction $\dfrac{\partial^2 f}{\partial x \partial y}$, which is computed first, $f_x$ or $f_y$?

## Problems

In Exercises 5 – 8, evaluate $f_x(x, y)$ and $f_y(x, y)$ at the indicated point.

5. $f(x, y) = x^2 y - x + 2y + 3$ at $(1, 2)$

6. $f(x, y) = x^3 - 3x + y^2 - 6y$ at $(-1, 3)$

7. $f(x, y) = \sin y \cos x$ at $(\pi/3, \pi/3)$

8. $f(x, y) = \ln(xy)$ at $(-2, -3)$

In Exercises 9 – 26, find $f_x$, $f_y$, $f_{xx}$, $f_{yy}$, $f_{xy}$ and $f_{yx}$.

9. $f(x, y) = x^2 y + 3x^2 + 4y - 5$

10. $f(x, y) = y^3 + 3xy^2 + 3x^2 y + x^3$

11. $f(x, y) = \dfrac{x}{y}$

12. $f(x, y) = \dfrac{4}{xy}$

13. $f(x, y) = e^{x^2 + y^2}$

14. $f(x, y) = e^{x + 2y}$

15. $f(x, y) = \sin x \cos y$

16. $f(x, y) = (x + y)^3$

17. $f(x, y) = \cos(5xy^3)$

18. $f(x, y) = \sin(5x^2 + 2y^3)$

19. $f(x, y) = \sqrt{4xy^2 + 1}$

20. $f(x, y) = (2x + 5y)\sqrt{y}$

21. $f(x, y) = \dfrac{1}{x^2 + y^2 + 1}$

22. $f(x, y) = 5x - 17y$

23. $f(x, y) = 3x^2 + 1$

24. $f(x, y) = \ln(x^2 + y)$

25. $f(x, y) = \dfrac{\ln x}{4y}$

26. $f(x, y) = 5e^x \sin y + 9$

In Exercises 27 – 30, form a function $z = f(x, y)$ such that $f_x$ and $f_y$ match those given.

27. $f_x = \sin y + 1, \quad f_y = x \cos y$

28. $f_x = x + y, \quad f_y = x + y$

29. $f_x = 6xy - 4y^2, \quad f_y = 3x^2 - 8xy + 2$

30. $f_x = \dfrac{2x}{x^2 + y^2}, \quad f_y = \dfrac{2y}{x^2 + y^2}$

In Exercises 31 – 34, find $f_x$, $f_y$, $f_z$, $f_{yz}$ and $f_{zy}$.

31. $f(x, y, z) = x^2 e^{2y - 3z}$

32. $f(x, y, z) = x^3 y^2 + x^3 z + y^2 z$

33. $f(x, y, z) = \dfrac{3x}{7y^2 z}$

34. $f(x, y, z) = \ln(xyz)$

## 12.4 Differentiability and the Total Differential

We studied **differentials** in Section 4.4, where Definition 4.4.1 states that if $y = f(x)$ and $f$ is differentiable, then $dy = f'(x)dx$. One important use of this differential is in Integration by Substitution. Another important application is approximation. Let $\Delta x = dx$ represent a change in $x$. When $dx$ is small, $dy \approx \Delta y$, the change in $y$ resulting from the change in $x$. Fundamental in this understanding is this: as $dx$ gets small, the difference between $\Delta y$ and $dy$ goes to 0. Another way of stating this: as $dx$ goes to 0, the *error* in approximating $\Delta y$ with $dy$ goes to 0.

We extend this idea to functions of two variables. Let $z = f(x, y)$, and let $\Delta x = dx$ and $\Delta y = dy$ represent changes in $x$ and $y$, respectively. Let $\Delta z = f(x + dx, y + dy) - f(x, y)$ be the change in $z$ over the change in $x$ and $y$. Recalling that $f_x$ and $f_y$ give the instantaneous rates of $z$-change in the $x$- and $y$-directions, respectively, we can approximate $\Delta z$ with $dz = f_x dx + f_y dy$; in words, the total change in $z$ is approximately the change caused by changing $x$ plus the change caused by changing $y$. In a moment we give an indication of whether or not this approximation is any good. First we give a name to $dz$.

**Note:** From Definition 12.4.1, we can write
$$dz = \langle f_x, f_y \rangle \cdot \langle dx, dy \rangle.$$
While not explored in this section, the vector $\langle f_x, f_y \rangle$ is seen again in the next section and fully defined in Section 12.6.

---

**Definition 12.4.1**     **Total Differential**

Let $z = f(x, y)$ be continuous on a set $S$. Let $dx$ and $dy$ represent changes in $x$ and $y$, respectively. Where the partial derivatives $f_x$ and $f_y$ exist, the **total differential of** $z$ is
$$dz = f_x(x, y)dx + f_y(x, y)dy.$$

---

**Example 12.4.1**     **Finding the total differential**
Let $z = x^4 e^{3y}$. Find $dz$.

**SOLUTION**     We compute the partial derivatives: $f_x = 4x^3 e^{3y}$ and $f_y = 3x^4 e^{3y}$. Following Definition 12.4.1, we have
$$dz = 4x^3 e^{3y} dx + 3x^4 e^{3y} dy.$$

We *can* approximate $\Delta z$ with $dz$, but as with all approximations, there is error involved. A good approximation is one in which the error is small. At a given point $(x_0, y_0)$, let $E_x$ and $E_y$ be functions of $dx$ and $dy$ such that $E_x dx + E_y dy$ describes this error. Then
$$\Delta z = dz + E_x dx + E_y dy$$
$$= f_x(x_0, y_0)dx + f_y(x_0, y_0)dy + E_x dx + E_y dy.$$

---

Notes:

## 12.4 Differentiability and the Total Differential

If the approximation of $\Delta z$ by $dz$ is good, then as $dx$ and $dy$ get small, so does $E_x dx + E_y dy$. The approximation of $\Delta z$ by $dz$ is even better if, as $dx$ and $dy$ go to 0, so do $E_x$ and $E_y$. This leads us to our definition of differentiability.

---

**Definition 12.4.2    Multivariable Differentiability**

Let $z = f(x, y)$ be defined on a set $S$ containing $(x_0, y_0)$ where $f_x(x_0, y_0)$ and $f_y(x_0, y_0)$ exist. Let $dz$ be the total differential of $z$ at $(x_0, y_0)$, let $\Delta z = f(x_0 + dx, y_0 + dy) - f(x_0, y_0)$, and let $E_x$ and $E_y$ be functions of $dx$ and $dy$ such that
$$\Delta z = dz + E_x dx + E_y dy.$$

1. We say $f$ is **differentiable at** $(x_0, y_0)$ if, given $\varepsilon > 0$, there is a $\delta > 0$ such that if $\| \langle dx, dy \rangle \| < \delta$, then $\| \langle E_x, E_y \rangle \| < \varepsilon$. That is, as $dx$ and $dy$ go to 0, so do $E_x$ and $E_y$.

2. We say $f$ is **differentiable on** $S$ if $f$ is differentiable at every point in $S$. If $f$ is differentiable on $\mathbb{R}^2$, we say that $f$ is **differentiable everywhere**.

---

**Example 12.4.2    Showing a function is differentiable**
Show $f(x, y) = xy + 3y^2$ is differentiable using Definition 12.4.2.

**SOLUTION**    We begin by finding $f(x + dx, y + dy)$, $\Delta z$, $f_x$ and $f_y$.

$$f(x + dx, y + dy) = (x + dx)(y + dy) + 3(y + dy)^2$$
$$= xy + xdy + ydx + dxdy + 3y^2 + 6ydy + 3dy^2.$$

$\Delta z = f(x + dx, y + dy) - f(x, y)$, so
$$\Delta z = xdy + ydx + dxdy + 6ydy + 3dy^2.$$

It is straightforward to compute $f_x = y$ and $f_y = x + 6y$. Consider once more $\Delta z$:

$$\Delta z = xdy + ydx + dxdy + 6ydy + 3dy^2 \quad \text{(now reorder)}$$
$$= ydx + xdy + 6ydy + dxdy + 3dy^2$$
$$= \underbrace{(y)}_{f_x} dx + \underbrace{(x + 6y)}_{f_y} dy + \underbrace{(dy)}_{E_x} dx + \underbrace{(3dy)}_{E_y} dy$$
$$= f_x dx + f_y dy + E_x dx + E_y dy.$$

With $E_x = dy$ and $E_y = 3dy$, it is clear that as $dx$ and $dy$ go to 0, $E_x$ and $E_y$ also go to 0. Since this did not depend on a specific point $(x_0, y_0)$, we can say that $f(x, y)$

---

Notes:

is differentiable for all pairs $(x, y)$ in $\mathbb{R}^2$, or, equivalently, that $f$ is differentiable everywhere.

Our intuitive understanding of differentiability of functions $y = f(x)$ of one variable was that the graph of $f$ was "smooth." A similar intuitive understanding of functions $z = f(x, y)$ of two variables is that the surface defined by $f$ is also "smooth," not containing cusps, edges, breaks, etc. The following theorem states that differentiable functions are continuous, followed by another theorem that provides a more tangible way of determining whether a great number of functions are differentiable or not.

---

**Theorem 12.4.1**     **Continuity and Differentiability of Multivariable Functions**

Let $z = f(x, y)$ be defined on a set $S$ containing $(x_0, y_0)$. If $f$ is differentiable at $(x_0, y_0)$, then $f$ is continuous at $(x_0, y_0)$.

---

**Theorem 12.4.2**     **Differentiability of Multivariable Functions**

Let $z = f(x, y)$ be defined on a set $S$. If $f_x$ and $f_y$ are both continuous on $S$, then $f$ is differentiable on $S$.

---

The theorems assure us that essentially all functions that we see in the course of our studies here are differentiable (and hence continuous) on their natural domains. There is a difference between Definition 12.4.2 and Theorem 12.4.2, though: it is possible for a function $f$ to be differentiable yet $f_x$ and/or $f_y$ is *not* continuous. Such strange behavior of functions is a source of delight for many mathematicians.

When $f_x$ and $f_y$ exist at a point but are not continuous at that point, we need to use other methods to determine whether or not $f$ is differentiable at that point.

For instance, consider the function

$$f(x, y) = \begin{cases} \frac{xy}{x^2+y^2} & (x, y) \neq (0, 0) \\ 0 & (x, y) = (0, 0) \end{cases}$$

---

Notes:

We can find $f_x(0,0)$ and $f_y(0,0)$ using Definition 12.3.1:

$$f_x(0,0) = \lim_{h \to 0} \frac{f(0+h, 0) - f(0,0)}{h}$$
$$= \lim_{h \to 0} \frac{0}{h^2} = 0;$$
$$f_y(0,0) = \lim_{h \to 0} \frac{f(0, 0+h) - f(0,0)}{h}$$
$$= \lim_{h \to 0} \frac{0}{h^2} = 0.$$

Both $f_x$ and $f_y$ *exist* at $(0,0)$, but they are not continuous at $(0,0)$, as

$$f_x(x,y) = \frac{y(y^2 - x^2)}{(x^2+y^2)^2} \quad \text{and} \quad f_y(x,y) = \frac{x(x^2-y^2)}{(x^2+y^2)^2}$$

are not continuous at $(0,0)$. (Take the limit of $f_x$ as $(x,y) \to (0,0)$ along the x- and y-axes; they give different results.) So even though $f_x$ and $f_y$ exist at every point in the x-y plane, they are not continuous. Therefore it is possible, by Theorem 12.4.2, for $f$ to not be differentiable.

Indeed, it is not. One can show that $f$ is not continuous at $(0,0)$ (see Example 12.2.4), and by Theorem 12.4.1, this means $f$ is not differentiable at $(0,0)$.

## Approximating with the Total Differential

By the definition, when $f$ is differentiable $dz$ is a good approximation for $\Delta z$ when $dx$ and $dy$ are small. We give some simple examples of how this is used here.

**Example 12.4.3    Approximating with the total differential**
Let $z = \sqrt{x} \sin y$. Approximate $f(4.1, 0.8)$.

**Solution**    Recognizing that $\pi/4 \approx 0.785 \approx 0.8$, we can approximate $f(4.1, 0.8)$ using $f(4, \pi/4)$. We can easily compute $f(4, \pi/4) = \sqrt{4} \sin(\pi/4) = 2\left(\frac{\sqrt{2}}{2}\right) = \sqrt{2} \approx 1.414$. Without calculus, this is the best approximation we could reasonably come up with. The total differential gives us a way of adjusting this initial approximation to hopefully get a more accurate answer.

We let $\Delta z = f(4.1, 0.8) - f(4, \pi/4)$. The total differential $dz$ is approximately equal to $\Delta z$, so

$$f(4.1, 0.8) - f(4, \pi/4) \approx dz \quad \Rightarrow \quad f(4.1, 0.8) \approx dz + f(4, \pi/4). \quad (12.1)$$

To find $dz$, we need $f_x$ and $f_y$.

Notes:

$$f_x(x,y) = \frac{\sin y}{2\sqrt{x}} \quad \Rightarrow \quad f_x(4, \pi/4) = \frac{\sin \pi/4}{2\sqrt{4}}$$
$$= \frac{\sqrt{2}/2}{4} = \sqrt{2}/8.$$
$$f_y(x,y) = \sqrt{x}\cos y \quad \Rightarrow \quad f_y(4, \pi/4) = \sqrt{4}\frac{\sqrt{2}}{2}$$
$$= \sqrt{2}.$$

Approximating 4.1 with 4 gives $dx = 0.1$; approximating 0.8 with $\pi/4$ gives $dy \approx 0.015$. Thus

$$dz(4, \pi/4) = f_x(4, \pi/4)(0.1) + f_y(4, \pi/4)(0.015)$$
$$= \frac{\sqrt{2}}{8}(0.1) + \sqrt{2}(0.015)$$
$$\approx 0.039.$$

Returning to Equation (12.1), we have

$$f(4.1, 0.8) \approx 0.039 + 1.414 = 1.4531.$$

We, of course, can compute the actual value of $f(4.1, 0.8)$ with a calculator; the actual value, accurate to 5 places after the decimal, is 1.45254. Obviously our approximation is quite good.

The point of the previous example was *not* to develop an approximation method for known functions. After all, we can very easily compute $f(4.1, 0.8)$ using readily available technology. Rather, it serves to illustrate how well this method of approximation works, and to reinforce the following concept:

"New position = old position + amount of change," so
"New position $\approx$ old position + approximate amount of change."

In the previous example, we could easily compute $f(4, \pi/4)$ and could approximate the amount of z-change when computing $f(4.1, 0.8)$, letting us approximate the new z-value.

It may be surprising to learn that it is not uncommon to know the values of $f$, $f_x$ and $f_y$ at a particular point without actually knowing the function $f$. The total differential gives a good method of approximating $f$ at nearby points.

**Example 12.4.4    Approximating an unknown function**
Given that $f(2, -3) = 6$, $f_x(2, -3) = 1.3$ and $f_y(2, -3) = -0.6$, approximate $f(2.1, -3.03)$.

Notes:

**Solution**    The total differential approximates how much $f$ changes from the point $(2, -3)$ to the point $(2.1, -3.03)$. With $dx = 0.1$ and $dy = -0.03$, we have

$$dz = f_x(2, -3)dx + f_y(2, -3)dy$$
$$= 1.3(0.1) + (-0.6)(-0.03)$$
$$= 0.148.$$

The change in $z$ is approximately 0.148, so we approximate $f(2.1, -3.03) \approx 6.148$.

## Error/Sensitivity Analysis

The total differential gives an approximation of the change in $z$ given small changes in $x$ and $y$. We can use this to approximate error propagation; that is, if the input is a little off from what it should be, how far from correct will the output be? We demonstrate this in an example.

**Example 12.4.5    Sensitivity analysis**
A cylindrical steel storage tank is to be built that is 10ft tall and 4ft across in diameter. It is known that the steel will expand/contract with temperature changes; is the overall volume of the tank more sensitive to changes in the diameter or in the height of the tank?

**Solution**    A cylindrical solid with height $h$ and radius $r$ has volume $V = \pi r^2 h$. We can view $V$ as a function of two variables, $r$ and $h$. We can compute partial derivatives of $V$:

$$\frac{\partial V}{\partial r} = V_r(r, h) = 2\pi rh \quad \text{and} \quad \frac{\partial V}{\partial h} = V_h(r, h) = \pi r^2.$$

The total differential is $dV = (2\pi rh)dr + (\pi r^2)dh$. When $h = 10$ and $r = 2$, we have $dV = 40\pi dr + 4\pi dh$. Note that the coefficient of $dr$ is $40\pi \approx 125.7$; the coefficient of $dh$ is a tenth of that, approximately 12.57. A small change in radius will be multiplied by 125.7, whereas a small change in height will be multiplied by 12.57. Thus the volume of the tank is more sensitive to changes in radius than in height.

The previous example showed that the volume of a particular tank was more sensitive to changes in radius than in height. Keep in mind that this analysis only applies to a tank of those dimensions. A tank with a height of 1ft and radius of 5ft would be more sensitive to changes in height than in radius.

Notes:

One could make a chart of small changes in radius and height and find exact changes in volume given specific changes. While this provides exact numbers, it does not give as much insight as the error analysis using the total differential.

### Differentiability of Functions of Three Variables

The definition of differentiability for functions of three variables is very similar to that of functions of two variables. We again start with the total differential.

---

**Definition 12.4.3**  **Total Differential**

Let $w = f(x,y,z)$ be continuous on a set $D$. Let $dx$, $dy$ and $dz$ represent changes in $x$, $y$ and $z$, respectively. Where the partial derivatives $f_x$, $f_y$ and $f_z$ exist, the **total differential of** $w$ is

$$dw = f_x(x,y,z)dx + f_y(x,y,z)dy + f_z(x,y,z)dz.$$

---

This differential can be a good approximation of the change in $w$ when $w = f(x,y,z)$ is **differentiable**.

---

**Definition 12.4.4**  **Multivariable Differentiability**

Let $w = f(x,y,z)$ be defined on a set $D$ containing $(x_0, y_0, z_0)$ where $f_x(x_0, y_0, z_0)$, $f_y(x_0, y_0, z_0)$ and $f_z(x_0, y_0, z_0)$ exist. Let $dw$ be the total differential of $w$ at $(x_0, y_0, z_0)$, let $\Delta w = f(x_0 + dx, y_0 + dy, z_0 + dz) - f(x_0, y_0, z_0)$, and let $E_x$, $E_y$ and $E_z$ be functions of $dx$, $dy$ and $dz$ such that

$$\Delta w = dw + E_x dx + E_y dy + E_z dz.$$

1. We say $f$ is **differentiable at** $(x_0, y_0, z_0)$ if, given $\varepsilon > 0$, there is a $\delta > 0$ such that if $||\langle dx, dy, dz\rangle|| < \delta$, then $||\langle E_x, E_y, E_z\rangle|| < \varepsilon$.

2. We say $f$ is **differentiable on** $B$ if $f$ is differentiable at every point in $B$. If $f$ is differentiable on $\mathbb{R}^3$, we say that $f$ is **differentiable everywhere**.

---

Just as before, this definition gives a rigorous statement about what it means to be differentiable that is not very intuitive. We follow it with a theorem similar to Theorem 12.4.2.

Notes:

> **Theorem 12.4.3** **Continuity and Differentiability of Functions of Three Variables**
>
> Let $w = f(x, y, z)$ be defined on a set $D$ containing $(x_0, y_0, z_0)$.
>
> 1. If $f$ is differentiable at $(x_0, y_0, z_0)$, then $f$ is continuous at $(x_0, y_0, z_0)$.
>
> 2. If $f_x$, $f_y$ and $f_z$ are continuous on $B$, then $f$ is differentiable on $B$.

This set of definition and theorem extends to functions of any number of variables. The theorem again gives us a simple way of verifying that most functions that we encounter are differentiable on their natural domains.

This section has given us a formal definition of what it means for a functions to be "differentiable," along with a theorem that gives a more accessible understanding. The following sections return to notions prompted by our study of partial derivatives that make use of the fact that most functions we encounter are differentiable.

Notes:

# Exercises 12.4

## Terms and Concepts

1. T/F: If $f(x,y)$ is differentiable on $S$, the $f$ is continuous on $S$.

2. T/F: If $f_x$ and $f_y$ are continuous on $S$, then $f$ is differentiable on $S$.

3. T/F: If $z = f(x,y)$ is differentiable, then the change in $z$ over small changes $dx$ and $dy$ in $x$ and $y$ is approximately $dz$.

4. Finish the sentence: "The new $z$-value is approximately the old $z$-value plus the approximate _____."

## Problems

**In Exercises 5 – 8, find the total differential $dz$.**

5. $z = x\sin y + x^2$

6. $z = (2x^2 + 3y)^2$

7. $z = 5x - 7y$

8. $z = xe^{x+y}$

**In Exercises 9 – 12, a function $z = f(x,y)$ is given. Give the indicated approximation using the total differential.**

9. $f(x,y) = \sqrt{x^2+y}$. Approximate $f(2.95, 7.1)$ knowing $f(3,7) = 4$.

10. $f(x,y) = \sin x \cos y$. Approximate $f(0.1, -0.1)$ knowing $f(0,0) = 0$.

11. $f(x,y) = x^2y - xy^2$. Approximate $f(2.04, 3.06)$ knowing $f(2,3) = -6$.

12. $f(x,y) = \ln(x-y)$. Approximate $f(5.1, 3.98)$ knowing $f(5,4) = 0$.

**Exercises 13 – 16 ask a variety of questions dealing with approximating error and sensitivity analysis.**

13. A cylindrical storage tank is to be 2ft tall with a radius of 1ft. Is the volume of the tank more sensitive to changes in the radius or the height?

14. **Projectile Motion:** The $x$-value of an object moving under the principles of projectile motion is $x(\theta, v_0, t) = (v_0 \cos\theta)t$. A particular projectile is fired with an initial velocity of $v_0 = 250$ft/s and an angle of elevation of $\theta = 60°$. It travels a distance of 375ft in 3 seconds.

Is the projectile more sensitive to errors in initial speed or angle of elevation?

15. The length $\ell$ of a long wall is to be approximated. The angle $\theta$, as shown in the diagram (not to scale), is measured to be $85°$, and the distance $x$ is measured to be 30'. Assume that the triangle formed is a right triangle.

Is the measurement of the length of $\ell$ more sensitive to errors in the measurement of $x$ or in $\theta$?

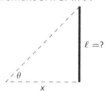

16. It is "common sense" that it is far better to measure a long distance with a long measuring tape rather than a short one. A measured distance $D$ can be viewed as the product of the length $\ell$ of a measuring tape times the number $n$ of times it was used. For instance, using a 3' tape 10 times gives a length of 30'. To measure the same distance with a 12' tape, we would use the tape 2.5 times. (I.e., $30 = 12 \times 2.5$.) Thus $D = n\ell$.

Suppose each time a measurement is taken with the tape, the recorded distance is within 1/16" of the actual distance. (I.e., $d\ell = 1/16'' \approx 0.005$ft). Using differentials, show why common sense proves correct in that it is better to use a long tape to measure long distances.

**In Exercises 17 – 18, find the total differential $dw$.**

17. $w = x^2yz^3$

18. $w = e^x \sin y \ln z$

**In Exercises 19 – 22, use the information provided and the total differential to make the given approximation.**

19. $f(3,1) = 7$, $f_x(3,1) = 9$, $f_y(3,1) = -2$. Approximate $f(3.05, 0.9)$.

20. $f(-4,2) = 13$, $f_x(-4,2) = 2.6$, $f_y(-4,2) = 5.1$. Approximate $f(-4.12, 2.07)$.

21. $f(2,4,5) = -1$, $f_x(2,4,5) = 2$, $f_y(2,4,5) = -3$, $f_z(2,4,5) = 3.7$. Approximate $f(2.5, 4.1, 4.8)$.

22. $f(3,3,3) = 5$, $f_x(3,3,3) = 2$, $f_y(3,3,3) = 0$, $f_z(3,3,3) = -2$. Approximate $f(3.1, 3.1, 3.1)$.

## 12.5 The Multivariable Chain Rule

Consider driving an off-road vehicle along a dirt road. As you drive, your elevation likely changes. What factors determine how quickly your elevation rises and falls? After some thought, generally one recognizes that one's *velocity* (speed and direction) and the *terrain* influence your rise and fall.

One can represent the terrain as the surface defined by a multivariable function $z = f(x, y)$; one can represent the path of the off-road vehicle, as seen from above, with a vector–valued function $\vec{r}(t) = \langle x(t), y(t)\rangle$; the velocity of the vehicle is thus $\vec{r}\,'(t) = \langle x'(t), y'(t)\rangle$.

Consider Figure 12.5.1 in which a surface $z = f(x, y)$ is drawn, along with a dashed curve in the *x-y* plane. Restricting $f$ to just the points on this circle gives the curve shown on the surface (i.e., "the path of the off-road vehicle.") The derivative $\frac{df}{dt}$ gives the instantaneous rate of change of $f$ with respect to $t$. If we consider an object traveling along this path, $\frac{df}{dt} = \frac{dz}{dt}$ gives the rate at which the object rises/falls (i.e., "the rate of elevation change" of the vehicle.) Conceptually, the Multivariable Chain Rule combines terrain and velocity information properly to compute this rate of elevation change.

Abstractly, let $z$ be a function of $x$ and $y$; that is, $z = f(x, y)$ for some function $f$, and let $x$ and $y$ each be functions of $t$. By choosing a $t$-value, $x$- and $y$-values are determined, which in turn determine $z$: this defines $z$ as a function of $t$. The Multivariable Chain Rule gives a method of computing $\frac{dz}{dt}$.

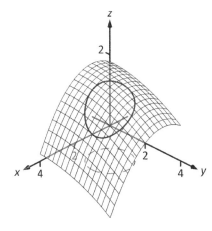

Figure 12.5.1: Understanding the application of the Multivariable Chain Rule.

---

**Theorem 12.5.1**     **Multivariable Chain Rule, Part I**

Let $z = f(x, y)$, $x = g(t)$ and $y = h(t)$, where $f$, $g$ and $h$ are differentiable functions. Then $z = f(x, y) = f\bigl(g(t), h(t)\bigr)$ is a function of $t$, and

$$\frac{dz}{dt} = \frac{df}{dt} = f_x(x,y)\frac{dx}{dt} + f_y(x,y)\frac{dy}{dt}$$
$$= \frac{\partial f}{\partial x}\frac{dx}{dt} + \frac{\partial f}{\partial y}\frac{dy}{dt}$$
$$= \langle f_x, f_y\rangle \cdot \langle x', y'\rangle.$$

---

The Chain Rule of Section 2.5 states that $\frac{d}{dx}\bigl(f(g(x))\bigr) = f'(g(x))g'(x)$. If $t = g(x)$, we can express the Chain Rule as

$$\frac{df}{dx} = \frac{df}{dt}\frac{dt}{dx};$$

recall that the derivative notation is deliberately chosen to reflect their fraction–

Notes:

like properties. A similar effect is seen in Theorem 12.5.1. In the second line of equations, one can think of the $dx$ and $\partial x$ as "sort of" canceling out, and likewise with $dy$ and $\partial y$.

Notice, too, the third line of equations in Theorem 12.5.1. The vector $\langle f_x, f_y \rangle$ contains information about the surface (terrain); the vector $\langle x', y' \rangle$ can represent velocity. In the context measuring the rate of elevation change of the off-road vehicle, the Multivariable Chain Rule states it can be found through a product of terrain and velocity information.

We now practice applying the Multivariable Chain Rule.

**Example 12.5.1  Using the Multivariable Chain Rule**
Let $z = x^2y + x$, where $x = \sin t$ and $y = e^{5t}$. Find $\dfrac{dz}{dt}$ using the Chain Rule.

**SOLUTION**  Following Theorem 12.5.1, we find

$$f_x(x,y) = 2xy + 1, \quad f_y(x,y) = x^2, \quad \frac{dx}{dt} = \cos t, \quad \frac{dy}{dt} = 5e^{5t}.$$

Applying the theorem, we have

$$\frac{dz}{dt} = (2xy + 1)\cos t + 5x^2 e^{5t}.$$

This may look odd, as it seems that $\frac{dz}{dt}$ is a function of $x$, $y$ and $t$. Since $x$ and $y$ are functions of $t$, $\frac{dz}{dt}$ is really just a function of $t$, and we can replace $x$ with $\sin t$ and $y$ with $e^{5t}$:

$$\frac{dz}{dt} = (2xy + 1)\cos t + 5x^2 e^{5t} = (2\sin(t)e^{5t} + 1)\cos t + 5e^{5t}\sin^2 t.$$

The previous example can make us wonder: if we substituted for $x$ and $y$ at the end to show that $\frac{dz}{dt}$ is really just a function of $t$, why not substitute before differentiating, showing clearly that $z$ is a function of $t$?

That is, $z = x^2y + x = (\sin t)^2 e^{5t} + \sin t$. Applying the Chain and Product Rules, we have

$$\frac{dz}{dt} = 2\sin t \cos t\, e^{5t} + 5\sin^2 t\, e^{5t} + \cos t,$$

which matches the result from the example.

This may now make one wonder "What's the point? If we could already find the derivative, why learn another way of finding it?" In some cases, applying this rule makes deriving simpler, but this is hardly the power of the Chain Rule. Rather, in the case where $z = f(x,y)$, $x = g(t)$ and $y = h(t)$, the Chain Rule is

Notes:

extremely powerful when *we do not know what f, g and/or h are*. It may be hard to believe, but often in "the real world" we know rate–of–change information (i.e., information about derivatives) without explicitly knowing the underlying functions. The Chain Rule allows us to combine several rates of change to find another rate of change. The Chain Rule also has theoretic use, giving us insight into the behavior of certain constructions (as we'll see in the next section).

We demonstrate this in the next example.

**Example 12.5.2    Applying the Multivarible Chain Rule**
An object travels along a path on a surface. The exact path and surface are not known, but at time $t = t_0$ it is known that:

$$\frac{\partial z}{\partial x} = 5, \quad \frac{\partial z}{\partial y} = -2, \quad \frac{dx}{dt} = 3 \quad \text{and} \quad \frac{dy}{dt} = 7.$$

Find $\frac{dz}{dt}$ at time $t_0$.

**SOLUTION**    The Multivariable Chain Rule states that

$$\frac{dz}{dt} = \frac{\partial z}{\partial x}\frac{dx}{dt} + \frac{\partial z}{\partial y}\frac{dy}{dt}$$
$$= 5(3) + (-2)(7)$$
$$= 1.$$

By knowing certain rates–of–change information about the surface and about the path of the particle in the *x-y* plane, we can determine how quickly the object is rising/falling.

We next apply the Chain Rule to solve a max/min problem.

**Example 12.5.3    Applying the Multivariable Chain Rule**
Consider the surface $z = x^2 + y^2 - xy$, a paraboloid, on which a particle moves with *x* and *y* coordinates given by $x = \cos t$ and $y = \sin t$. Find $\frac{dz}{dt}$ when $t = 0$, and find where the particle reaches its maximum/minimum *z*-values.

**SOLUTION**    It is straightforward to compute

$$f_x(x, y) = 2x - y, \quad f_y(x, y) = 2y - x, \quad \frac{dx}{dt} = -\sin t, \quad \frac{dy}{dt} = \cos t.$$

Combining these according to the Chain Rule gives:

$$\frac{dz}{dt} = -(2x - y)\sin t + (2y - x)\cos t.$$

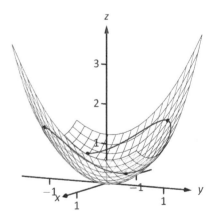

Figure 12.5.2: Plotting the path of a particle on a surface in Example 12.5.3.

Notes:

When $t = 0$, $x = 1$ and $y = 0$. Thus $\dfrac{dz}{dt} = -(2)(0) + (-1)(1) = -1$. When $t = 0$, the particle is moving down, as shown in Figure 12.5.2.

To find where $z$-value is maximized/minimized on the particle's path, we set $\dfrac{dz}{dt} = 0$ and solve for $t$:

$$\frac{dz}{dt} = 0 = -(2x - y)\sin t + (2y - x)\cos t$$

$$0 = -(2\cos t - \sin t)\sin t + (2\sin t - \cos t)\cos t$$

$$0 = \sin^2 t - \cos^2 t$$

$$\cos^2 t = \sin^2 t$$

$$t = n\frac{\pi}{4} \quad \text{(for odd } n\text{)}$$

We can use the First Derivative Test to find that on $[0, 2\pi]$, $z$ has reaches its absolute minimum at $t = \pi/4$ and $5\pi/4$; it reaches its absolute maximum at $t = 3\pi/4$ and $7\pi/4$, as shown in Figure 12.5.2.

We can extend the Chain Rule to include the situation where $z$ is a function of more than one variable, and each of these variables is also a function of more than one variable. The basic case of this is where $z = f(x, y)$, and $x$ and $y$ are functions of two variables, say $s$ and $t$.

---

**Theorem 12.5.2**     **Multivariable Chain Rule, Part II**

1. Let $z = f(x, y)$, $x = g(s, t)$ and $y = h(s, t)$, where $f$, $g$ and $h$ are differentiable functions. Then $z$ is a function of $s$ and $t$, and

   - $\dfrac{\partial z}{\partial s} = \dfrac{\partial f}{\partial x}\dfrac{\partial x}{\partial s} + \dfrac{\partial f}{\partial y}\dfrac{\partial y}{\partial s}$, and

   - $\dfrac{\partial z}{\partial t} = \dfrac{\partial f}{\partial x}\dfrac{\partial x}{\partial t} + \dfrac{\partial f}{\partial y}\dfrac{\partial y}{\partial t}.$

2. Let $z = f(x_1, x_2, \ldots, x_m)$ be a differentiable function of $m$ variables, where each of the $x_i$ is a differentiable function of the variables $t_1, t_2, \ldots, t_n$. Then $z$ is a function of the $t_i$, and

   $$\frac{\partial z}{\partial t_i} = \frac{\partial f}{\partial x_1}\frac{\partial x_1}{\partial t_i} + \frac{\partial f}{\partial x_2}\frac{\partial x_2}{\partial t_i} + \cdots + \frac{\partial f}{\partial x_m}\frac{\partial x_m}{\partial t_i}.$$

---

Notes:

## 12.5 The Multivariable Chain Rule

**Example 12.5.4  Using the Multivarible Chain Rule, Part II**

Let $z = x^2y + x$, $x = s^2 + 3t$ and $y = 2s - t$. Find $\frac{\partial z}{\partial s}$ and $\frac{\partial z}{\partial t}$, and evaluate each when $s = 1$ and $t = 2$.

**SOLUTION**  Following Theorem 12.5.2, we compute the following partial derivatives:

$$\frac{\partial f}{\partial x} = 2xy + 1 \qquad \frac{\partial f}{\partial y} = x^2,$$

$$\frac{\partial x}{\partial s} = 2s \qquad \frac{\partial x}{\partial t} = 3 \qquad \frac{\partial y}{\partial s} = 2 \qquad \frac{\partial y}{\partial t} = -1.$$

Thus

$$\frac{\partial z}{\partial s} = (2xy + 1)(2s) + (x^2)(2) = 4xys + 2s + 2x^2, \quad \text{and}$$

$$\frac{\partial z}{\partial t} = (2xy + 1)(3) + (x^2)(-1) = 6xy - x^2 + 3.$$

When $s = 1$ and $t = 2$, $x = 7$ and $y = 0$, so

$$\frac{\partial z}{\partial s} = 100 \quad \text{and} \quad \frac{\partial z}{\partial t} = -46.$$

**Example 12.5.5  Using the Multivarible Chain Rule, Part II**

Let $w = xy + z^2$, where $x = t^2 e^s$, $y = t\cos s$, and $z = s\sin t$. Find $\frac{\partial w}{\partial t}$ when $s = 0$ and $t = \pi$.

**SOLUTION**  Following Theorem 12.5.2, we compute the following partial derivatives:

$$\frac{\partial f}{\partial x} = y \qquad \frac{\partial f}{\partial y} = x \qquad \frac{\partial f}{\partial z} = 2z,$$

$$\frac{\partial x}{\partial t} = 2te^s \qquad \frac{\partial y}{\partial t} = \cos s \qquad \frac{\partial z}{\partial t} = s\cos t.$$

Thus

$$\frac{\partial w}{\partial t} = y(2te^s) + x(\cos s) + 2z(s\cos t).$$

When $s = 0$ and $t = \pi$, we have $x = \pi^2$, $y = \pi$ and $z = 0$. Thus

$$\frac{\partial w}{\partial t} = \pi(2\pi) + \pi^2 = 3\pi^2.$$

## Implicit Differentiation

We studied finding $\frac{dy}{dx}$ when $y$ is given as an implicit function of $x$ in detail in Section 2.6. We find here that the Multivariable Chain Rule gives a simpler method of finding $\frac{dy}{dx}$.

Notes:

For instance, consider the implicit function $x^2y - xy^3 = 3$. We learned to use the following steps to find $\frac{dy}{dx}$:

$$\frac{d}{dx}\left(x^2y - xy^3\right) = \frac{d}{dx}(3)$$

$$2xy + x^2\frac{dy}{dx} - y^3 - 3xy^2\frac{dy}{dx} = 0$$

$$\frac{dy}{dx} = -\frac{2xy - y^3}{x^2 - 3xy^2}. \tag{12.2}$$

Instead of using this method, consider $z = x^2y - xy^3$. The implicit function above describes the level curve $z = 3$. Considering $x$ and $y$ as functions of $x$, the Multivariable Chain Rule states that

$$\frac{dz}{dx} = \frac{\partial z}{\partial x}\frac{dx}{dx} + \frac{\partial z}{\partial y}\frac{dy}{dx}. \tag{12.3}$$

Since $z$ is constant (in our example, $z = 3$), $\frac{dz}{dx} = 0$. We also know $\frac{dx}{dx} = 1$. Equation (12.3) becomes

$$0 = \frac{\partial z}{\partial x}(1) + \frac{\partial z}{\partial y}\frac{dy}{dx} \quad \Rightarrow$$

$$\frac{dy}{dx} = -\frac{\partial z}{\partial x}\bigg/\frac{\partial z}{\partial y}$$

$$= -\frac{f_x}{f_y}.$$

Note how our solution for $\frac{dy}{dx}$ in Equation (12.2) is just the partial derivative of $z$ with respect to $x$, divided by the partial derivative of $z$ with respect to $y$, all multiplied by $(-1)$.

We state the above as a theorem.

---

**Theorem 12.5.3**     **Implicit Differentiation**

Let $f$ be a differentiable function of $x$ and $y$, where $f(x, y) = c$ defines $y$ as an implicit function of $x$, for some constant $c$. Then

$$\frac{dy}{dx} = -\frac{f_x(x, y)}{f_y(x, y)}.$$

---

We practice using Theorem 12.5.3 by applying it to a problem from Section 2.6.

Notes:

**Example 12.5.6    Implicit Differentiation**
Given the implicitly defined function $\sin(x^2y^2)+y^3 = x+y$, find $y'$. Note: this is the same problem as given in Example 2.6.4 of Section 2.6, where the solution took about a full page to find.

**SOLUTION**    Let $f(x,y) = \sin(x^2y^2) + y^3 - x - y$; the implicitly defined function above is equivalent to $f(x,y) = 0$. We find $\frac{dy}{dx}$ by applying Theorem 12.5.3. We find

$$f_x(x,y) = 2xy^2\cos(x^2y^2) - 1 \quad \text{and} \quad f_y(x,y) = 2x^2y\cos(x^2y^2) + 3y^2 - 1,$$

so

$$\frac{dy}{dx} = -\frac{2xy^2\cos(x^2y^2) - 1}{2x^2y\cos(x^2y^2) + 3y^2 - 1},$$

which matches our solution from Example 2.6.4.

In Section 12.3 we learned how partial derivatives give certain instantaneous rate of change information about a function $z = f(x,y)$. In that section, we measured the rate of change of $f$ by holding one variable constant and letting the other vary (such as, holding $y$ constant and letting $x$ vary gives $f_x$). We can visualize this change by considering the surface defined by $f$ at a point and moving parallel to the $x$-axis.

What if we want to move in a direction that is not parallel to a coordinate axis? Can we still measure instantaneous rates of change? Yes; we find out how in the next section. In doing so, we'll see how the Multivariable Chain Rule informs our understanding of these *directional derivatives*.

Notes:

# Exercises 12.5

## Terms and Concepts

1. Let a level curve of $z = f(x,y)$ be described by $x = g(t)$, $y = h(t)$. Explain why $\frac{dz}{dt} = 0$.

2. Fill in the blank: The single variable Chain Rule states $\frac{d}{dx}\big(f(g(x))\big) = f'(g(x)) \cdot$ _____.

3. Fill in the blank: The Multivariable Chain Rule states $\frac{df}{dt} = \frac{\partial f}{\partial x} \cdot$ _____ $+$ _____ $\cdot \frac{dy}{dt}$.

4. If $z = f(x,y)$, where $x = g(t)$ and $y = h(t)$, we can substitute and write $z$ as an explicit function of $t$.
T/F: Using the Multivariable Chain Rule to find $\frac{dz}{dt}$ is sometimes easier than first substituting and then taking the derivative.

5. T/F: The Multivariable Chain Rule is only useful when all the related functions are known explicitly.

6. The Multivariable Chain Rule allows us to compute implicit derivatives easily by just computing two _____ derivatives.

## Problems

In Exercises 7 – 12, functions $z = f(x,y)$, $x = g(t)$ and $y = h(t)$ are given.

(a) Use the Multivariable Chain Rule to compute $\frac{dz}{dt}$.

(b) Evaluate $\frac{dz}{dt}$ at the indicated $t$-value.

7. $z = 3x + 4y$, $\quad x = t^2$, $\quad y = 2t$; $\quad t = 1$

8. $z = x^2 - y^2$, $\quad x = t$, $\quad y = t^2 - 1$; $\quad t = 1$

9. $z = 5x + 2y$, $\quad x = 2\cos t + 1$, $\quad y = \sin t - 3$; $\quad t = \pi/4$

10. $z = \dfrac{x}{y^2+1}$, $\quad x = \cos t$, $\quad y = \sin t$; $\quad t = \pi/2$

11. $z = x^2 + 2y^2$, $\quad x = \sin t$, $\quad y = 3\sin t$; $\quad t = \pi/4$

12. $z = \cos x \sin y$, $\quad x = \pi t$, $\quad y = 2\pi t + \pi/2$; $\quad t = 3$

In Exercises 13 – 18, functions $z = f(x,y)$, $x = g(t)$ and $y = h(t)$ are given. Find the values of $t$ where $\frac{dz}{dt} = 0$. Note: these are the same surfaces/curves as found in Exercises 7 – 12.

13. $z = 3x + 4y$, $\quad x = t^2$, $\quad y = 2t$

14. $z = x^2 - y^2$, $\quad x = t$, $\quad y = t^2 - 1$

15. $z = 5x + 2y$, $\quad x = 2\cos t + 1$, $\quad y = \sin t - 3$

16. $z = \dfrac{x}{y^2+1}$, $\quad x = \cos t$, $\quad y = \sin t$

17. $z = x^2 + 2y^2$, $\quad x = \sin t$, $\quad y = 3\sin t$

18. $z = \cos x \sin y$, $\quad x = \pi t$, $\quad y = 2\pi t + \pi/2$

In Exercises 19 – 22, functions $z = f(x,y)$, $x = g(s,t)$ and $y = h(s,t)$ are given.

(a) Use the Multivariable Chain Rule to compute $\frac{\partial z}{\partial s}$ and $\frac{\partial z}{\partial t}$.

(b) Evaluate $\frac{\partial z}{\partial s}$ and $\frac{\partial z}{\partial t}$ at the indicated $s$ and $t$ values.

19. $z = x^2 y$, $\quad x = s - t$, $\quad y = 2s + 4t$; $\quad s = 1, t = 0$

20. $z = \cos\left(\pi x + \dfrac{\pi}{2}y\right)$, $\quad x = st^2$, $\quad y = s^2 t$; $\quad s = 1, t = 1$

21. $z = x^2 + y^2$, $\quad x = s\cos t$, $\quad y = s\sin t$; $\quad s = 2, t = \pi/4$

22. $z = e^{-(x^2+y^2)}$, $\quad x = t$, $\quad y = st^2$; $\quad s = 1, t = 1$

In Exercises 23 – 26, find $\frac{dy}{dx}$ using Implicit Differentiation and Theorem 12.5.3.

23. $x^2 \tan y = 50$

24. $(3x^2 + 2y^3)^4 = 2$

25. $\dfrac{x^2 + y}{x + y^2} = 17$

26. $\ln(x^2 + xy + y^2) = 1$

In Exercises 27 – 30, find $\frac{dz}{dt}$, or $\frac{\partial z}{\partial s}$ and $\frac{\partial z}{\partial t}$, using the supplied information.

27. $\dfrac{\partial z}{\partial x} = 2$, $\quad \dfrac{\partial z}{\partial y} = 1$, $\quad \dfrac{dx}{dt} = 4$, $\quad \dfrac{dy}{dt} = -5$

28. $\dfrac{\partial z}{\partial x} = 1$, $\quad \dfrac{\partial z}{\partial y} = -3$, $\quad \dfrac{dx}{dt} = 6$, $\quad \dfrac{dy}{dt} = 2$

29. $\dfrac{\partial z}{\partial x} = -4$, $\quad \dfrac{\partial z}{\partial y} = 9$,
$\dfrac{\partial x}{\partial s} = 5$, $\quad \dfrac{\partial x}{\partial t} = 7$, $\quad \dfrac{\partial y}{\partial s} = -2$, $\quad \dfrac{\partial y}{\partial t} = 6$

30. $\dfrac{\partial z}{\partial x} = 2$, $\quad \dfrac{\partial z}{\partial y} = 1$,
$\dfrac{\partial x}{\partial s} = -2$, $\quad \dfrac{\partial x}{\partial t} = 3$, $\quad \dfrac{\partial y}{\partial s} = 2$, $\quad \dfrac{\partial y}{\partial t} = -1$

## 12.6 Directional Derivatives

Partial derivatives give us an understanding of how a surface changes when we move in the $x$ and $y$ directions. We made the comparison to standing in a rolling meadow and heading due east: the amount of rise/fall in doing so is comparable to $f_x$. Likewise, the rise/fall in moving due north is comparable to $f_y$. The steeper the slope, the greater in magnitude $f_y$.

But what if we didn't move due north or east? What if we needed to move northeast and wanted to measure the amount of rise/fall? Partial derivatives alone cannot measure this. This section investigates **directional derivatives**, which do measure this rate of change.

We begin with a definition.

---

**Definition 12.6.1**    **Directional Derivatives**

Let $z = f(x, y)$ be continuous on a set $S$ and let $\vec{u} = \langle u_1, u_2 \rangle$ be a unit vector. For all points $(x, y)$, the **directional derivative of $f$ at $(x, y)$ in the direction of $\vec{u}$** is

$$D_{\vec{u}} f(x, y) = \lim_{h \to 0} \frac{f(x + hu_1, y + hu_2) - f(x, y)}{h}.$$

---

The partial derivatives $f_x$ and $f_y$ are defined with similar limits, but only $x$ or $y$ varies with $h$, not both. Here both $x$ and $y$ vary with a weighted $h$, determined by a particular unit vector $\vec{u}$. This may look a bit intimidating but in reality it is not too difficult to deal with; it often just requires extra algebra. However, the following theorem reduces this algebraic load.

---

**Theorem 12.6.1**    **Directional Derivatives**

Let $z = f(x, y)$ be differentiable on a set $S$ containing $(x_0, y_0)$, and let $\vec{u} = \langle u_1, u_2 \rangle$ be a unit vector. The directional derivative of $f$ at $(x_0, y_0)$ in the direction of $\vec{u}$ is

$$D_{\vec{u}} f(x_0, y_0) = f_x(x_0, y_0) u_1 + f_y(x_0, y_0) u_2.$$

---

**Example 12.6.1**    **Computing directional derivatives**

Let $z = 14 - x^2 - y^2$ and let $P = (1, 2)$. Find the directional derivative of $f$, at $P$, in the following directions:

1. toward the point $Q = (3, 4)$,

2. in the direction of $\langle 2, -1 \rangle$, and

---

Notes:

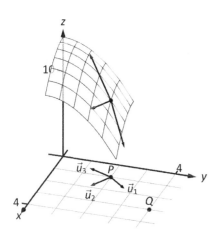

Figure 12.6.1: Understanding the directional derivative in Example 12.6.1.

3. toward the origin.

**SOLUTION** The surface is plotted in Figure 12.6.1, where the point $P = (1, 2)$ is indicated in the $x, y$-plane as well as the point $(1, 2, 9)$ which lies on the surface of $f$. We find that $f_x(x, y) = -2x$ and $f_x(1, 2) = -2$; $f_y(x, y) = -2y$ and $f_y(1, 2) = -4$.

1. Let $\vec{u}_1$ be the unit vector that points from the point $(1, 2)$ to the point $Q = (3, 4)$, as shown in the figure. The vector $\overrightarrow{PQ} = \langle 2, 2 \rangle$; the unit vector in this direction is $\vec{u}_1 = \langle 1/\sqrt{2}, 1/\sqrt{2} \rangle$. Thus the directional derivative of $f$ at $(1, 2)$ in the direction of $\vec{u}_1$ is

$$D_{\vec{u}_1} f(1, 2) = -2(1/\sqrt{2}) + (-4)(1/\sqrt{2}) = -6/\sqrt{2} \approx -4.24.$$

Thus the instantaneous rate of change in moving from the point $(1, 2, 9)$ on the surface in the direction of $\vec{u}_1$ (which points toward the point $Q$) is about $-4.24$. Moving in this direction moves one steeply downward.

2. We seek the directional derivative in the direction of $\langle 2, -1 \rangle$. The unit vector in this direction is $\vec{u}_2 = \langle 2/\sqrt{5}, -1/\sqrt{5} \rangle$. Thus the directional derivative of $f$ at $(1, 2)$ in the direction of $\vec{u}_2$ is

$$D_{\vec{u}_2} f(1, 2) = -2(2/\sqrt{5}) + (-4)(-1/\sqrt{5}) = 0.$$

Starting on the surface of $f$ at $(1, 2)$ and moving in the direction of $\langle 2, -1 \rangle$ (or $\vec{u}_2$) results in no instantaneous change in $z$-value. This is analogous to standing on the side of a hill and choosing a direction to walk that does not change the elevation. One neither walks up nor down, rather just "along the side" of the hill.

Finding these directions of "no elevation change" is important.

3. At $P = (1, 2)$, the direction towards the origin is given by the vector $\langle -1, -2 \rangle$; the unit vector in this direction is $\vec{u}_3 = \langle -1/\sqrt{5}, -2/\sqrt{5} \rangle$. The directional derivative of $f$ at $P$ in the direction of the origin is

$$D_{\vec{u}_3} f(1, 2) = -2(-1/\sqrt{5}) + (-4)(-2/\sqrt{5}) = 10/\sqrt{5} \approx 4.47.$$

Moving towards the origin means "walking uphill" quite steeply, with an initial slope of about 4.47.

As we study directional derivatives, it will help to make an important connection between the unit vector $\vec{u} = \langle u_1, u_2 \rangle$ that describes the direction and the partial derivatives $f_x$ and $f_y$. We start with a definition and follow this with a Key Idea.

Notes:

### Definition 12.6.2    Gradient

Let $z = f(x, y)$ be differentiable on a set $S$ that contains the point $(x_0, y_0)$.

1. The **gradient of** $f$ is $\nabla f(x,y) = \langle f_x(x,y), f_y(x,y) \rangle$.
2. The **gradient of** $f$ **at** $(x_0, y_0)$ **is** $\nabla f(x_0, y_0) = \langle f_x(x_0, y_0), f_y(x_0, y_0) \rangle$.

To simplify notation, we often express the gradient as $\nabla f = \langle f_x, f_y \rangle$. The gradient allows us to compute directional derivatives in terms of a dot product.

### Key Idea 12.6.1    The Gradient and Directional Derivatives

The directional derivative of $z = f(x, y)$ in the direction of $\vec{u}$ is

$$D_{\vec{u}} f = \nabla f \cdot \vec{u}.$$

**Note:** The symbol "$\nabla$" is named "nabla," derived from the Greek name of a Jewish harp. Oddly enough, in mathematics the expression $\nabla f$ is pronounced "del $f$."

The properties of the dot product previously studied allow us to investigate the properties of the directional derivative. Given that the directional derivative gives the instantaneous rate of change of $z$ when moving in the direction of $\vec{u}$, three questions naturally arise:

1. In what direction(s) is the change in $z$ the greatest (i.e., the "steepest uphill")?

2. In what direction(s) is the change in $z$ the least (i.e., the "steepest downhill")?

3. In what direction(s) is there no change in $z$?

Using the key property of the dot product, we have

$$\nabla f \cdot \vec{u} = ||\nabla f|| \, ||\vec{u}|| \cos\theta = ||\nabla f|| \cos\theta, \tag{12.4}$$

where $\theta$ is the angle between the gradient and $\vec{u}$. (Since $\vec{u}$ is a unit vector, $||\vec{u}|| = 1$.) This equation allows us to answer the three questions stated previously.

1. Equation 12.4 is maximized when $\cos\theta = 1$, i.e., when the gradient and $\vec{u}$ have the same direction. We conclude the gradient points in the direction of greatest $z$ change.

---

Notes:

2. Equation 12.4 is minimized when $\cos\theta = -1$, i.e., when the gradient and $\vec{u}$ have opposite directions. We conclude the gradient points in the opposite direction of the least $z$ change.

3. Equation 12.4 is 0 when $\cos\theta = 0$, i.e., when the gradient and $\vec{u}$ are orthogonal to each other. We conclude the gradient is orthogonal to directions of no $z$ change.

This result is rather amazing. Once again imagine standing in a rolling meadow and face the direction that leads you steepest uphill. Then the direction that leads steepest downhill is directly behind you, and side–stepping either left or right (i.e., moving perpendicularly to the direction you face) does not change your elevation at all.

Recall that a level curve is defined as a curve in the $x$-$y$ plane along which the $z$-values of a function do not change. Let a surface $z = f(x,y)$ be given, and let's represent one such level curve as a vector–valued function, $\vec{r}(t) = \langle x(t), y(t) \rangle$. As the output of $f$ does not change along this curve, $f(x(t), y(t)) = c$ for all $t$, for some constant $c$.

Since $f$ is constant for all $t$, $\frac{df}{dt} = 0$. By the Multivariable Chain Rule, we also know

$$\frac{df}{dt} = f_x(x,y)x'(t) + f_y(x,y)y'(t)$$
$$= \langle f_x(x,y), f_y(x,y) \rangle \cdot \langle x'(t), y'(t) \rangle$$
$$= \nabla f \cdot \vec{r}\,'(t)$$
$$= 0.$$

This last equality states $\nabla f \cdot \vec{r}\,'(t) = 0$: the gradient is orthogonal to the derivative of $\vec{r}$, meaning the gradient is orthogonal to the graph of $\vec{r}$. Our conclusion: at any point on a surface, the gradient at that point is orthogonal to the level curve that passes through that point.

We restate these ideas in a theorem, then use them in an example.

---

**Theorem 12.6.2**      **The Gradient and Directional Derivatives**

Let $z = f(x,y)$ be differentiable on a set $S$ with gradient $\nabla f$, let $P = (x_0, y_0)$ be a point in $S$ and let $\vec{u}$ be a unit vector.

1. The maximum value of $D_{\vec{u}}f(x_0, y_0)$ is $||\nabla f(x_0, y_0)||$; the direction of maximal $z$ increase is $\nabla f(x_0, y_0)$.

2. The minimum value of $D_{\vec{u}}f(x_0, y_0)$ is $-||\nabla f(x_0, y_0)||$; the direction of minimal $z$ increase is $-\nabla f(x_0, y_0)$.

3. At $P$, $\nabla f(x_0, y_0)$ is orthogonal to the level curve passing through $\big(x_0, y_0, f(x_0, y_0)\big)$.

---

Notes:

## 12.6 Directional Derivatives

**Example 12.6.2    Finding directions of maximal and minimal increase**

Let $f(x,y) = \sin x \cos y$ and let $P = (\pi/3, \pi/3)$. Find the directions of maximal/minimal increase, and find a direction where the instantaneous rate of $z$ change is 0.

**SOLUTION**    We begin by finding the gradient. $f_x = \cos x \cos y$ and $f_y = -\sin x \sin y$, thus

$$\nabla f = \langle \cos x \cos y, -\sin x \sin y \rangle \quad \text{and, at } P, \quad \nabla f\left(\frac{\pi}{3}, \frac{\pi}{3}\right) = \left\langle \frac{1}{4}, -\frac{3}{4} \right\rangle.$$

Thus the direction of maximal increase is $\langle 1/4, -3/4 \rangle$. In this direction, the instantaneous rate of $z$ change is $\| \langle 1/4, -3/4 \rangle \| = \sqrt{10}/4 \approx 0.79$.

Figure 12.6.2 shows the surface plotted from two different perspectives. In each, the gradient is drawn at $P$ with a dashed line (because of the nature of this surface, the gradient points "into" the surface). Let $\vec{u} = \langle u_1, u_2 \rangle$ be the unit vector in the direction of $\nabla f$ at $P$. Each graph of the figure also contains the vector $\langle u_1, u_2, \|\nabla f\| \rangle$. This vector has a "run" of 1 (because in the $x$-$y$ plane it moves 1 unit) and a "rise" of $\|\nabla f\|$, hence we can think of it as a vector with slope of $\|\nabla f\|$ in the direction of $\nabla f$, helping us visualize how "steep" the surface is in its steepest direction.

The direction of minimal increase is $\langle -1/4, 3/4 \rangle$; in this direction the instantaneous rate of $z$ change is $-\sqrt{10}/4 \approx -0.79$.

Any direction orthogonal to $\nabla f$ is a direction of no $z$ change. We have two choices: the direction of $\langle 3, 1 \rangle$ and the direction of $\langle -3, -1 \rangle$. The unit vector in the direction of $\langle 3, 1 \rangle$ is shown in each graph of the figure as well. The level curve at $z = \sqrt{3}/4$ is drawn: recall that along this curve the $z$-values do not change. Since $\langle 3, 1 \rangle$ is a direction of no $z$-change, this vector is tangent to the level curve at $P$.

**Example 12.6.3    Understanding when $\nabla f = \vec{0}$**

Let $f(x,y) = -x^2 + 2x - y^2 + 2y + 1$. Find the directional derivative of $f$ in any direction at $P = (1,1)$.

**SOLUTION**    We find $\nabla f = \langle -2x + 2, -2y + 2 \rangle$. At $P$, we have $\nabla f(1,1) = \langle 0, 0 \rangle$. According to Theorem 12.6.2, this is the direction of maximal increase. However, $\langle 0, 0 \rangle$ is directionless; it has no displacement. And regardless of the unit vector $\vec{u}$ chosen, $D_{\vec{u}} f = 0$.

Figure 12.6.3 helps us understand what this means. We can see that $P$ lies at the top of a paraboloid. In all directions, the instantaneous rate of change is 0.

So what is the direction of maximal increase? It is fine to give an answer of $\vec{0} = \langle 0, 0 \rangle$, as this indicates that all directional derivatives are 0.

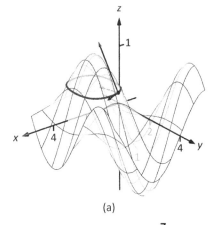

(a)

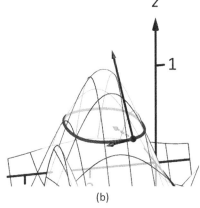

(b)

Figure 12.6.2: Graphing the surface and important directions in Example 12.6.2.

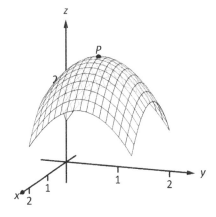

Figure 12.6.3: At the top of a paraboloid, all directional derivatives are 0.

Notes:

The fact that the gradient of a surface always points in the direction of steepest increase/decrease is very useful, as illustrated in the following example.

**Example 12.6.4    The flow of water downhill**
Consider the surface given by $f(x,y) = 20 - x^2 - 2y^2$. Water is poured on the surface at $(1, 1/4)$. What path does it take as it flows downhill?

**SOLUTION**    Let $\vec{r}(t) = \langle x(t), y(t) \rangle$ be the vector–valued function describing the path of the water in the x-y plane; we seek $x(t)$ and $y(t)$. We know that water will always flow downhill in the steepest direction; therefore, at any point on its path, it will be moving in the direction of $-\nabla f$. (We ignore the physical effects of momentum on the water.) Thus $\vec{r}'(t)$ will be parallel to $\nabla f$, and there is some constant $c$ such that $c\nabla f = \vec{r}'(t) = \langle x'(t), y'(t) \rangle$.

We find $\nabla f = \langle -2x, -4y \rangle$ and write $x'(t)$ as $\frac{dx}{dt}$ and $y'(t)$ as $\frac{dy}{dt}$. Then

$$c\nabla f = \langle x'(t), y'(t) \rangle$$
$$\langle -2cx, -4cy \rangle = \left\langle \frac{dx}{dt}, \frac{dy}{dt} \right\rangle.$$

This implies

$$-2cx = \frac{dx}{dt} \quad \text{and} \quad -4cy = \frac{dy}{dt}, \text{ i.e.,}$$
$$c = -\frac{1}{2x}\frac{dx}{dt} \quad \text{and} \quad c = -\frac{1}{4y}\frac{dy}{dt}.$$

As $c$ equals both expressions, we have

$$\frac{1}{2x}\frac{dx}{dt} = \frac{1}{4y}\frac{dy}{dt}.$$

To find an explicit relationship between $x$ and $y$, we can integrate both sides with respect to $t$. Recall from our study of differentials that $\frac{dx}{dt}dt = dx$. Thus:

$$\int \frac{1}{2x}\frac{dx}{dt}dt = \int \frac{1}{4y}\frac{dy}{dt}dt$$
$$\int \frac{1}{2x}dx = \int \frac{1}{4y}dy$$
$$\frac{1}{2}\ln|x| = \frac{1}{4}\ln|y| + C_1$$
$$2\ln|x| = \ln|y| + C_1$$
$$\ln|x^2| = \ln|y| + C_1$$

Notes:

Now raise both sides as a power of $e$:

$$x^2 = e^{\ln|y|+C_1}$$
$$x^2 = e^{\ln|y|}e^{C_1} \quad \text{(Note that } e^{C_1} \text{ is just a constant.)}$$
$$x^2 = yC_2$$
$$\frac{1}{C_2}x^2 = y \quad \text{(Note that } 1/C_2 \text{ is just a constant.)}$$
$$Cx^2 = y.$$

As the water started at the point $(1, 1/4)$, we can solve for $C$:

$$C(1)^2 = \frac{1}{4} \quad \Rightarrow \quad C = \frac{1}{4}.$$

Thus the water follows the curve $y = x^2/4$ in the x-y plane. The surface and the path of the water is graphed in Figure 12.6.4(a). In part (b) of the figure, the level curves of the surface are plotted in the x-y plane, along with the curve $y = x^2/4$. Notice how the path intersects the level curves at right angles. As the path follows the gradient downhill, this reinforces the fact that the gradient is orthogonal to level curves.

## Functions of Three Variables

The concepts of directional derivatives and the gradient are easily extended to three (and more) variables. We combine the concepts behind Definitions 12.6.1 and 12.6.2 and Theorem 12.6.1 into one set of definitions.

---

**Definition 12.6.3**    **Directional Derivatives and Gradient with Three Variables**

Let $w = F(x, y, z)$ be differentiable on a set $D$ and let $\vec{u}$ be a unit vector in $\mathbb{R}^3$.

1. The **gradient** of $F$ is $\nabla F = \langle F_x, F_y, F_z \rangle$.

2. The **directional derivative of $F$ in the direction of $\vec{u}$** is

$$D_{\vec{u}}F = \nabla F \cdot \vec{u}.$$

---

The same properties of the gradient given in Theorem 12.6.2, when $f$ is a

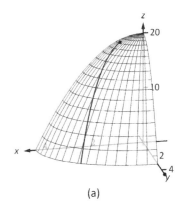

(a)

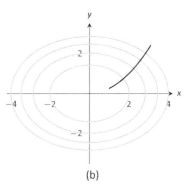

(b)

Figure 12.6.4: A graph of the surface described in Example 12.6.4 along with the path in the x-y plane with the level curves.

function of two variables, hold for F, a function of three variables.

> **Theorem 12.6.3  The Gradient and Directional Derivatives with Three Variables**
>
> Let $w = F(x, y, z)$ be differentiable on a set $D$, let $\nabla F$ be the gradient of $F$, and let $\vec{u}$ be a unit vector.
>
> 1. The maximum value of $D_{\vec{u}} F$ is $\|\nabla F\|$, obtained when the angle between $\nabla F$ and $\vec{u}$ is 0, i.e., the direction of maximal increase is $\nabla F$.
>
> 2. The minimum value of $D_{\vec{u}} F$ is $-\|\nabla F\|$, obtained when the angle between $\nabla F$ and $\vec{u}$ is $\pi$, i.e., the direction of minimal increase is $-\nabla F$.
>
> 3. $D_{\vec{u}} F = 0$ when $\nabla F$ and $\vec{u}$ are orthogonal.

We interpret the third statement of the theorem as "the gradient is orthogonal to level surfaces," the three–variable analogue to level curves.

**Example 12.6.5  Finding directional derivatives with functions of three variables**

If a point source $S$ is radiating energy, the intensity $I$ at a given point $P$ in space is inversely proportional to the square of the distance between $S$ and $P$. That is, when $S = (0, 0, 0)$, $I(x, y, z) = \dfrac{k}{x^2 + y^2 + z^2}$ for some constant $k$.

Let $k = 1$, let $\vec{u} = \langle 2/3, 2/3, 1/3 \rangle$ be a unit vector, and let $P = (2, 5, 3)$. Measure distances in inches. Find the directional derivative of $I$ at $P$ in the direction of $\vec{u}$, and find the direction of greatest intensity increase at $P$.

**Solution**  We need the gradient $\nabla I$, meaning we need $I_x$, $I_y$ and $I_z$. Each partial derivative requires a simple application of the Quotient Rule, giving

$$\nabla I = \left\langle \frac{-2x}{(x^2 + y^2 + z^2)^2}, \frac{-2y}{(x^2 + y^2 + z^2)^2}, \frac{-2z}{(x^2 + y^2 + z^2)^2} \right\rangle$$

$$\nabla I(2, 5, 3) = \left\langle \frac{-4}{1444}, \frac{-10}{1444}, \frac{-6}{1444} \right\rangle \approx \langle -0.003, -0.007, -0.004 \rangle$$

$$D_{\vec{u}} I = \nabla I(2, 5, 3) \cdot \vec{u}$$

$$= -\frac{17}{2166} \approx -0.0078.$$

The directional derivative tells us that moving in the direction of $\vec{u}$ from $P$ results in a decrease in intensity of about $-0.008$ units per inch. (The intensity is decreasing as $\vec{u}$ moves one farther from the origin than $P$.)

Notes:

The gradient gives the direction of greatest intensity increase. Notice that

$$\nabla I(2,5,3) = \left\langle \frac{-4}{1444}, \frac{-10}{1444}, \frac{-6}{1444} \right\rangle$$
$$= \frac{2}{1444} \langle -2, -5, -3 \rangle.$$

That is, the gradient at $(2, 5, 3)$ is pointing in the direction of $\langle -2, -5, -3 \rangle$, that is, towards the origin. That should make intuitive sense: the greatest increase in intensity is found by moving towards to source of the energy.

The directional derivative allows us to find the instantaneous rate of $z$ change in any direction at a point. We can use these instantaneous rates of change to define lines and planes that are *tangent* to a surface at a point, which is the topic of the next section.

Notes:

# Exercises 12.6

## Terms and Concepts

1. What is the difference between a directional derivative and a partial derivative?

2. For what $\vec{u}$ is $D_{\vec{u}}f = f_x$?

3. For what $\vec{u}$ is $D_{\vec{u}}f = f_y$?

4. The gradient is _____ to level curves.

5. The gradient points in the direction of _____ increase.

6. It is generally more informative to view the directional derivative not as the result of a limit, but rather as the result of a _____ product.

## Problems

In Exercises 7 – 12, a function $z = f(x, y)$ is given. Find $\nabla f$.

7. $f(x, y) = -x^2y + xy^2 + xy$

8. $f(x, y) = \sin x \cos y$

9. $f(x, y) = \dfrac{1}{x^2 + y^2 + 1}$

10. $f(x, y) = -4x + 3y$

11. $f(x, y) = x^2 + 2y^2 - xy - 7x$

12. $f(x, y) = x^2y^3 - 2x$

In Exercises 13 – 18, a function $z = f(x, y)$ and a point $P$ are given. Find the directional derivative of $f$ in the indicated directions. Note: these are the same functions as in Exercises 7 through 12.

13. $f(x, y) = -x^2y + xy^2 + xy$, $P = (2, 1)$
    (a) In the direction of $\vec{v} = \langle 3, 4 \rangle$.
    (b) In the direction toward the point $Q = (1, -1)$.

14. $f(x, y) = \sin x \cos y$, $P = \left(\dfrac{\pi}{4}, \dfrac{\pi}{3}\right)$
    (a) In the direction of $\vec{v} = \langle 1, 1 \rangle$.
    (b) In the direction toward the point $Q = (0, 0)$.

15. $f(x, y) = \dfrac{1}{x^2 + y^2 + 1}$, $P = (1, 1)$.
    (a) In the direction of $\vec{v} = \langle 1, -1 \rangle$.
    (b) In the direction toward the point $Q = (-2, -2)$.

16. $f(x, y) = -4x + 3y$, $P = (5, 2)$
    (a) In the direction of $\vec{v} = \langle 3, 1 \rangle$.
    (b) In the direction toward the point $Q = (2, 7)$.

17. $f(x, y) = x^2 + 2y^2 - xy - 7x$, $P = (4, 1)$
    (a) In the direction of $\vec{v} = \langle -2, 5 \rangle$
    (b) In the direction toward the point $Q = (4, 0)$.

18. $f(x, y) = x^2y^3 - 2x$, $P = (1, 1)$
    (a) In the direction of $\vec{v} = \langle 3, 3 \rangle$
    (b) In the direction toward the point $Q = (1, 2)$.

In Exercises 19 – 24, a function $z = f(x, y)$ and a point $P$ are given.

(a) Find the direction of maximal increase of $f$ at $P$.
(b) What is the maximal value of $D_{\vec{u}}f$ at $P$?
(c) Find the direction of minimal increase of $f$ at $P$.
(d) Give a direction $\vec{u}$ such that $D_{\vec{u}}f = 0$ at $P$.

Note: these are the same functions and points as in Exercises 13 through 18.

19. $f(x, y) = -x^2y + xy^2 + xy$, $P = (2, 1)$

20. $f(x, y) = \sin x \cos y$, $P = \left(\dfrac{\pi}{4}, \dfrac{\pi}{3}\right)$

21. $f(x, y) = \dfrac{1}{x^2 + y^2 + 1}$, $P = (1, 1)$.

22. $f(x, y) = -4x + 3y$, $P = (5, 4)$.

23. $f(x, y) = x^2 + 2y^2 - xy - 7x$, $P = (4, 1)$

24. $f(x, y) = x^2y^3 - 2x$, $P = (1, 1)$

In Exercises 25 – 28, a function $w = F(x, y, z)$, a vector $\vec{v}$ and a point $P$ are given.

(a) Find $\nabla F(x, y, z)$.
(b) Find $D_{\vec{u}}F$ at $P$, where $\vec{u}$ is the unit vector in the direction of $\vec{v}$.

25. $F(x, y, z) = 3x^2z^3 + 4xy - 3z^2$, $\vec{v} = \langle 1, 1, 1 \rangle$, $P = (3, 2, 1)$

26. $F(x, y, z) = \sin(x)\cos(y)e^z$, $\vec{v} = \langle 2, 2, 1 \rangle$, $P = (0, 0, 0)$

27. $F(x, y, z) = x^2y^2 - y^2z^2$, $\vec{v} = \langle -1, 7, 3 \rangle$, $P = (1, 0, -1)$

28. $F(x, y, z) = \dfrac{2}{x^2 + y^2 + z^2}$, $\vec{v} = \langle 1, 1, -2 \rangle$, $P = (1, 1, 1)$

## 12.7 Tangent Lines, Normal Lines, and Tangent Planes

Derivatives and tangent lines go hand-in-hand. Given $y = f(x)$, the line tangent to the graph of $f$ at $x = x_0$ is the line through $(x_0, f(x_0))$ with slope $f'(x_0)$; that is, the slope of the tangent line is the instantaneous rate of change of $f$ at $x_0$.

When dealing with functions of two variables, the graph is no longer a curve but a surface. At a given point on the surface, it seems there are many lines that fit our intuition of being "tangent" to the surface.

In Figure 12.7.1 we see lines that are tangent to curves in space. Since each curve lies on a surface, it makes sense to say that the lines are also tangent to the surface. The next definition formally defines what it means to be "tangent to a surface."

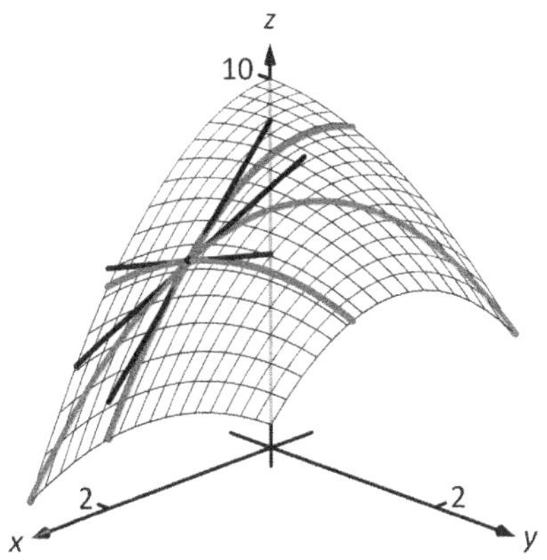

Figure 12.7.1: Showing various lines tangent to a surface.

---

**Definition 12.7.1      Directional Tangent Line**

Let $z = f(x, y)$ be differentiable on a set $S$ containing $(x_0, y_0)$ and let $\vec{u} = \langle u_1, u_2 \rangle$ be a unit vector.

1. The line $\ell_x$ through $(x_0, y_0, f(x_0, y_0))$ parallel to $\langle 1, 0, f_x(x_0, y_0) \rangle$ is the **tangent line to $f$ in the direction of $x$ at $(x_0, y_0)$**.

2. The line $\ell_y$ through $(x_0, y_0, f(x_0, y_0))$ parallel to $\langle 0, 1, f_y(x_0, y_0) \rangle$ is the **tangent line to $f$ in the direction of $y$ at $(x_0, y_0)$**.

3. The line $\ell_{\vec{u}}$ through $(x_0, y_0, f(x_0, y_0))$ parallel to $\langle u_1, u_2, D_{\vec{u}} f(x_0, y_0) \rangle$ is the **tangent line to $f$ in the direction of $\vec{u}$ at $(x_0, y_0)$**.

---

It is instructive to consider each of three directions given in the definition in terms of "slope." The direction of $\ell_x$ is $\langle 1, 0, f_x(x_0, y_0) \rangle$; that is, the "run" is one unit in the $x$-direction and the "rise" is $f_x(x_0, y_0)$ units in the $z$-direction. Note how the slope is just the partial derivative with respect to $x$. A similar statement can be made for $\ell_y$. The direction of $\ell_{\vec{u}}$ is $\langle u_1, u_2, D_{\vec{u}} f(x_0, y_0) \rangle$; the "run" is one unit in the $\vec{u}$ direction (where $\vec{u}$ is a unit vector) and the "rise" is the directional derivative of $z$ in that direction.

Definition 12.7.1 leads to the following parametric equations of directional tangent lines:

$$\ell_x(t) = \begin{cases} x = x_0 + t \\ y = y_0 \\ z = z_0 + f_x(x_0, y_0)t \end{cases}, \quad \ell_y(t) = \begin{cases} x = x_0 \\ y = y_0 + t \\ z = z_0 + f_y(x_0, y_0)t \end{cases} \text{ and } \quad \ell_{\vec{u}}(t) = \begin{cases} x = x_0 + u_1 t \\ y = y_0 + u_2 t \\ z = z_0 + D_{\vec{u}} f(x_0, y_0)t \end{cases}.$$

Notes:

## Chapter 12  Functions of Several Variables

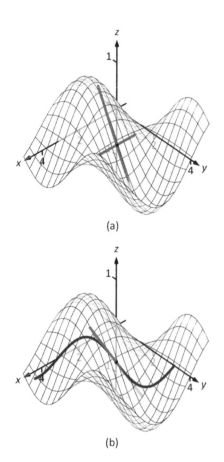

(a)

(b)

Figure 12.7.2: A surface and directional tangent lines in Example 12.7.1.

**Example 12.7.1    Finding directional tangent lines**
Find the lines tangent to the surface $z = \sin x \cos y$ at $(\pi/2, \pi/2)$ in the $x$ and $y$ directions and also in the direction of $\vec{v} = \langle -1, 1 \rangle$.

**SOLUTION**   The partial derivatives with respect to $x$ and $y$ are:
$$f_x(x,y) = \cos x \cos y \quad \Rightarrow \quad f_x(\pi/2, \pi/2) = 0$$
$$f_y(x,y) = -\sin x \sin y \quad \Rightarrow \quad f_y(\pi/2, \pi/2) = -1.$$

At $(\pi/2, \pi/2)$, the $z$-value is 0.

Thus the parametric equations of the line tangent to $f$ at $(\pi/2, \pi/2)$ in the directions of $x$ and $y$ are:

$$\ell_x(t) = \begin{cases} x = \pi/2 + t \\ y = \pi/2 \\ z = 0 \end{cases} \quad \text{and} \quad \ell_y(t) = \begin{cases} x = \pi/2 \\ y = \pi/2 + t \\ z = -t \end{cases}.$$

The two lines are shown with the surface in Figure 12.7.2(a). To find the equation of the tangent line in the direction of $\vec{v}$, we first find the unit vector in the direction of $\vec{v}$: $\vec{u} = \langle -1/\sqrt{2}, 1/\sqrt{2} \rangle$. The directional derivative at $(\pi/2, \pi, 2)$ in the direction of $\vec{u}$ is

$$D_{\vec{u}} f(\pi/2, \pi, 2) = \langle 0, -1 \rangle \cdot \langle -1/\sqrt{2}, 1/\sqrt{2} \rangle = -1/\sqrt{2}.$$

Thus the directional tangent line is

$$\ell_{\vec{u}}(t) = \begin{cases} x = \pi/2 - t/\sqrt{2} \\ y = \pi/2 + t/\sqrt{2} \\ z = -t/\sqrt{2} \end{cases}.$$

The curve through $(\pi/2, \pi/2, 0)$ in the direction of $\vec{v}$ is shown in Figure 12.7.2(b) along with $\ell_{\vec{u}}(t)$.

**Example 12.7.2   Finding directional tangent lines**
Let $f(x, y) = 4xy - x^4 - y^4$. Find the equations of *all* directional tangent lines to $f$ at $(1, 1)$.

**SOLUTION**   First note that $f(1, 1) = 2$. We need to compute directional derivatives, so we need $\nabla f$. We begin by computing partial derivatives.

$$f_x = 4y - 4x^3 \Rightarrow f_x(1,1) = 0; \quad f_y = 4x - 4y^3 \Rightarrow f_y(1,1) = 0.$$

Thus $\nabla f(1, 1) = \langle 0, 0 \rangle$. Let $\vec{u} = \langle u_1, u_2 \rangle$ be any unit vector. The directional derivative of $f$ at $(1, 1)$ will be $D_{\vec{u}} f(1, 1) = \langle 0, 0 \rangle \cdot \langle u_1, u_2 \rangle = 0$. It does not matter

Notes:

what direction we choose; the directional derivative is always 0. Therefore

$$\ell_{\vec{u}}(t) = \begin{cases} x = 1 + u_1 t \\ y = 1 + u_2 t \\ z = 2 \end{cases}.$$

Figure 12.7.3 shows a graph of $f$ and the point $(1, 1, 2)$. Note that this point comes at the top of a "hill," and therefore every tangent line through this point will have a "slope" of 0.

That is, consider any curve on the surface that goes through this point. Each curve will have a relative maximum at this point, hence its tangent line will have a slope of 0. The following section investigates the points on surfaces where all tangent lines have a slope of 0.

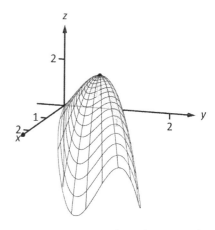

Figure 12.7.3: Graphing $f$ in Example 12.7.2.

## Normal Lines

When dealing with a function $y = f(x)$ of one variable, we stated that a line through $(c, f(c))$ was *tangent* to $f$ if the line had a slope of $f'(c)$ and was *normal* (or, *perpendicular, orthogonal*) to $f$ if it had a slope of $-1/f'(c)$. We extend the concept of normal, or orthogonal, to functions of two variables.

Let $z = f(x, y)$ be a differentiable function of two variables. By Definition 12.7.1, at $(x_0, y_0)$, $\ell_x(t)$ is a line parallel to the vector $\vec{d}_x = \langle 1, 0, f_x(x_0, y_0) \rangle$ and $\ell_y(t)$ is a line parallel to $\vec{d}_y = \langle 0, 1, f_y(x_0, y_0) \rangle$. Since lines in these directions through $(x_0, y_0, f(x_0, y_0))$ are tangent to the surface, a line through this point and orthogonal to these directions would be *orthogonal*, or *normal*, to the surface. We can use this direction to create a normal line.

The direction of the normal line is orthogonal to $\vec{d}_x$ and $\vec{d}_y$, hence the direction is parallel to $\vec{d}_n = \vec{d}_x \times \vec{d}_y$. It turns out this cross product has a very simple form:

$$\vec{d}_x \times \vec{d}_y = \langle 1, 0, f_x \rangle \times \langle 0, 1, f_y \rangle = \langle -f_x, -f_y, 1 \rangle.$$

It is often more convenient to refer to the opposite of this direction, namely $\langle f_x, f_y, -1 \rangle$. This leads to a definition.

Notes:

> **Definition 12.7.2  Normal Line**
>
> Let $z = f(x, y)$ be differentiable on a set $S$ containing $(x_0, y_0)$ where
> $$a = f_x(x_0, y_0) \quad \text{and} \quad b = f_y(x_0, y_0)$$
> are defined.
>
> 1. A nonzero vector parallel to $\vec{n} = \langle a, b, -1 \rangle$ is **orthogonal to $f$ at** $P = (x_0, y_0, f(x_0, y_0))$.
>
> 2. The line $\ell_n$ through $P$ with direction parallel to $\vec{n}$ is the **normal line to $f$ at $P$**.

Thus the parametric equations of the normal line to a surface $f$ at $(x_0, y_0, f(x_0, y_0))$ is:
$$\ell_n(t) = \begin{cases} x = x_0 + at \\ y = y_0 + bt \\ z = f(x_0, y_0) - t \end{cases}.$$

**Example 12.7.3  Finding a normal line**
Find the equation of the normal line to $z = -x^2 - y^2 + 2$ at $(0, 1)$.

**Solution**  We find $z_x(x, y) = -2x$ and $z_y(x, y) = -2y$; at $(0, 1)$, we have $z_x = 0$ and $z_y = -2$. We take the direction of the normal line, following Definition 12.7.2, to be $\vec{n} = \langle 0, -2, -1 \rangle$. The line with this direction going through the point $(0, 1, 1)$ is

$$\ell_n(t) = \begin{cases} x = 0 \\ y = -2t + 1 \\ z = -t + 1 \end{cases} \quad \text{or} \quad \ell_n(t) = \langle 0, -2, -1 \rangle t + \langle 0, 1, 1 \rangle.$$

The surface $z = -x^2 - y^2 + 2$, along with the found normal line, is graphed in Figure 12.7.4.

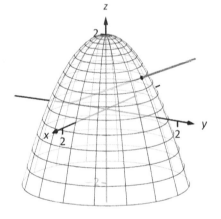

Figure 12.7.4: Graphing a surface with a normal line from Example 12.7.3.

The direction of the normal line has many uses, one of which is the definition of the **tangent plane** which we define shortly. Another use is in measuring distances from the surface to a point. Given a point $Q$ in space, it is a general geometric concept to define the distance from $Q$ to the surface as being the length of the shortest line segment $\overline{PQ}$ over all points $P$ on the surface. This, in turn, implies that $\overrightarrow{PQ}$ will be orthogonal to the surface at $P$. Therefore we can measure the distance from $Q$ to the surface $f$ by finding a point $P$ on the surface such that $\overrightarrow{PQ}$ is parallel to the normal line to $f$ at $P$.

---

Notes:

## 12.7 Tangent Lines, Normal Lines, and Tangent Planes

**Example 12.7.4   Finding the distance from a point to a surface**

Let $f(x, y) = 2 - x^2 - y^2$ and let $Q = (2, 2, 2)$. Find the distance from $Q$ to the surface defined by $f$.

**SOLUTION**   This surface is used in Example 12.7.2, so we know that at $(x, y)$, the direction of the normal line will be $\vec{d}_n = \langle -2x, -2y, -1 \rangle$. A point $P$ on the surface will have coordinates $(x, y, 2-x^2-y^2)$, so $\vec{PQ} = \langle 2 - x, 2 - y, x^2 + y^2 \rangle$. To find where $\vec{PQ}$ is parallel to $\vec{d}_n$, we need to find $x$, $y$ and $c$ such that $c\vec{PQ} = \vec{d}_n$.

$$c\vec{PQ} = \vec{d}_n$$
$$c\langle 2 - x, 2 - y, x^2 + y^2 \rangle = \langle -2x, -2y, -1 \rangle.$$

This implies

$$c(2 - x) = -2x$$
$$c(2 - y) = -2y$$
$$c(x^2 + y^2) = -1$$

In each equation, we can solve for $c$:

$$c = \frac{-2x}{2 - x} = \frac{-2y}{2 - y} = \frac{-1}{x^2 + y^2}.$$

The first two fractions imply $x = y$, and so the last fraction can be rewritten as $c = -1/(2x^2)$. Then

$$\frac{-2x}{2 - x} = \frac{-1}{2x^2}$$
$$-2x(2x^2) = -1(2 - x)$$
$$4x^3 = 2 - x$$
$$4x^3 + x - 2 = 0.$$

This last equation is a cubic, which is not difficult to solve with a numeric solver. We find that $x = 0.689$, hence $P = (0.689, 0.689, 1.051)$. We find the distance from $Q$ to the surface of $f$ is

$$\|\vec{PQ}\| = \sqrt{(2 - 0.689)^2 + (2 - 0.689)^2 + (2 - 1.051)^2} = 2.083.$$

We can take the concept of measuring the distance from a point to a surface to find a point $Q$ a particular distance from a surface at a given point $P$ on the surface.

Notes:

**Example 12.7.5**     **Finding a point a set distance from a surface**

Let $f(x,y) = x - y^2 + 3$. Let $P = (2, 1, f(2,1)) = (2, 1, 4)$. Find points $Q$ in space that are 4 units from the surface of $f$ at $P$. That is, find $Q$ such that $\|\overrightarrow{PQ}\| = 4$ and $\overrightarrow{PQ}$ is orthogonal to $f$ at $P$.

**SOLUTION**     We begin by finding partial derivatives:

$$f_x(x,y) = 1 \quad \Rightarrow \quad f_x(2,1) = 1$$
$$f_y(x,y) = -2y \quad \Rightarrow \quad f_y(2,1) = -2$$

The vector $\vec{n} = \langle 1, -2, -1 \rangle$ is orthogonal to $f$ at $P$. For reasons that will become more clear in a moment, we find the unit vector in the direction of $\vec{n}$:

$$\vec{u} = \frac{\vec{n}}{\|\vec{n}\|} = \langle 1/\sqrt{6}, -2/\sqrt{6}, -1/\sqrt{6} \rangle \approx \langle 0.408, -0.816, -0.408 \rangle.$$

Thus a the normal line to $f$ at $P$ can be written as

$$\ell_n(t) = \langle 2, 1, 4 \rangle + t \langle 0.408, -0.816, -0.408 \rangle.$$

An advantage of this parametrization of the line is that letting $t = t_0$ gives a point on the line that is $|t_0|$ units from $P$. (This is because the direction of the line is given in terms of a unit vector.) There are thus two points in space 4 units from $P$:

$$Q_1 = \ell_n(4) \qquad\qquad Q_2 = \ell_n(-4)$$
$$\approx \langle 3.63, -2.27, 2.37 \rangle \qquad\qquad \approx \langle 0.37, 4.27, 5.63 \rangle$$

width=150pt The surface is graphed along with points $P$, $Q_1$, $Q_2$ and a portion of the normal line to $f$ at $P$.

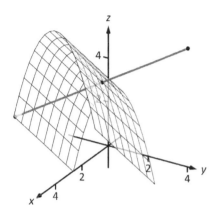

Figure 12.7.5: Graphing the surface in Example 12.7.5 along with points 4 units from the surface.

## Tangent Planes

We can use the direction of the normal line to define a plane. With $a = f_x(x_0, y_0)$, $b = f_y(x_0, y_0)$ and $P = (x_0, y_0, f(x_0, y_0))$, the vector $\vec{n} = \langle a, b, -1 \rangle$ is orthogonal to $f$ at $P$. The plane through $P$ with normal vector $\vec{n}$ is therefore **tangent** to $f$ at $P$.

Notes:

## 12.7 Tangent Lines, Normal Lines, and Tangent Planes

**Definition 12.7.3    Tangent Plane**

Let $z = f(x, y)$ be differentiable on a set $S$ containing $(x_0, y_0)$, where $a = f_x(x_0, y_0)$, $b = f_y(x_0, y_0)$, $\vec{n} = \langle a, b, -1 \rangle$ and $P = \bigl(x_0, y_0, f(x_0, y_0)\bigr)$.

The plane through $P$ with normal vector $\vec{n}$ is the **tangent plane to $f$ at $P$**. The standard form of this plane is

$$a(x - x_0) + b(y - y_0) - \bigl(z - f(x_0, y_0)\bigr) = 0.$$

**Example 12.7.6    Finding tangent planes**
Find the equation of the tangent plane to $z = -x^2 - y^2 + 2$ at $(0, 1)$.

**SOLUTION**    Note that this is the same surface and point used in Example 12.7.3. There we found $\vec{n} = \langle 0, -2, -1 \rangle$ and $P = (0, 1, 1)$. Therefore the equation of the tangent plane is

$$-2(y - 1) - (z - 1) = 0.$$

The surface $z = -x^2 - y^2 + 2$ and tangent plane are graphed in Figure 12.7.6.

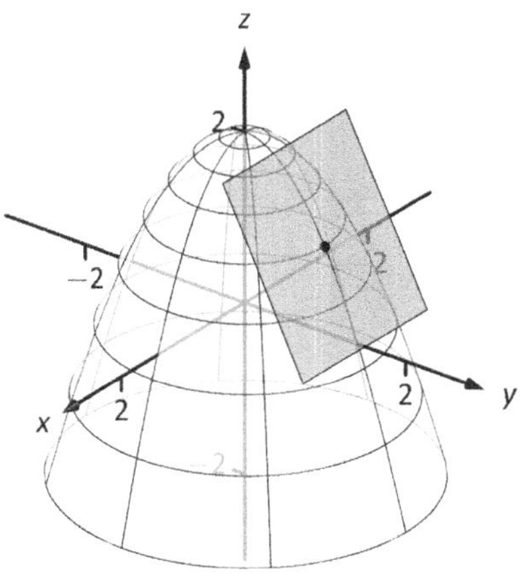

**Example 12.7.7    Using the tangent plane to approximate function values**
The point $(3, -1, 4)$ lies on the surface of an unknown differentiable function $f$ where $f_x(3, -1) = 2$ and $f_y(3, -1) = -1/2$. Find the equation of the tangent plane to $f$ at $P$, and use this to approximate the value of $f(2.9, -0.8)$.

**SOLUTION**    Knowing the partial derivatives at $(3, -1)$ allows us to form the normal vector to the tangent plane, $\vec{n} = \langle 2, -1/2, -1 \rangle$. Thus the equation of the tangent line to $f$ at $P$ is:

$$2(x-3) - 1/2(y+1) - (z-4) = 0 \quad \Rightarrow \quad z = 2(x-3) - 1/2(y+1) + 4. \quad (12.5)$$

Just as tangent lines provide excellent approximations of curves near their point of intersection, tangent planes provide excellent approximations of surfaces near their point of intersection. So $f(2.9, -0.8) \approx z(2.9, -0.8) = 3.7$.

This is not a new method of approximation. Compare the right hand expression for $z$ in Equation (12.5) to the total differential:

$$dz = f_x dx + f_y dy \quad \text{and} \quad z = \underbrace{\underbrace{2}_{f_x} \underbrace{(x-3)}_{dx} + \underbrace{-1/2}_{f_y} \underbrace{(y+1)}_{dy}}_{dz} + 4.$$

Figure 12.7.6: Graphing a surface with tangent plane from Example 12.7.6.

Notes:

Thus the "new z-value" is the sum of the change in z (i.e., $dz$) and the old z-value (4). As mentioned when studying the total differential, it is not uncommon to know partial derivative information about a unknown function, and tangent planes are used to give accurate approximations of the function.

### The Gradient and Normal Lines, Tangent Planes

The methods developed in this section so far give a straightforward method of finding equations of normal lines and tangent planes for surfaces with explicit equations of the form $z = f(x, y)$. However, they do not handle implicit equations well, such as $x^2 + y^2 + z^2 = 1$. There is a technique that allows us to find vectors orthogonal to these surfaces based on the **gradient**.

---

**Definition 12.7.4**   **Gradient**

Let $w = F(x, y, z)$ be differentiable on a set $D$ that contains the point $(x_0, y_0, z_0)$.

1. The **gradient of $F$** is $\nabla F(x, y, z) = \langle f_x(x, y, z), f_y(x, y, z), f_z(x, y, z) \rangle$.

2. The **gradient of $F$ at $(x_0, y_0, z_0)$** is

$$\nabla F(x_0, y_0, z_0) = \langle f_x(x_0, y_0, z_0), f_y(x_0, y_0, z_0), f_z(x_0, y_0, z_0) \rangle .$$

---

Recall that when $z = f(x, y)$, the gradient $\nabla f = \langle f_x, f_y \rangle$ is orthogonal to level curves of $f$. An analogous statement can be made about the gradient $\nabla F$, where $w = F(x, y, z)$. Given a point $(x_0, y_0, z_0)$, let $c = F(x_0, y_0, z_0)$. Then $F(x, y, z) = c$ is a **level surface** that contains the point $(x_0, y_0, z_0)$. The following theorem states that $\nabla F(x_0, y_0, z_0)$ is orthogonal to this level surface.

---

**Theorem 12.7.1**   **The Gradient and Level Surfaces**

Let $w = F(x, y, z)$ be differentiable on a set $D$ containing $(x_0, y_0, z_0)$ with gradient $\nabla F$, where $F(x_0, y_0, z_0) = c$.

The vector $\nabla F(x_0, y_0, z_0)$ is orthogonal to the level surface $F(x, y, z) = c$ at $(x_0, y_0, z_0)$.

---

The gradient at a point gives a vector orthogonal to the surface at that point. This direction can be used to find tangent planes and normal lines.

Notes:

### Example 12.7.8   Using the gradient to find a tangent plane

Find the equation of the plane tangent to the ellipsoid $\dfrac{x^2}{12} + \dfrac{y^2}{6} + \dfrac{z^2}{4} = 1$ at $P = (1, 2, 1)$.

**SOLUTION**   We consider the equation of the ellipsoid as a level surface of a function $F$ of three variables, where $F(x, y, z) = \dfrac{x^2}{12} + \dfrac{y^2}{6} + \dfrac{z^2}{4}$. The gradient is:

$$\nabla F(x, y, z) = \langle F_x, F_y, F_z \rangle$$
$$= \left\langle \frac{x}{6}, \frac{y}{3}, \frac{z}{2} \right\rangle.$$

At $P$, the gradient is $\nabla F(1, 2, 1) = \langle 1/6, 2/3, 1/2 \rangle$. Thus the equation of the plane tangent to the ellipsoid at $P$ is

$$\frac{1}{6}(x-1) + \frac{2}{3}(y-2) + \frac{1}{2}(z-1) = 0.$$

The ellipsoid and tangent plane are graphed in Figure 12.7.7.

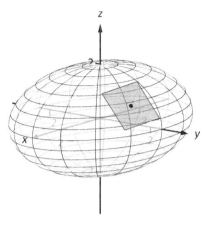

Figure 12.7.7: An ellipsoid and its tangent plane at a point.

Tangent lines and planes to surfaces have many uses, including the study of instantaneous rates of changes and making approximations. Normal lines also have many uses. In this section we focused on using them to measure distances from a surface. Another interesting application is in computer graphics, where the effects of light on a surface are determined using normal vectors.

The next section investigates another use of partial derivatives: determining relative extrema. When dealing with functions of the form $y = f(x)$, we found relative extrema by finding $x$ where $f'(x) = 0$. We can *start* finding relative extrema of $z = f(x, y)$ by setting $f_x$ and $f_y$ to 0, but it turns out that there is more to consider.

Notes:

# Exercises 12.7

## Terms and Concepts

1. Explain how the vector $\vec{v} = \langle 1, 0, 3 \rangle$ can be thought of as having a "slope" of 3.

2. Explain how the vector $\vec{v} = \langle 0.6, 0.8, -2 \rangle$ can be thought of as having a "slope" of $-2$.

3. T/F: Let $z = f(x, y)$ be differentiable at $P$. If $\vec{n}$ is a normal vector to the tangent plane of $f$ at $P$, then $\vec{n}$ is orthogonal to $\ell_x$ and $\ell_y$ at $P$.

4. Explain in your own words why we do not refer to *the* tangent line to a surface at a point, but rather to *directional* tangent lines to a surface at a point.

## Problems

In Exercises 5 – 8, a function $z = f(x, y)$, a vector $\vec{v}$ and a point $P$ are given. Give the parametric equations of the following directional tangent lines to $f$ at $P$:

(a) $\ell_x(t)$

(b) $\ell_y(t)$

(c) $\ell_{\vec{u}}(t)$, where $\vec{u}$ is the unit vector in the direction of $\vec{v}$.

5. $f(x, y) = 2x^2y - 4xy^2$, $\vec{v} = \langle 1, 3 \rangle$, $P = (2, 3)$.

6. $f(x, y) = 3 \cos x \sin y$, $\vec{v} = \langle 1, 2 \rangle$, $P = (\pi/3, \pi/6)$.

7. $f(x, y) = 3x - 5y$, $\vec{v} = \langle 1, 1 \rangle$, $P = (4, 2)$.

8. $f(x, y) = x^2 - 2x - y^2 + 4y$, $\vec{v} = \langle 1, 1 \rangle$, $P = (1, 2)$.

In Exercises 9 – 12, a function $z = f(x, y)$ and a point $P$ are given. Find the equation of the normal line to $f$ at $P$. Note: these are the same functions as in Exercises 5 – 8.

9. $f(x, y) = 2x^2y - 4xy^2$, $P = (2, 3)$.

10. $f(x, y) = 3 \cos x \sin y$, $P = (\pi/3, \pi/6)$.

11. $f(x, y) = 3x - 5y$, $P = (4, 2)$.

12. $f(x, y) = x^2 - 2x - y^2 + 4y$, $P = (1, 2)$.

In Exercises 13 – 16, a function $z = f(x, y)$ and a point $P$ are given. Find the two points that are 2 units from the surface $f$ at $P$. Note: these are the same functions as in Exercises 5 – 8.

13. $f(x, y) = 2x^2y - 4xy^2$, $P = (2, 3)$.

14. $f(x, y) = 3 \cos x \sin y$, $P = (\pi/3, \pi/6)$.

15. $f(x, y) = 3x - 5y$, $P = (4, 2)$.

16. $f(x, y) = x^2 - 2x - y^2 + 4y$, $P = (1, 2)$.

In Exercises 17 – 20, a function $z = f(x, y)$ and a point $P$ are given. Find the equation of the tangent plane to $f$ at $P$. Note: these are the same functions as in Exercises 5 – 8.

17. $f(x, y) = 2x^2y - 4xy^2$, $P = (2, 3)$.

18. $f(x, y) = 3 \cos x \sin y$, $P = (\pi/3, \pi/6)$.

19. $f(x, y) = 3x - 5y$, $P = (4, 2)$.

20. $f(x, y) = x^2 - 2x - y^2 + 4y$, $P = (1, 2)$.

In Exercises 21 – 24, an implicitly defined function of $x$, $y$ and $z$ is given along with a point $P$ that lies on the surface. Use the gradient $\nabla F$ to:

(a) find the equation of the normal line to the surface at $P$, and

(b) find the equation of the plane tangent to the surface at $P$.

21. $\dfrac{x^2}{8} + \dfrac{y^2}{4} + \dfrac{z^2}{16} = 1$, at $P = (1, \sqrt{2}, \sqrt{6})$

22. $z^2 - \dfrac{x^2}{4} - \dfrac{y^2}{9} = 0$, at $P = (4, -3, \sqrt{5})$

23. $xy^2 - xz^2 = 0$, at $P = (2, 1, -1)$

24. $\sin(xy) + \cos(yz) = 0$, at $P = (2, \pi/12, 4)$

## 12.8 Extreme Values

Given a function $z = f(x, y)$, we are often interested in points where $z$ takes on the largest or smallest values. For instance, if $z$ represents a cost function, we would likely want to know what $(x, y)$ values minimize the cost. If $z$ represents the ratio of a volume to surface area, we would likely want to know where $z$ is greatest. This leads to the following definition.

---

**Definition 12.8.1  Relative and Absolute Extrema**

Let $z = f(x, y)$ be defined on a set $S$ containing the point $P = (x_0, y_0)$.

1.  If $f(x_0, y_0) \geq f(x, y)$ for all $(x, y)$ in $S$, then $f$ has an **absolute maximum** at $P$

    If $f(x_0, y_0) \leq f(x, y)$ for all $(x, y)$ in $S$, then $f$ has an **absolute minimum** at $P$.

2.  If there is an open disk $D$ containing $P$ such that $f(x_0, y_0) \geq f(x, y)$ for all points $(x, y)$ that are in both $D$ and $S$, then $f$ has a **relative maximum** at $P$.

    If there is an open disk $D$ containing $P$ such that $f(x_0, y_0) \leq f(x, y)$ for all points $(x, y)$ that are in both $D$ and $S$, then $f$ has a **relative minimum** at $P$.

3.  If $f$ has an absolute maximum or minimum at $P$, then $f$ has an **absolute extrema** at $P$.

    If $f$ has a relative maximum or minimum at $P$, then $f$ has a **relative extrema** at $P$.

---

If $f$ has a relative or absolute maximum at $P = (x_0, y_0)$, it means every curve on the surface of $f$ through $P$ will also have a relative or absolute maximum at $P$. Recalling what we learned in Section 3.1, the slopes of the tangent lines to these curves at $P$ must be 0 or undefined. Since directional derivatives are computed using $f_x$ and $f_y$, we are led to the following definition and theorem.

---

**Definition 12.8.2  Critical Point**

Let $z = f(x, y)$ be continuous on a set $S$. A **critical point** $P = (x_0, y_0)$ of $f$ is a point in $S$ such that, at $P$,

- $f_x(x_0, y_0) = 0$ and $f_y(x_0, y_0) = 0$, or

- $f_x(x_0, y_0)$ and/or $f_y(x_0, y_0)$ is undefined.

---

Notes:

## Chapter 12 Functions of Several Variables

**Theorem 12.8.1  Critical Points and Relative Extrema**

Let $z = f(x,y)$ be defined on an open set $S$ containing $P = (x_0, y_0)$. If $f$ has a relative extrema at $P$, then $P$ is a critical point of $f$.

Therefore, to find relative extrema, we find the critical points of $f$ and determine which correspond to relative maxima, relative minima, or neither. The following examples demonstrate this process.

**Example 12.8.1  Finding critical points and relative extrema**
Let $f(x,y) = x^2 + y^2 - xy - x - 2$. Find the relative extrema of $f$.

**Solution**  We start by computing the partial derivatives of $f$:
$$f_x(x,y) = 2x - y - 1 \quad \text{and} \quad f_y(x,y) = 2y - x.$$

Each is never undefined. A critical point occurs when $f_x$ and $f_y$ are simultaneously 0, leading us to solve the following system of linear equations:
$$2x - y - 1 = 0 \quad \text{and} \quad -x + 2y = 0.$$

This solution to this system is $x = 2/3$, $y = 1/3$. (Check that at $(2/3, 1/3)$, both $f_x$ and $f_y$ are 0.)

The graph in Figure 12.8.1 shows $f$ along with this critical point. It is clear from the graph that this is a relative minimum; further consideration of the function shows that this is actually the absolute minimum.

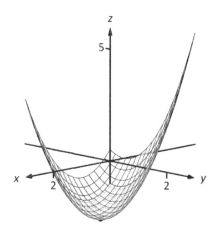

Figure 12.8.1: The surface in Example 12.8.1 with its absolute minimum indicated.

**Example 12.8.2  Finding critical points and relative extrema**
Let $f(x,y) = -\sqrt{x^2 + y^2} + 2$. Find the relative extrema of $f$.

**Solution**  We start by computing the partial derivatives of $f$:
$$f_x(x,y) = \frac{-x}{\sqrt{x^2+y^2}} \quad \text{and} \quad f_y(x,y) = \frac{-y}{\sqrt{x^2+y^2}}.$$

It is clear that $f_x = 0$ when $x = 0$ & $y \neq 0$, and that $f_y = 0$ when $y = 0$ & $x \neq 0$. At $(0,0)$, both $f_x$ and $f_y$ are *not* 0, but rather undefined. The point $(0,0)$ is still a critical point, though, because the partial derivatives are undefined. This is the only critical point of $f$.

The surface of $f$ is graphed in Figure 12.8.2 along with the point $(0,0,2)$. The graph shows that this point is the absolute maximum of $f$.

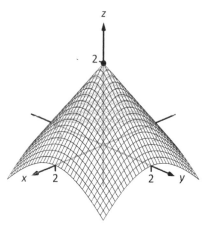

Figure 12.8.2: The surface in Example 12.8.2 with its absolute maximum indicated.

Notes:

In each of the previous two examples, we found a critical point of $f$ and then determined whether or not it was a relative (or absolute) maximum or minimum by graphing. It would be nice to be able to determine whether a critical point corresponded to a max or a min without a graph. Before we develop such a test, we do one more example that sheds more light on the issues our test needs to consider.

**Example 12.8.3  Finding critical points and relative extrema**
Let $f(x,y) = x^3 - 3x - y^2 + 4y$. Find the relative extrema of $f$.

**SOLUTION**  Once again we start by finding the partial derivatives of $f$:

$$f_x(x,y) = 3x^2 - 3 \quad \text{and} \quad f_y(x,y) = -2y + 4.$$

Each is always defined. Setting each equal to 0 and solving for $x$ and $y$, we find

$$f_x(x,y) = 0 \quad \Rightarrow x = \pm 1$$
$$f_y(x,y) = 0 \quad \Rightarrow y = 2.$$

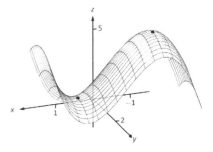

Figure 12.8.3: The surface in Example 12.8.3 with both critical points marked.

We have two critical points: $(-1, 2)$ and $(1, 2)$. To determine if they correspond to a relative maximum or minimum, we consider the graph of $f$ in Figure 12.8.3.

The critical point $(-1, 2)$ clearly corresponds to a relative maximum. However, the critical point at $(1, 2)$ is neither a maximum nor a minimum, displaying a different, interesting characteristic.

If one walks parallel to the $y$-axis towards this critical point, then this point becomes a relative maximum along this path. But if one walks towards this point parallel to the $x$-axis, this point becomes a relative minimum along this path. A point that seems to act as both a max and a min is a **saddle point**. A formal definition follows.

---

**Definition 12.8.3  Saddle Point**

Let $P = (x_0, y_0)$ be in the domain of $f$ where $f_x = 0$ and $f_y = 0$ at $P$. We say $P$ is a **saddle point** of $f$ if, for every open disk $D$ containing $P$, there are points $(x_1, y_1)$ and $(x_2, y_2)$ in $D$ such that $f(x_0, y_0) > f(x_1, y_1)$ and $f(x_0, y_0) < f(x_2, y_2)$.

---

At a saddle point, the instantaneous rate of change in all directions is 0 and there are points nearby with $z$-values both less than and greater than the $z$-value of the saddle point.

Notes:

Before Example 12.8.3 we mentioned the need for a test to differentiate between relative maxima and minima. We now recognize that our test also needs to account for saddle points. To do so, we consider the second partial derivatives of $f$.

Recall that with single variable functions, such as $y = f(x)$, if $f''(c) > 0$, then if $f$ is concave up at $c$, and if $f'(c) = 0$, then $f$ has a relative minimum at $x = c$. (We called this the Second Derivative Test.) Note that at a saddle point, it seems the graph is "both" concave up and concave down, depending on which direction you are considering.

It would be nice if the following were true:

$$\begin{aligned} f_{xx} \text{ and } f_{yy} > 0 &\Rightarrow \text{relative minimum} \\ f_{xx} \text{ and } f_{yy} < 0 &\Rightarrow \text{relative maximum} \\ f_{xx} \text{ and } f_{yy} \text{ have opposite signs} &\Rightarrow \text{saddle point.} \end{aligned}$$

However, this is not the case. Functions $f$ exist where $f_{xx}$ and $f_{yy}$ are both positive but a saddle point still exists. In such a case, while the concavity in the $x$-direction is up (i.e., $f_{xx} > 0$) and the concavity in the $y$-direction is also up (i.e., $f_{yy} > 0$), the concavity switches somewhere in between the $x$- and $y$-directions.

To account for this, consider $D = f_{xx}f_{yy} - f_{xy}f_{yx}$. Since $f_{xy}$ and $f_{yx}$ are equal when continuous (refer back to Theorem 12.3.1), we can rewrite this as $D = f_{xx}f_{yy} - f_{xy}^2$. $D$ can be used to test whether the concavity at a point changes depending on direction. If $D > 0$, the concavity does not switch (i.e., at that point, the graph is concave up or down in all directions). If $D < 0$, the concavity does switch. If $D = 0$, our test fails to determine whether concavity switches or not. We state the use of $D$ in the following theorem.

---

**Theorem 12.8.2**     **Second Derivative Test**

Let $R$ be an open set on which a function $z = f(x,y)$ and all its first and second partial derivatives are defined, let $P = (x_0, y_0)$ be a critical point of $f$ in $R$, and let

$$D = f_{xx}(x_0, y_0)f_{yy}(x_0, y_0) - f_{xy}^2(x_0, y_0).$$

1. If $D > 0$ and $f_{xx}(x_0, y_0) > 0$, then $f$ has a relative minimum at $P$.

2. If $D > 0$ and $f_{xx}(x_0, y_0) < 0$, then $f$ has a relative maximum at $P$.

3. If $D < 0$, then $f$ has a saddle point at $P$.

4. If $D = 0$, the test is inconclusive.

---

Notes:

We first practice using this test with the function in the previous example, where we visually determined we had a relative maximum and a saddle point.

**Example 12.8.4  Using the Second Derivative Test**
Let $f(x, y) = x^3 - 3x - y^2 + 4y$ as in Example 12.8.3. Determine whether the function has a relative minimum, maximum, or saddle point at each critical point.

**SOLUTION**  We determined previously that the critical points of $f$ are $(-1, 2)$ and $(1, 2)$. To use the Second Derivative Test, we must find the second partial derivatives of $f$:

$$f_{xx} = 6x; \qquad f_{yy} = -2; \qquad f_{xy} = 0.$$

Thus $D(x, y) = -12x$.

At $(-1, 2)$: $D(-1, 2) = 12 > 0$, and $f_{xx}(-1, 2) = -6$. By the Second Derivative Test, $f$ has a relative maximum at $(-1, 2)$.

At $(1, 2)$: $D(1, 2) = -12 < 0$. The Second Derivative Test states that $f$ has a saddle point at $(1, 2)$.

The Second Derivative Test confirmed what we determined visually.

**Example 12.8.5  Using the Second Derivative Test**
Find the relative extrema of $f(x, y) = x^2 y + y^2 + xy$.

**SOLUTION**  We start by finding the first and second partial derivatives of $f$:

$$f_x = 2xy + y \qquad f_y = x^2 + 2y + x$$
$$f_{xx} = 2y \qquad f_{yy} = 2$$
$$f_{xy} = 2x + 1 \qquad f_{yx} = 2x + 1.$$

We find the critical points by finding where $f_x$ and $f_y$ are simultaneously 0 (they are both never undefined). Setting $f_x = 0$, we have:

$$f_x = 0 \quad \Rightarrow \quad 2xy + y = 0 \quad \Rightarrow \quad y(2x + 1) = 0.$$

This implies that for $f_x = 0$, either $y = 0$ or $2x + 1 = 0$.

Assume $y = 0$ then consider $f_y = 0$:

$$f_y = 0$$
$$x^2 + 2y + x = 0, \quad \text{and since } y = 0, \text{ we have}$$
$$x^2 + x = 0$$
$$x(x + 1) = 0.$$

Thus if $y = 0$, we have either $x = 0$ or $x = -1$, giving two critical points: $(-1, 0)$ and $(0, 0)$.

Notes:

## Chapter 12  Functions of Several Variables

Going back to $f_x$, now assume $2x+1 = 0$, i.e., that $x = -1/2$, then consider $f_y = 0$:

$$f_y = 0$$
$$x^2 + 2y + x = 0,$$
$$1/4 + 2y - 1/2 = 0$$
$$y = 1/8.$$

and since $x = -1/2$, we have

Thus if $x = -1/2$, $y = 1/8$ giving the critical point $(-1/2, 1/8)$.

With $D = 4y - (2x+1)^2$, we apply the Second Derivative Test to each critical point.

At $(-1, 0)$, $D < 0$, so $(-1, 0)$ is a saddle point.
At $(0, 0)$, $D < 0$, so $(0, 0)$ is also a saddle point.
At $(-1/2, 1/8)$, $D > 0$ and $f_{xx} > 0$, so $(-1/2, 1/8)$ is a relative minimum.

Figure 12.8.4 shows a graph of $f$ and the three critical points. Note how this function does not vary much near the critical points – that is, visually it is difficult to determine whether a point is a saddle point or relative minimum (or even a critical point at all!). This is one reason why the Second Derivative Test is so important to have.

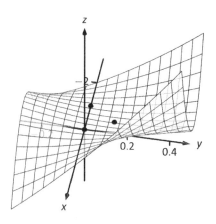

Figure 12.8.4: Graphing $f$ from Example 12.8.5 and its relative extrema.

## Constrained Optimization

When optimizing functions of one variable such as $y = f(x)$, we made use of Theorem 3.1.1, the Extreme Value Theorem, that said that over a closed interval $I$, a continuous function has both a maximum and minimum value. To find these maximum and minimum values, we evaluated $f$ at all critical points in the interval, as well as at the endpoints (the "boundary") of the interval.

A similar theorem and procedure applies to functions of two variables. A continuous function over a closed set also attains a maximum and minimum value (see the following theorem). We can find these values by evaluating the function at the critical values in the set and over the boundary of the set. After formally stating this extreme value theorem, we give examples.

---

**Theorem 12.8.3**  **Extreme Value Theorem**

Let $z = f(x, y)$ be a continuous function on a closed, bounded set $S$. Then $f$ has a maximum and minimum value on $S$.

---

**Example 12.8.6**  **Finding extrema on a closed set**

Let $f(x, y) = x^2 - y^2 + 5$ and let $S$ be the triangle with vertices $(-1, -2)$, $(0, 1)$ and $(2, -2)$. Find the maximum and minimum values of $f$ on $S$.

---

Notes:

**Solution**     It can help to see a graph of $f$ along with the set $S$. In Figure 12.8.5(a) the triangle defining $S$ is shown in the x-y plane in a dashed line. Above it is the surface of $f$; we are only concerned with the portion of $f$ enclosed by the "triangle" on its surface.

We begin by finding the critical points of $f$. With $f_x = 2x$ and $f_y = -2y$, we find only one critical point, at $(0, 0)$.

We now find the maximum and minimum values that $f$ attains along the boundary of $S$, that is, along the edges of the triangle. In Figure 12.8.5(b) we see the triangle sketched in the plane with the equations of the lines forming its edges labeled.

Start with the bottom edge, along the line $y = -2$. If $y$ is $-2$, then on the surface, we are considering points $f(x, -2)$; that is, our function reduces to $f(x, -2) = x^2 - (-2)^2 + 5 = x^2 + 1 = f_1(x)$. We want to maximize/minimize $f_1(x) = x^2 + 1$ on the interval $[-1, 2]$. To do so, we evaluate $f_1(x)$ at its critical points and at the endpoints.

The critical points of $f_1$ are found by setting its derivative equal to 0:

$$f_1'(x) = 0 \quad \Rightarrow x = 0.$$

Evaluating $f_1$ at this critical point, and at the endpoints of $[-1, 2]$ gives:

$$\begin{aligned} f_1(-1) = 2 &\quad \Rightarrow \quad f(-1, -2) = 2 \\ f_1(0) = 1 &\quad \Rightarrow \quad f(0, -2) = 1 \\ f_1(2) = 5 &\quad \Rightarrow \quad f(2, -2) = 5. \end{aligned}$$

Notice how evaluating $f_1$ at a point is the same as evaluating $f$ at its corresponding point.

We need to do this process twice more, for the other two edges of the triangle.

Along the left edge, along the line $y = 3x + 1$, we substitute $3x + 1$ in for $y$ in $f(x, y)$:

$$f(x, y) = f(x, 3x + 1) = x^2 - (3x + 1)^2 + 5 = -8x^2 - 6x + 4 = f_2(x).$$

We want the maximum and minimum values of $f_2$ on the interval $[-1, 0]$, so we evaluate $f_2$ at its critical points and the endpoints of the interval. We find the critical points:

$$f_2'(x) = -16x - 6 = 0 \quad \Rightarrow \quad x = -3/8.$$

Evaluate $f_2$ at its critical point and the endpoints of $[-1, 0]$:

$$\begin{aligned} f_2(-1) = 2 &\quad \Rightarrow \quad f(-1, -2) = 2 \\ f_2(-3/8) = 41/8 = 5.125 &\quad \Rightarrow \quad f(-3/8, -0.125) = 5.125 \\ f_2(0) = 4 &\quad \Rightarrow \quad f(0, 1) = 4. \end{aligned}$$

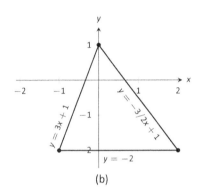

Figure 12.8.5: Plotting the surface of $f$ along with the restricted domain $S$ in Example 12.8.6.

Finally, we evaluate $f$ along the right edge of the triangle, where $y = -3/2x + 1$.

$$f(x, y) = f(x, -3/2x + 1) = x^2 - (-3/2x + 1)^2 + 5 = -\frac{5}{4}x^2 + 3x + 4 = f_3(x).$$

The critical points of $f_3(x)$ are:

$$f_3'(x) = 0 \quad \Rightarrow \quad x = 6/5 = 1.2.$$

We evaluate $f_3$ at this critical point and at the endpoints of the interval $[0, 2]$:

$$\begin{aligned} f_3(0) &= 4 & \Rightarrow & & f(0, 1) &= 4 \\ f_3(1.2) &= 5.8 & \Rightarrow & & f(1.2, -0.8) &= 5.8 \\ f_3(2) &= 5 & \Rightarrow & & f(2, -2) &= 5. \end{aligned}$$

One last point to test: the critical point of $f$, $(0, 0)$. We find $f(0, 0) = 5$.

We have evaluated $f$ at a total of 7 different places, all shown in Figure 12.8.6. We checked each vertex of the triangle twice, as each showed up as the endpoint of an interval twice. Of all the $z$-values found, the maximum is 5.8, found at $(1.2, -0.8)$; the minimum is 1, found at $(0, -2)$.

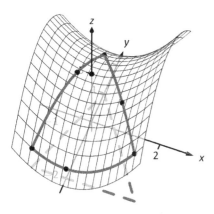

Figure 12.8.6: The surface of $f$ along with important points along the boundary of $S$ and the interior in Example 12.8.6.

This portion of the text is entitled "Constrained Optimization" because we want to optimize a function (i.e., find its maximum and/or minimum values) subject to a *constraint* – some limit to what values the function can attain. In the previous example, we constrained ourselves by considering a function only within the boundary of a triangle. This was largely arbitrary; the function and the boundary were chosen just as an example, with no real "meaning" behind the function or the chosen constraint.

However, solving constrained optimization problems is a very important topic in applied mathematics. The techniques developed here are the basis for solving larger problems, where more than two variables are involved.

We illustrate the technique once more with a classic problem.

### Example 12.8.7    Constrained Optimization

The U.S. Postal Service states that the girth+length of Standard Post Package must not exceed 130". Given a rectangular box, the "length" is the longest side, and the "girth" is twice the width+height.

Given a rectangular box where the width and height are equal, what are the dimensions of the box that give the maximum volume subject to the constraint of the size of a Standard Post Package?

**Solution**    Let $w$, $h$ and $\ell$ denote the width, height and length of a rectangular box; we assume here that $w = h$. The girth is then $2(w + h) = 4w$. The

Notes:

volume of the box is $V(w, \ell) = wh\ell = w^2\ell$. We wish to maximize this volume subject to the constraint $4w + \ell \leq 130$, or $\ell \leq 130 - 4w$. (Common sense also indicates that $\ell > 0, w > 0$.)

We begin by finding the critical values of $V$. We find that $V_w = 2w\ell$ and $V_\ell = w^2$; these are simultaneously 0 only at $(0, 0)$. This gives a volume of 0, so we can ignore this critical point.

We now consider the volume along the constraint $\ell = 130 - 4w$. Along this line, we have:

$$V(w, \ell) = V(w, 130 - 4w) = w^2(130 - 4w) = 130w^2 - 4w^3 = V_1(w).$$

The constraint is applicable on the $w$-interval $[0, 32.5]$ as indicated in the figure. Thus we want to maximize $V_1$ on $[0, 32.5]$.

Finding the critical values of $V_1$, we take the derivative and set it equal to 0:

$$V_1'(w) = 260w - 12w^2 = 0 \quad \Rightarrow \quad w(260 - 12w) = 0 \quad \Rightarrow \quad w = 0, \frac{260}{12} \approx 21.67.$$

We found two critical values: when $w = 0$ and when $w = 21.67$. We again ignore the $w = 0$ solution; the maximum volume, subject to the constraint, comes at $w = h = 21.67$, $\ell = 130 - 4(21.6) = 43.33$. This gives a volume of $V(21.67, 43.33) \approx 19,408 \text{in}^3$.

The volume function $V(w, \ell)$ is shown in Figure 12.8.7 along with the constraint $\ell = 130 - 4w$. As done previously, the constraint is drawn dashed in the x-y plane and also along the surface of the function. The point where the volume is maximized is indicated.

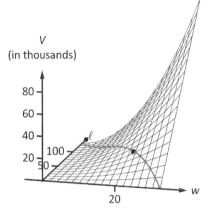

Figure 12.8.7: Graphing the volume of a box with girth $4w$ and length $\ell$, subject to a size constraint.

It is hard to overemphasize the importance of optimization. In "the real world," we routinely seek to make *something* better. By expressing the *something* as a mathematical function, "making *something* better" means "optimize *some function*."

The techniques shown here are only the beginning of an incredibly important field. Many functions that we seek to optimize are incredibly complex, making the step of "find the gradient and set it equal to $\vec{0}$" highly nontrivial. Mastery of the principles here are key to being able to tackle these more complicated problems.

Notes:

# Exercises 12.8

## Terms and Concepts

1. T/F: Theorem 12.8.1 states that if $f$ has a critical point at $P$, then $f$ has a relative extrema at $P$.

2. T/F: A point $P$ is a critical point of $f$ if $f_x$ and $f_y$ are both 0 at $P$.

3. T/F: A point $P$ is a critical point of $f$ if $f_x$ or $f_y$ are undefined at $P$.

4. Explain what it means to "solve a constrained optimization" problem.

## Problems

In Exercises 5 – 14, find the critical points of the given function. Use the Second Derivative Test to determine if each critical point corresponds to a relative maximum, minimum, or saddle point.

5. $f(x,y) = \frac{1}{2}x^2 + 2y^2 - 8y + 4x$

6. $f(x,y) = x^2 + 4x + y^2 - 9y + 3xy$

7. $f(x,y) = x^2 + 3y^2 - 6y + 4xy$

8. $f(x,y) = \dfrac{1}{x^2 + y^2 + 1}$

9. $f(x,y) = x^2 + y^3 - 3y + 1$

10. $f(x,y) = \frac{1}{3}x^3 - x + \frac{1}{3}y^3 - 4y$

11. $f(x,y) = x^2 y^2$

12. $f(x,y) = x^4 - 2x^2 + y^3 - 27y - 15$

13. $f(x,y) = \sqrt{16 - (x-3)^2 - y^2}$

14. $f(x,y) = \sqrt{x^2 + y^2}$

In Exercises 15 – 18, find the absolute maximum and minimum of the function subject to the given constraint.

15. $f(x,y) = x^2 + y^2 + y + 1$, constrained to the triangle with vertices $(0,1)$, $(-1,-1)$ and $(1,-1)$.

16. $f(x,y) = 5x - 7y$, constrained to the region bounded by $y = x^2$ and $y = 1$.

17. $f(x,y) = x^2 + 2x + y^2 + 2y$, constrained to the region bounded by the circle $x^2 + y^2 = 4$.

18. $f(x,y) = 3y - 2x^2$, constrained to the region bounded by the parabola $y = x^2 + x - 1$ and the line $y = x$.

# 13: MULTIPLE INTEGRATION

The previous chapter introduced multivariable functions and we applied concepts of differential calculus to these functions. We learned how we can view a function of two variables as a surface in space, and learned how partial derivatives convey information about how the surface is changing in any direction.

In this chapter we apply techniques of integral calculus to multivariable functions. In Chapter 5 we learned how the definite integral of a single variable function gave us "area under the curve." In this chapter we will see that integration applied to a multivariable function gives us "volume under a surface." And just as we learned applications of integration beyond finding areas, we will find applications of integration in this chapter beyond finding volume.

## 13.1 Iterated Integrals and Area

In Chapter 12 we found that it was useful to differentiate functions of several variables with respect to one variable, while treating all the other variables as constants or coefficients. We can integrate functions of several variables in a similar way. For instance, if we are told that $f_x(x,y) = 2xy$, we can treat $y$ as staying constant and integrate to obtain $f(x,y)$:

$$f(x,y) = \int f_x(x,y)\,dx$$
$$= \int 2xy\,dx$$
$$= x^2 y + C.$$

Make a careful note about the constant of integration, $C$. This "constant" is something with a derivative of 0 with respect to $x$, so it could be any expression that contains only constants and functions of $y$. For instance, if $f(x,y) = x^2 y + \sin y + y^3 + 17$, then $f_x(x,y) = 2xy$. To signify that $C$ is actually a function of $y$, we write:

$$f(x,y) = \int f_x(x,y)\,dx = x^2 y + C(y).$$

Using this process we can even evaluate definite integrals.

**Example 13.1.1**    **Integrating functions of more than one variable**

Evaluate the integral $\displaystyle\int_1^{2y} 2xy\,dx$.

**SOLUTION**    We find the indefinite integral as before, then apply the Fundamental Theorem of Calculus to evaluate the definite integral:

$$\int_1^{2y} 2xy\,dx = x^2 y \Big|_1^{2y}$$
$$= (2y)^2 y - (1)^2 y$$
$$= 4y^3 - y.$$

Chapter 13  Multiple Integration

We can also integrate with respect to y. In general,

$$\int_{h_1(y)}^{h_2(y)} f_x(x,y)\,dx = f(x,y)\Big|_{h_1(y)}^{h_2(y)} = f(h_2(y),y) - f(h_1(y),y),$$

and

$$\int_{g_1(x)}^{g_2(x)} f_y(x,y)\,dy = f(x,y)\Big|_{g_1(x)}^{g_2(x)} = f(x,g_2(x)) - f(x,g_1(x)).$$

Note that when integrating with respect to x, the bounds are functions of y (of the form $x = h_1(y)$ and $x = h_2(y)$) and the final result is also a function of y. When integrating with respect to y, the bounds are functions of x (of the form $y = g_1(x)$ and $y = g_2(x)$) and the final result is a function of x. Another example will help us understand this.

**Example 13.1.2**   **Integrating functions of more than one variable**
Evaluate $\int_{1}^{x} \left(5x^3 y^{-3} + 6y^2\right) dy$.

**SOLUTION**   We consider x as staying constant and integrate with respect to y:

$$\int_{1}^{x} \left(5x^3 y^{-3} + 6y^2\right) dy = \left(\frac{5x^3 y^{-2}}{-2} + \frac{6y^3}{3}\right)\Big|_{1}^{x}$$

$$= \left(-\frac{5}{2}x^3 x^{-2} + 2x^3\right) - \left(-\frac{5}{2}x^3 + 2\right)$$

$$= \frac{9}{2}x^3 - \frac{5}{2}x - 2.$$

Note how the bounds of the integral are from $y = 1$ to $y = x$ and that the final answer is a function of x.

In the previous example, we integrated a function with respect to y and ended up with a function of x. We can integrate this as well. This process is known as **iterated integration**, or **multiple integration.**

**Example 13.1.3**   **Integrating an integral**
Evaluate $\int_{1}^{2} \left( \int_{1}^{x} \left(5x^3 y^{-3} + 6y^2\right) dy \right) dx$.

**SOLUTION**   We follow a standard "order of operations" and perform the operations inside parentheses first (which is the integral evaluated in Example

Notes:

13.1.2.)

$$\int_1^2 \left(\int_1^x (5x^3y^{-3} + 6y^2)\, dy\right) dx = \int_1^2 \left(\left[\frac{5x^3y^{-2}}{-2} + \frac{6y^3}{3}\right]\Big|_1^x\right) dx$$

$$= \int_1^2 \left(\frac{9}{2}x^3 - \frac{5}{2}x - 2\right) dx$$

$$= \left(\frac{9}{8}x^4 - \frac{5}{4}x^2 - 2x\right)\Big|_1^2$$

$$= \frac{89}{8}.$$

Note how the bounds of $x$ were $x = 1$ to $x = 2$ and the final result was a number.

The previous example showed how we could perform something called an iterated integral; we do not yet know *why* we would be interested in doing so nor what the result, such as the number $89/8$, *means*. Before we investigate these questions, we offer some definitions.

---

**Definition 13.1.1**    **Iterated Integration**

**Iterated integration** is the process of repeatedly integrating the results of previous integrations. Integrating one integral is denoted as follows.

Let $a$, $b$, $c$ and $d$ be numbers and let $g_1(x)$, $g_2(x)$, $h_1(y)$ and $h_2(y)$ be functions of $x$ and $y$, respectively. Then:

1. $\displaystyle\int_c^d \int_{h_1(y)}^{h_2(y)} f(x,y)\, dx\, dy = \int_c^d \left(\int_{h_1(y)}^{h_2(y)} f(x,y)\, dx\right) dy.$

2. $\displaystyle\int_a^b \int_{g_1(x)}^{g_2(x)} f(x,y)\, dy\, dx = \int_a^b \left(\int_{g_1(x)}^{g_2(x)} f(x,y)\, dy\right) dx.$

---

Again make note of the bounds of these iterated integrals.

With $\displaystyle\int_c^d \int_{h_1(y)}^{h_2(y)} f(x,y)\, dx\, dy$, $x$ varies from $h_1(y)$ to $h_2(y)$, whereas $y$ varies from $c$ to $d$. That is, the bounds of $x$ are *curves*, the curves $x = h_1(y)$ and $x = h_2(y)$, whereas the bounds of $y$ are *constants*, $y = c$ and $y = d$. It is useful to remember that when setting up and evaluating such iterated integrals, we integrate "from

Notes:

# Chapter 13 Multiple Integration

curve to curve, then from point to point."

We now begin to investigate *why* we are interested in iterated integrals and *what* they mean.

## Area of a plane region

Consider the plane region $R$ bounded by $a \leq x \leq b$ and $g_1(x) \leq y \leq g_2(x)$, shown in Figure 13.1.1. We learned in Section 7.1 that the area of $R$ is given by

$$\int_a^b \big(g_2(x) - g_1(x)\big)\, dx.$$

We can view the expression $\big(g_2(x) - g_1(x)\big)$ as

$$\big(g_2(x) - g_1(x)\big) = \int_{g_1(x)}^{g_2(x)} 1\, dy = \int_{g_1(x)}^{g_2(x)} dy,$$

meaning we can express the area of $R$ as an iterated integral:

$$\text{area of } R = \int_a^b \big(g_2(x) - g_1(x)\big)\, dx = \int_a^b \left(\int_{g_1(x)}^{g_2(x)} dy\right) dx = \int_a^b \int_{g_1(x)}^{g_2(x)} dy\, dx.$$

In short: a certain iterated integral can be viewed as giving the area of a plane region.

A region $R$ could also be defined by $c \leq y \leq d$ and $h_1(y) \leq x \leq h_2(y)$, as shown in Figure 13.1.2. Using a process similar to that above, we have

$$\text{the area of } R = \int_c^d \int_{h_1(y)}^{h_2(y)} dx\, dy.$$

We state this formally in a theorem.

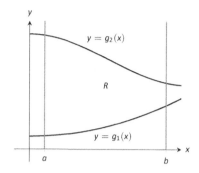

Figure 13.1.1: Calculating the area of a plane region $R$ with an iterated integral.

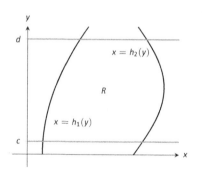

Figure 13.1.2: Calculating the area of a plane region $R$ with an iterated integral.

Notes:

## 13.1 Iterated Integrals and Area

> **Theorem 13.1.1  Area of a plane region**
>
> 1. Let $R$ be a plane region bounded by $a \leq x \leq b$ and $g_1(x) \leq y \leq g_2(x)$, where $g_1$ and $g_2$ are continuous functions on $[a, b]$. The area $A$ of $R$ is
> $$A = \int_a^b \int_{g_1(x)}^{g_2(x)} dy\, dx.$$
>
> 2. Let $R$ be a plane region bounded by $c \leq y \leq d$ and $h_1(y) \leq x \leq h_2(y)$, where $h_1$ and $h_2$ are continuous functions on $[c, d]$. The area $A$ of $R$ is
> $$A = \int_c^d \int_{h_1(y)}^{h_2(y)} dx\, dy.$$

The following examples should help us understand this theorem.

### Example 13.1.4  Area of a rectangle
Find the area $A$ of the rectangle with corners $(-1, 1)$ and $(3, 3)$, as shown in Figure 13.1.3.

**Solution**  Multiple integration is obviously overkill in this situation, but we proceed to establish its use.

The region $R$ is bounded by $x = -1$, $x = 3$, $y = 1$ and $y = 3$. Choosing to integrate with respect to $y$ first, we have

$$A = \int_{-1}^{3} \int_{1}^{3} 1\, dy\, dx = \int_{-1}^{3} \left( y \Big|_{1}^{3} \right) dx = \int_{-1}^{3} 2\, dx = 2x \Big|_{-1}^{3} = 8.$$

We could also integrate with respect to $x$ first, giving:

$$A = \int_{1}^{3} \int_{-1}^{3} 1\, dx\, dy = \int_{1}^{3} \left( x \Big|_{-1}^{3} \right) dy = \int_{1}^{3} 4\, dy = 4y \Big|_{1}^{3} = 8.$$

Clearly there are simpler ways to find this area, but it is interesting to note that this method works.

### Example 13.1.5  Area of a triangle
Find the area $A$ of the triangle with vertices at $(1, 1)$, $(3, 1)$ and $(5, 5)$, as shown in Figure 13.1.4.

**Solution**  The triangle is bounded by the lines as shown in the figure. Choosing to integrate with respect to $x$ first gives that $x$ is bounded by $x = y$

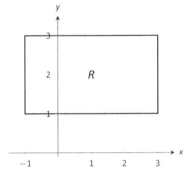

Figure 13.1.3: Calculating the area of a rectangle with an iterated integral in Example 13.1.4.

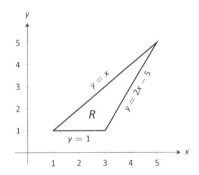

Figure 13.1.4: Calculating the area of a triangle with iterated integrals in Example 13.1.5.

Notes:

to $x = \frac{y+5}{2}$, while $y$ is bounded by $y = 1$ to $y = 5$. (Recall that since $x$-values increase from left to right, the leftmost curve, $x = y$, is the lower bound and the rightmost curve, $x = (y+5)/2$, is the upper bound.) The area is

$$A = \int_1^5 \int_y^{\frac{y+5}{2}} dx\, dy$$

$$= \int_1^5 \left( x \Big|_y^{\frac{y+5}{2}} \right) dy$$

$$= \int_1^5 \left( -\frac{1}{2}y + \frac{5}{2} \right) dy$$

$$= \left( -\frac{1}{4}y^2 + \frac{5}{2}y \right) \Big|_1^5$$

$$= 4.$$

We can also find the area by integrating with respect to $y$ first. In this situation, though, we have two functions that act as the lower bound for the region $R$, $y = 1$ and $y = 2x - 5$. This requires us to use two iterated integrals. Note how the $x$-bounds are different for each integral:

$$A = \int_1^3 \int_1^x 1\, dy\, dx \quad + \quad \int_3^5 \int_{2x-5}^x 1\, dy\, dx$$

$$= \int_1^3 (y)\Big|_1^x dx \quad + \quad \int_3^5 (y)\Big|_{2x-5}^x dx$$

$$= \int_1^3 (x - 1)\, dx \quad + \quad \int_3^5 (-x + 5)\, dx$$

$$= 2 \quad + \quad 2$$

$$= 4.$$

As expected, we get the same answer both ways.

### Example 13.1.6  Area of a plane region

Find the area of the region enclosed by $y = 2x$ and $y = x^2$, as shown in Figure 13.1.5.

**Solution**  Once again we'll find the area of the region using both orders of integration.

Using $dy\, dx$:

$$\int_0^2 \int_{x^2}^{2x} 1\, dy\, dx = \int_0^2 (2x - x^2)\, dx = \left( x^2 - \frac{1}{3}x^3 \right)\Big|_0^2 = \frac{4}{3}.$$

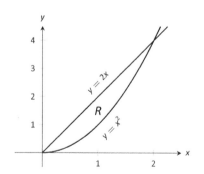

Figure 13.1.5: Calculating the area of a plane region with iterated integrals in Example 13.1.6.

Notes:

## 13.1 Iterated Integrals and Area

Using $dx\,dy$:

$$\int_0^4 \int_{y/2}^{\sqrt{y}} 1\,dx\,dy = \int_0^4 (\sqrt{y} - y/2)\,dy = \left(\frac{2}{3}y^{3/2} - \frac{1}{4}y^2\right)\Big|_0^4 = \frac{4}{3}.$$

### Changing Order of Integration

In each of the previous examples, we have been given a region $R$ and found the bounds needed to find the area of $R$ using both orders of integration. We integrated using both orders of integration to demonstrate their equality.

We now approach the skill of describing a region using both orders of integration from a different perspective. Instead of starting with a region and creating iterated integrals, we will start with an iterated integral and rewrite it in the other integration order. To do so, we'll need to understand the region over which we are integrating.

The simplest of all cases is when both integrals are bound by constants. The region described by these bounds is a rectangle (see Example 13.1.4), and so:

$$\int_a^b \int_c^d 1\,dy\,dx = \int_c^d \int_a^b 1\,dx\,dy.$$

When the inner integral's bounds are not constants, it is generally very useful to sketch the bounds to determine what the region we are integrating over looks like. From the sketch we can then rewrite the integral with the other order of integration.

Examples will help us develop this skill.

**Example 13.1.7**    **Changing the order of integration**

Rewrite the iterated integral $\displaystyle\int_0^6 \int_0^{x/3} 1\,dy\,dx$ with the order of integration $dx\,dy$.

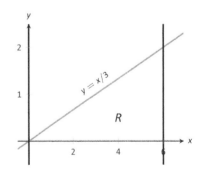

Figure 13.1.6: Sketching the region $R$ described by the iterated integral in Example 13.1.7.

**SOLUTION**    We need to use the bounds of integration to determine the region we are integrating over.

The bounds tell us that $y$ is bounded by $0$ and $x/3$; $x$ is bounded by $0$ and $6$. We plot these four curves: $y = 0$, $y = x/3$, $x = 0$ and $x = 6$ to find the region described by the bounds. Figure 13.1.6 shows these curves, indicating that $R$ is a triangle.

To change the order of integration, we need to consider the curves that bound the $x$-values. We see that the lower bound is $x = 3y$ and the upper bound is $x = 6$. The bounds on $y$ are $0$ to $2$. Thus we can rewrite the integral as

$$\int_0^2 \int_{3y}^6 1\,dx\,dy.$$

Notes:

## Chapter 13 Multiple Integration

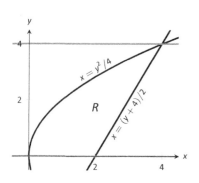

Figure 13.1.7: Drawing the region determined by the bounds of integration in Example 13.1.8.

**Example 13.1.8 Changing the order of integration**

Change the order of integration of $\displaystyle\int_0^4 \int_{y^2/4}^{(y+4)/2} 1\,dx\,dy$.

**SOLUTION** We sketch the region described by the bounds to help us change the integration order. $x$ is bounded below and above (i.e., to the left and right) by $x = y^2/4$ and $x = (y+4)/2$ respectively, and $y$ is bounded between 0 and 4. Graphing the previous curves, we find the region $R$ to be that shown in Figure 13.1.7.

To change the order of integration, we need to establish curves that bound $y$. The figure makes it clear that there are two lower bounds for $y$: $y = 0$ on $0 \leq x \leq 2$, and $y = 2x - 4$ on $2 \leq x \leq 4$. Thus we need two double integrals. The upper bound for each is $y = 2\sqrt{x}$. Thus we have

$$\int_0^4 \int_{y^2/4}^{(y+4)/2} 1\,dx\,dy = \int_0^2 \int_0^{2\sqrt{x}} 1\,dy\,dx + \int_2^4 \int_{2x-4}^{2\sqrt{x}} 1\,dy\,dx.$$

This section has introduced a new concept, the iterated integral. We developed one application for iterated integration: area between curves. However, this is not new, for we already know how to find areas bounded by curves.

In the next section we apply iterated integration to solve problems we currently do not know how to handle. The "real" goal of this section was not to learn a new way of computing area. Rather, our goal was to learn how to define a region in the plane using the bounds of an iterated integral. That skill is very important in the following sections.

Notes:

# Exercises 13.1

## Terms and Concepts

1. When integrating $f_x(x,y)$ with respect to $x$, the constant of integration $C$ is really which: $C(x)$ or $C(y)$? What does this mean?

2. Integrating an integral is called _____ _____.

3. When evaluating an iterated integral, we integrate from _____ to _____, then from _____ to _____.

4. One understanding of an iterated integral is that
$$\int_a^b \int_{g_1(x)}^{g_2(x)} dy\, dx$$
gives the _____ of a plane region.

## Problems

In Exercises 5 – 10, evaluate the integral and subsequent iterated integral.

5. (a) $\displaystyle\int_2^5 \left(6x^2 + 4xy - 3y^2\right) dy$

   (b) $\displaystyle\int_{-3}^{-2}\int_2^5 \left(6x^2 + 4xy - 3y^2\right) dy\, dx$

6. (a) $\displaystyle\int_0^\pi \left(2x\cos y + \sin x\right) dx$

   (b) $\displaystyle\int_0^{\pi/2}\int_0^\pi \left(2x\cos y + \sin x\right) dx\, dy$

7. (a) $\displaystyle\int_1^x \left(x^2 y - y + 2\right) dy$

   (b) $\displaystyle\int_0^2 \int_1^x \left(x^2 y - y + 2\right) dy\, dx$

8. (a) $\displaystyle\int_y^{y^2} (x-y)\, dx$

   (b) $\displaystyle\int_{-1}^1 \int_y^{y^2} (x-y)\, dx\, dy$

9. (a) $\displaystyle\int_0^y \left(\cos x \sin y\right) dx$

   (b) $\displaystyle\int_0^\pi \int_0^y \left(\cos x \sin y\right) dx\, dy$

10. (a) $\displaystyle\int_0^x \left(\frac{1}{1+x^2}\right) dy$

    (b) $\displaystyle\int_1^2 \int_0^x \left(\frac{1}{1+x^2}\right) dy\, dx$

In Exercises 11 – 16, a graph of a planar region $R$ is given. Give the iterated integrals, with both orders of integration $dy\, dx$ and $dx\, dy$, that give the area of $R$. Evaluate one of the iterated integrals to find the area.

11.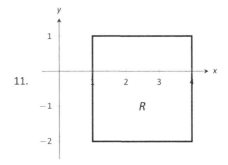

12.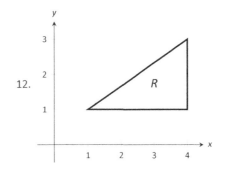

13.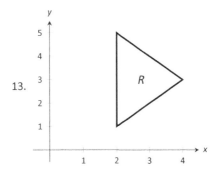

14.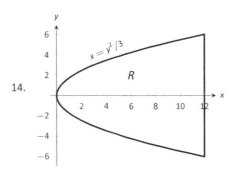

15.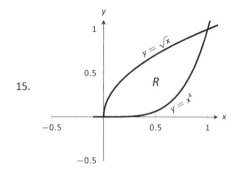

16.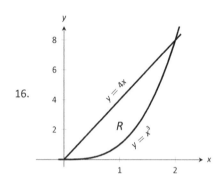

**In Exercises 17 – 22, iterated integrals are given that compute the area of a region $R$ in the x-y plane. Sketch the region $R$, and give the iterated integral(s) that give the area of $R$ with the opposite order of integration.**

17. $\displaystyle\int_{-2}^{2}\int_{0}^{4-x^2} dy\, dx$

18. $\displaystyle\int_{0}^{1}\int_{5-5x}^{5-5x^2} dy\, dx$

19. $\displaystyle\int_{-2}^{2}\int_{0}^{2\sqrt{4-y^2}} dx\, dy$

20. $\displaystyle\int_{-3}^{3}\int_{-\sqrt{9-x^2}}^{\sqrt{9-x^2}} dy\, dx$

21. $\displaystyle\int_{0}^{1}\int_{-\sqrt{y}}^{\sqrt{y}} dx\, dy + \int_{1}^{4}\int_{y-2}^{\sqrt{y}} dx\, dy$

22. $\displaystyle\int_{-1}^{1}\int_{(x-1)/2}^{(1-x)/2} dy\, dx$

## 13.2 Double Integration and Volume

The definite integral of $f$ over $[a, b]$, $\int_a^b f(x)\,dx$, was introduced as "the signed area under the curve." We approximated the value of this area by first subdividing $[a, b]$ into $n$ subintervals, where the $i^{\text{th}}$ subinterval has length $\Delta x_i$, and letting $c_i$ be any value in the $i^{\text{th}}$ subinterval. We formed rectangles that approximated part of the region under the curve with width $\Delta x_i$, height $f(c_i)$, and hence with area $f(c_i)\Delta x_i$. Summing all the rectangle's areas gave an approximation of the definite integral, and Theorem 5.3.2 stated that

$$\int_a^b f(x)\,dx = \lim_{\|\Delta x\| \to 0} \sum f(c_i)\Delta x_i,$$

connecting the area under the curve with sums of the areas of rectangles.

We use a similar approach in this section to find volume under a surface.

Let $R$ be a closed, bounded region in the x-y plane and let $z = f(x, y)$ be a continuous function defined on $R$. We wish to find the signed volume under the surface of $f$ over $R$. (We use the term "signed volume" to denote that space above the x-y plane, under $f$, will have a positive volume; space above $f$ and under the x-y plane will have a "negative" volume, similar to the notion of signed area used before.)

We start by partitioning $R$ into $n$ rectangular subregions as shown in Figure 13.2.1(a). For simplicity's sake, we let all widths be $\Delta x$ and all heights be $\Delta y$. Note that the sum of the areas of the rectangles is not equal to the area of $R$, but rather is a close approximation. Arbitrarily number the rectangles 1 through $n$, and pick a point $(x_i, y_i)$ in the $i^{\text{th}}$ subregion.

The volume of the rectangular solid whose base is the $i^{\text{th}}$ subregion and whose height is $f(x_i, y_i)$ is $V_i = f(x_i, y_i)\Delta x \Delta y$. Such a solid is shown in Figure 13.2.1(b). Note how this rectangular solid only approximates the true volume under the surface; part of the solid is above the surface and part is below.

For each subregion $R_i$ used to approximate $R$, create the rectangular solid with base area $\Delta x \Delta y$ and height $f(x_i, y_i)$. The sum of all rectangular solids is

$$\sum_{i=1}^{n} f(x_i, y_i)\Delta x \Delta y.$$

This approximates the signed volume under $f$ over $R$. As we have done before, to get a better approximation we can use more rectangles to approximate the region $R$.

In general, each rectangle could have a different width $\Delta x_j$ and height $\Delta y_k$, giving the $i^{\text{th}}$ rectangle an area $\Delta A_i = \Delta x_j \Delta y_k$ and the $i^{\text{th}}$ rectangular solid a

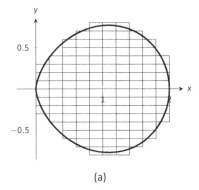

(a)

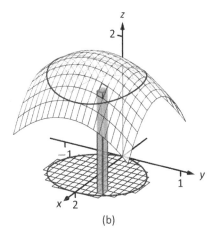

(b)

Figure 13.2.1: Developing a method for finding signed volume under a surface.

Notes:

## Chapter 13  Multiple Integration

volume of $f(x_i, y_i)\Delta A_i$. Let $||\Delta A||$ denote the length of the longest diagonal of all rectangles in the subdivision of $R$; $||\Delta A|| \to 0$ means each rectangle's width and height are both approaching 0. If $f$ is a continuous function, as $||\Delta A||$ shrinks (and hence $n \to \infty$) the summation $\sum_{i=1}^{n} f(x_i, y_i)\Delta A_i$ approximates the signed volume better and better. This leads to a definition.

**Note:** Recall that the integration symbol "$\int$" is an "elongated S," representing the word "sum." We interpreted $\int_a^b f(x)\,dx$ as "take the *sum* of the areas of rectangles over the interval $[a, b]$." The double integral uses two integration symbols to represent a "double sum." When adding up the volumes of rectangular solids over a partition of a region $R$, as done in Figure 13.2.1, one could first add up the volumes across each row (one type of sum), then add these totals together (another sum), as in

$$\sum_{j=1}^{n}\sum_{i=1}^{m} f(x_i, y_j)\Delta x_i \Delta y_j.$$

One can rewrite this as

$$\sum_{j=1}^{n}\left(\sum_{i=1}^{m} f(x_i, y_j)\Delta x_i\right)\Delta y_j.$$

The summation inside the parenthesis indicates the sum of heights × widths, which gives an area; multiplying these areas by the thickness $\Delta y_j$ gives a volume. The illustration in Figure 13.2.2 relates to this understanding.

---

**Definition 13.2.1**  **Double Integral, Signed Volume**

Let $z = f(x, y)$ be a continuous function defined over a closed, bounded region $R$ in the x-y plane. The **signed volume** $V$ under $f$ over $R$ is denoted by the **double integral**

$$V = \iint_R f(x, y)\, dA.$$

Alternate notations for the double integral are

$$\iint_R f(x, y)\, dA = \iint_R f(x, y)\, dx\, dy = \iint_R f(x, y)\, dy\, dx.$$

---

The definition above does not state how to find the signed volume, though the notation offers a hint. We need the next two theorems to evaluate double integrals to find volume.

---

**Theorem 13.2.1**  **Double Integrals and Signed Volume**

Let $z = f(x, y)$ be a continuous function defined over a closed, bounded region $R$ in the x-y plane. Then the signed volume $V$ under $f$ over $R$ is

$$V = \iint_R f(x, y)\, dA = \lim_{||\Delta A|| \to 0} \sum_{i=1}^{n} f(x_i, y_i)\Delta A_i.$$

---

This theorem states that we can find the exact signed volume using a limit of sums. The partition of the region $R$ is not specified, so any partitioning where the diagonal of each rectangle shrinks to 0 results in the same answer.

This does not offer a very satisfying way of computing volume, though. Our experience has shown that evaluating the limits of sums can be tedious. We seek a more direct method.

---

Notes:

Recall Theorem 7.2.1 in Section 7.2. This stated that if $A(x)$ gives the cross-sectional area of a solid at $x$, then $\int_a^b A(x)\,dx$ gave the volume of that solid over $[a, b]$.

Consider Figure 13.2.2, where a surface $z = f(x, y)$ is drawn over a region $R$. Fixing a particular $x$ value, we can consider the area under $f$ over $R$ where $x$ has that fixed value. That area can be found with a definite integral, namely

$$A(x) = \int_{g_1(x)}^{g_2(x)} f(x, y)\,dy.$$

Remember that though the integrand contains $x$, we are viewing $x$ as fixed. Also note that the bounds of integration are functions of $x$: the bounds depend on the value of $x$.

As $A(x)$ is a cross-sectional area function, we can find the signed volume $V$ under $f$ by integrating it:

$$V = \int_a^b A(x)\,dx = \int_a^b \left( \int_{g_1(x)}^{g_2(x)} f(x, y)\,dy \right) dx = \int_a^b \int_{g_1(x)}^{g_2(x)} f(x, y)\,dy\,dx.$$

This gives a concrete method for finding signed volume under a surface. We could do a similar procedure where we started with $y$ fixed, resulting in an iterated integral with the order of integration $dx\,dy$. The following theorem states that both methods give the same result, which is the value of the double integral. It is such an important theorem it has a name associated with it.

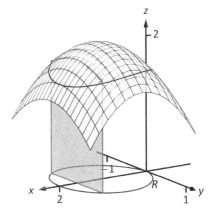

Figure 13.2.2: Finding volume under a surface by sweeping out a cross-sectional area.

---

**Theorem 13.2.2**     **Fubini's Theorem**

Let $R$ be a closed, bounded region in the $x$-$y$ plane and let $z = f(x, y)$ be a continuous function on $R$.

1. If $R$ is bounded by $a \leq x \leq b$ and $g_1(x) \leq y \leq g_2(x)$, where $g_1$ and $g_2$ are continuous functions on $[a, b]$, then

$$\iint_R f(x, y)\,dA = \int_a^b \int_{g_1(x)}^{g_2(x)} f(x, y)\,dy\,dx.$$

2. If $R$ is bounded by $c \leq y \leq d$ and $h_1(y) \leq x \leq h_2(y)$, where $h_1$ and $h_2$ are continuous functions on $[c, d]$, then

$$\iint_R f(x, y)\,dA = \int_c^d \int_{h_1(y)}^{h_2(y)} f(x, y)\,dx\,dy.$$

---

Notes:

Note that once again the bounds of integration follow the "curve to curve, point to point" pattern discussed in the previous section. In fact, one of the main points of the previous section is developing the skill of describing a region $R$ with the bounds of an iterated integral. Once this skill is developed, we can use double integrals to compute many quantities, not just signed volume under a surface.

**Example 13.2.1    Evaluating a double integral**

Let $f(x, y) = xy + e^y$. Find the signed volume under $f$ on the region $R$, which is the rectangle with corners $(3, 1)$ and $(4, 2)$ pictured in Figure 13.2.3, using Fubini's Theorem and both orders of integration.

**SOLUTION**    We wish to evaluate $\iint_R (xy + e^y)\, dA$. As $R$ is a rectangle, the bounds are easily described as $3 \leq x \leq 4$ and $1 \leq y \leq 2$.

Using the order $dy\, dx$:

$$\iint_R (xy + e^y)\, dA = \int_3^4 \int_1^2 (xy + e^y)\, dy\, dx$$

$$= \int_3^4 \left( \left[\frac{1}{2}xy^2 + e^y\right]\Big|_1^2 \right) dx$$

$$= \int_3^4 \left(\frac{3}{2}x + e^2 - e\right) dx$$

$$= \left(\frac{3}{4}x^2 + (e^2 - e)x\right)\Big|_3^4$$

$$= \frac{21}{4} + e^2 - e \approx 9.92.$$

Now we check the validity of Fubini's Theorem by using the order $dx\, dy$:

$$\iint_R (xy + e^y)\, dA = \int_1^2 \int_3^4 (xy + e^y)\, dx\, dy$$

$$= \int_1^2 \left( \left[\frac{1}{2}x^2 y + xe^y\right]\Big|_3^4 \right) dy$$

$$= \int_1^2 \left(\frac{7}{2}y + e^y\right) dy$$

$$= \left(\frac{7}{4}y^2 + e^y\right)\Big|_1^2$$

$$= \frac{21}{4} + e^2 - e \approx 9.92.$$

Both orders of integration return the same result, as expected.

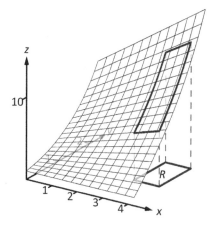

Figure 13.2.3: Finding the signed volume under a surface in Example 13.2.1.

Notes:

**Example 13.2.2  Evaluating a double integral**

Evaluate $\iint_R (3xy - x^2 - y^2 + 6)\, dA$, where $R$ is the triangle bounded by $x = 0$, $y = 0$ and $x/2 + y = 1$, as shown in Figure 13.2.4.

**SOLUTION**  While it is not specified which order we are to use, we will evaluate the double integral using both orders to help drive home the point that it does not matter which order we use.

Using the order $dy\, dx$: The bounds on $y$ go from "curve to curve," i.e., $0 \leq y \leq 1 - x/2$, and the bounds on $x$ go from "point to point," i.e., $0 \leq x \leq 2$.

$$\iint_R (3xy - x^2 - y^2 + 6)\, dA = \int_0^2 \int_0^{-\frac{x}{2}+1} (3xy - x^2 - y^2 + 6)\, dy\, dx$$

$$= \int_0^2 \left(\frac{3}{2}xy^2 - x^2y - \frac{1}{3}y^3 + 6y\right)\bigg|_0^{-\frac{x}{2}+1} dx$$

$$= \int_0^2 \left(\frac{11}{12}x^3 - \frac{11}{4}x^2 - x + \frac{17}{3}\right) dx$$

$$= \left(\frac{11}{48}x^4 - \frac{11}{12}x^3 - \frac{1}{2}x^2 + \frac{17}{3}x\right)\bigg|_0^2$$

$$= \frac{17}{3} = 5.\overline{6}.$$

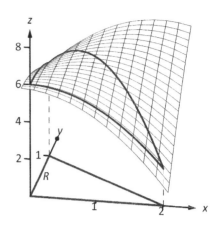

Figure 13.2.4: Finding the signed volume under the surface in Example 13.2.2.

Now lets consider the order $dx\, dy$. Here $x$ goes from "curve to curve," $0 \leq x \leq 2 - 2y$, and $y$ goes from "point to point," $0 \leq y \leq 1$:

$$\iint_R (3xy - x^2 - y^2 + 6)\, dA = \int_0^1 \int_0^{2-2y} (3xy - x^2 - y^2 + 6)\, dx\, dy$$

$$= \int_0^1 \left(\frac{3}{2}x^2 y - \frac{1}{3}x^3 - xy^2 + 6x\right)\bigg|_0^{2-2y} dy$$

$$= \int_0^1 \left(\frac{32}{3}y^3 - 22y^2 + 2y + \frac{28}{3}\right) dy$$

$$= \left(\frac{8}{3}y^4 - \frac{22}{3}y^3 + y^2 + \frac{28}{3}y\right)\bigg|_0^1$$

$$= \frac{17}{3} = 5.\overline{6}.$$

We obtained the same result using both orders of integration.

Note how in these two examples that the bounds of integration depend only on $R$; the bounds of integration have nothing to do with $f(x, y)$. This is an important concept, so we include it as a Key Idea.

Notes:

# Chapter 13  Multiple Integration

> **Key Idea 13.2.1  Double Integration Bounds**
>
> When evaluating $\iint_R f(x,y)\,dA$ using an iterated integral, the bounds of integration depend only on $R$. The surface $f$ does not determine the bounds of integration.

Before doing another example, we give some properties of double integrals. Each should make sense if we view them in the context of finding signed volume under a surface, over a region.

> **Theorem 13.2.3  Properties of Double Integrals**
>
> Let $f$ and $g$ be continuous functions over a closed, bounded plane region $R$, and let $c$ be a constant.
>
> 1. $\iint_R cf(x,y)\,dA = c\iint_R f(x,y)\,dA.$
> 2. $\iint_R \bigl(f(x,y) \pm g(x,y)\bigr)\,dA = \iint_R f(x,y)\,dA \pm \iint_R g(x,y)\,dA$
> 3. If $f(x,y) \geq 0$ on $R$, then $\iint_R f(x,y)\,dA \geq 0.$
> 4. If $f(x,y) \geq g(x,y)$ on $R$, then $\iint_R f(x,y)\,dA \geq \iint_R g(x,y)\,dA.$
> 5. Let $R$ be the union of two nonoverlapping regions, $R = R_1 \bigcup R_2$ (see Figure 13.2.5). Then
> $$\iint_R f(x,y)\,dA = \iint_{R_1} f(x,y)\,dA + \iint_{R_2} f(x,y)\,dA.$$

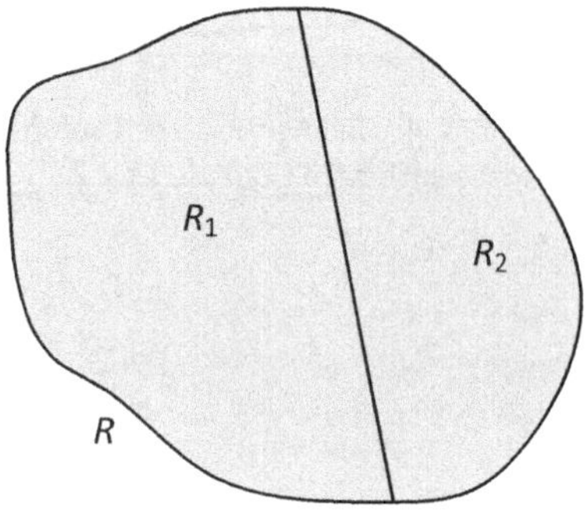

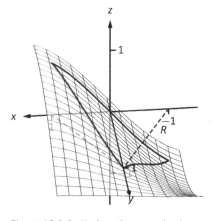

Figure 13.2.5: $R$ is the union of two nonoverlapping regions, $R_1$ and $R_2$.

**Example 13.2.3  Evaluating a double integral**
Let $f(x,y) = \sin x \cos y$ and $R$ be the triangle with vertices $(-1,0)$, $(1,0)$ and $(0,1)$ (see Figure 13.2.6). Evaluate the double integral $\iint_R f(x,y)\,dA$.

**Solution**  If we attempt to integrate using an iterated integral with the order $dy\,dx$, note how there are two upper bounds on $R$ meaning we'll need to use two iterated integrals. We would need to split the triangle into two regions along the $y$-axis, then use Theorem 13.2.3, part 5.

Instead, let's use the order $dx\,dy$. The curves bounding $x$ are $y - 1 \leq x \leq$

Figure 13.2.6: Finding the signed volume under a surface in Example 13.2.3.

Notes:

$1 - y$; the bounds on $y$ are $0 \leq y \leq 1$. This gives us:

$$\iint_R f(x, y)\, dA = \int_0^1 \int_{y-1}^{1-y} \sin x \cos y \, dx \, dy$$
$$= \int_0^1 \Big(-\cos x \cos y\Big)\Big|_{y-1}^{1-y} dy$$
$$= \int_0^1 \cos y \Big(-\cos(1-y) + \cos(y-1)\Big) dy.$$

Recall that the cosine function is an even function; that is, $\cos x = \cos(-x)$. Therefore, from the last integral above, we have $\cos(y-1) = \cos(1-y)$. Thus the integrand simplifies to 0, and we have

$$\iint_R f(x, y)\, dA = \int_0^1 0\, dy$$
$$= 0.$$

It turns out that over $R$, there is just as much volume above the $x$-$y$ plane as below (look again at Figure 13.2.6), giving a final signed volume of 0.

**Example 13.2.4  Evaluating a double integral**
Evaluate $\iint_R (4 - y)\, dA$, where $R$ is the region bounded by the parabolas $y^2 = 4x$ and $x^2 = 4y$, graphed in Figure 13.2.7.

**Solution**    Graphing each curve can help us find their points of intersection. Solving analytically, the second equation tells us that $y = x^2/4$. Substituting this value in for $y$ in the first equation gives us $x^4/16 = 4x$. Solving for $x$:

$$\frac{x^4}{16} = 4x$$
$$x^4 - 64x = 0$$
$$x(x^3 - 64) = 0$$
$$x = 0, 4.$$

Thus we've found analytically what was easy to approximate graphically: the regions intersect at $(0, 0)$ and $(4, 4)$, as shown in Figure 13.2.7.

We now choose an order of integration: $dy\, dx$ or $dx\, dy$? Either order works; since the integrand does not contain $x$, choosing $dx\, dy$ might be simpler – at least, the first integral is very simple.

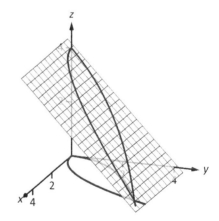

Figure 13.2.7: Finding the volume under the surface in Example 13.2.4.

Thus we have the following "curve to curve, point to point" bounds: $y^2/4 \leq x \leq 2\sqrt{y}$, and $0 \leq y \leq 4$.

$$\iint_R (4-y)\, dA = \int_0^4 \int_{y^2/4}^{2\sqrt{y}} (4-y)\, dx\, dy$$
$$= \int_0^4 \left(x(4-y)\right)\Big|_{y^2/4}^{2\sqrt{y}} dy$$
$$= \int_0^4 \left(\left(2\sqrt{y} - \frac{y^2}{4}\right)(4-y)\right) dy = \int_0^4 \left(\frac{y^3}{4} - y^2 - 2y^{3/2} + 8y^{1/2}\right) dy$$
$$= \left(\frac{y^4}{16} - \frac{y^3}{3} - \frac{4y^{5/2}}{5} + \frac{16y^{3/2}}{3}\right)\Big|_0^4$$
$$= \frac{176}{15} = 11.7\overline{3}.$$

The signed volume under the surface $f$ is about 11.7 cubic units.

In the previous section we practiced changing the order of integration of a given iterated integral, where the region $R$ was not explicitly given. Changing the bounds of an integral is more than just an test of understanding. Rather, there are cases where integrating in one order is really hard, if not impossible, whereas integrating with the other order is feasible.

**Example 13.2.5    Changing the order of integration**

Rewrite the iterated integral $\displaystyle\int_0^3 \int_y^3 e^{-x^2}\, dx\, dy$ with the order $dy\, dx$. Comment on the feasibility to evaluate each integral.

**SOLUTION**    Once again we make a sketch of the region over which we are integrating to facilitate changing the order. The bounds on $x$ are from $x = y$ to $x = 3$; the bounds on $y$ are from $y = 0$ to $y = 3$. These curves are sketched in Figure 13.2.8, enclosing the region $R$.

To change the bounds, note that the curves bounding $y$ are $y = 0$ up to $y = x$; the triangle is enclosed between $x = 0$ and $x = 3$. Thus the new bounds of integration are $0 \leq y \leq x$ and $0 \leq x \leq 3$, giving the iterated integral $\displaystyle\int_0^3 \int_0^x e^{-x^2}\, dy\, dx$.

How easy is it to evaluate each iterated integral? Consider the order of integrating $dx\, dy$, as given in the original problem. The first indefinite integral we need to evaluate is $\int e^{-x^2}\, dx$; we have stated before (see Section 5.5) that this integral cannot be evaluated in terms of elementary functions. We are stuck.

Changing the order of integration makes a big difference here. In the second

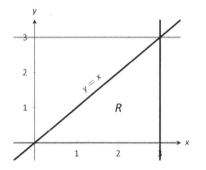

Figure 13.2.8: Determining the region $R$ determined by the bounds of integration in Example 13.2.5.

Notes:

iterated integral, we are faced with $\int e^{-x^2}\,dy$; integrating with respect to $y$ gives us $ye^{-x^2} + C$, and the first definite integral evaluates to
$$\int_0^x e^{-x^2}\,dy = xe^{-x^2}.$$
Thus
$$\int_0^3 \int_0^x e^{-x^2}\,dy\,dx = \int_0^3 \left(xe^{-x^2}\right)dx.$$
This last integral is easy to evaluate with substitution, giving a final answer of $\frac{1}{2}(1 - e^{-9}) \approx 0.5$. Figure 13.2.9 shows the surface over $R$.

In short, evaluating one iterated integral is impossible; the other iterated integral is relatively simple.

Definition 5.4.1 defines the average value of a single–variable function $f(x)$ on the interval $[a, b]$ as
$$\text{average value of } f(x) \text{ on } [a, b] = \frac{1}{b - a}\int_a^b f(x)\,dx;$$
that is, it is the "area under $f$ over an interval divided by the length of the interval." We make an analogous statement here: the average value of $z = f(x, y)$ over a region $R$ is the volume under $f$ over $R$ divided by the area of $R$.

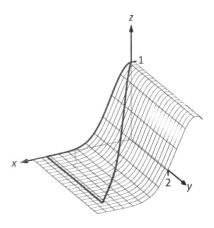

Figure 13.2.9: Showing the surface $f$ defined in Example 13.2.5 over its region $R$.

---

**Definition 13.2.2**   **The Average Value of $f$ on $R$**

Let $z = f(x, y)$ be a continuous function defined over a closed, bounded region $R$ in the $x$-$y$ plane. The **average value of $f$ on $R$** is
$$\text{average value of } f \text{ on } R = \frac{\iint_R f(x, y)\,dA}{\iint_R dA}.$$

---

**Example 13.2.6**   **Finding average value of a function over a region $R$**
Find the average value of $f(x, y) = 4 - y$ over the region $R$, which is bounded by the parabolas $y^2 = 4x$ and $x^2 = 4y$. Note: this is the same function and region as used in Example 13.2.4.

**Solution**   In Example 13.2.4 we found
$$\iint_R f(x, y)\,dA = \int_0^4 \int_{y^2/4}^{2\sqrt{y}} (4 - y)\,dx\,dy = \frac{176}{15}.$$

Notes:

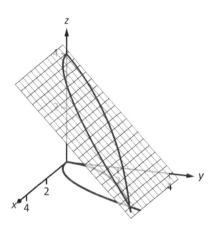

Figure 13.2.10: Finding the average value of $f$ in Example 13.2.6.

We find the area of $R$ by computing $\iint_R dA$:

$$\iint_R dA = \int_0^4 \int_{y^2/4}^{2\sqrt{y}} dx\, dy = \frac{16}{3}.$$

Dividing the volume under the surface by the area gives the average value:

$$\text{average value of } f \text{ on } R = \frac{176/15}{16/3} = \frac{11}{5} = 2.2.$$

While the surface, as shown in Figure 13.2.10, covers $z$-values from $z = 0$ to $z = 4$, the "average" $z$-value on $R$ is 2.2.

The previous section introduced the iterated integral in the context of finding the area of plane regions. This section has extended our understanding of iterated integrals; now we see they can be used to find the signed volume under a surface.

This new understanding allows us to revisit what we did in the previous section. Given a region $R$ in the plane, we computed $\iint_R 1\, dA$; again, our understanding at the time was that we were finding the area of $R$. However, we can now view the function $z = 1$ as a surface, a flat surface with constant $z$-value of 1. The double integral $\iint_R 1\, dA$ finds the volume, under $z = 1$, over $R$, as shown in Figure 13.2.11. Basic geometry tells us that if the base of a general right cylinder has area $A$, its volume is $A \cdot h$, where $h$ is the height. In our case, the height is 1. We were "actually" computing the volume of a solid, though we interpreted the number as an area.

The next section extends our abilities to find "volumes under surfaces." Currently, some integrals are hard to compute because either the region $R$ we are integrating over is hard to define with rectangular curves, or the integrand itself is hard to deal with. Some of these problems can be solved by converting everything into polar coordinates.

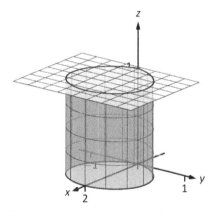

Figure 13.2.11: Showing how an iterated integral used to find area also finds a certain volume.

Notes:

# Exercises 13.2

## Terms and Concepts

1. An integral can be interpreted as giving the signed area over an interval; a double integral can be interpreted as giving the signed _____ over a region.

2. Explain why the following statement is false: "Fubini's Theorem states that $\int_a^b \int_{g_1(x)}^{g_2(x)} f(x,y) \, dy \, dx = \int_a^b \int_{g_1(y)}^{g_2(y)} f(x,y) \, dx \, dy$."

3. Explain why if $f(x,y) > 0$ over a region $R$, then $\iint_R f(x,y) \, dA > 0$.

4. If $\iint_R f(x,y) \, dA = \iint_R g(x,y) \, dA$, does this imply $f(x,y) = g(x,y)$?

## Problems

**In Exercises 5 – 10,**
(a) Evaluate the given iterated integral, and
(b) rewrite the integral using the other order of integration.

5. $\int_1^2 \int_{-1}^1 \left(\frac{x}{y} + 3\right) dx \, dy$

6. $\int_{-\pi/2}^{\pi/2} \int_0^\pi (\sin x \cos y) \, dx \, dy$

7. $\int_0^4 \int_0^{-x/2+2} (3x^2 - y + 2) \, dy \, dx$

8. $\int_1^3 \int_y^3 (x^2 y - xy^2) \, dx \, dy$

9. $\int_0^1 \int_{-\sqrt{1-y}}^{\sqrt{1-y}} (x + y + 2) \, dx \, dy$

10. $\int_0^9 \int_{y/3}^{\sqrt{y}} (xy^2) \, dx \, dy$

**In Exercises 11 – 18:**
(a) Sketch the region $R$ given by the problem.
(b) Set up the iterated integrals, in both orders, that evaluate the given double integral for the described region $R$.
(c) Evaluate one of the iterated integrals to find the signed volume under the surface $z = f(x,y)$ over the region $R$.

11. $\iint_R x^2 y \, dA$, where $R$ is bounded by $y = \sqrt{x}$ and $y = x^2$.

12. $\iint_R x^2 y \, dA$, where $R$ is bounded by $y = \sqrt[3]{x}$ and $y = x^3$.

13. $\iint_R x^2 - y^2 \, dA$, where $R$ is the rectangle with corners $(-1,-1)$, $(1,-1)$, $(1,1)$ and $(-1,1)$.

14. $\iint_R y e^x \, dA$, where $R$ is bounded by $x = 0$, $x = y^2$ and $y = 1$.

15. $\iint_R (6 - 3x - 2y) \, dA$, where $R$ is bounded by $x = 0$, $y = 0$ and $3x + 2y = 6$.

16. $\iint_R e^y \, dA$, where $R$ is bounded by $y = \ln x$ and $y = \frac{1}{e-1}(x - 1)$.

17. $\iint_R (x^3 y - x) \, dA$, where $R$ is the half of the circle $x^2 + y^2 = 9$ in the first and second quadrants.

18. $\iint_R (4 - 3y) \, dA$, where $R$ is bounded by $y = 0$, $y = x/e$ and $y = \ln x$.

**In Exercises 19 – 22, state why it is difficult/impossible to integrate the iterated integral in the given order of integration. Change the order of integration and evaluate the new iterated integral.**

19. $\int_0^4 \int_{y/2}^2 e^{x^2} dx \, dy$

20. $\int_0^{\sqrt{\pi/2}} \int_x^{\sqrt{\pi/2}} \cos(y^2) \, dy \, dx$

21. $\int_0^1 \int_y^1 \frac{2y}{x^2 + y^2} \, dx \, dy$

22. $\int_{-1}^1 \int_1^2 \frac{x \tan^2 y}{1 + \ln y} \, dy \, dx$

**In Exercises 23 – 26, find the average value of $f$ over the region $R$. Notice how these functions and regions are related to the iterated integrals given in Exercises 5 – 8.**

23. $f(x,y) = \frac{x}{y} + 3$; $R$ is the rectangle with opposite corners $(-1, 1)$ and $(1, 2)$.

24. $f(x,y) = \sin x \cos y$; $R$ is bounded by $x = 0$, $x = \pi$, $y = -\pi/2$ and $y = \pi/2$.

25. $f(x,y) = 3x^2 - y + 2$; $R$ is bounded by the lines $y = 0$, $y = 2 - x/2$ and $x = 0$.

26. $f(x,y) = x^2 y - xy^2$; $R$ is bounded by $y = x$, $y = 1$ and $x = 3$.

## 13.3 Double Integration with Polar Coordinates

We have used iterated integrals to evaluate double integrals, which give the signed volume under a surface, $z = f(x, y)$, over a region $R$ of the x-y plane. The integrand is simply $f(x, y)$, and the bounds of the integrals are determined by the region $R$.

Some regions $R$ are easy to describe using rectangular coordinates – that is, with equations of the form $y = f(x)$, $x = a$, etc. However, some regions are easier to handle if we represent their boundaries with polar equations of the form $r = f(\theta)$, $\theta = \alpha$, etc.

The basic form of the double integral is $\iint_R f(x, y)\, dA$. We interpret this integral as follows: over the region $R$, sum up lots of products of heights (given by $f(x_i, y_i)$) and areas (given by $\Delta A_i$). That is, $dA$ represents "a little bit of area." In rectangular coordinates, we can describe a small rectangle as having area $dx\, dy$ or $dy\, dx$ – the area of a rectangle is simply length×width – a small change in $x$ times a small change in $y$. Thus we replace $dA$ in the double integral with $dx\, dy$ or $dy\, dx$.

Now consider representing a region $R$ with polar coordinates. Consider Figure 13.3.1(a). Let $R$ be the region in the first quadrant bounded by the curve. We can approximate this region using the natural shape of polar coordinates: portions of sectors of circles. In the figure, one such region is shaded, shown again in part (b) of the figure.

As the area of a sector of a circle with radius $r$, subtended by an angle $\theta$, is $A = \frac{1}{2}r^2\theta$, we can find the area of the shaded region. The whole sector has area $\frac{1}{2}r_2^2\Delta\theta$, whereas the smaller, unshaded sector has area $\frac{1}{2}r_1^2\Delta\theta$. The area of the shaded region is the difference of these areas:

$$\Delta A_i = \frac{1}{2}r_2^2\Delta\theta - \frac{1}{2}r_1^2\Delta\theta = \frac{1}{2}\left(r_2^2 - r_1^2\right)(\Delta\theta) = \frac{r_2 + r_1}{2}(r_2 - r_1)\Delta\theta.$$

Note that $(r_2 + r_1)/2$ is just the average of the two radii.

To approximate the region $R$, we use many such subregions; doing so shrinks the difference $r_2 - r_1$ between radii to 0 and shrinks the change in angle $\Delta\theta$ also to 0. We represent these infinitesimal changes in radius and angle as $dr$ and $d\theta$, respectively. Finally, as $dr$ is small, $r_2 \approx r_1$, and so $(r_2 + r_1)/2 \approx r_1$. Thus, when $dr$ and $d\theta$ are small,

$$\Delta A_i \approx r_i\, dr\, d\theta.$$

Taking a limit, where the number of subregions goes to infinity and both $r_2 - r_1$ and $\Delta\theta$ go to 0, we get

$$dA = r\, dr\, d\theta.$$

So to evaluate $\iint_R f(x, y)\, dA$, replace $dA$ with $r\, dr\, d\theta$. Convert the function $z = f(x, y)$ to a function with polar coordinates with the substitutions $x = r\cos\theta$,

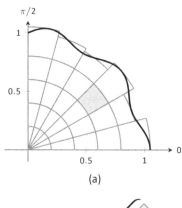

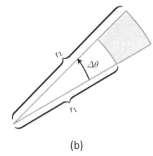

Figure 13.3.1: Approximating a region $R$ with portions of sectors of circles.

Notes:

## 13.3 Double Integration with Polar Coordinates

$y = r\sin\theta$. Finally, find bounds $g_1(\theta) \leq r \leq g_2(\theta)$ and $\alpha \leq \theta \leq \beta$ that describe $R$. This is the key principle of this section, so we restate it here as a Key Idea.

> **Key Idea 13.3.1  Evaluating Double Integrals with Polar Coordinates**
>
> Let $z = f(x, y)$ be a continuous function defined over a closed, bounded region $R$ in the x-y plane, where $R$ is bounded by the polar equations $\alpha \leq \theta \leq \beta$ and $g_1(\theta) \leq r \leq g_2(\theta)$. Then
> $$\iint_R f(x, y)\, dA = \int_\alpha^\beta \int_{g_1(\theta)}^{g_2(\theta)} f(r\cos\theta, r\sin\theta)\, r\, dr\, d\theta.$$

Examples will help us understand this Key Idea.

**Example 13.3.1  Evaluating a double integral with polar coordinates**
Find the signed volume under the plane $z = 4 - x - 2y$ over the disk bounded by the circle with equation $x^2 + y^2 = 1$.

**Solution**    The bounds of the integral are determined solely by the region $R$ over which we are integrating. In this case, it is a disk with boundary $x^2 + y^2 = 1$. We need to find polar bounds for this region. It may help to review Section 9.4; bounds for this disk are $0 \leq r \leq 1$ and $0 \leq \theta \leq 2\pi$.

We replace $f(x, y)$ with $f(r\cos\theta, r\sin\theta)$. That means we make the following substitutions:
$$4 - x - 2y \quad \Rightarrow \quad 4 - r\cos\theta - 2r\sin\theta.$$

Finally, we replace $dA$ in the double integral with $r\, dr\, d\theta$. This gives the final iterated integral, which we evaluate:

$$\begin{aligned}
\iint_R f(x, y)\, dA &= \int_0^{2\pi} \int_0^1 \left(4 - r\cos\theta - 2r\sin\theta\right) r\, dr\, d\theta \\
&= \int_0^{2\pi} \int_0^1 \left(4r - r^2(\cos\theta - 2\sin\theta)\right) dr\, d\theta \\
&= \int_0^{2\pi} \left(2r^2 - \frac{1}{3}r^3(\cos\theta - 2\sin\theta)\right)\bigg|_0^1 d\theta \\
&= \int_0^{2\pi} \left(2 - \frac{1}{3}(\cos\theta - 2\sin\theta)\right) d\theta \\
&= \left(2\theta - \frac{1}{3}(\sin\theta + 2\cos\theta)\right)\bigg|_0^{2\pi} \\
&= 4\pi \approx 12.566.
\end{aligned}$$

The surface and region $R$ are shown in Figure 13.3.2.

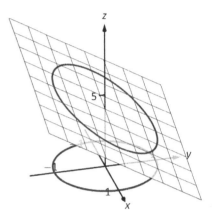

Figure 13.3.2: Evaluating a double integral with polar coordinates in Example 13.3.1.

## Chapter 13 Multiple Integration

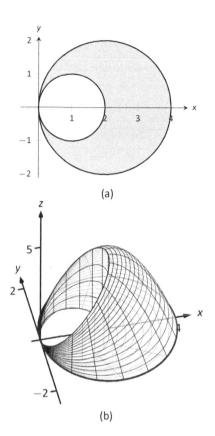

Figure 13.3.3: Showing the region $R$ and surface used in Example 13.3.2.

**Example 13.3.2  Evaluating a double integral with polar coordinates**
Find the volume under the paraboloid $z = 4 - (x-2)^2 - y^2$ over the region bounded by the circles $(x-1)^2 + y^2 = 1$ and $(x-2)^2 + y^2 = 4$.

**Solution**     At first glance, this seems like a very hard volume to compute as the region $R$ (shown in Figure 13.3.3(a)) has a hole in it, cutting out a strange portion of the surface, as shown in part (b) of the figure. However, by describing $R$ in terms of polar equations, the volume is not very difficult to compute. It is straightforward to show that the circle $(x-1)^2 + y^2 = 1$ has polar equation $r = 2\cos\theta$, and that the circle $(x-2)^2 + y^2 = 4$ has polar equation $r = 4\cos\theta$. Each of these circles is traced out on the interval $0 \le \theta \le \pi$. The bounds on $r$ are $2\cos\theta \le r \le 4\cos\theta$.

Replacing $x$ with $r\cos\theta$ in the integrand, along with replacing $y$ with $r\sin\theta$, prepares us to evaluate the double integral $\iint_R f(x,y)\,dA$:

$$\iint_R f(x,y)\,dA = \int_0^\pi \int_{2\cos\theta}^{4\cos\theta} \left(4 - (r\cos\theta - 2)^2 - (r\sin\theta)^2\right) r\,dr\,d\theta$$

$$= \int_0^\pi \int_{2\cos\theta}^{4\cos\theta} \left(-r^3 + 4r^2\cos\theta\right) dr\,d\theta$$

$$= \int_0^\pi \left(-\frac{1}{4}r^4 + \frac{4}{3}r^3\cos\theta\right)\bigg|_{2\cos\theta}^{4\cos\theta} d\theta$$

$$= \int_0^\pi \left(\left[-\frac{1}{4}(256\cos^4\theta) + \frac{4}{3}(64\cos^4\theta)\right] - \left[-\frac{1}{4}(16\cos^4\theta) + \frac{4}{3}(8\cos^4\theta)\right]\right) d\theta$$

$$= \int_0^\pi \frac{44}{3}\cos^4\theta\,d\theta.$$

To integrate $\cos^4\theta$, rewrite it as $\cos^2\theta\cos^2\theta$ and employ the power-reducing formula twice:

$$\cos^4\theta = \cos^2\theta\cos^2\theta$$
$$= \frac{1}{2}(1 + \cos(2\theta))\frac{1}{2}(1 + \cos(2\theta))$$
$$= \frac{1}{4}(1 + 2\cos(2\theta) + \cos^2(2\theta))$$
$$= \frac{1}{4}\left(1 + 2\cos(2\theta) + \frac{1}{2}(1 + \cos(4\theta))\right)$$
$$= \frac{3}{8} + \frac{1}{2}\cos(2\theta) + \frac{1}{8}\cos(4\theta).$$

Notes:

### 13.3 Double Integration with Polar Coordinates

Picking up from where we left off above, we have

$$= \int_0^\pi \frac{44}{3} \cos^4\theta \, d\theta$$

$$= \int_0^\pi \frac{44}{3}\left(\frac{3}{8} + \frac{1}{2}\cos(2\theta) + \frac{1}{8}\cos(4\theta)\right) d\theta$$

$$= \frac{44}{3}\left(\frac{3}{8}\theta + \frac{1}{4}\sin(2\theta) + \frac{1}{32}\sin(4\theta)\right)\Big|_0^\pi$$

$$= \frac{11}{2}\pi \approx 17.279.$$

While this example was not trivial, the double integral would have been *much* harder to evaluate had we used rectangular coordinates.

**Example 13.3.3  Evaluating a double integral with polar coordinates**

Find the volume under the surface $f(x,y) = \dfrac{1}{x^2+y^2+1}$ over the sector of the circle with radius $a$ centered at the origin in the first quadrant, as shown in Figure 13.3.4.

**SOLUTION**   The region $R$ we are integrating over is a circle with radius $a$, restricted to the first quadrant. Thus, in polar, the bounds on $R$ are $0 \leq r \leq a$, $0 \leq \theta \leq \pi/2$. The integrand is rewritten in polar as

$$\frac{1}{x^2+y^2+1} \Rightarrow \frac{1}{r^2\cos^2\theta + r^2\sin^2\theta + 1} = \frac{1}{r^2+1}.$$

We find the volume as follows:

$$\iint_R f(x,y) \, dA = \int_0^{\pi/2} \int_0^a \frac{r}{r^2+1} \, dr \, d\theta$$

$$= \int_0^{\pi/2} \frac{1}{2}\left(\ln|r^2+1|\right)\Big|_0^a d\theta$$

$$= \int_0^{\pi/2} \frac{1}{2}\ln(a^2+1) \, d\theta$$

$$= \left(\frac{1}{2}\ln(a^2+1)\theta\right)\Big|_0^{\pi/2}$$

$$= \frac{\pi}{4}\ln(a^2+1).$$

Figure 13.3.4 shows that $f$ shrinks to near 0 very quickly. Regardless, as $a$ grows, so does the volume, without bound.

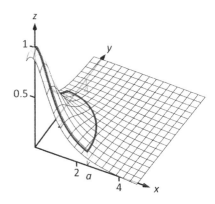

Figure 13.3.4: The surface and region $R$ used in Example 13.3.3.

**Note:** Previous work has shown that there is finite *area* under $\frac{1}{x^2+1}$ over the entire x-axis. However, Example 13.3.3 shows that there is infinite *volume* under $\frac{1}{x^2+y^2+1}$ over the entire x-y plane.

---

Notes:

**Example 13.3.4 Finding the volume of a sphere**
Find the volume of a sphere with radius $a$.

**Solution** The sphere of radius $a$, centered at the origin, has equation $x^2+y^2+z^2=a^2$; solving for $z$, we have $z=\sqrt{a^2-x^2-y^2}$. This gives the upper half of a sphere. We wish to find the volume under this top half, then double it to find the total volume.

The region we need to integrate over is the disk of radius $a$, centered at the origin. Polar bounds for this equation are $0\le r\le a$, $0\le\theta\le 2\pi$.

All together, the volume of a sphere with radius $a$ is:

$$\begin{aligned}2\iint_R \sqrt{a^2-x^2-y^2}\,dA &= 2\int_0^{2\pi}\int_0^a \sqrt{a^2-(r\cos\theta)^2-(r\sin\theta)^2}\,r\,dr\,d\theta\\ &= 2\int_0^{2\pi}\int_0^a r\sqrt{a^2-r^2}\,dr\,d\theta.\end{aligned}$$

We can evaluate this inner integral with substitution. With $u=a^2-r^2$, $du=-2r\,dr$. The new bounds of integration are $u(0)=a^2$ to $u(a)=0$. Thus we have:

$$\begin{aligned}&= \int_0^{2\pi}\int_{a^2}^0 \left(-u^{1/2}\right)du\,d\theta\\ &= \int_0^{2\pi}\left(-\frac{2}{3}u^{3/2}\right)\Bigg|_{a^2}^0 d\theta\\ &= \int_0^{2\pi}\left(\frac{2}{3}a^3\right)d\theta\\ &= \left(\frac{2}{3}a^3\theta\right)\Bigg|_0^{2\pi}\\ &= \frac{4}{3}\pi a^3.\end{aligned}$$

Generally, the formula for the volume of a sphere with radius $r$ is given as $4/3\pi r^3$; we have justified this formula with our calculation.

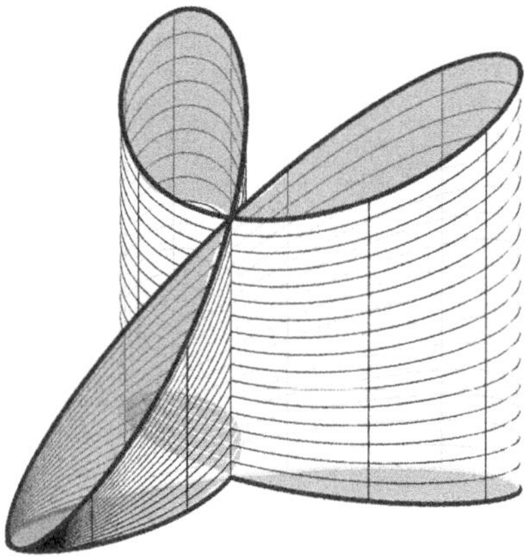

Figure 13.3.5: Visualizing the solid used in Example 13.3.5.

**Example 13.3.5 Finding the volume of a solid**
A sculptor wants to make a solid bronze cast of the solid shown in Figure 13.3.5, where the base of the solid has boundary, in polar coordinates, $r=\cos(3\theta)$, and the top is defined by the plane $z=1-x+0.1y$. Find the volume of the solid.

**Solution** From the outset, we should recognize that knowing *how to set up* this problem is probably more important than knowing *how to compute*

Notes:

*the integrals.* The iterated integral to come is not "hard" to evaluate, though it is long, requiring lots of algebra. Once the proper iterated integral is determined, one can use readily–available technology to help compute the final answer.

The region $R$ that we are integrating over is bound by $0 \leq r \leq \cos(3\theta)$, for $0 \leq \theta \leq \pi$ (note that this rose curve is traced out on the interval $[0, \pi]$, not $[0, 2\pi]$). This gives us our bounds of integration. The integrand is $z = 1-x+0.1y$; converting to polar, we have that the volume $V$ is:

$$V = \iint_R f(x,y)\, dA = \int_0^\pi \int_0^{\cos(3\theta)} \big(1 - r\cos\theta + 0.1r\sin\theta\big) r\, dr\, d\theta.$$

Distributing the $r$, the inner integral is easy to evaluate, leading to

$$\int_0^\pi \left( \frac{1}{2}\cos^2(3\theta) - \frac{1}{3}\cos^3(3\theta)\cos\theta + \frac{0.1}{3}\cos^3(3\theta)\sin\theta \right) d\theta.$$

This integral takes time to compute by hand; it is rather long and cumbersome. The powers of cosine need to be reduced, and products like $\cos(3\theta)\cos\theta$ need to be turned to sums using the Product To Sum formulas in the back cover of this text.

We rewrite $\frac{1}{2}\cos^2(3\theta)$ as $\frac{1}{4}(1+\cos(6\theta))$. We can also rewrite $\frac{1}{3}\cos^3(3\theta)\cos\theta$ as:

$$\frac{1}{3}\cos^3(3\theta)\cos\theta = \frac{1}{3}\cos^2(3\theta)\cos(3\theta)\cos\theta = \frac{1}{3}\frac{1+\cos(6\theta)}{2}\big(\cos(4\theta)+\cos(2\theta)\big).$$

This last expression still needs simplification, but eventually all terms can be reduced to the form $a\cos(m\theta)$ or $a\sin(m\theta)$ for various values of $a$ and $m$.

We forgo the algebra and recommend the reader employ technology, such as WolframAlpha®, to compute the numeric answer. Such technology gives:

$$\int_0^\pi \int_0^{\cos(3\theta)} \big(1 - r\cos\theta + 0.1r\sin\theta\big) r\, dr\, d\theta = \frac{\pi}{4} \approx 0.785 u^3.$$

Since the units were not specified, we leave the result as almost 0.8 cubic units (meters, feet, etc.) Should the artist want to scale the piece uniformly, so that each rose petal had a length other than 1, she should keep in mind that scaling by a factor of $k$ scales the volume by a factor of $k^3$.

We have used iterated integrals to find areas of plane regions and volumes under surfaces. Just as a single integral can be used to compute much more than "area under the curve," iterated integrals can be used to compute much more than we have thus far seen. The next two sections show two, among many, applications of iterated integrals.

Notes:

# Exercises 13.3

## Terms and Concepts

1. When evaluating $\iint_R f(x,y)\, dA$ using polar coordinates, $f(x,y)$ is replaced with _____ and $dA$ is replaced with _____.

2. Why would one be interested in evaluating a double integral with polar coordinates?

## Problems

In Exercises 3 – 10, a function $f(x,y)$ is given and a region $R$ of the x-y plane is described. Set up and evaluate $\iint_R f(x,y)\, dA$ using polar coordinates.

3. $f(x,y) = 3x - y + 4$; $R$ is the region enclosed by the circle $x^2 + y^2 = 1$.

4. $f(x,y) = 4x + 4y$; $R$ is the region enclosed by the circle $x^2 + y^2 = 4$.

5. $f(x,y) = 8 - y$; $R$ is the region enclosed by the circles with polar equations $r = \cos\theta$ and $r = 3\cos\theta$.

6. $f(x,y) = 4$; $R$ is the region enclosed by the petal of the rose curve $r = \sin(2\theta)$ in the first quadrant.

7. $f(x,y) = \ln(x^2 + y^2)$; $R$ is the annulus enclosed by the circles $x^2 + y^2 = 1$ and $x^2 + y^2 = 4$.

8. $f(x,y) = 1 - x^2 - y^2$; $R$ is the region enclosed by the circle $x^2 + y^2 = 1$.

9. $f(x,y) = x^2 - y^2$; $R$ is the region enclosed by the circle $x^2 + y^2 = 36$ in the first and fourth quadrants.

10. $f(x,y) = (x-y)/(x+y)$; $R$ is the region enclosed by the lines $y = x$, $y = 0$ and the circle $x^2 + y^2 = 1$ in the first quadrant.

In Exercises 11 – 14, an iterated integral in rectangular coordinates is given. Rewrite the integral using polar coordinates and evaluate the new double integral.

11. $\displaystyle\int_0^5 \int_{-\sqrt{25-x^2}}^{\sqrt{25-x^2}} \sqrt{x^2 + y^2}\, dy\, dx$

12. $\displaystyle\int_{-4}^4 \int_{-\sqrt{16-y^2}}^{0} (2y - x)\, dx\, dy$

13. $\displaystyle\int_0^2 \int_y^{\sqrt{8-y^2}} (x + y)\, dx\, dy$

14. $\displaystyle\int_{-2}^{-1} \int_0^{\sqrt{4-x^2}} (x+5)\, dy\, dx + \int_{-1}^{1} \int_{\sqrt{1-x^2}}^{\sqrt{4-x^2}} (x+5)\, dy\, dx + \int_1^2 \int_0^{\sqrt{4-x^2}} (x+5)\, dy\, dx$

Hint: draw the region of each integral carefully and see how they all connect.

In Exercises 15 – 16, special double integrals are presented that are especially well suited for evaluation in polar coordinates.

15. Consider $\displaystyle\iint_R e^{-(x^2+y^2)}\, dA$.

    (a) Why is this integral difficult to evaluate in rectangular coordinates, regardless of the region $R$?

    (b) Let $R$ be the region bounded by the circle of radius $a$ centered at the origin. Evaluate the double integral using polar coordinates.

    (c) Take the limit of your answer from (b), as $a \to \infty$. What does this imply about the volume under the surface of $e^{-(x^2+y^2)}$ over the entire x-y plane?

16. The surface of a right circular cone with height $h$ and base radius $a$ can be described by the equation $f(x,y) = h - h\sqrt{\dfrac{x^2}{a^2} + \dfrac{y^2}{a^2}}$, where the tip of the cone lies at $(0,0,h)$ and the circular base lies in the x-y plane, centered at the origin.

    Confirm that the volume of a right circular cone with height $h$ and base radius $a$ is $V = \dfrac{1}{3}\pi a^2 h$ by evaluating $\displaystyle\iint_R f(x,y)\, dA$ in polar coordinates.

## 13.4 Center of Mass

We have used iterated integrals to find areas of plane regions and signed volumes under surfaces. A brief recap of these uses will be useful in this section as we apply iterated integrals to compute the **mass** and **center of mass** of planar regions.

To find the area of a planar region, we evaluated the double integral $\iint_R dA$. That is, summing up the areas of lots of little subregions of $R$ gave us the total area. Informally, we think of $\iint_R dA$ as meaning "sum up lots of little areas over $R$."

To find the signed volume under a surface, we evaluated the double integral $\iint_R f(x, y)\, dA$. Recall that the "$dA$" is not just a "bookend" at the end of an integral; rather, it is multiplied by $f(x, y)$. We regard $f(x, y)$ as giving a height, and $dA$ still giving an area: $f(x, y)\, dA$ gives a volume. Thus, informally, $\iint_R f(x, y)\, dA$ means "sum up lots of little volumes over $R$."

We now extend these ideas to other contexts.

### Mass and Weight

Consider a thin sheet of material with constant thickness and finite area. Mathematicians (and physicists and engineers) call such a sheet a **lamina**. So consider a lamina, as shown in Figure 13.4.1(a), with the shape of some planar region $R$, as shown in part (b).

We can write a simple double integral that represents the mass of the lamina: $\iint_R dm$, where "$dm$" means "a little mass." That is, the double integral states the total mass of the lamina can be found by "summing up lots of little masses over $R$."

To evaluate this double integral, partition $R$ into $n$ subregions as we have done in the past. The $i^{\text{th}}$ subregion has area $\Delta A_i$. A fundamental property of mass is that "mass=density$\times$area." If the lamina has a constant density $\delta$, then the mass of this $i^{\text{th}}$ subregion is $\Delta m_i = \delta \Delta A_i$. That is, we can compute a small amount of mass by multiplying a small amount of area by the density.

If density is variable, with density function $\delta = \delta(x, y)$, then we can approximate the mass of the $i^{\text{th}}$ subregion of $R$ by multiplying $\Delta A_i$ by $\delta(x_i, y_i)$, where $(x_i, y_i)$ is a point in that subregion. That is, for a small enough subregion of $R$, the density across that region is almost constant.

The total mass $M$ of the lamina is approximately the sum of approximate masses of subregions:

$$M \approx \sum_{i=1}^{n} \Delta m_i = \sum_{i=1}^{n} \delta(x_i, y_i) \Delta A_i.$$

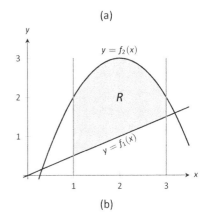

Figure 13.4.1: Illustrating the concept of a lamina.

**Note:** *Mass* and *weight* are different measures. Since they are scalar multiples of each other, it is often easy to treat them as the same measure. In this section we effectively treat them as the same, as our technique for finding mass is the same as for finding weight. The density functions used will simply have different units.

Taking the limit as the size of the subregions shrinks to 0 gives us the actual mass; that is, integrating $\delta(x, y)$ over $R$ gives the mass of the lamina.

---

**Definition 13.4.1**     **Mass of a Lamina with Vairable Density**

Let $\delta(x, y)$ be a continuous density function of a lamina corresponding to a closed, bounded plane region $R$. The mass $M$ of the lamina is

$$\text{mass } M = \iint_R dm = \iint_R \delta(x, y)\, dA.$$

---

**Example 13.4.1**     **Finding the mass of a lamina with constant density**

Find the mass of a square lamina, with side length 1, with a density of $\delta = 3\text{gm/cm}^2$.

**SOLUTION**     We represent the lamina with a square region in the plane as shown in Figure 13.4.2. As the density is constant, it does not matter where we place the square.

Following Definition 13.4.1, the mass $M$ of the lamina is

$$M = \iint_R 3\, dA = \int_0^1 \int_0^1 3\, dx\, dy = 3 \int_0^1 \int_0^1 dx\, dy = 3\text{gm}.$$

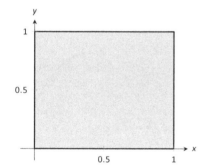

Figure 13.4.2: A region $R$ representing a lamina in Example 13.4.1.

This is all very straightforward; note that all we really did was find the area of the lamina and multiply it by the constant density of $3\text{gm/cm}^2$.

**Example 13.4.2**     **Finding the mass of a lamina with variable density**

Find the mass of a square lamina, represented by the unit square with lower lefthand corner at the origin (see Figure 13.4.2), with variable density $\delta(x, y) = (x + y + 2)\text{gm/cm}^2$.

**SOLUTION**     The variable density $\delta$, in this example, is very uniform, giving a density of 3 in the center of the square and changing linearly. A graph of $\delta(x, y)$ can be seen in Figure 13.4.3; notice how "same amount" of density is above $z = 3$ as below. We'll comment on the significance of this momentarily.

The mass $M$ is found by integrating $\delta(x, y)$ over $R$. The order of integration

---

Notes:

## 13.4 Center of Mass

is not important; we choose $dx\, dy$ arbitrarily. Thus:

$$M = \iint_R (x+y+2)\, dA = \int_0^1 \int_0^1 (x+y+2)\, dx\, dy$$
$$= \int_0^1 \left(\frac{1}{2}x^2 + x(y+2)\right)\bigg|_0^1 dy$$
$$= \int_0^1 \left(\frac{5}{2} + y\right) dy$$
$$= \left(\frac{5}{2}y + \frac{1}{2}y^2\right)\bigg|_0^1$$
$$= 3\text{gm}.$$

It turns out that since the density of the lamina is so uniformly distributed "above and below" $z=3$ that the mass of the lamina is the same as if it had a constant density of 3. The density functions in Examples 13.4.1 and 13.4.2 are graphed in Figure 13.4.3, which illustrates this concept.

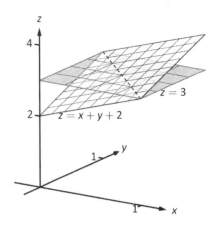

Figure 13.4.3: Graphing the density functions in Examples 13.4.1 and 13.4.2.

**Example 13.4.3   Finding the weight of a lamina with variable density**

Find the weight of the lamina represented by the disk with radius 2ft, centered at the origin, with density function $\delta(x,y) = (x^2+y^2+1)\text{lb/ft}^2$. Compare this to the weight of the lamina with the same shape and density $\delta(x,y) = (2\sqrt{x^2+y^2}+1)\text{lb/ft}^2$.

**SOLUTION**   A direct application of Definition 13.4.1 states that the weight of the lamina is $\iint_R \delta(x,y)\, dA$. Since our lamina is in the shape of a circle, it makes sense to approach the double integral using polar coordinates.

The density function $\delta(x,y) = x^2+y^2+1$ becomes $\delta(r,\theta) = (r\cos\theta)^2 + (r\sin\theta)^2 + 1 = r^2+1$. The circle is bounded by $0 \leq r \leq 2$ and $0 \leq \theta \leq 2\pi$. Thus the weight $W$ is:

$$W = \int_0^{2\pi} \int_0^2 (r^2+1)r\, dr\, d\theta$$
$$= \int_0^{2\pi} \left(\frac{1}{4}r^4 + \frac{1}{2}r^2\right)\bigg|_0^2 d\theta$$
$$= \int_0^{2\pi} (6)\, d\theta$$
$$= 12\pi \approx 37.70\text{lb}.$$

Now compare this with the density function $\delta(x,y) = 2\sqrt{x^2+y^2}+1$. Converting this to polar coordinates gives $\delta(r,\theta) = 2\sqrt{(r\cos\theta)^2+(r\sin\theta)^2}+1 =$

Notes:

$2r+1$. Thus the weight $W$ is:

$$\begin{aligned}W &= \int_0^{2\pi}\int_0^2 (2r+1)r\,dr\,d\theta\\ &= \int_0^{2\pi} \left(\tfrac{2}{3}r^3+\tfrac{1}{2}r^2\right)\Big|_0^2 d\theta\\ &= \int_0^{2\pi}\left(\frac{22}{3}\right)d\theta\\ &= \frac{44}{3}\pi \approx 46.08\text{lb}.\end{aligned}$$

One would expect different density functions to return different weights, as we have here. The density functions were chosen, though, to be similar: each gives a density of 1 at the origin and a density of 5 at the outside edge of the circle, as seen in Figure 13.4.4.

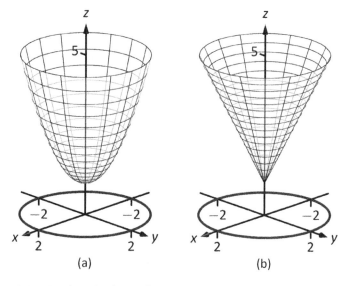

Figure 13.4.4: Graphing the density functions in Example 13.4.3. In (a) is the density function $\delta(x,y) = x^2+y^2+1$; in (b) is $\delta(x,y) = 2\sqrt{x^2+y^2}+1$.

Notice how $x^2+y^2+1 \leq 2\sqrt{x^2+y^2}+1$ over the circle; this results in less weight.

Plotting the density functions can be useful as our understanding of mass can be related to our understanding of "volume under a surface." We interpreted $\iint_R f(x,y)\,dA$ as giving the volume under $f$ over $R$; we can understand $\iint_R \delta(x,y)\,dA$ in the same way. The "volume" under $\delta$ over $R$ is actually mass;

Notes:

by compressing the "volume" under $\delta$ onto the x-y plane, we get "more mass" in some areas than others – i.e., areas of greater density.

Knowing the mass of a lamina is one of several important measures. Another is the **center of mass**, which we discuss next.

## Center of Mass

Consider a disk of radius 1 with uniform density. It is common knowledge that the disk will balance on a point if the point is placed at the center of the disk. What if the disk does not have a uniform density? Through trial-and-error, we should still be able to find a spot on the disk at which the disk will balance on a point. This balance point is referred to as the **center of mass**, or **center of gravity**. It is though all the mass is "centered" there. In fact, if the disk has a mass of 3kg, the disk will behave physically as though it were a point-mass of 3kg located at its center of mass. For instance, the disk will naturally spin with an axis through its center of mass (which is why it is important to "balance" the tires of your car: if they are "out of balance", their center of mass will be outside of the axle and it will shake terribly).

We find the center of mass based on the principle of a **weighted average**. Consider a college class in which your homework average is 90%, your test average is 73%, and your final exam grade is an 85%. Experience tells us that our final grade *is not* the *average* of these three grades: that is, it is not:

$$\frac{0.9 + 0.73 + 0.85}{3} \approx 0.837 = 83.7\%.$$

That is, you are probably not pulling a B in the course. Rather, your grades are *weighted*. Let's say the homework is worth 10% of the grade, tests are 60% and the exam is 30%. Then your final grade is:

$$(0.1)(0.9) + (0.6)(0.73) + (0.3)(0.85) = 0.783 = 78.3\%.$$

Each grade is multiplied by a **weight**.

In general, given values $x_1, x_2, \ldots, x_n$ and weights $w_1, w_2, \ldots, w_n$, the weighted average of the $n$ values is

$$\sum_{i=1}^{n} w_i x_i \bigg/ \sum_{i=1}^{n} w_i.$$

In the grading example above, the sum of the weights 0.1, 0.6 and 0.3 is 1, so we don't see the division by the sum of weights in that instance.

How this relates to center of mass is given in the following theorem.

Notes:

## Chapter 13 Multiple Integration

**Theorem 13.4.1**     **Center of Mass of Discrete Linear System**

Let point masses $m_1, m_2, \ldots, m_n$ be distributed along the x-axis at locations $x_1, x_2, \ldots, x_n$, respectively. The center of mass $\bar{x}$ of the system is located at

$$\bar{x} = \sum_{i=1}^{n} m_i x_i \bigg/ \sum_{i=1}^{n} m_i.$$

**Example 13.4.4**     **Finding the center of mass of a discrete linear system**

1. Point masses of 2gm are located at $x = -1$, $x = 2$ and $x = 3$ are connected by a thin rod of negligible weight. Find the center of mass of the system.

2. Point masses of 10gm, 2gm and 1gm are located at $x = -1$, $x = 2$ and $x = 3$, respectively, are connected by a thin rod of negligible weight. Find the center of mass of the system.

**Solution**

1. Following Theorem 13.4.1, we compute the center of mass as:

$$\bar{x} = \frac{2(-1) + 2(2) + 2(3)}{2 + 2 + 2} = \frac{4}{3} = 1.\bar{3}.$$

So the system would balance on a point placed at $x = 4/3$, as illustrated in Figure 13.4.5(a).

2. Again following Theorem 13.4.1, we find:

$$\bar{x} = \frac{10(-1) + 2(2) + 1(3)}{10 + 2 + 1} = \frac{-3}{13} \approx -0.23.$$

Placing a large weight at the left hand side of the system moves the center of mass left, as shown in Figure 13.4.5(b).

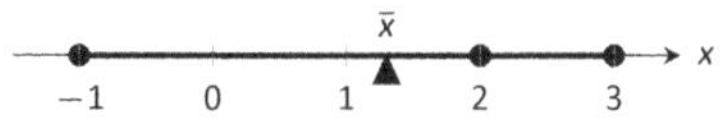

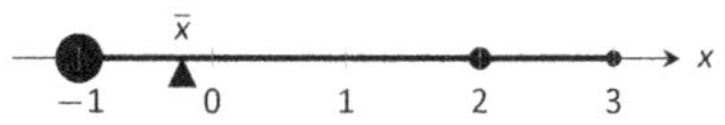

In a discrete system (i.e., mass is located at individual points, not along a continuum) we find the center of mass by dividing the mass into a **moment** of the system. In general, a moment is a weighted measure of distance from a particular point or line. In the case described by Theorem 13.4.1, we are finding a weighted measure of distances from the y-axis, so we refer to this as **the moment about the y-axis**, represented by $M_y$. Letting $M$ be the total mass of the system, we have $\bar{x} = M_y/M$.

Figure 13.4.5: Illustrating point masses along a thin rod and the center of mass.

Notes:

## 13.4 Center of Mass

We can extend the concept of the center of mass of discrete points along a line to the center of mass of discrete points in the plane rather easily. To do so, we define some terms then give a theorem.

> **Definition 13.4.2** **Moments about the x- and y- Axes.**
>
> Let point masses $m_1, m_2, \ldots, m_n$ be located at points $(x_1, y_1)$, $(x_2, y_2) \ldots, (x_n, y_n)$, respectively, in the x-y plane.
>
> 1. The **moment about the y-axis**, $M_y$, is $M_y = \sum_{i=1}^{n} m_i x_i$.
>
> 2. The **moment about the x-axis**, $M_x$, is $M_x = \sum_{i=1}^{n} m_i y_i$.

One can think that these definitions are "backwards" as $M_y$ sums up "x" distances. But remember, "x" distances are measurements of distance from the y-axis, hence defining the moment about the y-axis.

We now define the center of mass of discrete points in the plane.

> **Theorem 13.4.2** **Center of Mass of Discrete Planar System**
>
> Let point masses $m_1, m_2, \ldots, m_n$ be located at points $(x_1, y_1)$, $(x_2, y_2) \ldots, (x_n, y_n)$, respectively, in the x-y plane, and let $M = \sum_{i=1}^{n} m_i$.
> The center of mass of the system is at $(\bar{x}, \bar{y})$, where
>
> $$\bar{x} = \frac{M_y}{M} \quad \text{and} \quad \bar{y} = \frac{M_x}{M}.$$

**Example 13.4.5** **Finding the center of mass of a discrete planar system**
Let point masses of 1kg, 2kg and 5kg be located at points $(2, 0)$, $(1, 1)$ and $(3, 1)$, respectively, and are connected by thin rods of negligible weight. Find the center of mass of the system.

**Solution** We follow Theorem 13.4.2 and Definition 13.4.2 to find $M$, $M_x$ and $M_y$:
$M = 1 + 2 + 5 = 8\text{kg}.$

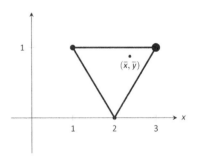

Figure 13.4.6: Illustrating the center of mass of a discrete planar system in Example 13.4.5.

Notes:

$$M_x = \sum_{i=1}^{n} m_i y_i \qquad\qquad M_y = \sum_{i=1}^{n} m_i x_i$$
$$= 1(0) + 2(1) + 5(1) \qquad\qquad = 1(2) + 2(1) + 5(3)$$
$$= 7. \qquad\qquad = 19.$$

Thus the center of mass is $(\bar{x}, \bar{y}) = \left(\dfrac{M_y}{M}, \dfrac{M_x}{M}\right) = \left(\dfrac{19}{8}, \dfrac{7}{8}\right) = (2.375, 0.875)$, illustrated in Figure 13.4.6.

We finally arrive at our true goal of this section: finding the center of mass of a lamina with variable density. While the above measurement of center of mass is interesting, it does not directly answer more realistic situations where we need to find the center of mass of a contiguous region. However, understanding the discrete case allows us to approximate the center of mass of a planar lamina; using calculus, we can refine the approximation to an exact value.

We begin by representing a planar lamina with a region $R$ in the $x$-$y$ plane with density function $\delta(x, y)$. Partition $R$ into $n$ subdivisions, each with area $\Delta A_i$. As done before, we can approximate the mass of the $i^{\text{th}}$ subregion with $\delta(x_i, y_i)\Delta A_i$, where $(x_i, y_i)$ is a point inside the $i^{\text{th}}$ subregion. We can approximate the moment of this subregion about the $y$-axis with $x_i \delta(x_i, y_i) \Delta A_i$ – that is, by multiplying the approximate mass of the region by its approximate distance from the $y$-axis. Similarly, we can approximate the moment about the $x$-axis with $y_i \delta(x_i, y_i) \Delta A_i$. By summing over all subregions, we have:

$$\text{mass: } M \approx \sum_{i=1}^{n} \delta(x_i, y_i) \Delta A_i \quad \text{(as seen before)}$$

$$\text{moment about the } x\text{-axis: } M_x \approx \sum_{i=1}^{n} y_i \delta(x_i, y_i) \Delta A_i$$

$$\text{moment about the } y\text{-axis: } M_y \approx \sum_{i=1}^{n} x_i \delta(x_i, y_i) \Delta A_i$$

By taking limits, where size of each subregion shrinks to 0 in both the $x$ and $y$ directions, we arrive at the double integrals given in the following theorem.

Notes:

## 13.4 Center of Mass

> **Theorem 13.4.3**     **Center of Mass of a Planar Lamina, Moments**
>
> Let a planar lamina be represented by a closed, bounded region $R$ in the $x$-$y$ plane with density function $\delta(x,y)$.
>
> 1. mass: $M = \iint_R \delta(x,y)\, dA$
>
> 2. moment about the $x$-axis: $M_x = \iint_R y\delta(x,y)\, dA$
>
> 3. moment about the $y$-axis: $M_y = \iint_R x\delta(x,y)\, dA$
>
> 4. The center of mass of the lamina is
> $$(\bar{x}, \bar{y}) = \left(\frac{M_y}{M}, \frac{M_x}{M}\right).$$

We start our practice of finding centers of mass by revisiting some of the lamina used previously in this section when finding mass. We will just set up the integrals needed to compute $M$, $M_x$ and $M_y$ and leave the details of the integration to the reader.

**Example 13.4.6**    **Finding the center of mass of a lamina**
Find the center mass of a square lamina, with side length 1, with a density of $\delta = 3\text{gm/cm}^2$. (Note: this is the lamina from Example 13.4.1.)

**Solution**    We represent the lamina with a square region in the plane as shown in Figure 13.4.7 as done previously.
Following Theorem 13.4.3, we find $M$, $M_x$ and $M_y$:

$$M = \iint_R 3\, dA = \int_0^1 \int_0^1 3\, dx\, dy = 3\text{gm}.$$
$$M_x = \iint_R 3y\, dA = \int_0^1 \int_0^1 3y\, dx\, dy = 3/2 = 1.5.$$
$$M_y = \iint_R 3x\, dA = \int_0^1 \int_0^1 3x\, dx\, dy = 3/2 = 1.5.$$

Thus the center of mass is $(\bar{x}, \bar{y}) = \left(\dfrac{M_y}{M}, \dfrac{M_x}{M}\right) = (1.5/3, 1.5/3) = (0.5, 0.5)$.
This is what we should have expected: the center of mass of a square with constant density is the center of the square.

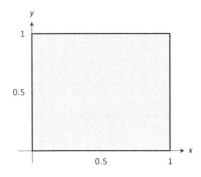

Figure 13.4.7: A region $R$ representing a lamina in Example 13.4.1.

Notes:

## Chapter 13 Multiple Integration

**Example 13.4.7**    **Finding the center of mass of a lamina**

Find the center of mass of a square lamina, represented by the unit square with lower lefthand corner at the origin (see Figure 13.4.7), with variable density $\delta(x,y) = (x+y+2)$gm/cm$^2$. (Note: this is the lamina from Example 13.4.2.)

**SOLUTION**    We follow Theorem 13.4.3, to find $M$, $M_x$ and $M_y$:

$$M = \iint_R (x+y+2)\,dA = \int_0^1 \int_0^1 (x+y+2)\,dx\,dy = 3\text{gm}.$$

$$M_x = \iint_R y(x+y+2)\,dA = \int_0^1 \int_0^1 y(x+y+2)\,dx\,dy = \frac{19}{12}.$$

$$M_y = \iint_R x(x+y+2)\,dA = \int_0^1 \int_0^1 x(x+y+2)\,dx\,dy = \frac{19}{12}.$$

Thus the center of mass is $(\bar{x}, \bar{y}) = \left(\dfrac{M_y}{M}, \dfrac{M_x}{M}\right) = \left(\dfrac{19}{36}, \dfrac{19}{36}\right) \approx (0.528, 0.528)$.

While the mass of this lamina is the same as the lamina in the previous example, the greater density found with greater $x$ and $y$ values pulls the center of mass from the center slightly towards the upper righthand corner.

**Example 13.4.8**    **Finding the center of mass of a lamina**

Find the center of mass of the lamina represented by the circle with radius 2ft, centered at the origin, with density function $\delta(x,y) = (x^2+y^2+1)$lb/ft$^2$. (Note: this is one of the lamina used in Example 13.4.3.)

**SOLUTION**    As done in Example 13.4.3, it is best to describe $R$ using polar coordinates. Thus when we compute $M_y$, we will integrate not $x\delta(x,y) = x(x^2+y^2+1)$, but rather $(r\cos\theta)\delta(r\cos\theta, r\sin\theta) = (r\cos\theta)(r^2+1)$. We compute $M$, $M_x$ and $M_y$:

$$M = \int_0^{2\pi} \int_0^2 (r^2+1)r\,dr\,d\theta = 12\pi \approx 37.7\text{lb}.$$

$$M_x = \int_0^{2\pi} \int_0^2 (r\sin\theta)(r^2+1)r\,dr\,d\theta = 0.$$

$$M_y = \int_0^{2\pi} \int_0^2 (r\cos\theta)(r^2+1)r\,dr\,d\theta = 0.$$

Since $R$ and the density of $R$ are both symmetric about the $x$ and $y$ axes, it should come as no big surprise that the moments about each axis is 0. Thus the center of mass is $(\bar{x}, \bar{y}) = (0,0)$.

---

Notes:

## 13.4 Center of Mass

**Example 13.4.9    Finding the center of mass of a lamina**

Find the center of mass of the lamina represented by the region $R$ shown in Figure 13.4.8, half an annulus with outer radius 6 and inner radius 5, with constant density 2lb/ft$^2$.

**SOLUTION**    Once again it will be useful to represent $R$ in polar coordinates. Using the description of $R$ and/or the illustration, we see that $R$ is bounded by $5 \leq r \leq 6$ and $0 \leq \theta \leq \pi$. As the lamina is symmetric about the y-axis, we should expect $M_y = 0$. We compute $M$, $M_x$ and $M_y$:

$$M = \int_0^\pi \int_5^6 (2) r \, dr \, d\theta = 11\pi \text{lb}.$$

$$M_x = \int_0^\pi \int_5^6 (r \sin\theta)(2) r \, dr \, d\theta = \frac{364}{3} \approx 121.33.$$

$$M_y = \int_0^\pi \int_5^6 (r \cos\theta)(2) r \, dr \, d\theta = 0.$$

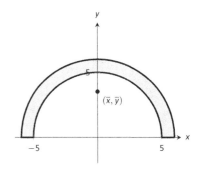

Figure 13.4.8: Illustrating the region $R$ in Example 13.4.9.

Thus the center of mass is $(\bar{x}, \bar{y}) = \left(0, \frac{364}{33\pi}\right) \approx (0, 3.51)$. The center of mass is indicated in Figure 13.4.8; note how it lies outside of $R$!

This section has shown us another use for iterated integrals beyond finding area or signed volume under the curve. While there are many uses for iterated integrals, we give one more application in the following section: computing surface area.

Notes:

# Exercises 13.4

## Terms and Concepts

1. Why is it easy to use "mass" and "weight" interchangeably, even though they are different measures?

2. Given a point $(x, y)$, the value of $x$ is a measure of distance from the _____-axis.

3. We can think of $\iint_R dm$ as meaning "sum up lots of _____"

4. What is a "discrete planar system?"

5. Why does $M_x$ use $\iint_R y\delta(x,y)\, dA$ instead of $\iint_R x\delta(x,y)\, dA$; that is, why do we use "$y$" and not "$x$"?

6. Describe a situation where the center of mass of a lamina does not lie within the region of the lamina itself.

## Problems

In Exercises 7 – 10, point masses are given along a line or in the plane. Find the center of mass $\bar{x}$ or $(\bar{x}, \bar{y})$, as appropriate. (All masses are in grams and distances are in cm.)

7. $m_1 = 4$ at $x = 1$;  $m_2 = 3$ at $x = 3$;  $m_3 = 5$ at $x = 10$

8. $m_1 = 2$ at $x = -3$;  $m_2 = 2$ at $x = -1$;  
   $m_3 = 3$ at $x = 0$;  $m_4 = 3$ at $x = 7$

9. $m_1 = 2$ at $(-2, -2)$;  $m_2 = 2$ at $(2, -2)$;  
   $m_3 = 20$ at $(0, 4)$

10. $m_1 = 1$ at $(-1, -1)$;  $m_2 = 2$ at $(-1, 1)$;  
    $m_3 = 2$ at $(1, 1)$;  $m_4 = 1$ at $(1, -1)$

In Exercises 11 – 18, find the mass/weight of the lamina described by the region $R$ in the plane and its density function $\delta(x, y)$.

11. $R$ is the rectangle with corners $(1, -3)$, $(1, 2)$, $(7, 2)$ and $(7, -3)$; $\delta(x, y) = 5\text{gm/cm}^2$

12. $R$ is the rectangle with corners $(1, -3)$, $(1, 2)$, $(7, 2)$ and $(7, -3)$; $\delta(x, y) = (x + y^2)\text{gm/cm}^2$

13. $R$ is the triangle with corners $(-1, 0)$, $(1, 0)$, and $(0, 1)$; $\delta(x, y) = 2\text{lb/in}^2$

14. $R$ is the triangle with corners $(0, 0)$, $(1, 0)$, and $(0, 1)$; $\delta(x, y) = (x^2 + y^2 + 1)\text{lb/in}^2$

15. $R$ is the disk centered at the origin with radius 2; $\delta(x, y) = (x + y + 4)\text{kg/m}^2$

16. $R$ is the circle sector bounded by $x^2 + y^2 = 25$ in the first quadrant; $\delta(x, y) = (\sqrt{x^2 + y^2} + 1)\text{kg/m}^2$

17. $R$ is the annulus in the first and second quadrants bounded by $x^2 + y^2 = 9$ and $x^2 + y^2 = 36$; $\delta(x, y) = 4\text{lb/ft}^2$

18. $R$ is the annulus in the first and second quadrants bounded by $x^2 + y^2 = 9$ and $x^2 + y^2 = 36$; $\delta(x, y) = \sqrt{x^2 + y^2}\text{lb/ft}^2$

In Exercises 19 – 26, find the center of mass of the lamina described by the region $R$ in the plane and its density function $\delta(x, y)$.
Note: these are the same lamina as in Exercises 11 – 18.

19. $R$ is the rectangle with corners $(1, -3)$, $(1, 2)$, $(7, 2)$ and $(7, -3)$; $\delta(x, y) = 5\text{gm/cm}^2$

20. $R$ is the rectangle with corners $(1, -3)$, $(1, 2)$, $(7, 2)$ and $(7, -3)$; $\delta(x, y) = (x + y^2)\text{gm/cm}^2$

21. $R$ is the triangle with corners $(-1, 0)$, $(1, 0)$, and $(0, 1)$; $\delta(x, y) = 2\text{lb/in}^2$

22. $R$ is the triangle with corners $(0, 0)$, $(1, 0)$, and $(0, 1)$; $\delta(x, y) = (x^2 + y^2 + 1)\text{lb/in}^2$

23. $R$ is the disk centered at the origin with radius 2; $\delta(x, y) = (x + y + 4)\text{kg/m}^2$

24. $R$ is the circle sector bounded by $x^2 + y^2 = 25$ in the first quadrant; $\delta(x, y) = (\sqrt{x^2 + y^2} + 1)\text{kg/m}^2$

25. $R$ is the annulus in the first and second quadrants bounded by $x^2 + y^2 = 9$ and $x^2 + y^2 = 36$; $\delta(x, y) = 4\text{lb/ft}^2$

26. $R$ is the annulus in the first and second quadrants bounded by $x^2 + y^2 = 9$ and $x^2 + y^2 = 36$; $\delta(x, y) = \sqrt{x^2 + y^2}\text{lb/ft}^2$

The *moment of inertia* $I$ is a measure of the tendency of a lamina to resist rotating about an axis or continue to rotate about an axis. $I_x$ is the moment of inertia about the x-axis, $I_y$ is the moment of inertia about the x-axis, and $I_O$ is the moment of inertia about the origin. These are computed as follows:

- $I_x = \iint_R y^2\, dm$

- $I_y = \iint_R x^2\, dm$

- $I_O = \iint_R (x^2 + y^2)\, dm$

In Exercises 27 – 30, a lamina corresponding to a planar region $R$ is given with a mass of 16 units. For each, compute $I_x$, $I_y$ and $I_O$.

27. $R$ is the $4 \times 4$ square with corners at $(-2, -2)$ and $(2, 2)$ with density $\delta(x, y) = 1$.

28. $R$ is the $8 \times 2$ rectangle with corners at $(-4, -1)$ and $(4, 1)$ with density $\delta(x, y) = 1$.

29. $R$ is the $4 \times 2$ rectangle with corners at $(-2, -1)$ and $(2, 1)$ with density $\delta(x, y) = 2$.

30. $R$ is the disk with radius 2 centered at the origin with density $\delta(x, y) = 4/\pi$.

## 13.5 Surface Area

In Section 7.4 we used definite integrals to compute the arc length of plane curves of the form $y = f(x)$. We later extended these ideas to compute the arc length of plane curves defined by parametric or polar equations.

The natural extension of the concept of "arc length over an interval" to surfaces is "surface area over a region."

Consider the surface $z = f(x, y)$ over a region $R$ in the $x$-$y$ plane, shown in Figure 13.5.1(a). Because of the domed shape of the surface, the surface area will be greater than that of the area of the region $R$. We can find this area using the same basic technique we have used over and over: we'll make an approximation, then using limits, we'll refine the approximation to the exact value.

As done to find the volume under a surface or the mass of a lamina, we subdivide $R$ into $n$ subregions. Here we subdivide $R$ into rectangles, as shown in the figure. One such subregion is outlined in the figure, where the rectangle has dimensions $\Delta x_i$ and $\Delta y_i$, along with its corresponding region on the surface.

In part (b) of the figure, we zoom in on this portion of the surface. When $\Delta x_i$ and $\Delta y_i$ are small, the function is approximated well by the tangent plane at any point $(x_i, y_i)$ in this subregion, which is graphed in part (b). In fact, the tangent plane approximates the function so well that in this figure, it is virtually indistinguishable from the surface itself! Therefore we can approximate the surface area $S_i$ of this region of the surface with the area $T_i$ of the corresponding portion of the tangent plane.

This portion of the tangent plane is a parallelogram, defined by sides $\vec{u}$ and $\vec{v}$, as shown. One of the applications of the cross product from Section 10.4 is that the area of this parallelogram is $\|\vec{u} \times \vec{v}\|$. Once we can determine $\vec{u}$ and $\vec{v}$, we can determine the area.

$\vec{u}$ is tangent to the surface in the direction of $x$, therefore, from Section 12.7, $\vec{u}$ is parallel to $\langle 1, 0, f_x(x_i, y_i) \rangle$. The $x$-displacement of $\vec{u}$ is $\Delta x_i$, so we know that $\vec{u} = \Delta x_i \langle 1, 0, f_x(x_i, y_i) \rangle$. Similar logic shows that $\vec{v} = \Delta y_i \langle 0, 1, f_y(x_i, y_i) \rangle$. Thus:

$$\text{surface area } S_i \approx \text{area of } T_i$$
$$= \|\vec{u} \times \vec{v}\|$$
$$= \|\Delta x_i \langle 1, 0, f_x(x_i, y_i) \rangle \times \Delta y_i \langle 0, 1, f_y(x_i, y_i) \rangle \|$$
$$= \sqrt{1 + f_x(x_i, y_i)^2 + f_y(x_i, y_i)^2}\, \Delta x_i \Delta y_i.$$

Note that $\Delta x_i \Delta y_i = \Delta A_i$, the area of the $i^{\text{th}}$ subregion.

Summing up all $n$ of the approximations to the surface area gives

$$\text{surface area over } R \approx \sum_{i=1}^{n} \sqrt{1 + f_x(x_i, y_i)^2 + f_y(x_i, y_i)^2}\, \Delta A_i.$$

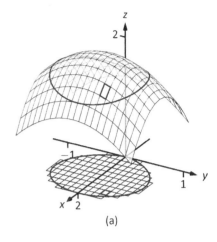

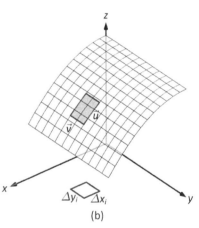

Figure 13.5.1: Developing a method of computing surface area.

Notes:

Once again take a limit as all of the $\Delta x_i$ and $\Delta y_i$ shrink to 0; this leads to a double integral.

**Note:** as done before, we think of "$\iint_R dS$" as meaning "sum up lots of little surface areas over $R$."

The concept of surface area is *defined* here, for while we already have a notion of the area of a region in the *plane*, we did not yet have a solid grasp of what "the area of a surface in *space*" means.

---

**Definition 13.5.1**  **Surface Area**

Let $z = f(x, y)$ where $f_x$ and $f_y$ are continuous over a closed, bounded region $R$. The surface area $S$ over $R$ is

$$S = \iint_R dS$$
$$= \iint_R \sqrt{1 + f_x(x,y)^2 + f_y(x,y)^2}\, dA.$$

---

We test this definition by using it to compute surface areas of known surfaces. We start with a triangle.

**Example 13.5.1**  **Finding the surface area of a plane over a triangle**
Let $f(x, y) = 4 - x - 2y$, and let $R$ be the region in the plane bounded by $x = 0$, $y = 0$ and $y = 2 - x/2$, as shown in Figure 13.5.2. Find the surface area of $f$ over $R$.

**SOLUTION**  We follow Definition 13.5.1. We start by noting that $f_x(x, y) = -1$ and $f_y(x, y) = -2$. To define $R$, we use bounds $0 \leq y \leq 2 - x/2$ and $0 \leq x \leq 4$. Therefore

$$S = \iint_R dS$$
$$= \int_0^4 \int_0^{2-x/2} \sqrt{1 + (-1)^2 + (-2)^2}\, dy\, dx$$
$$= \int_0^4 \sqrt{6}\left(2 - \frac{x}{2}\right) dx$$
$$= 4\sqrt{6}.$$

Because the surface is a triangle, we can figure out the area using geometry. Considering the base of the triangle to be the side in the x-y plane, we find the length of the base to be $\sqrt{20}$. We can find the height using our knowledge of vectors: let $\vec{u}$ be the side in the x-z plane and let $\vec{v}$ be the side in the x-y plane. The height is then $\|\vec{u} - \text{proj}_{\vec{v}}\vec{u}\| = 4\sqrt{6/5}$. Geometry states that the area is thus

$$\frac{1}{2} \cdot 4\sqrt{6/5} \cdot \sqrt{20} = 4\sqrt{6}.$$

We affirm the validity of our formula.

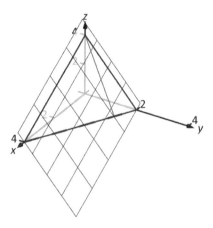

Figure 13.5.2: Finding the area of a triangle in space in Example 13.5.1.

It is "common knowledge" that the surface area of a sphere of radius $r$ is $4\pi r^2$. We confirm this in the following example, which involves using our formula with polar coordinates.

### Example 13.5.2  The surface area of a sphere.

Find the surface area of the sphere with radius $a$ centered at the origin, whose top hemisphere has equation $f(x,y) = \sqrt{a^2 - x^2 - y^2}$.

**SOLUTION**   We start by computing partial derivatives and find

$$f_x(x,y) = \frac{-x}{\sqrt{a^2 - x^2 - y^2}} \quad \text{and} \quad f_y(x,y) = \frac{-y}{\sqrt{a^2 - x^2 - y^2}}.$$

As our function $f$ only defines the top upper hemisphere of the sphere, we double our surface area result to get the total area:

$$S = 2\iint_R \sqrt{1 + f_x(x,y)^2 + f_y(x,y)^2}\, dA$$

$$= 2\iint_R \sqrt{1 + \frac{x^2 + y^2}{a^2 - x^2 - y^2}}\, dA.$$

The region $R$ that we are integrating over is bounded by the circle, centered at the origin, with radius $a$: $x^2 + y^2 = a^2$. Because of this region, we are likely to have greater success with our integration by converting to polar coordinates. Using the substitutions $x = r\cos\theta$, $y = r\sin\theta$, $dA = r\,dr\,d\theta$ and bounds $0 \leq \theta \leq 2\pi$ and $0 \leq r \leq a$, we have:

$$S = 2\int_0^{2\pi}\int_0^a \sqrt{1 + \frac{r^2\cos^2\theta + r^2\sin^2\theta}{a^2 - r^2\cos^2\theta - r^2\sin^2\theta}}\, r\,dr\,d\theta$$

$$= 2\int_0^{2\pi}\int_0^a r\sqrt{1 + \frac{r^2}{a^2 - r^2}}\, dr\,d\theta$$

$$= 2\int_0^{2\pi}\int_0^a r\sqrt{\frac{a^2}{a^2 - r^2}}\, dr\,d\theta. \tag{13.1}$$

Apply substitution $u = a^2 - r^2$ and integrate the inner integral, giving

$$= 2\int_0^{2\pi} a^2\, d\theta$$

$$= 4\pi a^2.$$

Our work confirms our previous formula.

**Note:** The inner integral in Equation (13.1) is an improper integral, as the integrand of $\int_0^a r\sqrt{\dfrac{a^2}{a^2 - r^2}}\, dr$ is not defined at $r = a$. To properly evaluate this integral, one must use the techniques of Section 6.8.

The reason this need arises is that the function $f(x,y) = \sqrt{a^2 - x^2 - y^2}$ fails the requirements of Definition 13.5.1, as $f_x$ and $f_y$ are not continuous on the boundary of the circle $x^2 + y^2 = a^2$.

The computation of the surface area is still valid. The definition makes stronger requirements than necessary in part to avoid the use of improper integration, as when $f_x$ and/or $f_y$ are not continuous, the resulting improper integral may not converge. Since the improper integral does converge in this example, the surface area is accurately computed.

Chapter 13  Multiple Integration

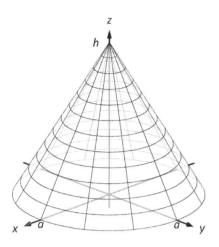

Figure 13.5.3: Finding the surface area of a cone in Example 13.5.3.

**Note:** Once again $f_x$ and $f_y$ are not continuous on the domain of $f$, as both are undefined at $(0,0)$. (A similar problem occurred in the previous example.) Once again the resulting improper integral converges and the computation of the surface area is valid.

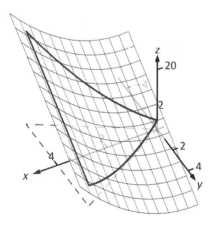

Figure 13.5.4: Graphing the surface in Example 13.5.4.

**Example 13.5.3**   **Finding the surface area of a cone**

The general formula for a right cone with height $h$ and base radius $a$ is
$$f(x,y) = h - \frac{h}{a}\sqrt{x^2+y^2},$$
shown in Figure 13.5.3. Find the surface area of this cone.

**SOLUTION**    We begin by computing partial derivatives.
$$f_x(x,y) = -\frac{xh}{a\sqrt{x^2+y^2}} \quad \text{and} \quad f_y(x,y) = -\frac{yh}{a\sqrt{x^2+y^2}}.$$

Since we are integrating over the disk bounded by $x^2+y^2=a^2$, we again use polar coordinates. Using the standard substitutions, our integrand becomes
$$\sqrt{1+\left(\frac{hr\cos\theta}{a\sqrt{r^2}}\right)^2+\left(\frac{hr\sin\theta}{a\sqrt{r^2}}\right)^2}.$$

This may look intimidating at first, but there are lots of simple simplifications to be done. It amazingly reduces to just
$$\sqrt{1+\frac{h^2}{a^2}} = \frac{1}{a}\sqrt{a^2+h^2}.$$

Our polar bounds are $0 \leq \theta \leq 2\pi$ and $0 \leq r \leq a$. Thus
$$\begin{aligned}
S &= \int_0^{2\pi}\int_0^a r\frac{1}{a}\sqrt{a^2+h^2}\,dr\,d\theta \\
&= \int_0^{2\pi}\left(\frac{1}{2}r^2\frac{1}{a}\sqrt{a^2+h^2}\right)\bigg|_0^a d\theta \\
&= \int_0^{2\pi}\frac{1}{2}a\sqrt{a^2+h^2}\,d\theta \\
&= \pi a\sqrt{a^2+h^2}.
\end{aligned}$$

This matches the formula found in the back of this text.

**Example 13.5.4**   **Finding surface area over a region**

Find the area of the surface $f(x,y) = x^2 - 3y + 3$ over the region $R$ bounded by $-x \leq y \leq x$, $0 \leq x \leq 4$, as pictured in Figure 13.5.4.

**SOLUTION**    It is straightforward to compute $f_x(x,y) = 2x$ and $f_y(x,y) = -3$. Thus the surface area is described by the double integral
$$\iint_R \sqrt{1+(2x)^2+(-3)^2}\,dA = \iint_R \sqrt{10+4x^2}\,dA.$$

Notes:

As with integrals describing arc length, double integrals describing surface area are in general hard to evaluate directly because of the square–root. This particular integral can be easily evaluated, though, with judicious choice of our order of integration.

Integrating with order $dx\,dy$ requires us to evaluate $\int \sqrt{10+4x^2}\,dx$. This can be done, though it involves Integration By Parts and $\sinh^{-1} x$. Integrating with order $dy\,dx$ has as its first integral $\int \sqrt{10+4x^2}\,dy$, which is easy to evaluate: it is simply $y\sqrt{10+4x^2}+C$. So we proceed with the order $dy\,dx$; the bounds are already given in the statement of the problem.

$$\iint_R \sqrt{10+4x^2}\,dA = \int_0^4 \int_{-x}^{x} \sqrt{10+4x^2}\,dy\,dx$$

$$= \int_0^4 \left(y\sqrt{10+4x^2}\right)\Big|_{-x}^{x} dx$$

$$= \int_0^4 \left(2x\sqrt{10+4x^2}\right) dx.$$

Apply substitution with $u = 10 + 4x^2$:

$$= \left(\frac{1}{6}(10+4x^2)^{3/2}\right)\Big|_0^4$$

$$= \frac{1}{3}\left(37\sqrt{74} - 5\sqrt{10}\right) \approx 100.825 \text{u}^2.$$

So while the region $R$ over which we integrate has an area of $16\text{u}^2$, the surface has a much greater area as its $z$-values change dramatically over $R$.

In practice, technology helps greatly in the evaluation of such integrals. High powered computer algebra systems can compute integrals that are difficult, or at least time consuming, by hand, and can at the least produce very accurate approximations with numerical methods. In general, just knowing *how* to set up the proper integrals brings one very close to being able to compute the needed value. Most of the work is actually done in just describing the region $R$ in terms of polar or rectangular coordinates. Once this is done, technology can usually provide a good answer.

We have learned how to integrate integrals; that is, we have learned to evaluate double integrals. In the next section, we learn how to integrate double integrals – that is, we learn to evaluate *triple integrals*, along with learning some uses for this operation.

---

Notes:

# Exercises 13.5

## Terms and Concepts

1. "Surface area" is analogous to what previously studied concept?

2. To approximate the area of a small portion of a surface, we computed the area of its _____ plane.

3. We interpret $\iint_R dS$ as "sum up lots of little _____ _____."

4. Why is it important to know how to set up a double integral to compute surface area, even if the resulting integral is hard to evaluate?

5. Why do $z = f(x, y)$ and $z = g(x, y) = f(x, y) + h$, for some real number $h$, have the same surface area over a region $R$?

6. Let $z = f(x, y)$ and $z = g(x, y) = 2f(x, y)$. Why is the surface area of $g$ over a region $R$ not twice the surface area of $f$ over $R$?

## Problems

In Exercises 7 – 10, *set up* the iterated integral that computes the surface area of the given surface over the region $R$.

7. $f(x, y) = \sin x \cos y$; $R$ is the rectangle with bounds $0 \leq x \leq 2\pi$, $0 \leq y \leq 2\pi$.

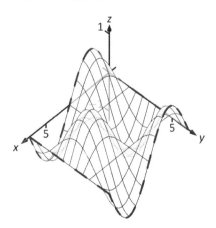

8. $f(x, y) = \dfrac{1}{x^2 + y^2 + 1}$; $R$ is bounded by the circle $x^2 + y^2 = 9$.

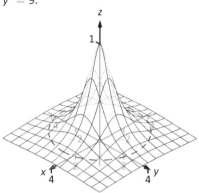

9. $f(x, y) = x^2 - y^2$; $R$ is the rectangle with opposite corners $(-1, -1)$ and $(1, 1)$.

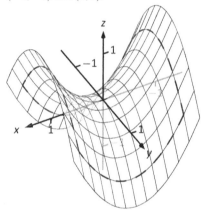

10. $f(x, y) = \dfrac{1}{e^{x^2} + 1}$; $R$ is the rectangle bounded by $-5 \leq x \leq 5$ and $0 \leq y \leq 1$.

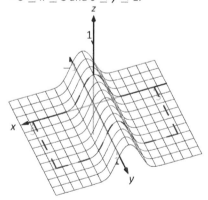

In Exercises 11 – 19, find the area of the given surface over the region $R$.

11. $f(x, y) = 3x - 7y + 2$; $R$ is the rectangle with opposite corners $(-1, 0)$ and $(1, 3)$.

12. $f(x, y) = 2x + 2y + 2$; $R$ is the triangle with corners $(0, 0)$,

(1, 0) and (0, 1).

13. $f(x,y) = x^2 + y^2 + 10$; R is bounded by the circle $x^2 + y^2 = 16$.

14. $f(x,y) = -2x + 4y^2 + 7$ over R, the triangle bounded by $y = -x, y = x, 0 \leq y \leq 1$.

15. $f(x,y) = x^2 + y$ over R, the triangle bounded by $y = 2x$, $y = 0$ and $x = 2$.

16. $f(x,y) = \frac{2}{3}x^{3/2} + 2y^{3/2}$ over R, the rectangle with opposite corners $(0,0)$ and $(1,1)$.

17. $f(x,y) = 10 - 2\sqrt{x^2 + y^2}$ over R, bounded by the circle $x^2 + y^2 = 25$. (This is the cone with height 10 and base radius 5; be sure to compare your result with the known formula.)

18. Find the surface area of the sphere with radius 5 by doubling the surface area of $f(x,y) = \sqrt{25 - x^2 - y^2}$ over R, bounded by the circle $x^2 + y^2 = 25$. (Be sure to compare your result with the known formula.)

19. Find the surface area of the ellipse formed by restricting the plane $f(x,y) = cx + dy + h$ to the region R, bounded by the circle $x^2 + y^2 = 1$, where c, d and h are some constants. Your answer should be given in terms of c and d; why does the value of h not matter?

## 13.6 Volume Between Surfaces and Triple Integration

We learned in Section 13.2 how to compute the signed volume $V$ under a surface $z = f(x, y)$ over a region $R$: $V = \iint_R f(x, y)\, dA$. It follows naturally that if $f(x, y) \geq g(x, y)$ on $R$, then the **volume between** $f(x, y)$ **and** $g(x, y)$ **on** $R$ is

$$V = \iint_R f(x, y)\, dA - \iint_R g(x, y)\, dA = \iint_R \left(f(x, y) - g(x, y)\right) dA.$$

> **Theorem 13.6.1**     **Volume Between Surfaces**
>
> Let $f$ and $g$ be continuous functions on a closed, bounded region $R$, where $f(x, y) \geq g(x, y)$ for all $(x, y)$ in $R$. The volume $V$ between $f$ and $g$ over $R$ is
>
> $$V = \iint_R \left(f(x, y) - g(x, y)\right) dA.$$

**Example 13.6.1**     **Finding volume between surfaces**

Find the volume of the space region bounded by the planes $z = 3x + y - 4$, $z = 8 - 3x - 2y$, $x = 0$ and $y = 0$. In Figure 13.6.1(a) the planes are drawn; in (b), only the defined region is given.

**SOLUTION**     We need to determine the region $R$ over which we will integrate. To do so, we need to determine where the planes intersect. They have common $z$-values when $3x + y - 4 = 8 - 3x - 2y$. Applying a little algebra, we have:

$$3x + y - 4 = 8 - 3x - 2y$$
$$6x + 3y = 12$$
$$2x + y = 4$$

The planes intersect along the line $2x + y = 4$. Therefore the region $R$ is bounded by $x = 0$, $y = 0$, and $y = 4 - 2x$; we can convert these bounds to integration bounds of $0 \leq x \leq 2$, $0 \leq y \leq 4 - 2x$. Thus

$$V = \iint_R \left(8 - 3x - 2y - (3x + y - 4)\right) dA$$
$$= \int_0^2 \int_0^{4-2x} \left(12 - 6x - 3y\right) dy\, dx$$
$$= 16 u^3.$$

The volume between the surfaces is 16 cubic units.

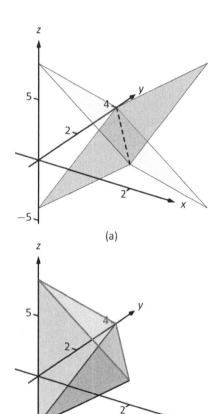

Figure 13.6.1: Finding the volume between the planes given in Example 13.6.1.

Notes:

## 13.6 Volume Between Surfaces and Triple Integration

In the preceding example, we found the volume by evaluating the integral

$$\int_0^2 \int_0^{4-2x} \big(8 - 3x - 2y - (3x + y - 4)\big)\, dy\, dx.$$

Note how we can rewrite the integrand as an integral, much as we did in Section 13.1:

$$8 - 3x - 2y - (3x + y - 4) = \int_{3x+y-4}^{8-3x-2y} dz.$$

Thus we can rewrite the double integral that finds volume as

$$\int_0^2 \int_0^{4-2x} \big(8-3x-2y-(3x+y-4)\big)\, dy\, dx = \int_0^2 \int_0^{4-2x} \left(\int_{3x+y-4}^{8-3x-2y} dz\right) dy\, dx.$$

This no longer looks like a "double integral," but more like a "triple integral." Just as our first introduction to double integrals was in the context of finding the area of a plane region, our introduction into triple integrals will be in the context of finding the volume of a space region.

To formally find the volume of a closed, bounded region $D$ in space, such as the one shown in Figure 13.6.2(a), we start with an approximation. Break $D$ into $n$ rectangular solids; the solids near the boundary of $D$ may possibly not include portions of $D$ and/or include extra space. In Figure 13.6.2(b), we zoom in on a portion of the boundary of $D$ to show a rectangular solid that contains space not in $D$; as this is an approximation of the volume, this is acceptable and this error will be reduced as we shrink the size of our solids.

The volume $\Delta V_i$ of the $i^{\text{th}}$ solid $D_i$ is $\Delta V_i = \Delta x_i \Delta y_i \Delta z_i$, where $\Delta x_i$, $\Delta y_i$ and $\Delta z_i$ give the dimensions of the rectangular solid in the $x$, $y$ and $z$ directions, respectively. By summing up the volumes of all $n$ solids, we get an approximation of the volume $V$ of $D$:

$$V \approx \sum_{i=1}^n \Delta V_i = \sum_{i=1}^n \Delta x_i \Delta y_i \Delta z_i.$$

Let $\|\Delta D\|$ represent the length of the longest diagonal of rectangular solids in the subdivision of $D$. As $\|\Delta D\| \to 0$, the volume of each solid goes to 0, as do each of $\Delta x_i$, $\Delta y_i$ and $\Delta z_i$, for all $i$. Our calculus experience tells us that taking a limit as $\|\Delta D\| \to 0$ turns our approximation of $V$ into an exact calculation of $V$. Before we state this result in a theorem, we use a definition to define some terms.

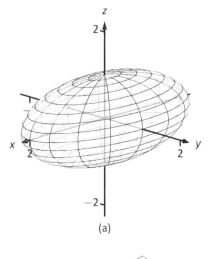

(a)

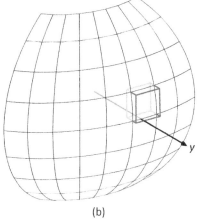

(b)

Figure 13.6.2: Approximating the volume of a region $D$ in space.

Notes:

## Chapter 13 Multiple Integration

**Definition 13.6.1**    **Triple Integrals, Iterated Integration (Part I)**

Let $D$ be a closed, bounded region in space. Let $a$ and $b$ be real numbers, let $g_1(x)$ and $g_2(x)$ be continuous functions of $x$, and let $f_1(x,y)$ and $f_2(x,y)$ be continuous functions of $x$ and $y$.

1. The volume $V$ of $D$ is denoted by a **triple integral**,
$$V = \iiint_D dV.$$

2. The iterated integral $\displaystyle\int_a^b \int_{g_1(x)}^{g_2(x)} \int_{f_1(x,y)}^{f_2(x,y)} dz\,dy\,dx$ is evaluated as
$$\int_a^b \int_{g_1(x)}^{g_2(x)} \int_{f_1(x,y)}^{f_2(x,y)} dz\,dy\,dx = \int_a^b \int_{g_1(x)}^{g_2(x)} \left( \int_{f_1(x,y)}^{f_2(x,y)} dz \right) dy\,dx.$$

Evaluating the above iterated integral is **triple integration.**

Our informal understanding of the notation $\iiint_D dV$ is "sum up lots of little volumes over $D$," analogous to our understanding of $\iint_R dA$ and $\iint_R dm$.
We now state the major theorem of this section.

**Theorem 13.6.2**    **Triple Integration (Part I)**

Let $D$ be a closed, bounded region in space and let $\Delta D$ be any subdivision of $D$ into $n$ rectangular solids, where the $i^{\text{th}}$ subregion $D_i$ has dimensions $\Delta x_i \times \Delta y_i \times \Delta z_i$ and volume $\Delta V_i$.

1. The volume $V$ of $D$ is
$$V = \iiint_D dV = \lim_{\|\Delta D\| \to 0} \sum_{i=1}^n \Delta V_i = \lim_{\|\Delta D\| \to 0} \sum_{i=1}^n \Delta x_i \Delta y_i \Delta z_i.$$

2. If $D$ is defined as the region bounded by the planes $x = a$ and $x = b$, the cylinders $y = g_1(x)$ and $y = g_2(x)$, and the surfaces $z = f_1(x,y)$ and $z = f_2(x,y)$, where $a < b$, $g_1(x) \leq g_2(x)$ and $f_1(x,y) \leq f_2(x,y)$ on $D$, then
$$\iiint_D dV = \int_a^b \int_{g_1(x)}^{g_2(x)} \int_{f_1(x,y)}^{f_2(x,y)} dz\,dy\,dx.$$

3. $V$ can be determined using iterated integration with other orders of integration (there are 6 total), as long as $D$ is defined by the region enclosed by a pair of planes, a pair of cylinders, and a pair of surfaces.

Notes:

## 13.6 Volume Between Surfaces and Triple Integration

We evaluated the area of a plane region R by iterated integration, where the bounds were "from curve to curve, then from point to point." Theorem 13.6.2 allows us to find the volume of a space region with an iterated integral with bounds "from surface to surface, then from curve to curve, then from point to point." In the iterated integral

$$\int_a^b \int_{g_1(x)}^{g_2(x)} \int_{f_1(x,y)}^{f_2(x,y)} dz\, dy\, dx,$$

the bounds $a \leq x \leq b$ and $g_1(x) \leq y \leq g_2(x)$ define a region R in the x-y plane over which the region D exists in space. However, these bounds are also defining surfaces in space; $x = a$ is a plane and $y = g_1(x)$ is a cylinder. The combination of these 6 surfaces enclose, and define, D.

Examples will help us understand triple integration, including integrating with various orders of integration.

**Example 13.6.2 Finding the volume of a space region with triple integration**
Find the volume of the space region in the first octant bounded by the plane $z = 2 - y/3 - 2x/3$, shown in Figure 13.6.3(a), using the order of integration $dz\, dy\, dx$. Set up the triple integrals that give the volume in the other 5 orders of integration.

**SOLUTION** Starting with the order of integration $dz\, dy\, dx$, we need to first find bounds on z. The region D is bounded below by the plane $z = 0$ (because we are restricted to the first octant) and above by $z = 2 - y/3 - 2x/3$; $0 \leq z \leq 2 - y/3 - 2x/3$.

To find the bounds on y and x, we "collapse" the region onto the x-y plane, giving the triangle shown in Figure 13.6.3(b). (We know the equation of the line $y = 6 - 2x$ in two ways. First, by setting $z = 0$, we have $0 = 2 - y/3 - 2x/3 \Rightarrow y = 6 - 2x$. Secondly, we know this is going to be a straight line between the points $(3, 0)$ and $(0, 6)$ in the x-y plane.)

We define that region R, in the integration order of $dy\, dx$, with bounds $0 \leq y \leq 6 - 2x$ and $0 \leq x \leq 3$. Thus the volume V of the region D is:

$$V = \iiint_D dV$$
$$= \int_0^3 \int_0^{6-2x} \int_0^{2-\frac{1}{3}y-\frac{2}{3}x} dz\, dy\, dx$$

Notes:

**Note:** Example 13.6.2 uses the term "first octant." Recall how the x-, y- and z-axes divide space into eight *octants*; the octant in which x, y and z are all positive is called the *first octant*.

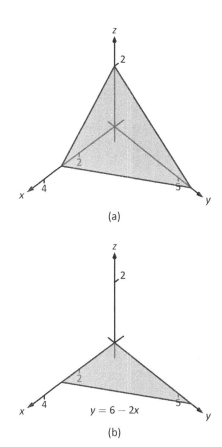

Figure 13.6.3: The region D used in Example 13.6.2 in (a); in (b), the region found by collapsing D onto the x-y plane.

# Chapter 13  Multiple Integration

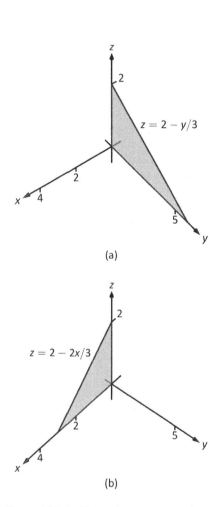

Figure 13.6.4: The region $D$ in Example 13.6.2 is collapsed onto the $y$-$z$ plane in (a); in (b), the region is collapsed onto the $x$-$z$ plane.

$$= \int_0^3 \int_0^{6-2x} \left( \int_0^{2-\frac{1}{3}y-\frac{2}{3}x} dz \right) dy\, dx$$

$$= \int_0^3 \int_0^{6-2x} z \Big|_0^{2-\frac{1}{3}y-\frac{2}{3}x} dy\, dx$$

$$= \int_0^3 \int_0^{6-2x} \left( 2 - \frac{1}{3}y - \frac{2}{3}x \right) dy\, dx.$$

From this step on, we are evaluating a double integral as done many times before. We skip these steps and give the final volume,

$$= 6u^3.$$

The order $dz\, dx\, dy$:

Now consider the volume using the order of integration $dz\, dx\, dy$. The bounds on $z$ are the same as before, $0 \leq z \leq 2-y/3-2x/3$. Collapsing the space region on the $x$-$y$ plane as shown in Figure 13.6.3(b), we now describe this triangle with the order of integration $dx\, dy$. This gives bounds $0 \leq x \leq 3-y/2$ and $0 \leq y \leq 6$. Thus the volume is given by the triple integral

$$V = \int_0^6 \int_0^{3-\frac{1}{2}y} \int_0^{2-\frac{1}{3}y-\frac{2}{3}x} dz\, dx\, dy.$$

The order $dx\, dy\, dz$:

Following our "surface to surface..." strategy, we need to determine the $x$-surfaces that bound our space region. To do so, approach the region "from behind," in the direction of increasing $x$. The first surface we hit as we enter the region is the $y$-$z$ plane, defined by $x = 0$. We come out of the region at the plane $z = 2-y/3-2x/3$; solving for $x$, we have $x = 3-y/2-3z/2$. Thus the bounds on $x$ are: $0 \leq x \leq 3-y/2-3z/2$.

Now collapse the space region onto the $y$-$z$ plane, as shown in Figure 13.6.4(a). (Again, we find the equation of the line $z = 2-y/3$ by setting $x = 0$ in the equation $x = 3-y/2-3z/2$.) We need to find bounds on this region with the order $dy\, dz$. The *curves* that bound $y$ are $y = 0$ and $y = 6-3z$; the *points* that bound $z$ are 0 and 2. Thus the triple integral giving volume is:

$$\begin{array}{c} 0 \leq x \leq 3-y/2-3z/2 \\ 0 \leq y \leq 6-3z \\ 0 \leq z \leq 2 \end{array} \quad \Rightarrow \quad \int_0^2 \int_0^{6-3z} \int_0^{3-y/2-3z/2} dx\, dy\, dz.$$

Notes:

The order *dx dz dy*:

The *x*-bounds are the same as the order above. We now consider the triangle in Figure 13.6.4(a) and describe it with the order *dz dy*: $0 \leq z \leq 2 - y/3$ and $0 \leq y \leq 6$. Thus the volume is given by:

$$\begin{array}{l} 0 \leq x \leq 3 - y/2 - 3z/2 \\ 0 \leq z \leq 2 - y/3 \\ 0 \leq y \leq 6 \end{array} \Rightarrow \int_0^6 \int_0^{2-y/3} \int_0^{3-y/2-3z/2} dx\, dz\, dy.$$

The order *dy dz dx*:

We now need to determine the *y*-surfaces that determine our region. Approaching the space region from "behind" and moving in the direction of increasing *y*, we first enter the region at $y = 0$, and exit along the plane $z = 2 - y/3 - 2x/3$. Solving for *y*, this plane has equation $y = 6 - 2x - 3z$. Thus *y* has bounds $0 \leq y \leq 6 - 2x - 3z$.

Now collapse the region onto the *x*-*z* plane, as shown in Figure 13.6.4(b). The curves bounding this triangle are $z = 0$ and $z = 2 - 2x/3$; *x* is bounded by the points $x = 0$ to $x = 3$. Thus the triple integral giving volume is:

$$\begin{array}{l} 0 \leq y \leq 6 - 2x - 3z \\ 0 \leq z \leq 2 - 2x/3 \\ 0 \leq x \leq 3 \end{array} \Rightarrow \int_0^3 \int_0^{2-2x/3} \int_0^{6-2x-3z} dy\, dz\, dx.$$

The order *dy dx dz*:

The *y*-bounds are the same as in the order above. We now determine the bounds of the triangle in Figure 13.6.4(b) using the order *dy dx dz*. *x* is bounded by $x = 0$ and $x = 3 - 3z/2$; *z* is bounded between $z = 0$ and $z = 2$. This leads to the triple integral:

$$\begin{array}{l} 0 \leq y \leq 6 - 2x - 3z \\ 0 \leq x \leq 3 - 3z/2 \\ 0 \leq z \leq 2 \end{array} \Rightarrow \int_0^2 \int_0^{3-3z/2} \int_0^{6-2x-3z} dy\, dx\, dz.$$

This problem was long, but hopefully useful, demonstrating how to determine bounds with every order of integration to describe the region *D*. In practice, we only need 1, but being able to do them all gives us flexibility to choose the order that suits us best.

Notes:

In the previous example, we collapsed the surface into the x-y, x-z, and y-z planes as we determined the "curve to curve, point to point" bounds of integration. Since the surface was a triangular portion of a plane, this collapsing, or *projecting*, was simple: the *projection* of a straight line in space onto a coordinate plane is a line.

The following example shows us how to do this when dealing with more complicated surfaces and curves.

**Example 13.6.3** **Finding the projection of a curve in space onto the coordinate planes**

Consider the surfaces $z = 3 - x^2 - y^2$ and $z = 2y$, as shown in Figure 13.6.5(a). The curve of their intersection is shown, along with the projection of this curve into the coordinate planes, shown dashed. Find the equations of the projections into the coordinate planes.

**SOLUTION** The two surfaces are $z = 3 - x^2 - y^2$ and $z = 2y$. To find where they intersect, it is natural to set them equal to each other: $3 - x^2 - y^2 = 2y$. This is an implicit function of $x$ and $y$ that gives all points $(x, y)$ in the x-y plane where the $z$ values of the two surfaces are equal.

We can rewrite this implicit function by completing the square:

$$3 - x^2 - y^2 = 2y \quad \Rightarrow \quad y^2 + 2y + x^2 = 3 \quad \Rightarrow \quad (y+1)^2 + x^2 = 4.$$

Thus in the x-y plane the projection of the intersection is a circle with radius 2, centered at $(0, -1)$.

To project onto the x-z plane, we do a similar procedure: find the $x$ and $z$ values where the $y$ values on the surface are the same. We start by solving the equation of each surface for $y$. In this particular case, it works well to actually solve for $y^2$:

$z = 3 - x^2 - y^2 \quad \Rightarrow \quad y^2 = 3 - x^2 - z$
$z = 2y \quad \Rightarrow \quad y^2 = z^2/4.$

Thus we have (after again completing the square):

$$3 - x^2 - z = z^2/4 \quad \Rightarrow \quad \frac{(z+2)^2}{16} + \frac{x^2}{4} = 1,$$

and ellipse centered at $(0, -2)$ in the x-z plane with a major axis of length 8 and a minor axis of length 4.

Finally, to project the curve of intersection into the y-z plane, we solve equation for $x$. Since $z = 2y$ is a cylinder that lacks the variable $x$, it becomes our equation of the projection in the y-z plane.

All three projections are shown in Figure 13.6.5(b).

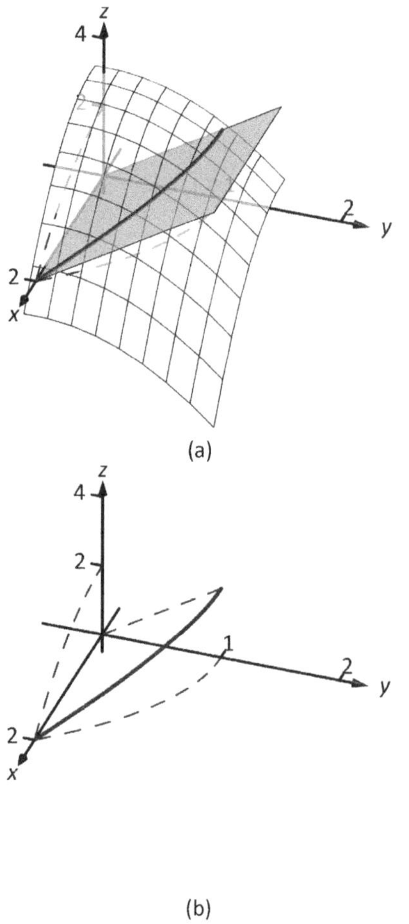

Figure 13.6.5: Finding the projections of the curve of intersection in Example 13.6.3.

---

Notes:

## 13.6 Volume Between Surfaces and Triple Integration

**Example 13.6.4  Finding the volume of a space region with triple integration**
Set up the triple integrals that find the volume of the space region $D$ bounded by the surfaces $x^2 + y^2 = 1$, $z = 0$ and $z = -y$, as shown in Figure 13.6.6(a), with the orders of integration $dz\,dy\,dx$, $dy\,dx\,dz$ and $dx\,dz\,dy$.

**SOLUTION**  The order $dz\,dy\,dx$:

The region $D$ is bounded below by the plane $z = 0$ and above by the plane $z = -y$. The cylinder $x^2 + y^2 = 1$ does not offer any bounds in the $z$-direction, as that surface is parallel to the $z$-axis. Thus $0 \leq z \leq -y$.

Collapsing the region into the $x$-$y$ plane, we get part of the disk bounded by the circle with equation $x^2 + y^2 = 1$ as shown in Figure 13.6.6(b). As a function of $x$, this half circle has equation $y = -\sqrt{1-x^2}$. Thus $y$ is bounded below by $-\sqrt{1-x^2}$ and above by $y = 0$: $-\sqrt{1-x^2} \leq y \leq 0$. The $x$ bounds of the half circle are $-1 \leq x \leq 1$. All together, the bounds of integration and triple integral are as follows:

$$\begin{array}{c} 0 \leq z \leq -y \\ -\sqrt{1-x^2} \leq y \leq 0 \\ -1 \leq x \leq 1 \end{array} \Rightarrow \int_{-1}^{1} \int_{-\sqrt{1-x^2}}^{0} \int_{0}^{-y} dz\,dy\,dx.$$

We evaluate this triple integral:

$$\int_{-1}^{1} \int_{-\sqrt{1-x^2}}^{0} \int_{0}^{-y} dz\,dy\,dx = \int_{-1}^{1} \int_{-\sqrt{1-x^2}}^{0} (-y)\,dy\,dx$$

$$= \int_{-1}^{1} \left(-\frac{1}{2}y^2\right)\Big|_{-\sqrt{1-x^2}}^{0} dx$$

$$= \int_{-1}^{1} \frac{1}{2}(1-x^2)\,dx$$

$$= \left(\frac{1}{2}\left(x - \frac{1}{3}x^3\right)\right)\Big|_{-1}^{1}$$

$$= \frac{2}{3}\text{ units}^3.$$

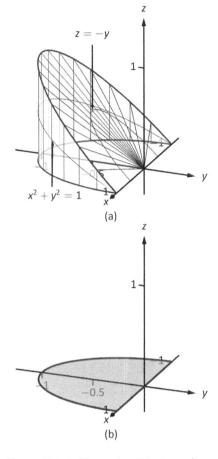

Figure 13.6.6: The region $D$ in Example 13.6.4 is shown in (a); in (b), it is collapsed onto the $x$-$y$ plane.

With the order $dy\,dx\,dz$:

The region is bounded "below" in the $y$-direction by the surface $x^2 + y^2 = 1 \Rightarrow y = -\sqrt{1-x^2}$ and "above" by the surface $y = -z$. Thus the $y$ bounds are $-\sqrt{1-x^2} \leq y \leq -z$.

Collapsing the region onto the $x$-$z$ plane gives the region shown in Figure 13.6.7(a); this half disk is bounded by $z = 0$ and $x^2 + z^2 = 1$. (We find this curve

## Chapter 13 Multiple Integration

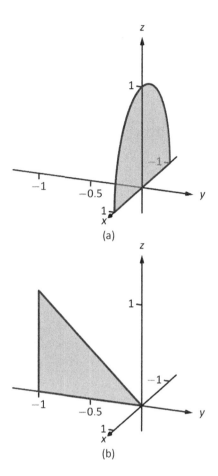

Figure 13.6.7: The region $D$ in Example 13.6.4 is shown collapsed onto the $x$-$z$ plane in (a); in (b), it is collapsed onto the $y$-$z$ plane.

by solving each surface for $y^2$, then setting them equal to each other. We have $y^2 = 1 - x^2$ and $y = -z \Rightarrow y^2 = z^2$. Thus $x^2 + z^2 = 1$.) It is bounded below by $x = -\sqrt{1-z^2}$ and above by $x = \sqrt{1-z^2}$, where $z$ is bounded by $0 \leq z \leq 1$. All together, we have:

$$\begin{array}{c} -\sqrt{1-x^2} \leq y \leq -z \\ -\sqrt{1-z^2} \leq x \leq \sqrt{1-z^2} \\ 0 \leq z \leq 1 \end{array} \Rightarrow \int_0^1 \int_{-\sqrt{1-z^2}}^{\sqrt{1-z^2}} \int_{-\sqrt{1-x^2}}^{-z} dy\, dx\, dz.$$

With the order $dx\, dz\, dy$:

$D$ is bounded below by the surface $x = -\sqrt{1-y^2}$ and above by $\sqrt{1-y^2}$. We then collapse the region onto the $y$-$z$ plane and get the triangle shown in Figure 13.6.7(b). (The hypotenuse is the line $z = -y$, just as the plane.) Thus $z$ is bounded by $0 \leq z \leq -y$ and $y$ is bounded by $-1 \leq y \leq 0$. This gives:

$$\begin{array}{c} -\sqrt{1-y^2} \leq x \leq \sqrt{1-y^2} \\ 0 \leq z \leq -y \\ -1 \leq y \leq 0 \end{array} \Rightarrow \int_{-1}^0 \int_0^{-y} \int_{-\sqrt{1-y^2}}^{\sqrt{1-y^2}} dx\, dz\, dy.$$

The following theorem states two things that should make "common sense" to us. First, using the triple integral to find volume of a region $D$ should always return a positive number; we are computing *volume* here, not *signed volume*. Secondly, to compute the volume of a "complicated" region, we could break it up into subregions and compute the volumes of each subregion separately, summing them later to find the total volume.

---

**Theorem 13.6.3  Properties of Triple Integrals**

Let $D$ be a closed, bounded region in space, and let $D_1$ and $D_2$ be non-overlapping regions such that $D = D_1 \bigcup D_2$.

1. $\iiint_D dV \geq 0$

2. $\iiint_D dV = \iiint_{D_1} dV + \iiint_{D_2} dV.$

---

Notes:

## 13.6 Volume Between Surfaces and Triple Integration

We use this latter property in the next example.

**Example 13.6.5**  **Finding the volume of a space region with triple integration**
Find the volume of the space region $D$ bounded by the coordinate planes, $z = 1 - x/2$ and $z = 1 - y/4$, as shown in Figure 13.6.8(a). Set up the triple integrals that find the volume of $D$ in all 6 orders of integration.

**SOLUTION**  Following the bounds–determining strategy of "surface to surface, curve to curve, and point to point," we can see that the most difficult orders of integration are the two in which we integrate with respect to $z$ first, for there are two "upper" surfaces that bound $D$ in the $z$-direction. So we start by noting that we have

$$0 \leq z \leq 1 - \frac{1}{2}x \quad \text{and} \quad 0 \leq z \leq 1 - \frac{1}{4}y.$$

We now collapse the region $D$ onto the $x$-$y$ axis, as shown in Figure 13.6.8(b). The boundary of $D$, the line from $(0, 0, 1)$ to $(2, 4, 0)$, is shown in part (b) of the figure as a dashed line; it has equation $y = 2x$. (We can recognize this in two ways: one, in collapsing the line from $(0, 0, 1)$ to $(2, 4, 0)$ onto the $x$-$y$ plane, we simply ignore the $z$-values, meaning the line now goes from $(0, 0)$ to $(2, 4)$. Secondly, the two surfaces meet where $z = 1 - x/2$ is equal to $z = 1 - y/4$: thus $1 - x/2 = 1 - y/4 \Rightarrow y = 2x$.)

We use the second property of Theorem 13.6.3 to state that

$$\iiint_D dV = \iiint_{D_1} dV + \iiint_{D_2} dV,$$

where $D_1$ and $D_2$ are the space regions above the plane regions $R_1$ and $R_2$, respectively. Thus we can say

$$\iiint_D dV = \iint_{R_1} \left( \int_0^{1-x/2} dz \right) dA + \iint_{R_2} \left( \int_0^{1-y/4} dz \right) dA.$$

All that is left is to determine bounds of $R_1$ and $R_2$, depending on whether we are integrating with order $dx\,dy$ or $dy\,dx$. We give the final integrals here, leaving it to the reader to confirm these results.

$dz\,dy\,dx$:

$$\begin{array}{ll} 0 \leq z \leq 1 - x/2 & 0 \leq z \leq 1 - y/4 \\ 0 \leq y \leq 2x & 2x \leq y \leq 4 \\ 0 \leq x \leq 2 & 0 \leq x \leq 2 \end{array}$$

$$\iiint_D dV = \int_0^2 \int_0^{2x} \int_0^{1-x/2} dz\,dy\,dx + \int_0^2 \int_{2x}^4 \int_0^{1-y/4} dz\,dy\,dx$$

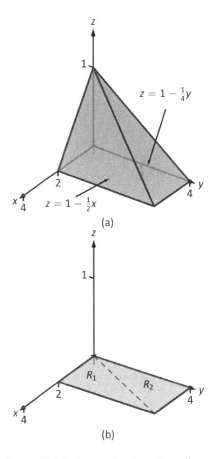

Figure 13.6.8: The region $D$ in Example 13.6.5 is shown in (a); in (b), it is collapsed onto the $x$-$y$ plane.

## Chapter 13  Multiple Integration

*dz dx dy*:

$$0 \leq z \leq 1 - x/2 \qquad\qquad 0 \leq z \leq 1 - y/4$$
$$y/2 \leq x \leq 2 \qquad\qquad 0 \leq x \leq y/2$$
$$0 \leq y \leq 4 \qquad\qquad 0 \leq y \leq 4$$

$$\iiint_D dV = \int_0^4 \int_{y/2}^2 \int_0^{1-x/2} dz\, dx\, dy + \int_0^4 \int_0^{y/2} \int_0^{1-y/4} dz\, dx\, dy$$

The remaining four orders of integration do not require a sum of triple integrals. In Figure 13.6.9 we show $D$ collapsed onto the other two coordinate planes. Using these graphs, we give the final orders of integration here, again leaving it to the reader to confirm these results.

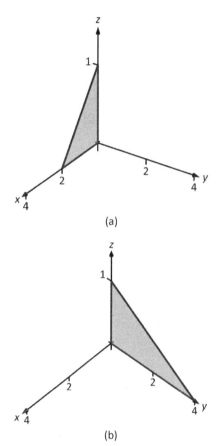

Figure 13.6.9: The region $D$ in Example 13.6.5 is shown collapsed onto the $x$-$z$ plane in (a); in (b), it is collapsed onto the $y$-$z$ plane.

*dy dx dz*:

$$0 \leq y \leq 4 - 4z$$
$$0 \leq x \leq 2 - 2z \quad \Rightarrow \quad \int_0^1 \int_0^{2-2z} \int_0^{4-4z} dy\, dx\, dz$$
$$0 \leq z \leq 1$$

*dy dz dx*:

$$0 \leq y \leq 4 - 4z$$
$$0 \leq z \leq 1 - x/2 \quad \Rightarrow \quad \int_0^2 \int_0^{1-x/2} \int_0^{4-4z} dy\, dx\, dz$$
$$0 \leq x \leq 2$$

*dx dy dz*:

$$0 \leq x \leq 2 - 2z$$
$$0 \leq y \leq 4 - 4z \quad \Rightarrow \quad \int_0^1 \int_0^{4-4z} \int_0^{2-2z} dx\, dy\, dz$$
$$0 \leq z \leq 1$$

*dx dz dy*:

$$0 \leq x \leq 2 - 2z$$
$$0 \leq z \leq 1 - y/4 \quad \Rightarrow \quad \int_0^4 \int_0^{1-y/4} \int_0^{2-2z} dx\, dz\, dy$$
$$0 \leq y \leq 4$$

Notes:

## 13.6 Volume Between Surfaces and Triple Integration

We give one more example of finding the volume of a space region.

**Example 13.6.6    Finding the volume of a space region**
Set up a triple integral that gives the volume of the space region $D$ bounded by $z = 2x^2 + 2$ and $z = 6 - 2x^2 - y^2$. These surfaces are plotted in Figure 13.6.10(a) and (b), respectively; the region $D$ is shown in part (c) of the figure.

**SOLUTION**    The main point of this example is this: integrating with respect to $z$ first is rather straightforward; integrating with respect to $x$ first is not.

The order $dz\, dy\, dx$:

The bounds on $z$ are clearly $2x^2 + 2 \leq z \leq 6 - 2x^2 - y^2$. Collapsing $D$ onto the x-y plane gives the ellipse shown in Figure 13.6.10(c). The equation of this ellipse is found by setting the two surfaces equal to each other:

$$2x^2 + 2 = 6 - 2x^2 - y^2 \quad\Rightarrow\quad 4x^2 + y^2 = 4 \quad\Rightarrow\quad x^2 + \frac{y^2}{4} = 1.$$

We can describe this ellipse with the bounds

$$-\sqrt{4 - 4x^2} \leq y \leq \sqrt{4 - 4x^2} \quad \text{and} \quad -1 \leq x \leq 1.$$

Thus we find volume as

$$\begin{array}{c} 2x^2 + 2 \leq z \leq 6 - 2x^2 - y^2 \\ -\sqrt{4-4x^2} \leq y \leq \sqrt{4-4x^2} \\ -1 \leq x \leq 1 \end{array} \quad\Rightarrow\quad \int_{-1}^{1}\int_{-\sqrt{4-4x^2}}^{\sqrt{4-4x^2}}\int_{2x^2+2}^{6-2x^2-y^2} dz\, dy\, dx.$$

The order $dy\, dz\, dx$:

Integrating with respect to $y$ is not too difficult. Since the surface $z = 2x^2 + 2$ is a cylinder whose directrix is the y-axis, it does not create a border for $y$. The paraboloid $z = 6 - 2x^2 - y^2$ does; solving for $y$, we get the bounds

$$-\sqrt{6 - 2x^2 - z} \leq y \leq \sqrt{6 - 2x^2 - z}.$$

Collapsing $D$ onto the x-z axes gives the region shown in Figure 13.6.11(a); the lower curve is from the cylinder, with equation $z = 2x^2 + 2$. The upper curve is from the paraboloid; with $y = 0$, the curve is $z = 6 - 2x^2$. Thus bounds on $z$ are $2x^2 + 2 \leq z \leq 6 - 2x^2$; the bounds on $x$ are $-1 \leq x \leq 1$. Thus we have:

$$\begin{array}{c} -\sqrt{6-2x^2-z} \leq y \leq \sqrt{6-2x^2-z} \\ 2x^2+2 \leq z \leq 6-2x^2 \\ -1 \leq x \leq 1 \end{array} \quad\Rightarrow\quad \int_{-1}^{1}\int_{2x^2+2}^{6-2x^2}\int_{-\sqrt{6-2x^2-z}}^{\sqrt{6-2x^2-z}} dy\, dz\, dx.$$

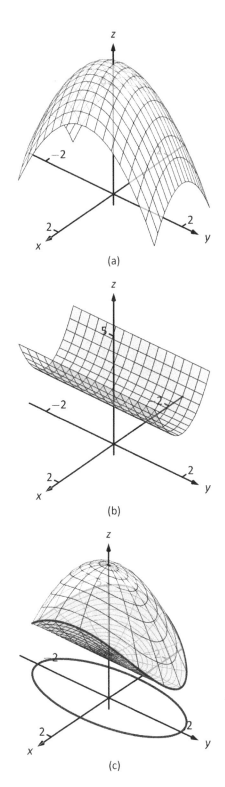

(a)

(b)

(c)

Figure 13.6.10: The region $D$ is bounded by the surfaces shown in (a) and (b); $D$ is shown in (c).

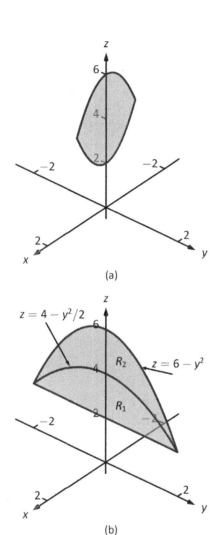

Figure 13.6.11: The region $D$ in Example 13.6.6 is collapsed onto the $x$-$z$ plane in (a); in (b), it is collapsed onto the $y$-$z$ plane.

The order $dx\,dz\,dy$:

This order takes more effort as $D$ must be split into two subregions. The two surfaces create two sets of upper/lower bounds in terms of $x$; the cylinder creates bounds

$$-\sqrt{z/2-1} \leq x \leq \sqrt{z/2-1}$$

for region $D_1$ and the paraboloid creates bounds

$$-\sqrt{3-y^2/2-z^2/2} \leq x \leq \sqrt{3-y^2/2-z^2/2}$$

for region $D_2$.

Collapsing $D$ onto the $y$-$z$ axes gives the regions shown in Figure 13.6.11(b). We find the equation of the curve $z = 4 - y^2/2$ by noting that the equation of the ellipse seen in Figure 13.6.10(c) has equation

$$x^2 + y^2/4 = 1 \quad \Rightarrow \quad x = \sqrt{1 - y^2/4}.$$

Substitute this expression for $x$ in either surface equation, $z = 6 - 2x^2 - y^2$ or $z = 2x^2 + 2$. In both cases, we find

$$z = 4 - \frac{1}{2}y^2.$$

Region $R_1$, corresponding to $D_1$, has bounds

$$2 \leq z \leq 4 - y^2/2, \quad -2 \leq y \leq 2$$

and region $R_2$, corresponding to $D_2$, has bounds

$$4 - y^2/2 \leq z \leq 6 - y^2, \quad -2 \leq y \leq 2.$$

Thus the volume of $D$ is given by:

$$\int_{-2}^{2} \int_{2}^{4-y^2/2} \int_{-\sqrt{z/2-1}}^{\sqrt{z/2-1}} dx\,dz\,dy + \int_{-2}^{2} \int_{4-y^2/2}^{6-y^2} \int_{-\sqrt{3-y^2/2-z^2/2}}^{\sqrt{3-y^2/2-z^2/2}} dx\,dz\,dy.$$

If all one wanted to do in Example 13.6.6 was find the volume of the region $D$, one would have likely stopped at the first integration setup (with order $dz\,dy\,dx$) and computed the volume from there. However, we included the other two methods 1) to show that it could be done, "messy" or not, and 2) because

Notes:

sometimes we "have" to use a less desirable order of integration in order to actually integrate.

## Triple Integration and Functions of Three Variables

There are uses for triple integration beyond merely finding volume, just as there are uses for integration beyond "area under the curve." These uses start with understanding how to integrate functions of three variables, which is effectively no different than integrating functions of two variables. This leads us to a definition, followed by an example.

---

**Definition 13.6.2**  **Iterated Integration, (Part II)**

Let $D$ be a closed, bounded region in space, over which $g_1(x)$, $g_2(x)$, $f_1(x,y)$, $f_2(x,y)$ and $h(x,y,z)$ are all continuous, and let $a$ and $b$ be real numbers.

The **iterated integral** $\int_a^b \int_{g_1(x)}^{g_2(x)} \int_{f_1(x,y)}^{f_2(x,y)} h(x,y,z)\, dz\, dy\, dx$ is evaluated as

$$\int_a^b \int_{g_1(x)}^{g_2(x)} \int_{f_1(x,y)}^{f_2(x,y)} h(x,y,z)\, dz\, dy\, dx = \int_a^b \int_{g_1(x)}^{g_2(x)} \left( \int_{f_1(x,y)}^{f_2(x,y)} h(x,y,z)\, dz \right) dy\, dx.$$

---

**Example 13.6.7**  Evaluating a triple integral of a function of three variables

Evaluate $\int_0^1 \int_{x^2}^{x} \int_{x^2-y}^{2x+3y} (xy + 2xz)\, dz\, dy\, dx$.

**SOLUTION**  We evaluate this integral according to Definition 13.6.2.

$$\int_0^1 \int_{x^2}^{x} \int_{x^2-y}^{2x+3y} (xy + 2xz)\, dz\, dy\, dx$$

$$= \int_0^1 \int_{x^2}^{x} \left( \int_{x^2-y}^{2x+3y} (xy + 2xz)\, dz \right) dy\, dx$$

$$= \int_0^1 \int_{x^2}^{x} \left( (xyz + xz^2) \Big|_{x^2-y}^{2x+3y} \right) dy\, dx$$

$$= \int_0^1 \int_{x^2}^{x} \left( xy(2x+3y) + x(2x+3y)^2 - \left( xy(x^2-y) + x(x^2-y)^2 \right) \right) dy\, dx$$

$$= \int_0^1 \int_{x^2}^{x} \left( -x^5 + x^3 y + 4x^3 + 14x^2 y + 12xy^2 \right) dy\, dx.$$

---

Notes:

We continue as we have in the past, showing fewer steps.

$$= \int_0^1 \left( -\frac{7}{2}x^7 - 8x^6 - \frac{7}{2}x^5 + 15x^4 \right) dx$$

$$= \frac{281}{336} \approx 0.836.$$

We now know *how* to evaluate a triple integral of a function of three variables; we do not yet understand what it *means*. We build up this understanding in a way very similar to how we have understood integration and double integration.

Let $h(x,y,z)$ be a continuous function of three variables, defined over some space region $D$. We can partition $D$ into $n$ rectangular–solid subregions, each with dimensions $\Delta x_i \times \Delta y_i \times \Delta z_i$. Let $(x_i, y_i, z_i)$ be some point in the $i^{\text{th}}$ subregion, and consider the product $h(x_i, y_i, z_i)\Delta x_i \Delta y_i \Delta z_i$. It is the product of a function value (that's the $h(x_i, y_i, z_i)$ part) and a small volume $\Delta V_i$ (that's the $\Delta x_i \Delta y_i \Delta z_i$ part). One of the simplest understanding of this type of product is when $h$ describes the density of an object, for then $h \times$ volume $=$ mass.

We can sum up all $n$ products over $D$. Again letting $||\Delta D||$ represent the length of the longest diagonal of the $n$ rectangular solids in the partition, we can take the limit of the sums of products as $||\Delta D|| \to 0$. That is, we can find

$$S = \lim_{||\Delta D|| \to 0} \sum_{i=1}^n h(x_i, y_i, z_i) \Delta V_i = \lim_{||\Delta D|| \to 0} \sum_{i=1}^n h(x_i, y_i, z_i) \Delta x_i \Delta y_i \Delta z_i.$$

While this limit has lots of interpretations depending on the function $h$, in the case where $h$ describes density, $S$ is the total mass of the object described by the region $D$.

We now use the above limit to define the **triple integral**, give a theorem that relates triple integrals to iterated iteration, followed by the application of triple integrals to find the centers of mass of solid objects.

---

**Definition 13.6.3**   **Triple Integral**

Let $w = h(x, y, z)$ be a continuous function over a closed, bounded region $D$ in space, and let $\Delta D$ be any partition of $D$ into $n$ rectangular solids with volume $\Delta V_i$. The **triple integral of $h$ over $D$** is

$$\iiint_D h(x,y,z)\, dV = \lim_{||\Delta D|| \to 0} \sum_{i=1}^n h(x_i, y_i, z_i) \Delta V_i.$$

---

Notes:

## 13.6 Volume Between Surfaces and Triple Integration

The following theorem assures us that the above limit exists for continuous functions $h$ and gives us a method of evaluating the limit.

---

**Theorem 13.6.4**     **Triple Integration (Part II)**

Let $w = h(x, y, z)$ be a continuous function over a closed, bounded region $D$ in space, and let $\Delta D$ be any partition of $D$ into $n$ rectangular solids with volume $V_i$.

1. The limit $\displaystyle\lim_{||\Delta D|| \to 0} \sum_{i=1}^{n} h(x_i, y_i, z_i) \Delta V_i$ exists.

2. If $D$ is defined as the region bounded by the planes $x = a$ and $x = b$, the cylinders $y = g_1(x)$ and $y = g_2(x)$, and the surfaces $z = f_1(x, y)$ and $z = f_2(x, y)$, where $a < b$, $g_1(x) \leq g_2(x)$ and $f_1(x, y) \leq f_2(x, y)$ on $D$, then

$$\iiint_D h(x, y, z)\, dV = \int_a^b \int_{g_1(x)}^{g_2(x)} \int_{f_1(x,y)}^{f_2(x,y)} h(x, y, z)\, dz\, dy\, dx.$$

---

**Note:** In the marginal note on page 770, we showed how the summation of rectangles over a region $R$ in the plane could be viewed as a double sum, leading to the double integral. Likewise, we can view the sum

$$\sum_{i=1}^{n} h(x_i, y_i, z_i) \Delta x_i \Delta y_i \Delta z_i$$

as a triple sum,

$$\sum_{k=1}^{p} \sum_{j=1}^{n} \sum_{i=1}^{m} h(x_i, y_j, z_k) \Delta x_i \Delta y_j \Delta z_k,$$

which we evaluate as

$$\sum_{k=1}^{p} \left( \sum_{j=1}^{n} \left( \sum_{i=1}^{m} h(x_i, y_j, z_k) \Delta x_i \right) \Delta y_j \right) \Delta z_k.$$

Here we fix a $k$ value, which establishes the $z$-height of the rectangular solids on one "level" of all the rectangular solids in the space region $D$. The inner double summation adds up all the volumes of the rectangular solids on this level, while the outer summation adds up the volumes of each level.

This triple summation understanding leads to the $\iiint_D$ notation of the triple integral, as well as the method of evaluation shown in Theorem 13.6.4.

---

Notes:

## Chapter 13 Multiple Integration

We now apply triple integration to find the centers of mass of solid objects.

### Mass and Center of Mass

One may wish to review Section 13.4 for a reminder of the relevant terms and concepts.

---

**Definition 13.6.4**     **Mass, Center of Mass of Solids**

Let a solid be represented by a closed, bounded region $D$ in space with variable density function $\delta(x, y, z)$.

1. The **mass** of the object is $M = \iiint_D dm = \iiint_D \delta(x, y, z)\, dV$.

2. The **moment about the y-z plane** is $M_{yz} = \iiint_D x\delta(x, y, z)\, dV$.

3. The **moment about the x-z plane** is $M_{xz} = \iiint_D y\delta(x, y, z)\, dV$.

4. The **moment about the x-y plane** is $M_{xy} = \iiint_D z\delta(x, y, z)\, dV$.

5. The **center of mass** of the object is
$$(\bar{x}, \bar{y}, \bar{z}) = \left(\frac{M_{yz}}{M}, \frac{M_{xz}}{M}, \frac{M_{xy}}{M}\right).$$

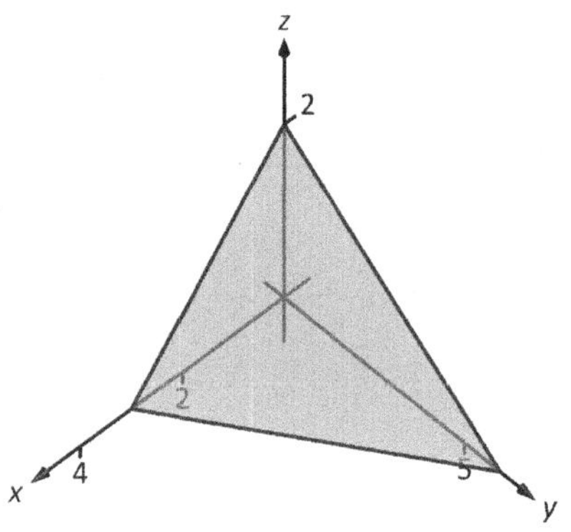

---

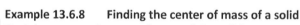

Figure 13.6.12: Finding the center of mass of this solid in Example 13.6.8.

**Example 13.6.8**    **Finding the center of mass of a solid**

Find the mass and center of mass of the solid represented by the space region bounded by the coordinate planes and $z = 2 - y/3 - 2x/3$, shown in Figure 13.6.12, with constant density $\delta(x, y, z) = 3\text{gm/cm}^3$. (Note: this space region was used in Example 13.6.2.)

**SOLUTION**    We apply Definition 13.6.4. In Example 13.6.2, we found bounds for the order of integration $dz\, dy\, dx$ to be $0 \leq z \leq 2 - y/3 - 2x/3$,

---

Notes:

$0 \leq y \leq 6 - 2x$ and $0 \leq x \leq 3$. We find the mass of the object:

$$M = \iiint_D \delta(x,y,z)\, dV$$
$$= \int_0^3 \int_0^{6-2x} \int_0^{2-y/3-2x/3} (3)\, dz\, dy\, dx$$
$$= 3 \int_0^3 \int_0^{6-2x} \int_0^{2-y/3-2x/3} dz\, dy\, dx$$
$$= 3(6) = 18\text{gm}.$$

The evaluation of the triple integral is done in Example 13.6.2, so we skipped those steps above. Note how the mass of an object with constant density is simply "density×volume."

We now find the moments about the planes.

$$M_{xy} = \iiint_D 3z\, dV$$
$$= \int_0^3 \int_0^{6-2x} \int_0^{2-y/3-2x/3} (3z)\, dz\, dy\, dx$$
$$= \int_0^3 \int_0^{6-2x} \frac{3}{2}(2 - y/3 - 2x/3)^2\, dy\, dx$$
$$= \int_0^3 -\frac{4}{9}(x-3)^3\, dx$$
$$= 9.$$

We omit the steps of integrating to find the other moments.

$$M_{yz} = \iiint_D 3x\, dV$$
$$= \frac{27}{2}.$$
$$M_{xz} = \iiint_D 3y\, dV$$
$$= 27.$$

The center of mass is

$$(\bar{x}, \bar{y}, \bar{z}) = \left(\frac{27/2}{18}, \frac{27}{18}, \frac{9}{18}\right) = (0.75, 1.5, 0.5).$$

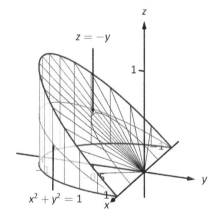

Figure 13.6.13: Finding the center of mass of this solid in Example 13.6.9.

Notes:

## Chapter 13 Multiple Integration

**Example 13.6.9  Finding the center of mass of a solid**

Find the center of mass of the solid represented by the region bounded by the planes $z = 0$ and $z = -y$ and the cylinder $x^2 + y^2 = 1$, shown in Figure 13.6.13, with density function $\delta(x, y, z) = 10 + x^2 + 5y - 5z$. (Note: this space region was used in Example 13.6.4.)

**SOLUTION**   As we start, consider the density function. It is symmetric about the y-z plane, and the farther one moves from this plane, the denser the object is. The symmetry indicates that $\bar{x}$ should be 0.

As one moves away from the origin in the y or z directions, the object becomes less dense, though there is more volume in these regions.

Though none of the integrals needed to compute the center of mass are particularly hard, they do require a number of steps. We emphasize here the importance of knowing how to set up the proper integrals; in complex situations we can appeal to technology for a good approximation, if not the exact answer. We use the order of integration dz dy dx, using the bounds found in Example 13.6.4. (As these are the same for all four triple integrals, we explicitly show the bounds only for M.)

$$M = \iiint_D (10 + x^2 + 5y - 5z)\, dV$$

$$= \int_{-1}^{1} \int_{-\sqrt{1-x^2}}^{0} \int_{0}^{-y} (10 + x^2 + 5y - 5z)\, dV$$

$$= \frac{64}{5} - \frac{15\pi}{16} \approx 3.855.$$

$$M_{yz} = \iiint_D x(10 + x^2 + 5y - 5z)\, dV$$

$$= 0.$$

$$M_{xz} = \iiint_D y(10 + x^2 + 5y - 5z)\, dV$$

$$= 2 - \frac{61\pi}{48} \approx -1.99.$$

$$M_{xy} = \iiint_D z(10 + x^2 + 5y - 5z)\, dV$$

$$= \frac{61\pi}{96} - \frac{10}{9} \approx 0.885.$$

Note how $M_{yz} = 0$, as expected. The center of mass is

$$(\bar{x}, \bar{y}, \bar{z}) = \left(0, \frac{-1.99}{3.855}, \frac{0.885}{3.855}\right) \approx (0, -0.516, 0.230).$$

---

Notes:

As stated before, there are many uses for triple integration beyond finding volume. When $h(x,y,z)$ describes a rate of change function over some space region $D$, then $\iiint_D h(x,y,z)\,dV$ gives the total change over $D$. Our one specific example of this was computing mass; a density function is simply a "rate of mass change per volume" function. Integrating density gives total mass.

While knowing *how to integrate* is important, it is arguably much more important to know *how to set up* integrals. It takes skill to create a formula that describes a desired quantity; modern technology is very useful in evaluating these formulas quickly and accurately.

In the next section, we learn about two new coordinate systems (each related to polar coordinates) that allow us to integrate over closed regions in space more easily than when using rectangular coordinates.

Notes:

# Exercises 13.6

## Terms and Concepts

1. The strategy for establishing bounds for triple integrals is "_____ to _____, _____ to _____ and _____ to _____."

2. Give an informal interpretation of what "$\iiint_D dV$" means.

3. Give two uses of triple integration.

4. If an object has a constant density $\delta$ and a volume $V$, what is its mass?

## Problems

**In Exercises 5 – 8, two surfaces $f_1(x, y)$ and $f_2(x, y)$ and a region $R$ in the $x, y$ plane are given. Set up and evaluate the double integral that finds the volume between these surfaces over $R$.**

5. $f_1(x, y) = 8 - x^2 - y^2, f_2(x, y) = 2x + y$;
   $R$ is the square with corners $(-1, -1)$ and $(1, 1)$.

6. $f_1(x, y) = x^2 + y^2, f_2(x, y) = -x^2 - y^2$;
   $R$ is the square with corners $(0, 0)$ and $(2, 3)$.

7. $f_1(x, y) = \sin x \cos y, f_2(x, y) = \cos x \sin y + 2$;
   $R$ is the triangle with corners $(0, 0)$, $(\pi, 0)$ and $(\pi, \pi)$.

8. $f_1(x, y) = 2x^2 + 2y^2 + 3, f_2(x, y) = 6 - x^2 - y^2$;
   $R$ is the disk bounded by $x^2 + y^2 = 1$.

**In Exercises 9 – 16, a domain $D$ is described by its bounding surfaces, along with a graph. Set up the triple integrals that give the volume of $D$ in all 6 orders of integration, and find the volume of $D$ by evaluating the indicated triple integral.**

9. $D$ is bounded by the coordinate planes and $z = 2 - 2x/3 - 2y$.

   Evaluate the triple integral with order $dz\,dy\,dx$.

   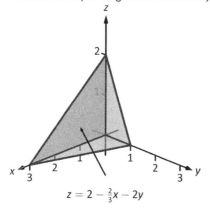
   $z = 2 - \frac{2}{3}x - 2y$

10. $D$ is bounded by the planes $y = 0, y = 2, x = 1, z = 0$ and $z = (3 - x)/2$.

    Evaluate the triple integral with order $dx\,dy\,dz$.

    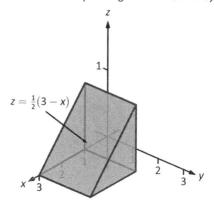
    $z = \frac{1}{2}(3 - x)$

11. $D$ is bounded by the planes $x = 0, x = 2, z = -y$ and by $z = y^2/2$.

    Evaluate the triple integral with the order $dy\,dz\,dx$.

    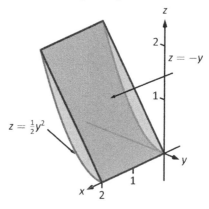
    $z = \frac{1}{2}y^2$ ; $z = -y$

12. $D$ is bounded by the planes $z = 0, y = 9, x = 0$ and by $z = \sqrt{y^2 - 9x^2}$.

    Do not evaluate any triple integral.

    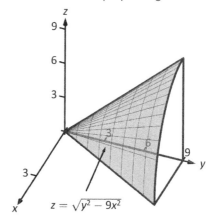
    $z = \sqrt{y^2 - 9x^2}$

826

13. $D$ is bounded by the planes $x = 2$, $y = 1$, $z = 0$ and $z = 2x + 4y - 4$.

    Evaluate the triple integral with the order $dx\,dy\,dz$.

    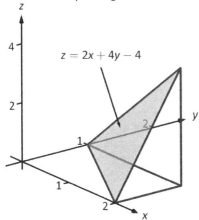

14. $D$ is bounded by the plane $z = 2y$ and by $y = 4 - x^2$.

    Evaluate the triple integral with the order $dz\,dy\,dx$.

    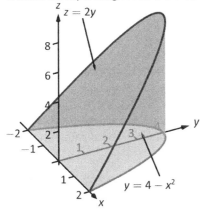

15. $D$ is bounded by the coordinate planes and by $y = 1 - x^2$ and $y = 1 - z^2$.

    Do not evaluate any triple integral. Which order is easier to evaluate: $dz\,dy\,dx$ or $dy\,dz\,dx$? Explain why.

    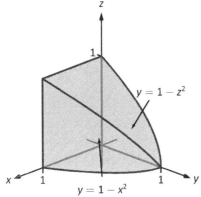

16. $D$ is bounded by the coordinate planes and by $z = 1 - y/3$ and $z = 1 - x$.

    Evaluate the triple integral with order $dx\,dy\,dz$.

    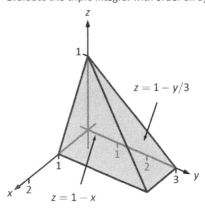

**In Exercises 17 – 20, evaluate the triple integral.**

17. $\displaystyle\int_{-\pi/2}^{\pi/2}\int_0^{\pi}\int_0^{\pi} \left(\cos x \sin y \sin z\right)\,dz\,dy\,dx$

18. $\displaystyle\int_0^1\int_0^x\int_0^{x+y} (x + y + z)\,dz\,dy\,dx$

19. $\displaystyle\int_0^{\pi}\int_0^1\int_0^z \left(\sin(yz)\right)\,dx\,dy\,dz$

20. $\displaystyle\int_{\pi}^{\pi^2}\int_x^{x^3}\int_{-y^2}^{y^2} \left(z\frac{x^2y + y^2x}{e^{x^2+y^2}}\right)\,dz\,dy\,dx$

**In Exercises 21 – 24, find the center of mass of the solid represented by the indicated space region $D$ with density function $\delta(x, y, z)$.**

21. $D$ is bounded by the coordinate planes and $z = 2 - 2x/3 - 2y$; $\delta(x, y, z) = 10\text{gm/cm}^3$.
    (Note: this is the same region as used in Exercise 9.)

22. $D$ is bounded by the planes $y = 0$, $y = 2$, $x = 1$, $z = 0$ and $z = (3 - x)/2$; $\delta(x, y, z) = 2\text{gm/cm}^3$.
    (Note: this is the same region as used in Exercise 10.)

23. $D$ is bounded by the planes $x = 2$, $y = 1$, $z = 0$ and $z = 2x + 4y - 4$; $\delta(x, y, z) = x^2\text{lb/in}^3$.
    (Note: this is the same region as used in Exercise 13.)

24. $D$ is bounded by the plane $z = 2y$ and by $y = 4 - x^2$. $\delta(x, y, z) = y^2\text{lb/in}^3$.
    (Note: this is the same region as used in Exercise 14.)

## 13.7 Triple Integration with Cylindrical and Spherical Coordinates

Just as polar coordinates gave us a new way of describing curves in the plane, in this section we will see how *cylindrical* and *spherical* coordinates give us new ways of desribing surfaces and regions in space.

### Cylindrical Coordinates

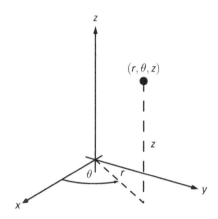

Figure 13.7.1: Illustrating the principles behind cylindrical coordinates.

In short, cylindrical coordinates can be thought of as a combination of the polar and rectangular coordinate systems. One can identify a point $(x_0, y_0, z_0)$, given in rectangular coordinates, with the point $(r_0, \theta_0, z_0)$, given in cylindrical coordinates, where the z-value in both systems is the same, and the point $(x_0, y_0)$ in the x-y plane is identified with the polar point $P(r_0, \theta_0)$; see Figure 13.7.1. So that each point in space that does not lie on the z-axis is defined uniquely, we will restrict $r \geq 0$ and $0 \leq \theta \leq 2\pi$.

We use the identity $z = z$ along with the identities found in Key Idea 9.4.1 to convert between the rectangular coordinate $(x, y, z)$ and the cylindrical coordinate $(r, \theta, z)$, namely:

**Note:** Our rectangular to polar conversion formulas used $r^2 = x^2 + y^2$, allowing for negative $r$ values. Since we now restrict $r \geq 0$, we can use $r = \sqrt{x^2 + y^2}$.

From rectangular to cylindrical: $r = \sqrt{x^2 + y^2}$, $\tan \theta = y/x$ and $z = z$;
From cylindrical to rectangular: $x = r \cos \theta$  $y = r \sin \theta$ and $z = z$.

These identities, along with conversions related to spherical coordinates, are given later in Key Idea 13.7.1.

**Example 13.7.1**     **Converting between rectangular and cylindrical coordinates**
Convert the rectangular point $(2, -2, 1)$ to cylindrical coordinates, and convert the cylindrical point $(4, 3\pi/4, 5)$ to rectangular.

    **Solution**      Following the identities given above (and, later in Key Idea 13.7.1), we have $r = \sqrt{2^2 + (-2)^2} = 2\sqrt{2}$. Using $\tan \theta = y/x$, we find $\theta = \tan^{-1}(-2/2) = -\pi/4$. As we restrict $\theta$ to being between 0 and $2\pi$, we set $\theta = 7\pi/4$. Finally, $z = 1$, giving the cylindrical point $(2\sqrt{2}, 7\pi/4, 1)$.

In converting the cylindrical point $(4, 3\pi/4, 5)$ to rectangular, we have $x = 4\cos(3\pi/4) = -2\sqrt{2}$, $y = 4\sin(3\pi/4) = 2\sqrt{2}$ and $z = 5$, giving the rectangular point $(-2\sqrt{2}, 2\sqrt{2}, 5)$.

Setting each of $r$, $\theta$ and $z$ equal to a constant defines a surface in space, as illustrated in the following example.

Notes:

## 13.7 Triple Integration with Cylindrical and Spherical Coordinates

**Example 13.7.2  Canonical surfaces in cylindrical coordinates**
Describe the surfaces $r = 1$, $\theta = \pi/3$ and $z = 2$, given in cylindrical coordinates.

**SOLUTION**   The equation $r = 1$ describes all points in space that are 1 unit away from the $z$-axis. This surface is a "tube" or "cylinder" of radius 1, centered on the $z$-axis, as graphed in Figure 10.1.8 (which describes the cylinder $x^2 + y^2 = 1$ in space).

The equation $\theta = \pi/3$ describes the plane formed by extending the line $\theta = \pi/3$, as given by polar coordinates in the $x$-$y$ plane, parallel to the $z$-axis.

The equation $z = 2$ describes the plane of all points in space that are 2 units above the $x$-$y$ plane. This plane is the same as the plane described by $z = 2$ in rectangular coordinates.

All three surfaces are graphed in Figure 13.7.2. Note how their intersection uniquely defines the point $P = (1, \pi/3, 2)$.

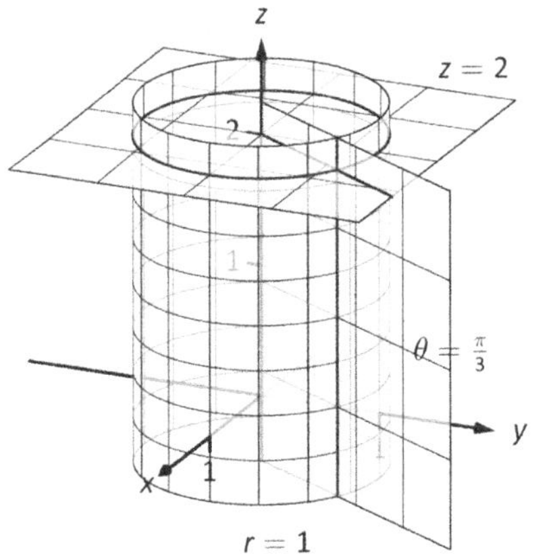

Figure 13.7.2: Graphing the canoncial surfaces in cylindrical coordinates from Example 13.7.2.

Cylindrical coordinates are useful when describing certain domains in space, allowing us to evaluate triple integrals over these domains more easily than if we used rectangular coordinates.

Theorem 13.6.4 shows how to evaluate $\iiint_D h(x, y, z) \, dV$ using rectangular coordinates. In that evaluation, we use $dV = dz \, dy \, dx$ (or one of the other five orders of integration). Recall how, in this order of integration, the bounds on $y$ are "curve to curve" and the bounds on $x$ are "point to point": these bounds describe a region $R$ in the $x$-$y$ plane. We could describe $R$ using polar coordinates as done in Section 13.3. In that section, we saw how we used $dA = r \, dr \, d\theta$ instead of $dA = dy \, dx$.

Considering the above thoughts, we have $dV = dz(r \, dr \, d\theta) = r \, dz \, dr \, d\theta$. We set bounds on $z$ as "surface to surface" as done in the previous section, and then use "curve to curve" and "point to point" bounds on $r$ and $\theta$, respectively. Finally, using the identities given above, we change the integrand $h(x, y, z)$ to $h(r, \theta, z)$.

This process should sound plausible; the following theorem states it is truly a way of evaluating a triple integral.

---

**Theorem 13.7.1   Triple Integration in Cylindrical Coordinates**

Let $w = h(r, \theta, z)$ be a continuous function on a closed, bounded region $D$ in space, bounded in cylindrical coordinates by $\alpha \leq \theta \leq \beta$, $g_1(\theta) \leq r \leq g_2(\theta)$ and $f_1(r, \theta) \leq z \leq f_2(r, \theta)$. Then

$$\iiint_D h(r, \theta, z) \, dV = \int_\alpha^\beta \int_{g_1(\theta)}^{g_2(\theta)} \int_{f_1(r,\theta)}^{f_2(r,\theta)} h(r, \theta, z) r \, dz \, dr \, d\theta.$$

---

Notes:

Chapter 13  Multiple Integration

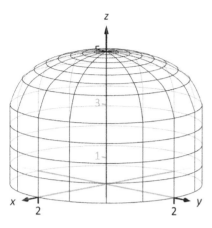

Figure 13.7.3: Visualizing the solid used in Example 13.7.3.

**Example 13.7.3**  **Evaluating a triple integral with cylindrical coordinates**
Find the mass of the solid represented by the region in space bounded by $z = 0$, $z = \sqrt{4 - x^2 - y^2} + 3$ and the cylinder $x^2 + y^2 = 4$ (as shown in Figure 13.7.3), with density function $\delta(x, y, z) = x^2 + y^2 + z + 1$, using a triple integral in cylindrical coordinates. Distances are measured in centimeters and density is measured in grams/cm$^3$.

**SOLUTION**  We begin by describing this region of space with cylindrical coordinates. The plane $z = 0$ is left unchanged; with the identity $r = \sqrt{x^2 + y^2}$, we convert the hemisphere of radius 2 to the equation $z = \sqrt{4 - r^2}$; the cylinder $x^2 + y^2 = 4$ is converted to $r^2 = 4$, or, more simply, $r = 2$. We also convert the density function: $\delta(r, \theta, z) = r^2 + z + 1$.

To describe this solid with the bounds of a triple integral, we bound $z$ with $0 \leq z \leq \sqrt{4 - r^2} + 3$; we bound $r$ with $0 \leq r \leq 2$; we bound $\theta$ with $0 \leq \theta \leq 2\pi$.

Using Definition 13.6.4 and Theorem 13.7.1, we have the mass of the solid is

$$M = \iiint_D \delta(x,y,z)\, dV = \int_0^{2\pi} \int_0^2 \int_0^{\sqrt{4-r^2}+3} (r^2 + z + 1) r\, dz\, dr\, d\theta$$
$$= \int_0^{2\pi} \int_0^2 \left( (r^3 + 4r)\sqrt{4-r^2} + \frac{5}{2}r^3 + \frac{19}{2}r \right) dr\, d\theta$$
$$= \frac{1318\pi}{15} \approx 276.04 \text{ gm},$$

where we leave the details of the remaining double integral to the reader.

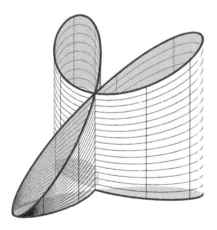

Figure 13.7.4: Visualizing the solid used in Example 13.7.4.

**Example 13.7.4**  **Finding the center of mass using cylindrical coordinates**
Find the center of mass of the solid with constant density whose base can be described by the polar curve $r = \cos(3\theta)$ and whose top is defined by the plane $z = 1 - x + 0.1y$, where distances are measured in feet, as seen in Figure 13.7.4. (The volume of this solid was found in Example 13.3.5.)

**SOLUTION**  We convert the equation of the plane to use cylindrical coordinates: $z = 1 - r\cos\theta + 0.1r\sin\theta$. Thus the region is space is bounded by $0 \leq z \leq 1 - r\cos\theta + 0.1r\sin\theta$, $0 \leq r \leq \cos(3\theta)$, $0 \leq \theta \leq \pi$ (recall that the rose curve $r = \cos(3\theta)$ is traced out once on $[0, \pi]$.

Since density is constant, we set $\delta = 1$ and finding the mass is equivalent to finding the volume of the solid. We set up the triple integral to compute this but do not evaluate it; we leave it to the reader to confirm it evaluates to the same

Notes:

result found in Example 13.3.5.

$$M = \iiint_D \delta\, dV = \int_0^\pi \int_0^{\cos(3\theta)} \int_0^{1-r\cos\theta+0.1r\sin\theta} r\, dz\, dr\, d\theta \approx 0.785.$$

From Definition 13.6.4 we set up the triple integrals to compute the moments about the three coordinate planes. The computation of each is left to the reader (using technology is recommended):

$$M_{yz} = \iiint_D x\, dV = \int_0^\pi \int_0^{\cos(3\theta)} \int_0^{1-r\cos\theta+0.1r\sin\theta} (r\cos\theta) r\, dz\, dr\, d\theta$$
$$= -0.147.$$

$$M_{xz} = \iiint_D y\, dV = \int_0^\pi \int_0^{\cos(3\theta)} \int_0^{1-r\cos\theta+0.1r\sin\theta} (r\sin\theta) r\, dz\, dr\, d\theta$$
$$= 0.015.$$

$$M_{xy} = \iiint_D z\, dV = \int_0^\pi \int_0^{\cos(3\theta)} \int_0^{1-r\cos\theta+0.1r\sin\theta} (z) r\, dz\, dr\, d\theta$$
$$= 0.467.$$

The center of mass, in rectangular coordinates, is located at $(-0.147, 0.015, 0.467)$, which lies outside the bounds of the solid.

## Spherical Coordinates

In short, spherical coordinates can be thought of as a "double application" of the polar coordinate system. In spherical coordinates, a point $P$ is identified with $(\rho, \theta, \varphi)$, where $\rho$ is the distance from the origin to $P$, $\theta$ is the same angle as would be used to describe $P$ in the cylindrical coordinate system, and $\varphi$ is the angle between the positive $z$-axis and the ray from the origin to $P$; see Figure 13.7.5. So that each point in space that does not lie on the $z$-axis is defined uniquely, we will restrict $\rho \geq 0$, $0 \leq \theta \leq 2\pi$ and $0 \leq \varphi \leq \pi$.

The following Key Idea gives conversions to/from our three spatial coordinate systems.

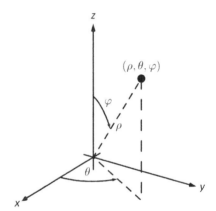

Figure 13.7.5: Illustrating the principles behind spherical coordinates.

**Note:** The symbol $\rho$ is the Greek letter "rho." Traditionally it is used in the spherical coordinate system, while $r$ is used in the polar and cylindrical coordinate systems.

**Note:** The role of $\theta$ and $\varphi$ in spherical coordinates differs between mathematicians and physicists. When reading about physics in spherical coordinates, be careful to note how that particular author uses these variables and recognize that these identities will may no longer be valid.

> **Key Idea 13.7.1  Converting Between Rectangular, Cylindrical and Spherical Coordinates**
>
> **Rectangular and Cylindrical**
> $$r^2 = x^2 + y^2, \quad \tan\theta = y/x, \quad z = z$$
> $$x = r\cos\theta, \quad y = r\sin\theta, \quad z = z$$
>
> **Rectangular and Spherical**
> $$\rho = \sqrt{x^2 + y^2 + z^2}, \quad \tan\theta = y/x, \quad \cos\varphi = z/\sqrt{x^2 + y^2 + z^2}$$
> $$x = \rho\sin\varphi\cos\theta, \quad y = \rho\sin\varphi\sin\theta, \quad z = \rho\cos\varphi$$
>
> **Cylindrical and Spherical**
> $$\rho = \sqrt{r^2 + z^2}, \quad \theta = \theta, \quad \tan\varphi = r/z$$
> $$r = \rho\sin\varphi, \quad \theta = \theta, \quad z = \rho\cos\varphi$$

**Example 13.7.5  Converting between rectangular and spherical coordinates**
Convert the rectangular point $(2, -2, 1)$ to spherical coordinates, and convert the spherical point $(6, \pi/3, \pi/2)$ to rectangular and cylindrical coordinates.

**SOLUTION**  This rectangular point is the same as used in Example 13.7.1. Using Key Idea 13.7.1, we find $\rho = \sqrt{2^2 + (-1)^2 + 1^2} = 3$. Using the same logic as in Example 13.7.1, we find $\theta = 7\pi/4$. Finally, $\cos\varphi = 1/3$, giving $\varphi = \cos^{-1}(1/3) \approx 1.23$, or about $70.53°$. Thus the spherical coordinates are approximately $(3, 7\pi/4, 1.23)$.

Converting the spherical point $(6, \pi/3, \pi/2)$ to rectangular, we have $x = 6\sin(\pi/2)\cos(\pi/3) = 3$, $y = 6\sin(\pi/2)\sin(\pi/3) = 3\sqrt{3}$ and $z = 6\cos(\pi/2) = 0$. Thus the rectangular coordinates are $(3, 3\sqrt{3}, 0)$.

To convert this spherical point to cylindrical, we have $r = 6\sin(\pi/2) = 6$, $\theta = \pi/3$ and $z = 6\cos(\pi/2) = 0$, giving the cylindrical point $(6, \pi/3, 0)$.

**Example 13.7.6  Canonical surfaces in spherical coordinates**
Describe the surfaces $\rho = 1$, $\theta = \pi/3$ and $\varphi = \pi/6$, given in spherical coordinates.

**SOLUTION**  The equation $\rho = 1$ describes all points in space that are 1 unit away from the origin: this is the sphere of radius 1, centered at the origin.

The equation $\theta = \pi/3$ describes the same surface in spherical coordinates as it does in cylindrical coordinates: beginning with the line $\theta = \pi/3$ in the x-y plane as given by polar coordinates, extend the line parallel to the z-axis, forming

Notes:

## 13.7 Triple Integration with Cylindrical and Spherical Coordinates

a plane.

The equation $\varphi = \pi/6$ describes all points $P$ in space where the ray from the origin to $P$ makes an angle of $\pi/6$ with the positive $z$-axis. This describes a cone, with the positive $z$-axis its axis of symmetry, with point at the origin.

All three surfaces are graphed in Figure 13.7.6. Note how their intersection uniquely defines the point $P = (1, \pi/3, \pi/6)$.

Spherical coordinates are useful when describing certain domains in space, allowing us to evaluate triple integrals over these domains more easily than if we used rectangular coordinates or cylindrical coordinates. The crux of setting up a triple integral in spherical coordinates is appropriately describing the "small amount of volume," $dV$, used in the integral.

Considering Figure 13.7.7, we can make a small "spherical wedge" by varying $\rho$, $\theta$ and $\varphi$ each a small amount, $\Delta\rho$, $\Delta\theta$ and $\Delta\varphi$, respectively. This wedge is approximately a rectangular solid when the change in each coordinate is small, giving a volume of about

$$\Delta V \approx \Delta\rho \times \rho\Delta\varphi \times \rho\sin(\varphi)\Delta\theta.$$

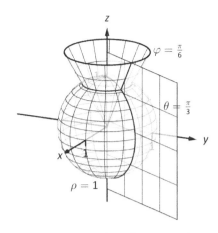

Figure 13.7.6: Graphing the canonical surfaces in spherical coordinates from Example 13.7.6.

Given a region $D$ in space, we can approximate the volume of $D$ with many such wedges. As the size of each of $\Delta\rho$, $\Delta\theta$ and $\Delta\varphi$ goes to zero, the number of wedges increases to infinity and the volume of $D$ is more accurately approximated, giving

$$dV = d\rho \times \rho\, d\varphi \times \rho\sin(\varphi)d\theta = \rho^2 \sin(\varphi)\, d\rho\, d\theta\, d\varphi.$$

Again, this development of $dV$ should sound reasonable, and the following theorem states it is the appropriate manner by which triple integrals are to be evaluated in spherical coordinates.

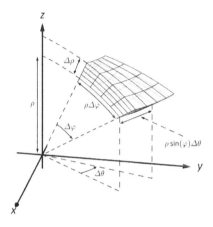

Figure 13.7.7: Approximating the volume of a standard region in space using spherical coordinates.

---

**Theorem 13.7.2    Triple Integration in Spherical Coordinates**

Let $w = h(\rho, \theta, \varphi)$ be a continuous function on a closed, bounded region $D$ in space, bounded in spherical coordinates by $\alpha_1 \leq \varphi \leq \alpha_2$, $\beta_1 \leq \theta \leq \beta_2$ and $f_1(\theta, \varphi) \leq \rho \leq f_2(\theta, \varphi)$. Then

$$\iiint_D h(\rho, \theta, \varphi)\, dV = \int_{\alpha_1}^{\alpha_2} \int_{\beta_1}^{\beta_2} \int_{f_1(\theta,\varphi)}^{f_2(\theta,\varphi)} h(\rho, \theta, \varphi)\rho^2 \sin(\varphi)\, d\rho\, d\theta\, d\varphi.$$

---

**Example 13.7.7    Establishing the volume of a sphere**

Let $D$ be the region in space bounded by the sphere, centered at the origin, of radius $r$. Use a triple integral in spherical coordinates to find the volume $V$ of $D$.

**Note:** It is generally most intuitive to evaluate the triple integral in Theorem 13.7.2 by integrating with respect to $\rho$ first; it often does not matter whether we next integrate with respect to $\theta$ or $\varphi$. Different texts present different standard orders, some preferring $d\varphi\, d\theta$ instead of $d\theta\, d\varphi$. As the bounds for these variables are usually constants in practice, it generally is a matter of preference.

Notes:

SOLUTION  The sphere of radius r, centered at the origin, has equation $\rho = r$. To obtain the full sphere, the bounds on $\theta$ and $\varphi$ are $0 \leq \theta \leq 2\pi$ and $0 \leq \varphi \leq \pi$. This leads us to:

$$V = \iiint_D dV$$
$$= \int_0^\pi \int_0^{2\pi} \int_0^r (\rho^2 \sin(\varphi)) \, d\rho \, d\theta \, d\varphi$$
$$= \int_0^\pi \int_0^{2\pi} \left(\frac{1}{3}\rho^3 \sin(\varphi) \Big|_0^r\right) d\theta \, d\varphi$$
$$= \int_0^\pi \int_0^{2\pi} \left(\frac{1}{3}r^3 \sin(\varphi)\right) d\theta \, d\varphi$$
$$= \int_0^\pi \left(\frac{2\pi}{3}r^3 \sin(\varphi)\right) d\varphi$$
$$= \left(-\frac{2\pi}{3}r^3 \cos(\varphi)\right)\Big|_0^\pi$$
$$= \frac{4\pi}{3}r^3,$$

the familiar formula for the volume of a sphere. Note how the integration steps were easy, not using square–roots nor integration steps such as Substitution.

**Example 13.7.8  Finding the center of mass using spherical coordinates**
Find the center of mass of the solid with constant density enclosed above by $\rho = 4$ and below by $\varphi = \pi/6$, as illustrated in Figure 13.7.8.

SOLUTION  We will set up the four triple integrals needed to find the center of mass (i.e., to compute $M$, $M_{yz}$, $M_{xz}$ and $M_{xy}$) and leave it to the reader to evaluate each integral. Because of symmetry, we expect the x- and y- coordinates of the center of mass to be 0.

While the surfaces describing the solid are given in the statement of the problem, to describe the full solid D, we use the following bounds: $0 \leq \rho \leq 4$, $0 \leq \theta \leq 2\pi$ and $0 \leq \varphi \leq \pi/6$. Since density $\delta$ is constant, we assume $\delta = 1$.

The mass of the solid:

$$M = \iiint_D dm = \iiint_D dV$$
$$= \int_0^{\pi/6} \int_0^{2\pi} \int_0^4 (\rho^2 \sin(\varphi)) \, d\rho \, d\theta \, d\varphi$$
$$= \frac{64}{3}(2 - \sqrt{3})\pi \approx 17.958.$$

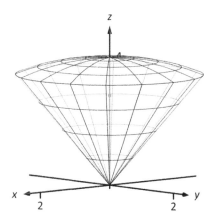

Figure 13.7.8: Graphing the solid, and its center of mass, from Example 13.7.8.

Notes:

## 13.7 Triple Integration with Cylindrical and Spherical Coordinates

To compute $M_{yz}$, the integrand is $x$; using Key Idea 13.7.1, we have $x = \rho \sin\varphi \cos\theta$. This gives:

$$M_{yz} = \iiint_D x\, dm$$
$$= \int_0^{\pi/6} \int_0^{2\pi} \int_0^4 \left((\rho \sin(\varphi) \cos(\theta))\rho^2 \sin(\varphi)\right) d\rho\, d\theta\, d\varphi$$
$$= \int_0^{\pi/6} \int_0^{2\pi} \int_0^4 \left(\rho^3 \sin^2(\varphi) \cos(\theta)\right) d\rho\, d\theta\, d\varphi$$
$$= 0,$$

which we expected as we expect $\bar{x} = 0$.

To compute $M_{xz}$, the integrand is $y$; using Key Idea 13.7.1, we have $y = \rho \sin\varphi \sin\theta$. This gives:

$$M_{xz} = \iiint_D y\, dm$$
$$= \int_0^{\pi/6} \int_0^{2\pi} \int_0^4 \left((\rho \sin(\varphi) \sin(\theta))\rho^2 \sin(\varphi)\right) d\rho\, d\theta\, d\varphi$$
$$= \int_0^{\pi/6} \int_0^{2\pi} \int_0^4 \left(\rho^3 \sin^2(\varphi) \sin(\theta)\right) d\rho\, d\theta\, d\varphi$$
$$= 0,$$

which we also expected as we expect $\bar{y} = 0$.

To compute $M_{xy}$, the integrand is $z$; using Key Idea 13.7.1, we have $z = \rho \cos\varphi$. This gives:

$$M_{xy} = \iiint_D z\, dm$$
$$= \int_0^{\pi/6} \int_0^{2\pi} \int_0^4 \left((\rho \cos(\varphi))\rho^2 \sin(\varphi)\right) d\rho\, d\theta\, d\varphi$$
$$= \int_0^{\pi/6} \int_0^{2\pi} \int_0^4 \left(\rho^3 \cos(\varphi) \sin(\varphi)\right) d\rho\, d\theta\, d\varphi$$
$$= 16\pi \approx 50.266.$$

Thus the center of mass is $(0, 0, M_{xy}/M) \approx (0, 0, 2.799)$, as indicated in Figure 13.7.8.

This section has provided a brief introduction into two new coordinate systems useful for identifying points in space. Each can be used to define a variety

Notes:

of surfaces in space beyond the canonical surfaces graphed as each system was introduced.

However, the usefulness of these coordinate systems does not lie in the variety of surfaces that they can describe nor the regions in space these surfaces may enclose. Rather, cylindrical coordinates are mostly used to describe cylinders and spherical coordinates are mostly used to describe spheres. These shapes are of special interest in the sciences, especially in physics, and computations on/inside these shapes is difficult using rectangular coordinates. For instance, in the study of electricity and magnetism, one often studies the effects of an electrical current passing through a wire; that wire is essentially a cylinder, described well by cylindrical coordinates.

This chapter investigated the natural follow–on to partial derivatives: iterated integration. We learned how to use the bounds of a double integral to describe a region in the plane using both rectangular and polar coordinates, then later expanded to use the bounds of a triple integral to describe a region in space. We used double integrals to find volumes under surfaces, surface area, and the center of mass of lamina; we used triple integrals as an alternate method of finding volumes of space regions and also to find the center of mass of a region in space.

Integration does not stop here. We could continue to iterate our integrals, next investigating "quadruple integrals" whose bounds describe a region in 4–dimensional space (which are very hard to visualize). We can also look back to "regular" integration where we found the area under a curve in the plane. A natural analogue to this is finding the "area under a curve," where the curve is in space, not in a plane. These are just two of many avenues to explore under the heading of "integration."

Notes:

# Exercises 13.7

## Terms and Concepts

1. Explain the difference between the roles $r$, in cylindrical coordinates, and $\rho$, in spherical coordinates, play in determining the location of a point.

2. Why are points on the $z$-axis not determined uniquely when using cylindrical and spherical coordinates?

3. What surfaces are naturally defined using cylindrical coordinates?

4. What surfaces are naturally defined using spherical coordinates?

## Problems

In Exercises 5 – 6, points are given in either the rectangular, cylindrical or spherical coordinate systems. Find the coordinates of the points in the other systems.

5. (a) Points in rectangular coordinates:
   $(2, 2, 1)$ and $(-\sqrt{3}, 1, 0)$

   (b) Points in cylindrical coordinates:
   $(2, \pi/4, 2)$ and $(3, 3\pi/2, -4)$

   (c) Points in spherical coordinates:
   $(2, \pi/4, \pi/4)$ and $(1, 0, 0)$

6. (a) Points in rectangular coordinates:
   $(0, 1, 1)$ and $(-1, 0, 1)$

   (b) Points in cylindrical coordinates:
   $(0, \pi, 1)$ and $(2, 4\pi/3, 0)$

   (c) Points in spherical coordinates:
   $(2, \pi/6, \pi/2)$ and $(3, \pi, \pi)$

In Exercises 7 – 8, describe the curve, surface or region in space determined by the given bounds.

7. Bounds in cylindrical coordinates:

   (a) $r = 1$, $\quad 0 \leq \theta \leq 2\pi$, $\quad 0 \leq z \leq 1$

   (b) $1 \leq r \leq 2$, $\quad 0 \leq \theta \leq \pi$, $\quad 0 \leq z \leq 1$

   Bounds in spherical coordinates:

   (c) $\rho = 3$, $\quad 0 \leq \theta \leq 2\pi$, $\quad 0 \leq \varphi \leq \pi/2$

   (d) $2 \leq \rho \leq 3$, $\quad 0 \leq \theta \leq 2\pi$, $\quad 0 \leq \varphi \leq \pi$

8. Bounds in cylindrical coordinates:

   (a) $1 \leq r \leq 2$, $\quad \theta = \pi/2$, $\quad 0 \leq z \leq 1$

   (b) $r = 2$, $\quad 0 \leq \theta \leq 2\pi$, $\quad z = 5$

   Bounds in spherical coordinates:

   (c) $0 \leq \rho \leq 2$, $\quad 0 \leq \theta \leq \pi$, $\quad \varphi = \pi/4$

   (d) $\rho = 2$, $\quad 0 \leq \theta \leq 2\pi$, $\quad \varphi = \pi/6$

In Exercises 9 – 10, standard regions in space, as defined by cylindrical and spherical coordinates, are shown. Set up the triple integral that integrates the given function over the graphed region.

9. Cylindrical coordinates, integrating $h(r, \theta, z)$:

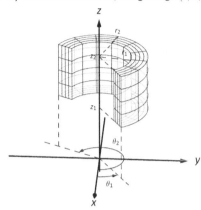

10. Cylindrical coordinates, integrating $h(\rho, \theta, \varphi)$:

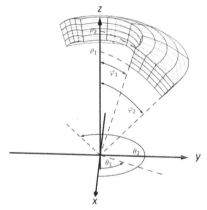

In Exercises 11 – 16, a triple integral in cylindrical coordinates is given. Describe the region in space defined by the bounds of the integral.

11. $\displaystyle\int_0^{\pi/2} \int_0^2 \int_0^2 r\, dz\, dr\, d\theta$

12. $\displaystyle\int_0^{2\pi} \int_3^4 \int_0^5 r\, dz\, dr\, d\theta$

13. $\displaystyle\int_0^{2\pi} \int_0^1 \int_0^{1-r} r\, dz\, dr\, d\theta$

14. $\displaystyle\int_0^{\pi} \int_0^1 \int_0^{2-r} r\, dz\, dr\, d\theta$

15. $\displaystyle\int_0^{\pi} \int_0^3 \int_0^{\sqrt{9-r^2}} r\, dz\, dr\, d\theta$

16. $\int_0^{2\pi} \int_0^a \int_0^{\sqrt{a^2-r^2}+b} r \, dz \, dr \, d\theta$

**In Exercises 17 – 22, a triple integral in spherical coordinates is given. Describe the region in space defined by the bounds of the integral.**

17. $\int_0^{\pi/2} \int_0^{\pi} \int_0^1 \rho^2 \sin(\varphi) \, d\rho \, d\theta \, d\varphi$

18. $\int_0^{\pi} \int_0^{\pi} \int_1^{1.1} \rho^2 \sin(\varphi) \, d\rho \, d\theta \, d\varphi$

19. $\int_0^{2\pi} \int_0^{\pi/4} \int_0^2 \rho^2 \sin(\varphi) \, d\rho \, d\theta \, d\varphi$

20. $\int_0^{2\pi} \int_{\pi/6}^{\pi/4} \int_0^2 \rho^2 \sin(\varphi) \, d\rho \, d\theta \, d\varphi$

21. $\int_0^{2\pi} \int_0^{\pi/6} \int_0^{\sec\varphi} \rho^2 \sin(\varphi) \, d\rho \, d\theta \, d\varphi$

22. $\int_0^{2\pi} \int_0^{\pi/6} \int_0^{a\sec\varphi} \rho^2 \sin(\varphi) \, d\rho \, d\theta \, d\varphi$

**In Exercises 23 – 26, a solid is described along with its density function. Find the mass of the solid using cylindrical coordinates.**

23. Bounded by the cylinder $x^2 + y^2 = 4$ and the planes $z = 0$ and $z = 4$ with density function $\delta(x,y,z) = \sqrt{x^2+y^2} + 1$.

24. Bounded by the cylinders $x^2 + y^2 = 4$ and $x^2 + y^2 = 9$, between the planes $z = 0$ and $z = 10$ with density function $\delta(x,y,z) = z$.

25. Bounded by $y \geq 0$, the cylinder $x^2 + y^2 = 1$, and between the planes $z = 0$ and $z = 4 - y$ with density function $\delta(x,y,z) = 1$.

26. The upper half of the unit ball, bounded between $z = 0$ and $z = \sqrt{1 - x^2 - y^2}$, with density function $\delta(x,y,z) = 1$.

**In Exercises 27 – 30, a solid is described along with its density function. Find the center of mass of the solid using cylindrical coordinates. (Note: these are the same solids and density functions as found in Exercises 23 through 26.)**

27. Bounded by the cylinder $x^2 + y^2 = 4$ and the planes $z = 0$ and $z = 4$ with density function $\delta(x,y,z) = \sqrt{x^2+y^2} + 1$.

28. Bounded by the cylinders $x^2 + y^2 = 4$ and $x^2 + y^2 = 9$, between the planes $z = 0$ and $z = 10$ with density function $\delta(x,y,z) = z$.

29. Bounded by $y \geq 0$, the cylinder $x^2 + y^2 = 1$, and between the planes $z = 0$ and $z = 4 - y$ with density function $\delta(x,y,z) = 1$.

30. The upper half of the unit ball, bounded between $z = 0$ and $z = \sqrt{1 - x^2 - y^2}$, with density function $\delta(x,y,z) = 1$.

**In Exercises 31 – 34, a solid is described along with its density function. Find the mass of the solid using spherical coordinates.**

31. The upper half of the unit ball, bounded between $z = 0$ and $z = \sqrt{1 - x^2 - y^2}$, with density function $\delta(x,y,z) = 1$.

32. The spherical shell bounded between $x^2 + y^2 + z^2 = 16$ and $x^2 + y^2 + z^2 = 25$ with density function $\delta(x,y,z) = \sqrt{x^2+y^2+z^2}$.

33. The conical region bounded above $z = \sqrt{x^2+y^2}$ and below the sphere $x^2 + y^2 + z^2 = 1$ with density function $\delta(x,y,z) = z$.

34. The cone bounded above $z = \sqrt{x^2+y^2}$ and below the plane $z = 1$ with density function $\delta(x,y,z) = z$.

**In Exercises 35 – 38, a solid is described along with its density function. Find the center of mass of the solid using spherical coordinates. (Note: these are the same solids and density functions as found in Exercises 31 through 34.)**

35. The upper half of the unit ball, bounded between $z = 0$ and $z = \sqrt{1 - x^2 - y^2}$, with density function $\delta(x,y,z) = 1$.

36. The spherical shell bounded between $x^2 + y^2 + z^2 = 16$ and $x^2 + y^2 + z^2 = 25$ with density function $\delta(x,y,z) = \sqrt{x^2+y^2+z^2}$.

37. The conical region bounded above $z = \sqrt{x^2+y^2}$ and below the sphere $x^2 + y^2 + z^2 = 1$ with density function $\delta(x,y,z) = z$.

38. The cone bounded above $z = \sqrt{x^2+y^2}$ and below the plane $z = 1$ with density function $\delta(x,y,z) = z$.

**In Exercises 39 – 42, a region is space is described. Set up the triple integrals that find the volume of this region using rectangular, cylindrical and spherical coordinates, then comment on which of the three appears easiest to evaluate.**

39. The region enclosed by the unit sphere, $x^2 + y^2 + z^2 = 1$.

40. The region enclosed by the cylinder $x^2 + y^2 = 1$ and planes $z = 0$ and $z = 1$.

41. The region enclosed by the cone $z = \sqrt{x^2+y^2}$ and plane $z = 1$.

42. The cube enclosed by the planes $x = 0$, $x = 1$, $y = 0$, $y = 1$, $z = 0$ and $z = 1$. (Hint: in spherical, use order of integration $d\rho \, d\varphi \, d\theta$.)

# 14: VECTOR ANALYSIS

In previous chapters we have explored a relationship between vectors and integration. Our most tangible result: if $\vec{v}(t)$ is the vector–valued velocity function of a moving object, then integrating $\vec{v}(t)$ from $t = a$ to $t = b$ gives the displacement of that object over that time interval.

This chapter explores completely different relationships between vectors and integration. These relationships will enable us to compute the work done by a magnetic field in moving an object along a path and find how much air moves through an oddly–shaped screen in space, among other things.

Our upcoming work with integration will benefit from a review. We are not concerned here with techniques of integration, but rather what an integral "does" and how that relates to the notation we use to describe it.

## Integration Review

Recall from Section 13.1 that when $R$ is a region in the x-y plane, $\iint_R dA$ gives the area of the region $R$. The integral symbols are "elongated esses" meaning "sum" and $dA$ represents "a small amount of area." Taken together, $\iint_R dA$ means "sum up, over $R$, small amounts of area." This sum then gives the total area of $R$. We use two integral symbols since $R$ is a two–dimensional region.

Now let $z = f(x, y)$ represent a surface. The double integral $\iint_R f(x, y)\, dA$ means "sum up, over $R$, function values (heights) given by $f$ times small amounts of area." Since "height $\times$ area = volume," we are summing small amounts of volume over $R$, giving the total signed volume under the surface $z = f(x, y)$ and above the x-y plane.

This notation does not directly inform us *how* to evaluate the double integrals to find an area or a volume. With additional work, we recognize that a small amount of area $dA$ can be measured as the area of a small rectangle, with one side length a small change in $x$ and the other side length a small change in $y$. That is, $dA = dx\,dy$ or $dA = dy\,dx$. We could also compute a small amount of area by thinking in terms of polar coordinates, where $dA = r\,dr\,d\theta$. These understandings lead us to the iterated integrals we used in Chapter 13.

Let us back our review up farther. Note that $\int_1^3 dx = x\big|_1^3 = 3 - 1 = 2$. We have simply measured the length of the interval $[1, 3]$. We could rewrite the above integral using syntax similar to the double integral syntax above:

$$\int_1^3 dx = \int_I dx, \quad \text{where } I = [1, 3].$$

We interpret "$\int_I dx$" as meaning "sum up, over the interval $I$, small changes in $x$." A change in $x$ is a length along the x-axis, so we are adding up along $I$ small

# Chapter 14 Vector Analysis

lengths, giving the total length of $I$.

We could also write $\int_1^3 f(x)\,dx$ as $\int_I f(x)\,dx$, interpreted as "sum up, over $I$, heights given by $y = f(x)$ times small changes in $x$." Since "height×length = area," we are summing up areas and finding the total signed area between $y = f(x)$ and the $x$-axis.

This method of referring to the process of integration can be very powerful. It is the core of our notion of the Riemann Sum. When faced with a quantity to compute, if one can think of a way to approximate its value through a sum, the one is well on their way to constructing an integral (or, double or triple integral) that computes the desired quantity. We will demonstrate this process throughout this chapter, starting with the next section.

## 14.1 Introduction to Line Integrals

We first used integration to find "area under a curve." In this section, we learn to do this (again), but in a different context.

Consider the surface and curve shown in Figure 14.1.1(a). The surface is given by $f(x, y) = 1 - \cos(x)\sin(y)$. The dashed curve lies in the $x$-$y$ plane and is the familiar $y = x^2$ parabola from $-1 \leq x \leq 1$; we'll call this curve $C$. The curve drawn with a solid line in the graph is the curve in space that lies on our surface with $x$ and $y$ values that lie on $C$.

The question we want to answer is this: what is the area that lies below the curve drawn with the solid line? In other words, what is the area of the region above $C$ and under the the surface $f$? This region is shown in Figure 14.1.1(b).

We suspect the answer can be found using an integral, but before trying to figure out what that integral is, let us first try to approximate its value.

In Figure 14.1.1(c), four rectangles have been drawn over the curve $C$. The bottom corners of each rectangle lie on $C$, and each rectangle has a height given by the function $f(x, y)$ for some $(x, y)$ pair along $C$ between the rectangle's bottom corners.

As we know how to find the area of each rectangle, we are able to approximate the area above $C$ and under $f$. Clearly, our approximation will be *an approximation*. The heights of the rectangles do not match exactly with the surface $f$, nor does the base of each rectangle follow perfectly the path of $C$.

In typical calculus fashion, our approximation can be improved by using more rectangles. The sum of the areas of these rectangles gives an approximate value of the true area above $C$ and under $f$. As the area of each rectangle is "height × width", we assert that the

$$\text{area above } C \approx \sum (\text{heights} \times \text{widths}).$$

When first learning of the integral, and approximating areas with "heights ×

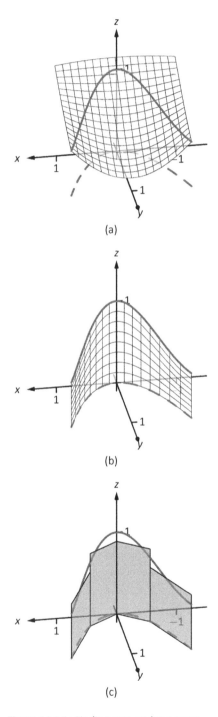

Figure 14.1.1: Finding area under a curve in space.

Notes:

## 14.1 Introduction to Line Integrals

widths", the width was a small change in x: dx. That will not suffice in this context. Rather, each width of a rectangle is actually approximating the arc length of a small portion of C. In Section 11.5, we used s to represent the arc–length parameter of a curve. A small amount of arc length will thus be represented by ds.

The height of each rectangle will be determined in some way by the surface f. If we parametrize C by s, an s-value corresponds to an (x, y) pair that lies on the parabola C. Since f is a function of x and y, and x and y are functions of s, we can say that f is a function of s. Given a value s, we can compute $f(s)$ and find a height. Thus

$$\text{area under } f \text{ and above } C \approx \sum (\text{heights} \times \text{widths});$$
$$\text{area under } f \text{ and above } C = \lim_{||\Delta s|| \to 0} \sum f(c_i) \Delta s_i$$
$$= \int_C f(s)\, ds. \tag{14.1}$$

Here we have introduce a new notation, the integral symbol with a subscript of C. It is reminiscent of our usage of $\iint_R$. Using the train of thought found in the Integration Review preceding this section, we interpret "$\int_C f(s)\, ds$" as meaning "sum up, along a curve C, function values $f(s) \times$ small arc lengths." It is understood here that s represents the arc–length parameter.

All this leads us to a definition. The integral found in Equation 14.1 is called a **line integral**. We formally define it below, but note that the definition is very abstract. On one hand, one is apt to say "the defintion makes sense," while on the other, one is equally apt to say "but I don't know what I'm supposed to do with this definition." We'll address that after the definition, and actually find an answer to the area problem we posed at the beginning of this section.

---

**Definition 14.1.1**     **Line Integral Over A Scalar Field**

Let C be a smooth curve parametrized by s, the arc–length parameter, and let f be a continuous function of s. A **line integral** is an integral of the form

$$\int_C f(s)\, ds = \lim_{||\Delta s|| \to 0} \sum_{i=1}^{n} f(c_i) \Delta s_i,$$

where $s_1 < s_2 < \ldots < s_n$ is any partition of the s-interval over which C is defined, $c_i$ is any value in the $i^{\text{th}}$ subinterval, $\Delta s_i$ is the width of the $i^{\text{th}}$ subinterval, and $||\Delta s||$ is the length of the longest subinterval in the partition.

---

**Note:** Definition 14.1.1 uses the term **scalar field** which has not yet been defined. Its meaning is discussed in the paragraph preceding Definition 14.3.1 when it is compared to a **vector field**.

Notes:

When $C$ is a **closed** curve, i.e., a curve that ends at the same point at which it starts, we use

$$\oint_C f(s)\, ds \quad \text{instead of} \quad \int_C f(s)\, ds.$$

The definition of the line integral does not specify whether $C$ is a curve in the plane or space (or hyperspace), as the definition holds regardless. For now, we'll assume $C$ lies in the $x$-$y$ plane.

This definition of the line integral doesn't really say anything new. If $C$ is a curve and $s$ is the arc–length parameter of $C$ on $a \leq s \leq b$, then

$$\int_C f(s)\, ds = \int_a^b f(s)\, ds.$$

The real difference with this integral from the standard "$\int_a^b f(x)\, dx$" we used in the past is that of context. Our previous integrals naturally summed up values over an interval on the $x$-axis, whereas now we are summing up values over a curve. *If* we can parametrize the curve with the arc–length parameter, we can evaluate the line integral just as before. Unfortunately, parametrizing a curve in terms of the arc–length parameter is usually very difficult, so we must develop a method of evaluating line integrals using a different parametrization.

Given a curve $C$, find any parametrization of $C$: $x = g(t)$ and $y = h(t)$, for continuous functions $g$ and $h$, where $a \leq t \leq b$. We can represent this parametrization with a vector–valued function, $\vec{r}(t) = \langle g(t), h(t) \rangle$.

In Section 11.5, we defined the arc–length parameter in Equation 11.1 as

$$s(t) = \int_0^t \|\vec{r}'(u)\|\, du.$$

By the Fundamental Theorem of Calculus, $ds = \|\vec{r}'(t)\|\, dt$. We can substitute the right hand side of this equation for $ds$ in the line integral definition.

We can view $f$ as being a function of $x$ and $y$ since it is a function of $s$. Thus $f(s) = f(x, y) = f(g(t), h(t))$. This gives us a concrete way to evaluate a line integral:

$$\int_C f(s)\, ds = \int_a^b f(g(t), h(t)) \|\vec{r}'(t)\|\, dt.$$

We restate this as a theorem, along with its three–dimensional analogue, followed by an example where we finally evaluate an integral and find an area.

---

Notes:

## 14.1 Introduction to Line Integrals

**Theorem 14.1.1     Evaluating a Line Integral Over A Scalar Field**

- Let $C$ be a curve parametrized by $\vec{r}(t) = \langle g(t), h(t) \rangle$, $a \leq t \leq b$, where $g$ and $h$ are continuously differentiable, and let $z = f(x, y)$, where $f$ is continuous over $C$. Then
$$\int_C f(s)\, ds = \int_a^b f\big(g(t), h(t)\big) \|\vec{r}\,'(t)\|\, dt.$$

- Let $C$ be a curve parametrized by $\vec{r}(t) = \langle g(t), h(t), k(t) \rangle$, $a \leq t \leq b$, where $g$, $h$ and $k$ are continuously differentiable, and let $w = f(x, y, z)$, where $f$ is continuous over $C$. Then
$$\int_C f(s)\, ds = \int_a^b f\big(g(t), h(t), k(t)\big) \|\vec{r}\,'(t)\|\, dt.$$

To be clear, the first point of Theorem 14.1.1 can be used to find the area under a surface $z = f(x, y)$ and above a curve $C$. We will later give an understanding of the line integral when $C$ is a curve in space.

Let's do an example where we actually compute an area.

**Example 14.1.1     Evaluating a line integral: area under a surface over a curve.**
Find the area under the surface $f(x, y) = \cos(x) + \sin(y) + 2$ over the curve $C$, which is the segment of the line $y = 2x + 1$ on $-1 \leq x \leq 1$, as shown in Figure 14.1.2.

**SOLUTION**     Our first step is to represent $C$ with a vector–valued function. Since $C$ is a simple line, and we have a explicit relationship between $y$ and $x$ (namely, that $y$ is $2x+1$), we can let $x = t$, $y = 2t+1$, and write $\vec{r}(t) = \langle t, 2t+1 \rangle$ for $-1 \leq t \leq 1$.

We find the values of $f$ over $C$ as $f(x, y) = f(t, 2t+1) = \cos(t) + \sin(2t+1) + 2$. We also need $\|\vec{r}\,'(t)\|$; with $\vec{r}\,'(t) = \langle 1, 2 \rangle$, we have $\|\vec{r}\,'(t)\| = \sqrt{5}$. Thus $ds = \sqrt{5}\, dt$.

The area we seek is
$$\int_C f(s)\, ds = \int_{-1}^{1} \big(\cos(t) + \sin(2t+1) + 2\big) \sqrt{5}\, dt$$
$$= \sqrt{5}\left(\sin(t) - \frac{1}{2}\cos(2t+1) + 2t\right)\Big|_{-1}^{1}$$
$$\approx 14.418 \text{ units}^2.$$

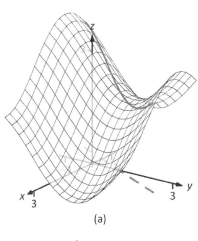

(a)

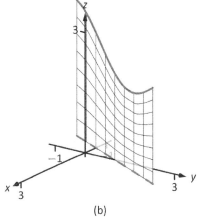

(b)

Figure 14.1.2: Finding area under a curve in Example 14.1.1.

Notes:

## Chapter 14 Vector Analysis

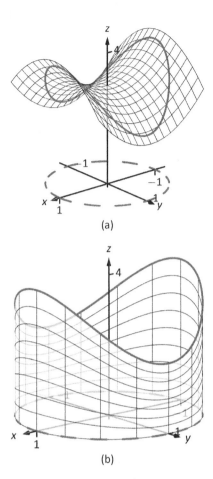

(a)

(b)

Figure 14.1.3: Finding area under a curve in Example 14.1.2.

We will practice setting up and evaluating a line integral in another example, then find the area described at the beginning of this section.

**Example 14.1.2  Evaluating a line integral: area under a surface over a curve.**
Find the area over the unit circle in the *x-y* plane and under the surface $f(x,y) = x^2 - y^2 + 3$, shown in Figure 14.1.3.

**SOLUTION**  The curve $C$ is the unit circle, which we will describe with the parametrization $\vec{r}(t) = \langle \cos t, \sin t \rangle$ for $0 \le t \le 2\pi$. We find $||\vec{r}'(t)|| = 1$, so $ds = 1 dt$.

We find the values of $f$ over $C$ as $f(x,y) = f(\cos t, \sin t) = \cos^2 t - \sin^2 t + 3$. Thus the area we seek is (note the use of the $\oint f(s)ds$ notation):

$$\oint_C f(s)\,ds = \int_0^{2\pi} \left(\cos^2 t - \sin^2 t + 3\right)\,dt$$
$$= 6\pi.$$

(Note: we may have approximated this answer from the start. The unit circle has a circumference of $2\pi$, and we may have guessed that due to the apparent symmetry of our surface, the average height of the surface is 3.)

We now consider the example that introduced this section.

**Example 14.1.3  Evaluating a line integral: area under a surface over a curve.**
Find the area under $f(x,y) = 1 - \cos(x)\sin(y)$ and over the parabola $y = x^2$, from $-1 \le x \le 1$.

**SOLUTION**  We parametrize our curve $C$ as $\vec{r}(t) = \langle t, t^2 \rangle$ for $-1 \le t \le 1$; we find $||\vec{r}'(t)|| = \sqrt{1+4t^2}$, so $ds = \sqrt{1+4t^2}\,dt$.

Replacing $x$ and $y$ with their respective functions of $t$, we have $f(x,y) = f(t,t^2) = 1 - \cos(t)\sin(t^2)$. Thus the area under $f$ and over $C$ is found to be

$$\int_C f(s)\,ds = \int_{-1}^{1} \left(1 - \cos(t)\sin\left(t^2\right)\right)\sqrt{1+t^2}\,dt.$$

This integral is impossible to evaluate using the techniques developed in this text. We resort to a numerical approximation; accurate to two places after the decimal, we find the area is

$$= 2.17.$$

Notes:

844

We give one more example of finding area.

**Example 14.1.4** **Evaluating a line integral: area under a curve in space.**
Find the area above the x-y plane and below the helix parametrized by $\vec{r}(t) = \langle \cos t, 2\sin t, t/\pi \rangle$, for $0 \leq t \leq 2\pi$, as shown in Figure 14.1.4.

**SOLUTION** Note how this is problem is different than the previous examples: here, the height is not given by a surface, but by the curve itself.

We use the given vector-valued function $\vec{r}(t)$ to determine the curve $C$ in the x-y plane by simply using the first two components of $\vec{r}(t)$: $\vec{c}(t) = \langle \cos t, 2\sin t \rangle$. Thus $ds = ||\vec{c}'(t)|| \, dt = \sqrt{\sin^2 t + 4\cos^2 t} \, dt$.

The height is not found by evaluating a surface over $C$, but rather it is given directly by the third component of $\vec{r}(t)$: $t/\pi$. Thus

$$\oint_C f(s)\, ds = \int_0^{2\pi} \frac{t}{\pi} \sqrt{\sin^2 t + 4\cos^2 t} \, dt \approx 9.69,$$

where the approximation was obtained using numerical methods.

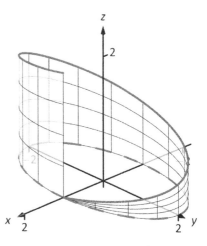

Figure 14.1.4: Finding area under a curve in Example 14.1.4.

Note how in each of the previous examples we are effectively finding "area under a curve", just as we did when first learning of integration. We have used the phrase "area *over* a curve $C$ and under a surface," but that is because of the important role $C$ plays in the integral. The figures show how the curve $C$ defines another curve on the surface $z = f(x,y)$, and we are finding the area under that curve.

## Properties of Line Integrals

Many properties of line integrals can be inferred from general integration properties. For instance, if $k$ is a scalar, then $\int_C kf(s)\,ds = k\int_C f(s)\,ds$.

One property in particular of line integrals is worth noting. If $C$ is a curve composed of subcurves $C_1$ and $C_2$, where they share only one point in common (see Figure 14.1.5(a)), then the line integral over $C$ is the sum of the line integrals over $C_1$ and $C_2$:

$$\int_C f(s)\, ds = \int_{C_1} f(s)\, ds + \int_{C_2} f(s)\, ds.$$

This property allows us to evaluate line integrals over some curves $C$ that are not smooth. Note how in Figure 14.1.5(b) the curve is not smooth at $D$, so by our definition of the line integral we cannot evaluate $\int_C f(s)\,ds$. However, one can evaluate line integrals over $C_1$ and $C_2$ and their sum will be the desired quantity.

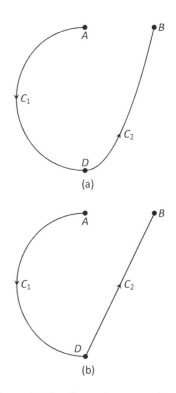

Figure 14.1.5: Illustrating properties of line integrals.

A curve C that is composed of two or more smooth curves is said to be **piecewise smooth**. In this chapter, any statement that is made about smooth curves also holds for piecewise smooth curves.

We state these properties as a theorem.

---

**Theorem 14.1.2**      **Properties of Line Integrals Over Scalar Fields**

1. Let $C$ be a smooth curve parametrized by the arc–length parameter $s$, let $f$ and $g$ be continuous functions of $s$, and let $k_1$ and $k_2$ be scalars. Then

$$\int_C \left(k_1 f(s) + k_2 g(s)\right) ds = k_1 \int_C f(s)\, ds + k_2 \int_C g(s)\, ds.$$

2. Let $C$ be piecewise smooth, composed of smooth components $C_1$ and $C_2$. Then

$$\int_C f(s)\, ds = \int_{C_1} f(s)\, ds + \int_{C_2} f(s)\, ds.$$

---

**Mass and Center of Mass**

We first learned integration as a method to find area under a curve, then later used integration to compute a variety of other quantities, such as arc length, volume, force, etc. In this section, we also introduced line integrals as a method to find area under a curve, and now we explore one more application.

Let a curve $C$ (either in the plane or in space) represent a thin wire with variable density $\delta(s)$. We can approximate the mass of the wire by dividing the wire (i.e., the curve) into small segments of length $\Delta s_i$ and assume the density is constant across these small segments. The mass of each segment is density of the segment $\times$ its length; by summing up the approximate mass of each segment we can approximate the total mass:

$$\text{Total Mass of Wire } = \sum \delta(s_i) \Delta s_i.$$

By taking the limit as the length of the segments approaches 0, we have the definition of the line integral as seen in Definition 14.1.1. When learning of the line integral, we let $f(s)$ represent a height; now we let $f(s) = \delta(s)$ represent a density.

We can extend this understanding of computing mass to also compute the center of mass of a thin wire. (As a reminder, the center of mass can be a useful

---

Notes:

piece of information as objects rotate about that center.) We give the relevant formulas in the next definition, followed by an example. Note the similarities between this definition and Definition 13.6.4, which gives similar properties of solids in space.

---

**Definition 14.1.2     Mass, Center of Mass of Thin Wire**

Let a thin wire lie along a smooth curve C with continuous density function $\delta(s)$, where s is the arc length parameter.

1. The **mass** of the thin wire is $M = \int_C \delta(s)\,ds$.

2. The **moment about the y-z plane** is $M_{yz} = \int_C x\delta(s)\,ds$.

3. The **moment about the x-z plane** is $M_{xz} = \int_C y\delta(s)\,ds$.

4. The **moment about the x-y plane** is $M_{xy} = \int_C z\delta(s)\,ds$.

5. The **center of mass** of the wire is
$$(\bar{x}, \bar{y}, \bar{z}) = \left(\frac{M_{yz}}{M}, \frac{M_{xz}}{M}, \frac{M_{xy}}{M}\right).$$

---

**Example 14.1.5     Evaluating a line integral: calculating mass.**

A thin wire follows the path $\vec{r}(t) = \langle 1+\cos t, 1+\sin t, 1+\sin(2t)\rangle$, $0 \leq t \leq 2\pi$. The density of the wire is determined by its position in space: $\delta(x,y,z) = y + z$ gm/cm. The wire is shown in Figure 14.1.6, where a light color indicates low density and a dark color represents high density. Find the mass and center of mass of the wire.

**SOLUTION**   We compute the density of the wire as
$$\delta(x,y,z) = \delta(1+\cos t, 1+\sin t, 1+\sin(2t)) = 2 + \sin t + \sin(2t).$$

We compute ds as
$$ds = \|\vec{r}'(t)\|\,dt = \sqrt{\sin^2 t + \cos^2 t + 4\cos^2(2t)}\,dt = \sqrt{1 + 4\cos^2(2t)}\,dt.$$

Thus the mass is
$$M = \oint_C \delta(s)\,ds = \int_0^{2\pi} (2 + \sin t + \sin(2t))\sqrt{1 + 4\cos^2(2t)}\,dt \approx 21.08 \text{gm}.$$

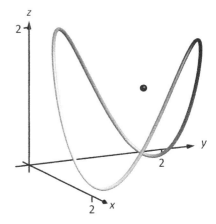

Figure 14.1.6: Finding the mass of a thin wire in Example 14.1.5.

Notes:

We compute the moments about the coordinate planes:

$$M_{yz} = \oint_C x\delta(s)\,ds = \int_0^{2\pi} (1+\cos t)(2+\sin t + \sin(2t))\sqrt{1+4\cos^2(2t)}\,dt \approx 21.08.$$

$$M_{xz} = \oint_C y\delta(s)\,ds = \int_0^{2\pi} (1+\sin t)(2+\sin t + \sin(2t))\sqrt{1+4\cos^2(2t)}\,dt \approx 26.35$$

$$M_{xy} = \oint_C z\delta(s)\,ds = \int_0^{2\pi} (1+\sin(2t))(2+\sin t + \sin(2t))\sqrt{1+4\cos^2(2t)}\,dt \approx 25.40$$

Thus the center of mass of the wire is located at

$$(\bar{x},\bar{y},\bar{z}) = \left(\frac{M_{yz}}{M},\frac{M_{xz}}{M},\frac{M_{xy}}{M}\right) \approx (1, 1.25, 1.20),$$

as indicated by the dot in Figure 14.1.6. Note how in this example, the curve $C$ is "centered" about the point $(1,1,1)$, though the variable density of the wire pulls the center of mass out along the $y$ and $z$ axes.

We end this section with a callback to the Integration Review that preceded this section. A line integral looks like: $\int_C f(s)\,ds$. As stated before the definition of the line integral, this means "sum up, along a curve $C$, function values $f(s) \times$ small arc lengths." When $f(s)$ represents a height, we have "height $\times$ length = area." When $f(s)$ is a density (and we use $\delta(s)$ by convention), we have "density (mass per unit length) $\times$ length = mass."

In the next section, we investigate a new mathematical object, the *vector field*. The remaining sections of this chapter are devoted to understanding integration in the context of vector fields.

Notes:

# Exercises 14.1

## Terms and Concepts

1. Explain how a line integral can be used to find the area under a curve.

2. How does the evaluation of a line integral given as $\int_C f(s)\,ds$ differ from a line integral given as $\oint_C f(s)\,ds$?

3. Why are most line integrals evaluated using Key Idea 14.1.1 instead of "directly" as $\int_C f(s)\,ds$?

4. Sketch a closed, piecewise smooth curve composed of three subcurves.

## Problems

In Exercises 5 – 10, a planar curve $C$ is given along with a surface $f$ that is defined over $C$. Evaluate the line integral $\int_C f(s)\,ds$.

5. $C$ is the line segment joining the points $(-2, -1)$ and $(1, 2)$; the surface is $f(x, y) = x^2 + y^2 + 2$.

6. $C$ is the segment of $y = 3x + 2$ on $[1, 2]$; the surface is $f(x, y) = 5x + 2y$.

7. $C$ is the circle with radius 2 centered at the point $(4, 2)$; the surface is $f(x, y) = 3x - y$.

8. $C$ is the curve given by $\vec{r}(t) = \langle \cos t + t \sin t, \sin t - t \cos t \rangle$ on $[0, 2\pi]$; the surface is $f(x, y) = 5$.

9. $C$ is the piecewise curve composed of the line segments that connect $(0, 1)$ to $(1, 1)$, then connect $(1, 1)$ to $(1, 0)$; the surface is $f(x, y) = x + y^2$.

10. $C$ is the piecewise curve composed of the line segment joining the points $(0, 0)$ and $(1, 1)$, along with the quarter-circle parametrized by $\langle \cos t, -\sin t + 1 \rangle$ on $[0, \pi/2]$ (which starts at the point $(1, 1)$ and ends at $(0, 0)$; the surface is $f(x, y) = x^2 + y^2$.

In Exercises 11 – 14, a planar curve $C$ is given along with a surface $f$ that is defined over $C$. Set up the line integral $\int_C f(s)\,ds$, then approximate its value using technology.

11. $C$ is the portion of the parabola $y = 2x^2 + x + 1$ on $[0, 1]$; the surface is $f(x, y) = x^2 + 2y$.

12. $C$ is the portion of the curve $y = \sin x$ on $[0, \pi]$; the surface is $f(x, y) = x$.

13. $C$ is the ellipse given by $\vec{r}(t) = \langle 2\cos t, \sin t \rangle$ on $[0, 2\pi]$; the surface is $f(x, y) = 10 - x^2 - y^2$.

14. $C$ is the portion of $y = x^3$ on $[-1, 1]$; the surface is $f(x, y) = 2x + 3y + 5$.

In Exercises 15 – 18, a parametrized curve $C$ in space is given. Find the area above the x-y plane that is under $C$.

15. $C$: $\vec{r}(t) = \langle 5t, t, t^2 \rangle$ for $1 \le t \le 2$.

16. $C$: $\vec{r}(t) = \langle \cos t, \sin t, \sin(2t) + 1 \rangle$ for $0 \le t \le 2\pi$.

17. $C$: $\vec{r}(t) = \langle 3\cos t, 3\sin t, t^2 \rangle$ for $0 \le t \le 2\pi$.

18. $C$: $\vec{r}(t) = \langle 3t, 4t, t \rangle$ for $0 \le t \le 1$.

In Exercises 19 – 20, a parametrized curve $C$ is given that represents a thin wire with density $\delta$. Find the mass and center of mass of the thin wire.

19. $C$: $\vec{r}(t) = \langle \cos t, \sin t, t \rangle$ for $0 \le t \le 4\pi$; $\delta(x, y, z) = z$.

20. $C$: $\vec{r}(t) = \langle t - t^2, t^2 - t^3, t^3 - t^4 \rangle$ for $0 \le t \le 1$; $\delta(x, y, z) = x + 2y + 2z$. Use technology to approximate the value of each integral.

## 14.2 Vector Fields

We have studied functions of two and three variables, where the input of such functions is a point (either a point in the plane or in space) and the output is a number.

We could also create functions where the input is a point (again, either in the plane or in space), but the output is a *vector*. For instance, we could create the following function: $\vec{F}(x,y) = \langle x+y, x-y \rangle$, where $\vec{F}(2,3) = \langle 5, -1 \rangle$. We are to think of $\vec{F}$ assigning the vector $\langle 5, -1 \rangle$ to the point $(2,3)$; in some sense, the vector $\langle 5, -1 \rangle$ lies at the point $(2,3)$.

Such functions are extremely useful in any context where magnitude and direction are important. For instance, we could create a function $\vec{F}$ that represents the electromagnetic force exerted at a point by a electromagnetic field, or the velocity of air as it moves across an airfoil.

Because these functions are so important, we need to formally define them.

---

**Definition 14.2.1**      **Vector Field**

1. A **vector field in the plane** is a function $\vec{F}(x,y)$ whose domain is a subset of $\mathbb{R}^2$ and whose output is a two–dimensional vector:
$$\vec{F}(x,y) = \langle M(x,y), N(x,y) \rangle.$$

2. A **vector field in space** is a function $\vec{F}(x,y,z)$ whose domain is a subset of $\mathbb{R}^3$ and whose output is a three–dimensional vector:
$$\vec{F}(x,y,z) = \langle M(x,y,z), N(x,y,z), P(x,y,z) \rangle.$$

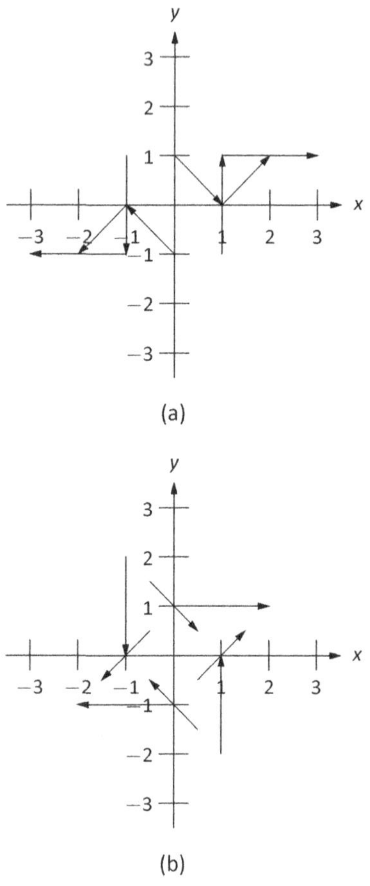

---

This definition may seem odd at first, as a special type of function is called a "field." However, as the function determines a "field of vectors", we can say the field is *defined by* the function, and thus the field *is* a function.

Visualizing vector fields helps cement this connection. When graphing a vector field in the plane, the general idea is to draw the vector $\vec{F}(x,y)$ at the point $(x,y)$. For instance, using $\vec{F}(x,y) = \langle x+y, x-y \rangle$ as before, at $(1,1)$ we would draw $\langle 2, 0 \rangle$.

In Figure 14.2.1(a), one can see that the vector $\langle 2, 0 \rangle$ is drawn *starting from* the point $(1,1)$. A total of 8 vectors are drawn, with the *x*- and *y*-values of $-1, 0, 1$. In many ways, the resulting graph is a mess; it is hard to tell what this field "looks like."

In Figure 14.2.1(b), the same field is redrawn with each vector $\vec{F}(x,y)$ drawn *centered on* the point $(x,y)$. This makes for a better looking image, though the

Figure 14.2.1: Demonstrating methods of graphing vector fields.

Notes:

long vectors can cause confusion: when one vector intersects another, the image looks cluttered.

A common way to address this problem is limit the length of each arrow, and represent long vectors with thick arrows, as done in Figure 14.2.2(a). Usually we do not use a graph of a vector field to determine exactly the magnitude of a particular vector. Rather, we are more concerned with the relative magnitudes of vectors: which are bigger than others? Thus limiting the length of the vectors is not problematic.

Drawing arrows with variable thickness is best done with technology; search the documentation of your favorite graphing program for terms like "vector fields" or "slope fields" to learn how. Technology obviously allows us to plot many vectors in a vector field nicely; in Figure 14.2.2(b), we see the same vector field drawn with many vectors, and finally get a clear picture of how this vector field behaves. (If this vector field represented the velocity of air moving across a flat surface, we could see that the air tends to move either to the upper–right or lower–left, and moves very slowly near the origin.)

We can similarly plot vector fields in space, as shown in Figure 14.2.3, though it is not often done. The plots get very busy very quickly, as there are lots of arrows drawn in a small amount of space. In Figure 14.2.3 the field $\vec{F} = \langle -y, x, z \rangle$ is graphed. If one could view the graph from above, one could see the arrows point in a cirlce about the $z$-axis. One should also note how the arrows far from the origin are larger than those close to the origin.

It is good practice to try to visualize certain vector fields in one's head. For instance, consider a point mass at the origin and the vector field that represents the gravitational force exerted by the mass at any point in the room. The field would consist of arrows pointing toward the origin, increasing in size as they near the origin (as the gravitational pull is strongest near the point mass).

## Vector Field Notation and Del Operator

Definition 14.2.1 defines a vector field $\vec{F}$ using the notation

$$\vec{F}(x,y) = \langle M(x,y), N(x,y) \rangle \quad \text{and} \quad \vec{F}(x,y,z) = \langle M(x,y,z), N(x,y,z), P(x,y,z) \rangle.$$

That is, the components of $\vec{F}$ are each functions of $x$ and $y$ (and also $z$ in space). As done in other contexts, we will drop the "of $x$, $y$ and $z$" portions of the notation and refer to vector fields in the plane and in space as

$$\vec{F} = \langle M, N \rangle \quad \text{and} \quad \vec{F} = \langle M, N, P \rangle,$$

respectively, as this shorthand is quite convenient.

Another item of notation will become useful: the "del operator." Recall in Section 12.6 how we used the symbol $\nabla$ (pronounced "del") to represent the

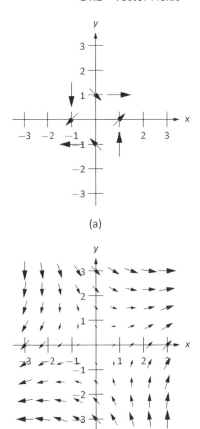

Figure 14.2.2: Demonstrating methods of graphing vector fields.

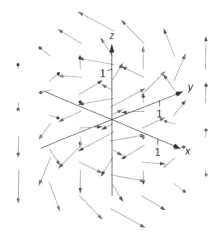

Figure 14.2.3: Graphing a vector field in space.

Notes:

gradient of a function of two variables. That is, if $z = f(x,y)$, then "del $f$" = $\nabla f = \langle f_x, f_y \rangle$.

We now define $\nabla$ to be the "del operator." It is a vector whose components are partial derivative operations.

In the plane, $\nabla = \left\langle \dfrac{\partial}{\partial x}, \dfrac{\partial}{\partial y} \right\rangle$; in space, $\nabla = \left\langle \dfrac{\partial}{\partial x}, \dfrac{\partial}{\partial y}, \dfrac{\partial}{\partial z} \right\rangle$.

With this definition of $\nabla$, we can better understand the gradient $\nabla f$. As $f$ returns a scalar, the properties of scalar and vector multiplication gives

$$\nabla f = \left\langle \dfrac{\partial}{\partial x}, \dfrac{\partial}{\partial y} \right\rangle f = \left\langle \dfrac{\partial}{\partial x} f, \dfrac{\partial}{\partial y} f \right\rangle = \langle f_x, f_y \rangle.$$

Now apply the del operator $\nabla$ to vector fields. Let $\vec{F} = \langle x + \sin y, y^2 + z, x^2 \rangle$. We can use vector operations and find the dot product of $\nabla$ and $\vec{F}$:

$$\nabla \cdot \vec{F} = \left\langle \dfrac{\partial}{\partial x}, \dfrac{\partial}{\partial y}, \dfrac{\partial}{\partial z} \right\rangle \cdot \langle x + \sin y, y^2 + z, x^2 \rangle$$

$$= \dfrac{\partial}{\partial x}(x + \sin y) + \dfrac{\partial}{\partial y}(y^2 + z) + \dfrac{\partial}{\partial z}(x^2)$$

$$= 1 + 2y.$$

We can also compute their cross products:

$$\nabla \times \vec{F} = \left\langle \dfrac{\partial}{\partial y}(x^2) - \dfrac{\partial}{\partial z}(y^2 + z), \dfrac{\partial}{\partial z}(x + \sin y) - \dfrac{\partial}{\partial x}(x^2), \dfrac{\partial}{\partial x}(y^2 + z) - \dfrac{\partial}{\partial y}(x + \sin y) \right\rangle$$

$$= \langle -1, -2x, -\cos y \rangle.$$

We do not yet know why we would want to compute the above. However, as we next learn about properties of vector fields, we will see how these dot and cross products with the del operator are quite useful.

### Divergence and Curl

Two properties of vector fields will prove themselves to be very important: divergence and curl. Each is a special "derivative" of a vector field; that is, each measures an instantaneous rate of change of a vector field.

If the vector field represents the velocity of a fluid or gas, then the **divergence** of the field is a measure of the "compressibility" of the fluid. If the divergence is negative at a point, it means that the fluid is compressing: more fluid is going into the point than is going out. If the divergence is positive, it means the fluid is expanding: more fluid is going out at that point than going in. A divergence of zero means the same amount of fluid is going in as is going out. If the divergence is zero at all points, we say the field is **incompressible**.

Notes:

It turns out that the proper measure of divergence is simply $\nabla \cdot \vec{F}$, as stated in the following definition.

---

**Definition 14.2.2**      **Divergence of a Vector Field**

The **divergence** of a vector field $\vec{F}$ is

$$\text{div}\,\vec{F} = \nabla \cdot \vec{F}.$$

- In the plane, with $\vec{F} = \langle M, N \rangle$, $\text{div}\,\vec{F} = M_x + N_y$.
- In space, with $\vec{F} = \langle M, N, P \rangle$, $\text{div}\,\vec{F} = M_x + N_y + P_z$.

---

**Curl** is a measure of the spinning action of the field. Let $\vec{F}$ represent the flow of water over a flat surface. If a small round cork were held in place at a point in the water, would the water cause the cork to spin? No spin corresponds to zero curl; counterclockwise spin corresponds to positive curl and clockwise spin corresponds to negative curl.

In space, things are a bit more complicated. Again let $\vec{F}$ represent the flow of water, and imagine suspending a tennis ball in one location in this flow. The water may cause the ball to spin along an axis. If so, the curl of the vector field is a *vector* (not a *scalar*, as before), parallel to the axis of rotation, following a right hand rule: when the thumb of one's right hand points in the direction of the curl, the ball will spin in the direction of the curling fingers of the hand.

In space, it turns out the proper measure of curl is $\nabla \times \vec{F}$, as stated in the following definition. To find the curl of a planar vector field $\vec{F} = \langle M, N \rangle$, embed it into space as $\vec{F} = \langle M, N, 0 \rangle$ and apply the cross product definition. Since $M$ and $N$ are functions of just $x$ and $y$ (and not $z$), all partial derivatives with respect to $z$ become 0 and the result is simply $\langle 0, 0, N_x - M_y \rangle$. The third component is the measure of curl of a planar vector field.

---

**Definition 14.2.3**      **Curl of a Vector Field**

- Let $\vec{F} = \langle M, N \rangle$ be a vector field in the plane. The **curl** of $\vec{F}$ is $\text{curl}\,\vec{F} = N_x - M_y$.
- Let $\vec{F} = \langle M, N, P \rangle$ be a vector field in space. The **curl** of $\vec{F}$ is $\text{curl}\,\vec{F} = \nabla \times \vec{F} = \langle P_y - N_z, M_z - P_x, N_x - M_y \rangle$.

---

We adopt the convention of referring to curl as $\nabla \times \vec{F}$, regardless of whether

$\vec{F}$ is a vector field in two or three dimensions.

We now practice computing these quantities.

**Example 14.2.1   Computing divergence and curl of planar vector fields**

For each of the planar vector fields given below, view its graph and try to visually determine if its divergence and curl are 0. Then compute the divergence and curl.

1. $\vec{F} = \langle y, 0 \rangle$ (see Figure 14.2.4(a))
2. $\vec{F} = \langle -y, x \rangle$ (see Figure 14.2.4(b))
3. $\vec{F} = \langle x, y \rangle$ (see Figure 14.2.5(a))
4. $\vec{F} = \langle \cos y, \sin x \rangle$ (see Figure 14.2.5(b))

**SOLUTION**

1. The arrow sizes are constant along any horizontal line, so if one were to draw a small box anywhere on the graph, it would seem that the same amount of fluid would enter the box as exit. Therefore it seems the divergence is zero; it is, as

$$\text{div}\,\vec{F} = \nabla \cdot \vec{F} = M_x + N_y = \frac{\partial}{\partial x}(y) + \frac{\partial}{\partial y}(0) = 0.$$

At any point on the x-axis, arrows above it move to the right and arrows below it move to the left, indicating that a cork placed on the axis would spin clockwise. A cork placed anywhere above the x-axis would have water above it moving to the right faster than the water below it, also creating a clockwise spin. A clockwise spin also appears to be created at points below the x-axis. Thus it seems the curl should be negative (and not zero). Indeed, it is:

$$\text{curl}\,\vec{F} = \nabla \times \vec{F} = N_x - M_y = \frac{\partial}{\partial x}(0) - \frac{\partial}{\partial y}(y) = -1.$$

2. It appears that all vectors that lie on a circle of radius $r$, centered at the origin, have the same length (and indeed this is true). That implies that the divergence should be zero: draw any box on the graph, and any fluid coming in will lie along a circle that takes the same amount of fluid out. Indeed, the divergence is zero, as

$$\text{div}\,\vec{F} = \nabla \cdot \vec{F} = M_x + N_y = \frac{\partial}{\partial x}(-y) + \frac{\partial}{\partial y}(x) = 0.$$

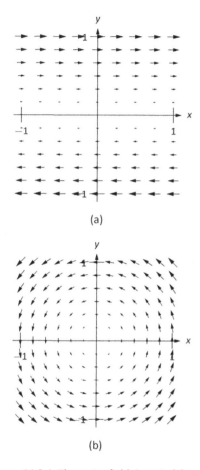

Figure 14.2.4: The vector fields in parts (a) and (b) in Example 14.2.1.

Notes:

Clearly this field moves objects in a circle, but would it induce a cork to spin? It appears that yes, it would: place a cork anywhere in the flow, and the point of the cork closest to the origin would feel less flow than the point on the cork farthest from the origin, which would induce a counter-clockwise flow. Indeed, the curl is positive:

$$\text{curl}\,\vec{F} = \nabla \times \vec{F} = N_x - M_y = \frac{\partial}{\partial x}(x) - \frac{\partial}{\partial y}(-y) = 1 - (-1) = 2.$$

Since the curl is constant, we conclude the induced spin is the same no matter where one is in this field.

3. At the origin, there are many arrows pointing out but no arrows pointing in. We conclude that at the origin, the divergence must be positive (and not zero). If one were to draw a box anywhere in the field, the edges farther from the origin would have larger arrows passing through them than the edges close to the origin, indicating that more is going from a point than going in. This indicates a positive (and not zero) divergence. This is correct:

$$\text{div}\,\vec{F} = \nabla \cdot \vec{F} = M_x + N_y = \frac{\partial}{\partial x}(x) + \frac{\partial}{\partial y}(y) = 1 + 1 = 2.$$

One may find this curl to be harder to determine visually than previous examples. One might note that any arrow that induces a clockwise spin on a cork will have an equally sized arrow inducing a counterclockwise spin on the other side, indicating no spin and no curl. This is correct, as

$$\text{curl}\,\vec{F} = \nabla \times \vec{F} = N_x - M_y = \frac{\partial}{\partial x}(y) - \frac{\partial}{\partial y}(x) = 0.$$

4. One might find this divergence hard to determine visually as large arrows appear in close proximity to small arrows, each pointing in different directions. Instead of trying to rationalize a guess, we compute the divergence:

$$\text{div}\,\vec{F} = \nabla \cdot \vec{F} = M_x + N_y = \frac{\partial}{\partial x}(\cos y) + \frac{\partial}{\partial y}(\sin x) = 0.$$

Perhaps surprisingly, the divergence is 0.

Will all the loops of different directions in the field, one is apt to reason the curl is variable. Indeed, it is:

$$\text{curl}\,\vec{F} = \nabla \times \vec{F} = N_x - M_y = \frac{\partial}{\partial x}(\sin x) - \frac{\partial}{\partial y}(\cos y) = \cos x + \sin y.$$

Depending on the values of $x$ and $y$, the curl may be positive, negative, or zero.

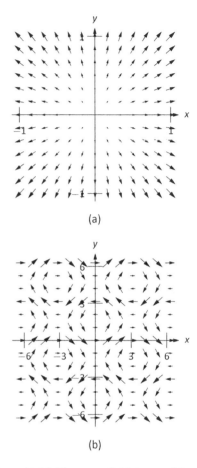

Figure 14.2.5: The vector fields in parts (c) and (d) in Example 14.2.1.

Notes:

Chapter 14  Vector Analysis

**Example 14.2.2  Computing divergence and curl of vector fields in space**
Compute the divergence and curl of each of the following vector fields.

1. $\vec{F} = \langle x^2 + y + z, -x - z, x + y \rangle$
2. $\vec{F} = \langle e^{xy}, \sin(x + z), x^2 + y \rangle$

**SOLUTION**  We compute the divergence and curl of each field following the definitions.

1. $\operatorname{div} \vec{F} = \nabla \cdot \vec{F} = M_x + N_y + P_z = 2x + 0 + 0 = 2x$.
   $\operatorname{curl} \vec{F} = \nabla \times \vec{F} = \langle P_y - N_z, M_z - P_x, N_x - M_y \rangle$
   $= \langle 1 - (-1), 1 - 1, -1 - (1) \rangle = \langle 2, 0, -2 \rangle$.

   For this particular field, no matter the location in space, a spin is induced with axis parallel to $\langle 2, 0, -2 \rangle$.

2. $\operatorname{div} \vec{F} = \nabla \cdot \vec{F} = M_x + N_y + P_z = ye^{xy} + 0 + 0 = ye^{xy}$.
   $\operatorname{curl} \vec{F} = \nabla \times \vec{F} = \langle P_y - N_z, M_z - P_x, N_x - M_y \rangle$
   $= \langle 1 - \cos(x + z), -2x, \cos(x + z) - xe^{xy} \rangle$.

**Example 14.2.3  Creating a field representing gravitational force**
The force of gravity between two objects is inversely proportional to the square of the distance between the objects. Locate a point mass at the origin. Create a vector field $\vec{F}$ that represents the gravitational pull of the point mass at any point $(x, y, z)$. Find the divergence and curl of this field.

**SOLUTION**  The point mass pulls toward the origin, so at $(x, y, z)$, the force will pull in the direction of $\langle -x, -y, -z \rangle$. To get the proper magnitude, it will be useful to find the unit vector in this direction. Dividing by its magnitude, we have

$$\vec{u} = \left\langle \frac{-x}{\sqrt{x^2 + y^2 + z^2}}, \frac{-y}{\sqrt{x^2 + y^2 + z^2}}, \frac{-z}{\sqrt{x^2 + y^2 + z^2}} \right\rangle.$$

The magnitude of the force is inversely proportional to the square of the distance between the two points. Letting $k$ be the constant of proportionality, we have the magnitude as $\dfrac{k}{x^2 + y^2 + z^2}$. Multiplying this magnitude by the unit vector above, we have the desired vector field:

$$\vec{F} = \left\langle \frac{-kx}{(x^2 + y^2 + z^2)^{3/2}}, \frac{-ky}{(x^2 + y^2 + z^2)^{3/2}}, \frac{-kz}{(x^2 + y^2 + z^2)^{3/2}} \right\rangle.$$

Notes:

## 14.2 Vector Fields

We leave it to the reader to confirm that div $\vec{F} = 0$ and curl $\vec{F} = \vec{0}$.

The analogous planar vector field is given in Figure 14.2.6. Note how all arrows point to the origin, and the magnitude gets very small when "far" from the origin.

A function $z = f(x,y)$ naturally induces a vector field, $\vec{F} = \nabla f = \langle f_x, f_y \rangle$. Given what we learned of the gradient in Section 12.6, we know that the vectors of $\vec{F}$ point in the direction of greatest increase of $f$. Because of this, $f$ is said to be the **potential function** of $\vec{F}$. Vector fields that are the gradient of potential functions will play an important role in the next section.

**Example 14.2.4     A vector field that is the gradient of a potential function**
Let $f(x,y) = 3 - x^2 - 2y^2$ and let $\vec{F} = \nabla f$. Graph $\vec{F}$, and find the divergence and curl of $\vec{F}$.

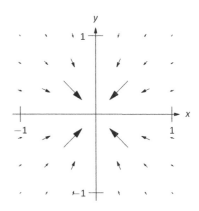

Figure 14.2.6: A vector field representing a planar gravitational force.

**Solution**     Given $f$, we find $\vec{F} = \nabla f = \langle -2x, -4y \rangle$. A graph of $\vec{F}$ is given in Figure 14.2.7(a). In part (b) of the figure, the vector field is given along with a graph of the surface itself; one can see how each vector is pointing in the direction of "steepest uphill", which, in this case, is not simply just "toward the origin."

We leave it to the reader to confirm that div $\vec{F} = -6$ and curl $\vec{F} = 0$.

There are some important concepts visited in this section that will be revisited in subsequent sections and again at the very end of this chapter. One is: given a vector field $\vec{F}$, both div $\vec{F}$ and curl $\vec{F}$ are measures of rates of change of $\vec{F}$. The divergence measures how much the field spreads (diverges) at a point, and the curl measures how much the field twists (curls) at a point. Another important concept is this: given $z = f(x, y)$, the gradient $\nabla f$ is also a measure of a rate of change of $f$. We will see how the integrals of these rates of change produce meaningful results.

This section introduces the concept of a vector field. The next section "applies calculus" to vector fields. A common application is this: let $\vec{F}$ be a vector field representing a force (hence it is called a "force field," though this name has a decidedly comic-book feel) and let a particle move along a curve $C$ under the influence of this force. What work is performed by the field on this particle? The solution lies in correctly applying the concepts of line integrals in the context of vector fields.

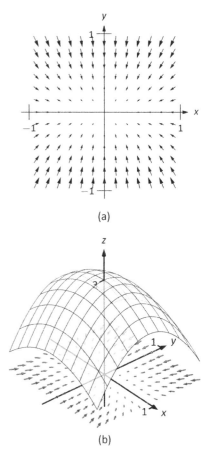

Figure 14.2.7: A graph of a function $z = f(x,y)$ and the vector field $\vec{F} = \nabla f$ in Example 14.2.4.

Notes:

# Exercises 14.2

## Terms and Concepts

1. Give two quantities that can be represented by a vector field in the plane or in space.

2. In your own words, describe what it means for a vector field to have a negative divergence at a point.

3. In your own words, describe what it means for a vector field to have a negative curl at a point.

4. The divergence of a vector field $\vec{F}$ at a particular point is 0. Does this mean that $\vec{F}$ is incompressible? Why/why not?

## Problems

In Exercises 5 – 8, sketch the given vector field over the rectangle with opposite corners $(-2, -2)$ and $(2, 2)$, sketching one vector for every point with integer coordinates (i.e., at $(0, 0)$, $(1, 2)$, etc.).

5. $\vec{F} = \langle x, 0 \rangle$

6. $\vec{F} = \langle 0, x \rangle$

7. $\vec{F} = \langle 1, -1 \rangle$

8. $\vec{F} = \langle y^2, 1 \rangle$

In Exercises 9 – 18, find the divergence and curl of the given vector field.

9. $\vec{F} = \langle x, y^2 \rangle$

10. $\vec{F} = \langle -y^2, x \rangle$

11. $\vec{F} = \langle \cos(xy), \sin(xy) \rangle$

12. $\vec{F} = \left\langle \dfrac{-2x}{(x^2 + y^2)^2}, \dfrac{-2y}{(x^2 + y^2)^2} \right\rangle$

13. $\vec{F} = \langle x + y, y + z, x + z \rangle$

14. $\vec{F} = \langle x^2 + z^2, x^2 + y^2, y^2 + z^2 \rangle$

15. $\vec{F} = \nabla f$, where $f(x, y) = \frac{1}{2}x^2 + \frac{1}{3}y^3$.

16. $\vec{F} = \nabla f$, where $f(x, y) = x^2 y$.

17. $\vec{F} = \nabla f$, where $f(x, y, z) = x^2 y + \sin z$.

18. $\vec{F} = \nabla f$, where $f(x, y, z) = \dfrac{1}{x^2 + y^2 + z^2}$.

## 14.3 Line Integrals over Vector Fields

Suppose a particle moves along a curve $C$ under the influence of an electromagnetic force described by a vector field $\vec{F}$. Since a force is inducing motion, work is performed. How can we calculate how much work is performed?

Recall that when moving in a straight line, if $\vec{F}$ represents a constant force and $\vec{d}$ represents the direction and length of travel, then work is simply $W = \vec{F} \cdot \vec{d}$. However, we generally want to be able to calculate work even if $\vec{F}$ is not constant and $C$ is not a straight line.

As we have practiced many times before, we can calculate work by first approximating, then refining our approximation through a limit that leads to integration.

Assume as we did in Section 14.1 that $C$ can be parametrized by the arc length parameter $s$. Over a short piece of the curve with length $ds$, the curve is approximately straight and our force is approximately constant. The straight–line direction of this short length of curve is given by $\vec{T}$, the unit tangent vector; let $\vec{d} = \vec{T}\,ds$, which gives the direction and magnitude of a small section of $C$. Thus work over this small section of $C$ is $\vec{F} \cdot \vec{d} = \vec{F} \cdot \vec{T}\,ds$.

Summing up all the work over these small segments gives an approximation of the work performed. By taking the limit as $ds$ goes to zero, and hence the number of segments approaches infinity, we can obtain the exact amount of work. Following the logic presented at the beginning of this chapter in the Integration Review, we see that

$$W = \int_C \vec{F} \cdot \vec{T}\,ds,$$

a line integral.

This line integral is beautiful in its simplicity, yet is not so useful in making actual computations (largely because the arc length parameter is so difficult to work with). To compute actual work, we need to parametrize $C$ with another parameter $t$ via a vector–valued function $\vec{r}(t)$. As stated in Section 14.1, $ds = \|\vec{r}'(t)\|\,dt$, and recall that $\vec{T} = \vec{r}'(t)/\|\vec{r}'(t)\|$. Thus

$$W = \int_C \vec{F} \cdot \vec{T}\,ds = \int_C \vec{F} \cdot \frac{\vec{r}'(t)}{\|\vec{r}'(t)\|} \|\vec{r}'(t)\|\,dt = \int_C \vec{F} \cdot \vec{r}'(t)\,dt = \int_C \vec{F} \cdot d\vec{r}, \quad (14.2)$$

where the final integral uses the differential $d\vec{r}$ for $\vec{r}'(t)\,dt$.

These integrals are known as **line integrals over vector fields**. By contrast, the line integrals we dealt with in Section 14.1 are sometimes referred to as **line integrals over scalar fields**. Just as a vector field is defined by a function that returns a vector, a scalar field is a function that returns a scalar, such as $z = f(x, y)$. We waited until now to introduce this terminology so we could contrast the concept with vector fields.

Notes:

We formally define this line integral, then give examples and applications.

---

**Definition 14.3.1**     **Line Integral Over A Vector Field**

Let $\vec{F}$ be a vector field with continuous components defined on a smooth curve $C$, parametrized by $\vec{r}(t)$, and let $\vec{T}$ be the unit tangent vector of $\vec{r}(t)$. The **line integral over $\vec{F}$ along $C$** is

$$\int_C \vec{F} \cdot d\vec{r} = \int_C \vec{F} \cdot \vec{T}\, ds.$$

---

In Definition 14.3.1, note how the dot product $\vec{F} \cdot \vec{T}$ is just a scalar. Therefore, this new line integral is really just a special kind of line integral found in Section 14.1; letting $f(s) = \vec{F}(s) \cdot \vec{T}(s)$, the right–hand side simply becomes $\int_C f(s)\, ds$, and we can use the techniques of that section to evaluate the integral. We combine those techniques, along with parts of Equation (14.2), to clearly state how to evaluate a line integral over a vector field in the following Key Idea.

---

**Key Idea 14.3.1**     **Evaluating a Line Integral Over A Vector Field**

Let $\vec{F}$ be a vector field with continuous components defined on a smooth curve $C$, parametrized by $\vec{r}(t)$, $a \leq t \leq b$, where $\vec{r}$ is continuously differentiable. Then

$$\int_C \vec{F} \cdot \vec{T}\, ds = \int_C \vec{F} \cdot d\vec{r} = \int_a^b \vec{F}(\vec{r}(t)) \cdot \vec{r}\,'(t)\, dt.$$

---

An important concept implicit in this Key Idea: we can use any continuously differentiable parametrization $\vec{r}(t)$ of $C$ that preserves the orientation of $C$: there isn't a "right" one. In practice, choose one that seems easy to work with.

**Notation note:** the above Definition and Key Idea implicitly evaluate $\vec{F}$ along the curve $C$, which is parametrized by $\vec{r}(t)$. For instance, if $\vec{F} = \langle x + y, x - y \rangle$ and $\vec{r}(t) = \langle t^2, \cos t \rangle$, then evaluating $\vec{F}$ along $C$ means substituting the $x$- and $y$-components of $\vec{r}(t)$ in for $x$ and $y$, respectively, in $\vec{F}$. Therefore, along $C$, $\vec{F} = \langle x + y, x - y \rangle = \langle t^2 + \cos t, t^2 - \cos t \rangle$. Since we are substituting the *output* of $\vec{r}(t)$ for the *input* of $\vec{F}$, we write this as $\vec{F}(\vec{r}(t))$. This is a slight abuse of notation as technically the input of $\vec{F}$ is to be a *point*, not a *vector*, but this shorthand is useful.

We use an example to practice evaluating line integrals over vector fields.

---

Notes:

## 14.3 Line Integrals over Vector Fields

**Example 14.3.1**     **Evaluating a line integral over a vector field: computing work**

Two particles move from $(0,0)$ to $(1,1)$ under the influence of the force field $\vec{F} = \langle x, x+y \rangle$. One particle follows $C_1$, the line $y = x$; the other follows $C_2$, the curve $y = x^4$, as shown in Figure 14.3.1. Force is measured in newtons and distance is measured in meters. Find the work performed by each particle.

**Solution**     To compute work, we need to parametrize each path. We use $\vec{r}_1(t) = \langle t, t \rangle$ to parametrize $y = x$, and let $\vec{r}_2(t) = \langle t, t^4 \rangle$ parametrize $y = x^4$; for each, $0 \leq t \leq 1$.

Along the straight-line path, $\vec{F}(\vec{r}_1(t)) = \langle x, x+y \rangle = \langle t, t+t \rangle = \langle t, 2t \rangle$. We find $\vec{r}_1'(t) = \langle 1, 2 \rangle$. The integral that computes work is:

$$\int_{C_1} \vec{F} \cdot d\vec{r} = \int_0^1 \langle t, 2t \rangle \cdot \langle 1, 1 \rangle \, dt$$
$$= \int_0^1 3t \, dt$$
$$= \frac{3}{2} t^2 \Big|_0^1 = 1.5 \text{ joules.}$$

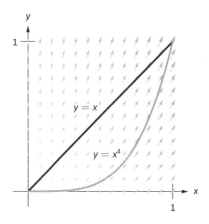

Figure 14.3.1: Paths through a vector field in Example 14.3.1.

Along the curve $y = x^4$, $\vec{F}(\vec{r}_2(t)) = \langle x, x+y \rangle = \langle t, t+t^4 \rangle$. We find $\vec{r}_2'(t) = \langle 1, 4t^3 \rangle$. The work performed along this path is

$$\int_{C_2} \vec{F} \cdot d\vec{r} = \int_0^1 \langle t, t+t^4 \rangle \cdot \langle 1, 4t^3 \rangle \, dt$$
$$= \int_0^1 \left( t + 4t^4 + 4t^7 \right) dt$$
$$= \left( \frac{1}{2} t^2 + \frac{4}{5} t^5 + \frac{1}{2} t^8 \right) \Big|_0^1 = 1.8 \text{ joules.}$$

Note how differing amounts of work are performed along the different paths. This should not be too surprising: the force is variable, one path is longer than the other, etc.

**Example 14.3.2**     **Evaluating a line integral over a vector field: computing work**

Two particles move from $(-1, 1)$ to $(1, 1)$ under the influence of a force field $\vec{F} = \langle y, x \rangle$. One moves along the curve $C_1$, the parabola defined by $y = 2x^2 - 1$. The other particle moves along the curve $C_2$, the bottom half of the circle defined by $x^2 + (y-1)^2 = 1$, as shown in Figure 14.3.2. Force is measured in pounds

Notes:

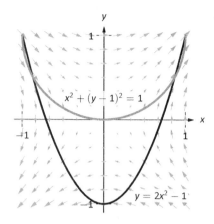

Figure 14.3.2: Paths through a vector field in Example 14.3.2.

and distances are measured in feet. Find the work performed by moving each particle along its path.

**SOLUTION** We start by parametrizing $C_1$: the parametrization $\vec{r}_1(t) = \langle t, 2t^2 - 1 \rangle$ is straightforward, giving $\vec{r}_1' = \langle 1, 4t \rangle$. On $C_1$, $\vec{F}(\vec{r}_1(t)) = \langle y, x \rangle = \langle 2t^2 - 1, t \rangle$.

Computing the work along $C_1$, we have:

$$\int_{C_1} \vec{F} \cdot d\vec{r}_1 = \int_{-1}^{1} \langle 2t^2 - 1, t \rangle \cdot \langle 1, 4t \rangle \, dt$$
$$= \int_{-1}^{1} \left(2t^2 - 1 + 4t^2\right) dt = 2 \text{ ft-lbs}.$$

For $C_2$, it is probably simplest to parametrize the half circle using sine and cosine. Recall that $\vec{r}(t) = \langle \cos t, \sin t \rangle$ is a parametrization of the unit circle on $0 \leq t \leq 2\pi$; we add 1 to the second component to shift the circle up one unit, then restrict the domain to $\pi \leq t \leq 2\pi$ to obtain only the lower half, giving $\vec{r}_2(t) = \langle \cos t, \sin t + 1 \rangle$, $\pi \leq t \leq 2\pi$, and hence $\vec{r}_2'(t) = \langle -\sin t, \cos t \rangle$ and $\vec{F}(\vec{r}_2(t)) = \langle y, x \rangle = \langle \sin t + 1, \cos t \rangle$.

Computing the work along $C_2$, we have:

$$\int_{C_2} \vec{F} \cdot d\vec{r}_2 = \int_{\pi}^{2\pi} \langle \sin t + 1, \cos t \rangle \cdot \langle -\sin t, \cos t \rangle \, dt$$
$$= \int_{\pi}^{2\pi} \left(-\sin^2 t - \sin t + \cos^2 t\right) dt = 2 \text{ ft-lbs}.$$

Note how the work along $C_1$ and $C_2$ in this example is the same. We'll address why later in this section when *conservative fields* and *path independence* are discussed.

## Properties of Line Integrals Over Vector Fields

Line integrals over vector fields share the same properties as line integrals over scalar fields, with one important distinction. The orientation of the curve $C$ matters with line integrals over vector fields, whereas it did not matter with line integrals over scalar fields.

It is relatively easy to see why. Let $C$ be the unit circle. The area under a surface over $C$ is the same whether we traverse the circle in a clockwise or counterclockwise fashion, hence the line integral over a scalar field on $C$ is the same irrespective of orientation. On the other hand, if we are computing work done by a force field, direction of travel definitely matters. Opposite directions create

Notes:

opposite signs when computing dot products, so traversing the circle in opposite directions will create line integrals that differ by a factor of $-1$.

> **Theorem 14.3.1  Properties of Line Integrals Over Vector Fields**
>
> 1. Let $\vec{F}$ and $\vec{G}$ be vector fields with continuous components defined on a smooth curve $C$, parametrized by $\vec{r}(t)$, and let $k_1$ and $k_2$ be scalars. Then
>
> $$\int_C \left(k_1\vec{F} + k_2\vec{G}\right) \cdot d\vec{r} = k_1\int_C \vec{F} \cdot d\vec{r} + k_2\int_C \vec{G} \cdot d\vec{r}.$$
>
> 2. Let $C$ be piecewise smooth, composed of smooth components $C_1$ and $C_2$. Then
>
> $$\int_C \vec{F} \cdot d\vec{r} = \int_{C_1} \vec{F} \cdot d\vec{r} + \int_{C_2} \vec{F} \cdot d\vec{r}.$$
>
> 3. Let $C^*$ be the curve $C$ with opposite orientation, parametrized by $\vec{r}^{\,*}$. Then
>
> $$\int_C \vec{F} \cdot d\vec{r} = -\int_{C^*} \vec{F} \cdot d\vec{r}^{\,*}.$$

We demonstrate using these properties in the following example.

**Example 14.3.3  Using properties of line integrals over vector fields**
Let $\vec{F} = \langle 3(y - 1/2), 1\rangle$ and let $C$ be the path that starts at $(0,0)$, goes to $(1,1)$ along the curve $y = x^3$, then returns to $(0,0)$ along the line $y = x$, as shown in Figure 14.3.3. Evaluate $\oint_C \vec{F} \cdot d\vec{r}$.

**Solution**    As $C$ is piecewise smooth, we break it into two components $C_1$ and $C_2$, where $C_1$ follows the curve $y = x^3$ and $C_2$ follows the curve $y = x$.

We parametrize $C_1$ with $\vec{r}_1(t) = \langle t, t^3\rangle$ on $0 \leq t \leq 1$, with $\vec{r}_1'(t) = \langle 1, 3t^2\rangle$. We will use $\vec{F}(\vec{r}_1(t)) = \langle 3(t^3 - 1/2), 1\rangle$.

While we always have unlimited ways in which to parametrize a curve, there are 2 "direct" methods to choose from when parametrizing $C_2$. The parametrization $\vec{r}_2(t) = \langle t, t\rangle$, $0 \leq t \leq 1$ traces the correct line segment but with the wrong orientation. Using Property 3 of Theorem 14.3.1, we can use this parametrization and negate the result.

Another choice is to use the techniques of Section 10.5 to create the line with the orientation we desire. We wish to start at $(1, 1)$ and travel in the

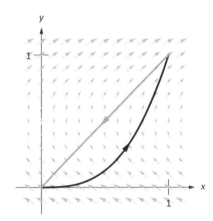

Figure 14.3.3: The vector field and curve in Example 14.3.3.

## Chapter 14 Vector Analysis

$\vec{d} = \langle -1, -1 \rangle$ direction for one length of $\vec{d}$, giving equation $\vec{\ell}(t) = \langle 1, 1 \rangle + t\langle -1, -1 \rangle = \langle 1-t, 1-t \rangle$ on $0 \leq t \leq 1$.

Either choice is fine; we choose $\vec{r}_2(t)$ to practice using line integral properties. We find $\vec{r}_2'(t) = \langle 1, 1 \rangle$ and $\vec{F}(\vec{r}_2(t)) = \langle 3(t-1/2), 1 \rangle$.

Evaluating the line integral (note how we subtract the integral over $C_2$ as the orientation of $\vec{r}_2(t)$ is opposite):

$$\oint_C \vec{F} \cdot d\vec{r} = \int_{C_1} \vec{F} \cdot d\vec{r}_1 - \int_{C_2} \vec{F} \cdot d\vec{r}_2$$

$$= \int_0^1 \langle 3(t^3 - 1/2), 1 \rangle \cdot \langle 1, 3t^2 \rangle \, dt - \int_0^1 \langle 3(t-1/2), 1 \rangle \cdot \langle 1, 1 \rangle \, dt$$

$$= \int_0^1 \left( 3t^3 + 3t^2 - 3/2 \right) dt - \int_0^1 \left( 3t - 1/2 \right) dt$$

$$= (1/4) - (1)$$

$$= -3/4.$$

If we interpret this integral as computing work, the negative work implies that the motion is mostly *against* the direction of the force, which seems plausible when we look at Figure 14.3.3.

**Example 14.3.4**    **Evaluating a line integral over a vector field in space**
Let $\vec{F} = \langle -y, x, 1 \rangle$, and let $C$ be the portion of the helix given by $\vec{r}(t) = \langle \cos t, \sin t, t/(2\pi) \rangle$ on $[0, 2\pi]$, as shown in Figure 14.3.4. Evaluate $\int_C \vec{F} \cdot d\vec{r}$.

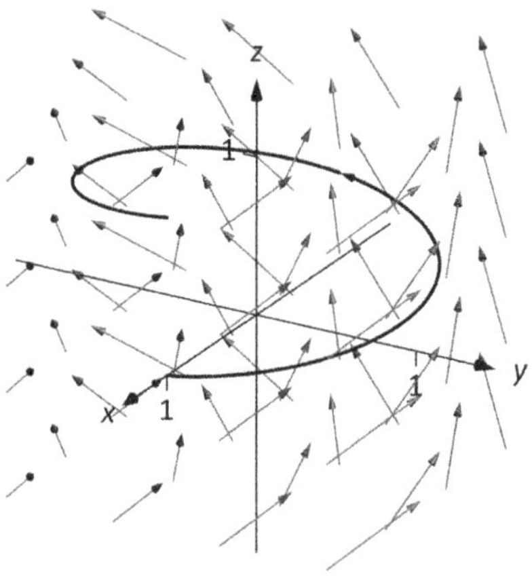

**Solution**    A parametrization is already given for $C$, so we just need to find $\vec{F}(\vec{r}(t))$ and $\vec{r}'(t)$.

We have $\vec{F}(\vec{r}(t)) = \langle -\sin t, \cos t, 1 \rangle$ and $\vec{r}'(t) = \langle -\sin t, \cos t, 1/(2\pi) \rangle$. Thus

$$\int_C \vec{F} \cdot d\vec{r} = \int_0^{2\pi} \langle -\sin t, \cos t, 1 \rangle \cdot \langle -\sin t, \cos t, 1/(2\pi) \rangle \, dt$$

$$= \int_0^{2\pi} \left( \sin^2 t + \cos^2 t + \frac{1}{2\pi} \right) dt$$

$$= 2\pi + 1 \approx 7.28.$$

Figure 14.3.4: The graph of $\vec{r}(t)$ in Example 14.3.4.

### The Fundamental Theorem of Line Integrals

We are preparing to make important statements about the value of certain line integrals over special vector fields. Before we can do that, we need to define some terms that describe the domains over which a vector field is defined.

Notes:

A region in the plane is **connected** if any two points in the region can be joined by a piecewise smooth curve that lies entirely in the region. In Figure 14.3.5, sets $R_1$ and $R_2$ are connected; set $R_3$ is not connected, though it is composed of two connected subregions.

A region is **simply connected** if every simple closed curve that lies entirely in the region can be continuously deformed (shrunk) to a single point without leaving the region. (A curve is **simple** if it does not cross itself.) In Figure 14.3.5, only set $R_1$ is simply connected. Region $R_2$ is not simply connected as any closed curve that goes around the "hole" in $R_2$ cannot be continously shrunk to a single point. As $R_3$ is not even connected, it cannot be simply connected, though again it consists of two simply connected subregions.

We have applied these terms to regions of the plane, but they can be extended intuitively to domains in space (and hyperspace). In Figure 14.3.6(a), the domain bounded by the sphere (at left) and the domain with a subsphere removed (at right) are both simply connected. Any simple closed path that lies entirely within these domains can be continuously deformed into a single point. In Figure 14.3.6(b), neither domain is simply connected. A left, the ball has a hole that extends its length and the pictured closed path cannot be deformed to a point. At right, two paths are illustrated on the torus that cannot be shrunk to a point.

We will use the terms connected and simply connected in subsequent definitions and theorems.

Recall how in Example 14.3.2 particles moved from $A = (-1, 1)$ to $B = (1, 1)$ along two different paths, wherein the same amount of work was performed along each path. It turns out that regardless of the choice of path from $A$ to $B$, the amount of work performed under the field $\vec{F} = \langle y, x \rangle$ is the same. Since our expectation is that differing amounts of work are performed along different paths, we give such special fields a name.

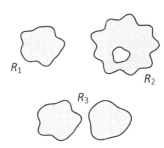

Figure 14.3.5: $R_1$ is simply connected; $R_2$ is connected, but not simply connected; $R_3$ is not connected.

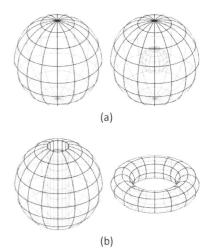

(a)

(b)

Figure 14.3.6: The domains in (a) are simply connected, while the domains in (b) are not.

---

**Definition 14.3.2    Conservative Field, Path Independent**

Let $\vec{F}$ be a vector field defined on an open, connected domain $D$ in the plane or in space containing points $A$ and $B$. If the line integral $\int_C \vec{F} \cdot d\vec{r}$ has the same value for all choices of paths $C$ starting at $A$ and ending at $B$, then

- $\vec{F}$ is a **conservative field** and

- The line integral $\int_C \vec{F} \cdot d\vec{r}$ is **path independent** and can be written as

$$\int_C \vec{F} \cdot d\vec{r} = \int_A^B \vec{F} \cdot d\vec{r}.$$

---

Notes:

When $\vec{F}$ is a conservative field, the line integral from points A to B is sometimes written as $\int_A^B \vec{F} \cdot d\vec{r}$ to emphasize the independence of its value from the choice of path; all that matters are the beginning and ending points of the path.

How can we tell if a field is conservative? To show a field $\vec{F}$ is conservative using the definition, we need to show that *all* line integrals from points A to B have the same value. It is equivalent to show that *all* line integrals over closed paths C are 0. Each of these tasks are generally nontrivial.

There is a simpler method. Consider the surface defined by $z = f(x, y) = xy$. We can compute the gradient of this function: $\nabla f = \langle f_x, f_y \rangle = \langle y, x \rangle$. Note that this is the field from Example 14.3.2, which we have claimed is conservative. We will soon give a theorem that states that a field $\vec{F}$ is conservative if, and only if, it is the gradient of some scalar function $f$. To show $\vec{F}$ is conservative, we need to determine whether or not $\vec{F} = \nabla f$ for some function $f$. (We'll later see that there is a yet simpler method). To recognize the special relationship between $\vec{F}$ and $f$ in this situation, $f$ is given a name.

---

**Definition 14.3.3**     **Potential Function**

Let $f$ be a differentiable function defined on a domain $D$ in the plane or in space (i.e., $z = f(x, y)$ or $w = f(x, y, z)$) and let $\vec{F} = \nabla f$, the gradient of $f$. Then $f$ is a **potential function** of $\vec{F}$.

---

We now state the Fundamental Theorem of Line Integrals, which connects conservative fields and path independence to fields with potential functions.

---

**Theorem 14.3.2**     **Fundamental Theorem of Line Integrals**

Let $\vec{F}$ be a vector field whose components are continuous on a connected domain $D$ in the plane or in space, let $A$ and $B$ be any points in $D$, and let $C$ be any path in $D$ starting at $A$ and ending at $B$.

1. $\vec{F}$ is conservative if and only if there exists a differentiable function $f$ such that $\vec{F} = \nabla f$.

2. If $\vec{F}$ is conservative, then

$$\int_C \vec{F} \cdot d\vec{r} = \int_A^B \vec{F} \cdot d\vec{r} = f(B) - f(A).$$

---

Once again considering Example 14.3.2, we have $A = (-1, 1)$, $B = (1, 1)$

---

Notes:

and $\vec{F} = \langle y, x \rangle$. In that example, we evaluated two line integrals from $A$ to $B$ and found the value of each was 2. Note that $f(x, y) = xy$ is a potential function for $\vec{F}$. Following the Fundamental Theorem of Line Integrals, consider $f(B) - f(A)$:

$$f(B) - f(A) = f(1, 1) - f(-1, 1) = 1 - (-1) = 2,$$

the same value given by the line integrals.

We practice using this theorem again in the next example.

**Example 14.3.5    Using the Fundamental Theorem of Line Integrals**
Let $\vec{F} = \langle 3x^2y + 2x, x^3 + 1 \rangle$, $A = (0, 1)$ and $B = (1, 4)$. Use the first part of the Fundamental Theorem of Line Integrals to show that $\vec{F}$ is conservative, then choose any path from $A$ to $B$ and confirm the second part of the theorem.

**Solution**    To show $\vec{F}$ is conservative, we need to find $z = f(x, y)$ such that $\vec{F} = \nabla f = \langle f_x, f_y \rangle$. That is, we need to find $f$ such that $f_x = 3x^2y + 2x$ and $f_y = x^3 + 1$. As all we know about $f$ are its partial derivatives, we recover $f$ by integration:

$$\int \frac{\partial f}{\partial x} \, dx = f(x, y) + C(y).$$

Note how the constant of integration is more than "just a constant": it is anything that acts as a constant when taking a derivative with respect to $x$. Any function that is a function of $y$ (containing no $x$'s) acts as a constant when deriving with respect to $x$.

Integrating $f_x$ in this example gives:

$$\int \frac{\partial f}{\partial x} \, dx = \int (3x^2y + 2x) \, dx = x^3y + x^2 + C(y).$$

Likewise, integrating $f_y$ with respect to $y$ gives:

$$\int \frac{\partial f}{\partial y} \, dy = \int (x^3 + 1) \, dy = x^3y + y + C(x).$$

These two results should be equal with appropriate choices of $C(x)$ and $C(y)$:

$$x^3y + x^2 + C(y) = x^3y + y + C(x) \quad \Rightarrow \quad C(x) = x^2 \quad \text{and} \quad C(y) = y.$$

We find $f(x, y) = x^3y + x^2 + y$, a potential function of $\vec{F}$. (If $\vec{F}$ were not conservative, no choice of $C(x)$ and $C(y)$ would give equality.)

By the Fundamental Theorem of Line Integrals, regardless of the path from $A$ to $B$,

$$\int_A^B \vec{F} \cdot d\vec{r} = f(B) - f(A)$$
$$= f(1, 4) - f(0, 1)$$
$$= 9 - 1 = 8.$$

Notes:

To illustrate the validity of the Fundamental Theorem, we pick a path from $A$ to $B$. The line between these two points would be simple to construct; we choose a slightly more complicated path by choosing the parabola $y = x^2 + 2x + 1$. This leads to the parametrization $\vec{r}(t) = \langle t, t^2 + 2t + 1 \rangle$, $0 \leq t \leq 1$, with $\vec{r}'(t) = \langle t, 2t + 2 \rangle$. Thus

$$\int_C \vec{F} \cdot d\vec{r} = \int_C \vec{F}(\vec{r}(t)) \cdot \vec{r}'(t)\, dt$$
$$= \int_0^1 \langle 3(t)(t^2 + 2t + 1) + 2t, t^3 + 1 \rangle \cdot \langle t, 2t + 2 \rangle\, dt$$
$$= \int_0^1 \left(5t^4 + 8t^3 + 3t^2 + 4t + 2\right) dt$$
$$= \left(t^5 + 2t^4 + t^3 + 2t^2 + 2t\right)\Big|_0^1$$
$$= 8,$$

which matches our previous result.

The Fundamental Theorem of Line Integrals states that we can determine whether or not $\vec{F}$ is conservative by determining whether or not $\vec{F}$ has a potential function. This can be difficult. A simpler method exists if the domain of $\vec{F}$ is simply connected (not just connected as needed in the Fundamental Theorem of Line Integrals), which is a reasonable requirement. We state this simpler method as a theorem.

---

**Theorem 14.3.3**      **Curl of Conservative Fields**

Let $\vec{F}$ be a vector field whose components are continuous on a simply connected domain $D$ in the plane or in space. Then $\vec{F}$ is conservative if and only if curl $\vec{F} = 0$ or $\vec{0}$, respectively.

---

In Example 14.3.5, we showed that $\vec{F} = \langle 3x^2y + 2x, x^3 + 1 \rangle$ is conservative by finding a potential function for $\vec{F}$. Using the above theorem, we can show that $\vec{F}$ is conservative much more easily by computing its curl:

$$\text{curl}\, \vec{F} = N_x - M_y = 3x^2 - 3x^2 = 0.$$

Notes:

# Exercises 14.3

## Terms and Concepts

1. T/F: In practice, the evaluation of line integrals over vector fields involves computing the magnitude of a vector–valued function.

2. Let $\vec{F}(x, y)$ be a vector field in the plane and let $\vec{r}(t)$ be a two–dimensional vector–valued function. Why is "$\vec{F}(\vec{r}(t))$" an "abuse of notation"?

3. T/F: The orientation of a curve C matters when computing a line integral over a vector field.

4. T/F: The orientation of a curve C matters when computing a line integral over a scalar field.

5. Under "reasonable conditions," if curl $\vec{F} = \vec{0}$, what can we conclude about the vector field $\vec{F}$?

6. Let $\vec{F}$ be a conservative field and let C be a closed curve. Why are we able to conclude that $\oint_C \vec{F} \cdot d\vec{r} = 0$?

## Problems

In Exercises 7 – 12, a vector field $\vec{F}$ and a curve C are given. Evaluate $\int_C \vec{F} \cdot d\vec{r}$.

7. $\vec{F} = \langle y, y^2 \rangle$; C is the line segment from $(0, 0)$ to $(3, 1)$.

8. $\vec{F} = \langle x, x + y \rangle$; C is the portion of the parabola $y = x^2$ from $(0, 0)$ to $(1, 1)$.

9. $\vec{F} = \langle y, x \rangle$; C is the top half of the unit circle, beginning at $(1, 0)$ and ending at $(-1, 0)$.

10. $\vec{F} = \langle xy, x \rangle$; C is the portion of the curve $y = x^3$ on $-1 \leq x \leq 1$.

11. $\vec{F} = \langle z, x^2, y \rangle$; C is the line segment from $(1, 2, 3)$ to $(4, 3, 2)$.

12. $\vec{F} = \langle y + z, x + z, x + y \rangle$; C is the helix $\vec{r}(t) = \langle \cos t, \sin t, t/(2\pi) \rangle$ on $0 \leq t \leq 2\pi$.

In Exercises 13 – 16, find the work performed by the force field $\vec{F}$ moving a particle along the path C.

13. $\vec{F} = \langle y, x^2 \rangle$ N; C is the segment of the line $y = x$ from $(0, 0)$ to $(1, 1)$, where distances are measured in meters.

14. $\vec{F} = \langle y, x^2 \rangle$ N; C is the portion of $y = \sqrt{x}$ from $(0, 0)$ to $(1, 1)$, where distances are measured in meters.

15. $\vec{F} = \langle 2xy, x^2, 1 \rangle$ lbs; C is the path from $(0, 0, 0)$ to $(2, 4, 8)$ via $\vec{r}(t) = \langle t, t^2, t^3 \rangle$ on $0 \leq t \leq 2$, where distance are measured in feet.

16. $\vec{F} = \langle 2xy, x^2, 1 \rangle$ lbs; C is the path from $(0, 0, 0)$ to $(2, 4, 8)$ via $\vec{r}(t) = \langle t, 2t, 4t \rangle$ on $0 \leq t \leq 2$, where distance are measured in feet.

In Exercises 17 – 20, a conservative vector field $\vec{F}$ and a curve C are given.

1. Find a potential function $f$ for $\vec{F}$.

2. Compute curl $\vec{F}$.

3. Evaluate $\int_C \vec{F} \cdot d\vec{r}$ directly, i.e., using Key Idea 14.3.1.

4. Evaluate $\int_C \vec{F} \cdot d\vec{r}$ using the Fundamental Theorem of Line Integrals.

17. $\vec{F} = \langle y + 1, x \rangle$, C is the line segment from $(0, 1)$ to $(1, 0)$.

18. $\vec{F} = \langle 2x + y, 2y + x \rangle$, C is curve parametrized by $\vec{r}(t) = \langle t^2 - t, t^3 - t \rangle$ on $0 \leq t \leq 1$.

19. $\vec{F} = \langle 2xyz, x^2 z, x^2 y \rangle$, C is curve parametrized by $\vec{r}(t) = \langle 2t + 1, 3t - 1, t \rangle$ on $0 \leq t \leq 2$.

20. $\vec{F} = \langle 2x, 2y, 2z \rangle$, C is curve parametrized by $\vec{r}(t) = \langle \cos t, \sin t, \sin(2t) \rangle$ on $0 \leq t \leq 2\pi$.

21. Prove part of Theorem 14.3.3: let $\vec{F} = \langle M, N, P \rangle$ be a conservative vector field. Show that curl $\vec{F} = 0$.

## 14.4 Flow, Flux, Green's Theorem and the Divergence Theorem

### Flow and Flux

Line integrals over vector fields have the natural interpretation of computing work when $\vec{F}$ represents a force field. It is also common to use vector fields to represent velocities. In these cases, the line integral $\int_C \vec{F} \cdot d\vec{r}$ is said to represent **flow**.

Let the vector field $\vec{F} = \langle 1, 0 \rangle$ represent the velocity of water as it moves across a smooth surface, depicted in Figure 14.4.1. A line integral over $C$ will compute "how much water is moving *along* the path $C$."

In the figure, "all" of the water above $C_1$ is moving along that curve, whereas "none" of the water above $C_2$ is moving along that curve (the curve and the flow of water are at right angles to each other). Because $C_3$ has nonzero horizontal and vertical components, "some" of the water above that curve is moving along the curve.

When $C$ is a closed curve, we call flow **circulation**, represented by $\oint_C \vec{F} \cdot d\vec{r}$.

The "opposite" of flow is **flux**, a measure of "how much water is moving *across* the path $C$." If a curve represents a filter in flowing water, flux measures how much water will pass through the filter. Considering again Figure 14.4.1, we see that a screen along $C_1$ will not filter any water as no water passes across that curve. Because of the nature of this field, $C_2$ and $C_3$ each filter the same amount of water per second.

The terms "flow" and "flux" are used apart from velocity fields, too. Flow is measured by $\int_C \vec{F} \cdot d\vec{r}$, which is the same as $\int_C \vec{F} \cdot \vec{T}\, ds$ by Definition 14.3.1. That is, flow is a summation of the amount of $\vec{F}$ that is *tangent* to the curve $C$.

By contrast, flux is a summation of the amount of $\vec{F}$ that is *orthogonal* to the direction of travel. To capture this orthogonal amount of $\vec{F}$, we use $\int_C \vec{F} \cdot \vec{n}\, ds$ to measure flux, where $\vec{n}$ is a unit vector orthogonal to the curve $C$. (Later, we'll measure flux across surfaces, too. For example, in physics it is useful to measure the amount of a magnetic field that passes through a surface.)

How is $\vec{n}$ determined? We'll later see that if $C$ is a closed curve, we'll want $\vec{n}$ to point to the outside of the curve (measuring how much is "going out"). We'll also adopt the convention that closed curves should be traversed counterclockwise.

(If $C$ is a complicated closed curve, it can be difficult to determine what "counterclockwise" means. Consider Figure 14.4.2. Seeing the curve as a whole, we know which way "counterclockwise" is. If we zoom in on point $A$, one might incorrectly choose to traverse the path in the wrong direction. So we offer this definition: *a closed curve is being traversed counterclockwise if the outside is to the right of the path and the inside is to the left.*)

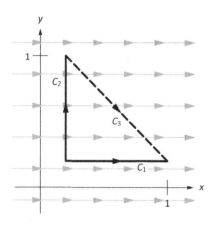

Figure 14.4.1: Illustrating the principles of flow and flux.

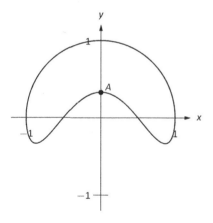

Figure 14.4.2: Determining "counterclockwise" is not always simple without a good definition.

Notes:

## 14.4 Flow, Flux, Green's Theorem and the Divergence Theorem

When a curve $C$ is traversed counterclockwise by $\vec{r}(t) = \langle f(t), g(t) \rangle$, we rotate $\vec{T}$ clockwise $90°$ to obtain $\vec{n}$:

$$\vec{T} = \frac{\langle f'(t), g'(t) \rangle}{\|\vec{r}'(t)\|} \quad \Rightarrow \quad \vec{n} = \frac{\langle g'(t), -f'(t) \rangle}{\|\vec{r}'(t)\|}.$$

Letting $\vec{F} = \langle M, N \rangle$, we calculate flux as:

$$\begin{aligned}
\int_C \vec{F} \cdot \vec{n}\, ds &= \int_C \vec{F} \cdot \frac{\langle g'(t), -f'(t) \rangle}{\|\vec{r}'(t)\|} \|\vec{r}'(t)\|\, dt \\
&= \int_C \langle M, N \rangle \cdot \langle g'(t), -f'(t) \rangle\, dt \\
&= \int_C \Big(M g'(t) - N f'(t)\Big) dt \\
&= \int_C M g'(t)\, dt - \int_C N f'(t)\, dt.
\end{aligned}$$

As the $x$ and $y$ components of $\vec{r}(t)$ are $f(t)$ and $g(t)$ respectively, the differentials of $x$ and $y$ are $dx = f'(t)dt$ and $dy = g'(t)dt$. We can then write the above integrals as:

$$= \int_C M\, dy - \int_C N\, dx.$$

This is often written as one integral (not incorrectly, though somewhat confusingly, as this one integral has two "$d$'s"):

$$= \int_C M\, dy - N\, dx.$$

We summarize the above in the following definition.

---

Notes:

## Chapter 14 Vector Analysis

**Definition 14.4.1**     **Flow, Flux**

Let $\vec{F} = \langle M, N \rangle$ be a vector field with continuous components defined on a smooth curve $C$, parametrized by $\vec{r}(t) = \langle f(t), g(t) \rangle$, let $\vec{T}$ be the unit tangent vector of $\vec{r}(t)$, and let $\vec{n}$ be the clockwise 90°degree rotation of $\vec{T}$.

- The **flow** of $\vec{F}$ along $C$ is

$$\int_C \vec{F} \cdot \vec{T}\, ds = \int_C \vec{F} \cdot d\vec{r}.$$

- The **flux** of $\vec{F}$ across $C$ is

$$\int_C \vec{F} \cdot \vec{n}\, ds = \int_C M\, dy - N\, dx = \int_C \Big(M g'(t) - N f'(t)\Big)\, dt.$$

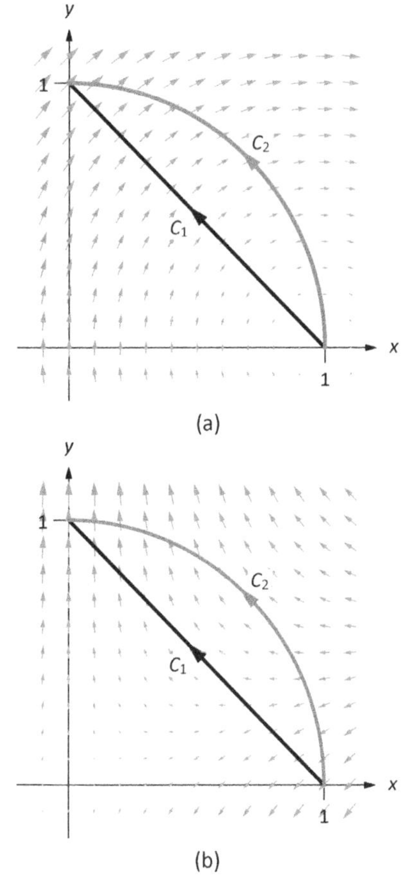

This definition of flow also holds for curves in space, though it does not make sense to measure "flux across a curve" in space.

Measuring flow is essentially the same as finding work performed by a force as done in the previous examples. Therefore we practice finding only flux in the following example.

**Example 14.4.1**     **Finding flux across curves in the plane**
Curves $C_1$ and $C_2$ each start at $(1, 0)$ and end at $(0, 1)$, where $C_1$ follows the line $y = 1 - x$ and $C_2$ follows the unit circle, as shown in Figure 14.4.3. Find the flux across both curves for the vector fields $\vec{F}_1 = \langle y, -x + 1 \rangle$ and $\vec{F}_2 = \langle -x, 2y - x \rangle$.

**Solution**     We begin by finding parametrizations of $C_1$ and $C_2$. As done in Example 14.3.3, parametrize $C_1$ by creating the line that starts at $(1, 0)$ and moves in the $\langle -1, 1 \rangle$ direction: $\vec{r}_1(t) = \langle 1, 0 \rangle + t \langle -1, 1 \rangle = \langle 1 - t, t \rangle$, for $0 \leq t \leq 1$. We parametrize $C_2$ with the familiar $\vec{r}_2(t) = \langle \cos t, \sin t \rangle$ on $0 \leq t \leq \pi/2$. For reference later, we give each function and its derivative below:

$$\vec{r}_1(t) = \langle 1 - t, t \rangle, \quad \vec{r}_1'(t) = \langle -1, 1 \rangle.$$

$$\vec{r}_2(t) = \langle \cos t, \sin t \rangle, \quad \vec{r}_2'(t) = \langle -\sin t, \cos t \rangle.$$

When $\vec{F} = \vec{F}_1 = \langle y, -x + 1 \rangle$ (as shown in Figure 14.4.3(a)), over $C_1$ we have $M = y = t$ and $N = -x + 1 = -(1 - t) + 1 = t$. Using Definition 14.4.1, we compute the flux:

Figure 14.4.3: Illustrating the curves and vector fields in Example 14.4.1. In (a) the vector field is $\vec{F}_1$, and in (b) the vector field is $\vec{F}_2$.

Notes:

$$\int_{C_1} \vec{F}\cdot\vec{n}\,ds = \int_{C_1}\Big(Mg'(t) - Nf'(t)\Big)\,dt$$
$$= \int_0^1 \Big(t(1) - t(-1)\Big)\,dt$$
$$= \int_0^1 2t\,dt$$
$$= 1.$$

Over $C_2$, we have $M = y = \sin t$ and $N = -x + 1 = 1 - \cos t$. Thus the flux across $C_2$ is:

$$\int_{C_1} \vec{F}\cdot\vec{n}\,ds = \int_{C_1}\Big(Mg'(t) - Nf'(t)\Big)\,dt$$
$$= \int_0^{\pi/2} \Big((\sin t)(\cos t) - (1 - \cos t)(-\sin t)\Big)\,dt$$
$$= \int_0^{\pi/2} \sin t\,dt$$
$$= 1.$$

Notice how the flux was the same across both curves. This won't hold true when we change the vector field.

When $\vec{F} = \vec{F}_2 = \langle -x, 2y - x\rangle$ (as shown in Figure 14.4.3(b)), over $C_1$ we have $M = -x = t - 1$ and $N = 2y - x = 2t - (1 - t) = 3t - 1$. Computing the flux across $C_1$:

$$\int_{C_1} \vec{F}\cdot\vec{n}\,ds = \int_{C_1}\Big(Mg'(t) - Nf'(t)\Big)\,dt$$
$$= \int_0^1 \Big((t - 1)(1) - (3t - 1)(-1)\Big)\,dt$$
$$= \int_0^1 (4t - 2)\,dt$$
$$= 0.$$

Over $C_2$, we have $M = -x = -\cos t$ and $N = 2y - x = 2\sin t - \cos t$. Thus the flux across $C_2$ is:

---

Notes:

$$\int_{C_1} \vec{F} \cdot \vec{n}\, ds = \int_{C_1} \left( M g'(t) - N f'(t) \right) dt$$
$$= \int_0^{\pi/2} \left( (-\cos t)(\cos t) - (2\sin t - \cos t)(-\sin t) \right) dt$$
$$= \int_0^{\pi/2} \left( 2\sin^2 t - \sin t \cos t - \cos^2 t \right) dt$$
$$= \pi/4 - 1/2 \approx 0.285.$$

We analyze the results of this example below.

In Example 14.4.1, we saw that the flux across the two curves was the same when the vector field was $\vec{F}_1 = \langle y, -x+1 \rangle$. This is not a coincidence. We show why they are equal in Example 14.4.6. In short, the reason is this: the divergence of $\vec{F}_1$ is 0, and when $\operatorname{div}\vec{F} = 0$, the flux across any two paths with common beginning and ending points will be the same.

We also saw in the example that the flux across $C_1$ was 0 when the field was $\vec{F}_2 = \langle -x, 2y - x \rangle$. Flux measures "how much" of the field crosses the path from left to right (following the conventions established before). Positive flux means most of the field is crossing from left to right; negative flux means most of the field is crossing from right to left; zero flux means the same amount crosses from each side. When we consider Figure 14.4.3(b), it seems plausible that the same amount of $\vec{F}_2$ was crossing $C_1$ from left to right as from right to left.

## Green's Theorem

There is an important connection between the circulation around a closed region $R$ and the curl of the vector field inside of $R$, as well as a connection between the flux across the boundary of $R$ and the divergence of the field inside $R$. These connections are described by Green's Theorem and the Divergence Theorem, respectively. We'll explore each in turn.

Green's Theorem states "the counterclockwise circulation around a closed region $R$ is equal to the sum of the curls over $R$."

---

**Theorem 14.4.1**     **Green's Theorem**

Let $R$ be a closed, bounded region of the plane whose boundary $C$ is composed of finitely many smooth curves, let $\vec{r}(t)$ be a counterclockwise parametrization of $C$, and let $\vec{F} = \langle M, N \rangle$ where $N_x$ and $M_y$ are continuous over $R$. Then

$$\oint_C \vec{F} \cdot d\vec{r} = \iint_R \operatorname{curl} \vec{F}\, dA.$$

---

Notes:

We'll explore Green's Theorem through an example.

**Example 14.4.2  Confirming Green's Theorem**
Let $\vec{F} = \langle -y, x^2 + 1\rangle$ and let $R$ be the region of the plane bounded by the triangle with vertices $(-1, 0)$, $(1, 0)$ and $(0, 2)$, shown in Figure 14.4.4. Verify Green's Theorem; that is, find the circulation of $\vec{F}$ around the boundary of $R$ and show that is equal to the double integral of curl $\vec{F}$ over $R$.

**Solution**  The curve $C$ that bounds $R$ is composed of 3 lines. While we need to traverse the boundary of $R$ in a counterclockwise fashion, we may start anywhere we choose. We arbitrarily choose to start at $(-1, 0)$, move to $(1, 0)$, etc., with each line parametrized by $\vec{r}_1(t)$, $\vec{r}_2(t)$ and $\vec{r}_3(t)$, respectively.

We leave it to the reader to confirm that the following parametrizations of the three lines are accurate:
$\vec{r}_1(t) = \langle 2t - 1, 0\rangle$,   for $0 \leq t \leq 1$,  with $\vec{r}_1'(t) = \langle 2, 0\rangle$,
$\vec{r}_2(t) = \langle 1 - t, 2t\rangle$,   for $0 \leq t \leq 1$,  with $\vec{r}_2'(t) = \langle -1, 2\rangle$, and
$\vec{r}_3(t) = \langle -t, 2 - 2t\rangle$,  for $0 \leq t \leq 1$,  with $\vec{r}_3'(t) = \langle -1, -2\rangle$.

The circulation around $C$ is found by summing the flow along each of the sides of the triangle. We again leave it to the reader to confirm the following computations:

$$\int_{C_1} \vec{F} \cdot d\vec{r}_1 = \int_0^1 \langle 0, (2t-1)^2 + 1\rangle \cdot \langle 2, 0\rangle \, dt = 0,$$

$$\int_{C_2} \vec{F} \cdot d\vec{r}_2 = \int_0^1 \langle -2t, (1-t)^2 + 1\rangle \cdot \langle -1, 2\rangle \, dt = 11/3, \text{ and}$$

$$\int_{C_3} \vec{F} \cdot d\vec{r}_3 = \int_0^1 \langle 2t - 2, t^2 + 1\rangle \cdot \langle -1, -2\rangle \, dt = -5/3.$$

The circulation is the sum of the flows: 2.

We confirm Green's Theorem by computing $\iint_R \text{curl } \vec{F} \, dA$. We find curl $\vec{F} = 2x + 1$. The region $R$ is bounded by the lines $y = 2x + 2$, $y = -2x + 2$ and $y = 0$. Integrating with the order $dx\,dy$ is most straightforward, leading to

$$\int_0^2 \int_{y/2-1}^{1-y/2} (2x+1) \, dx \, dy = \int_0^2 (2-y) \, dy = 2,$$

which matches our previous measurement of circulation.

**Example 14.4.3  Using Green's Theorem**
Let $\vec{F} = \langle \sin x, \cos y\rangle$ and let $R$ be the region enclosed by the curve $C$ parametrized by $\vec{r}(t) = \langle 2\cos t + \frac{1}{10}\cos(10t), 2\sin t + \frac{1}{10}\sin(10t)\rangle$ on $0 \leq t \leq 2\pi$, as shown in Figure 14.4.5. Find the circulation around $C$.

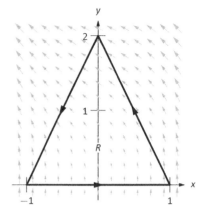

Figure 14.4.4: The vector field and planar region used in Example 14.4.2.

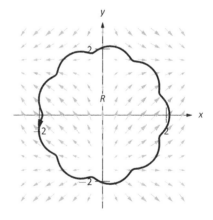

Figure 14.4.5: The vector field and planar region used in Example 14.4.3.

Notes:

## Chapter 14 Vector Analysis

**SOLUTION** Computing the circulation directly using the line integral looks difficult, as the integrand will include terms like "$\sin\left(2\cos t + \frac{1}{10}\cos(10t)\right)$."

Green's Theorem states that $\oint_C \vec{F} \cdot d\vec{r} = \iint_R \text{curl}\,\vec{F}\, dA$; since curl $\vec{F} = 0$ in this example, the double integral is simply 0 and hence the circulation is 0.

Since curl $\vec{F} = 0$, we can conclude that the circulation is 0 in two ways. One method is to employ Green's Theorem as done above. The second way is to recognize that $\vec{F}$ is a conservative field, hence there is a function $z = f(x,y)$ wherein $\vec{F} = \nabla f$. Let $A$ be any point on the curve $C$; since $C$ is closed, we can say that $C$ "begins" and "ends" at $A$. By the Fundamental Theorem of Line Integrals, $\oint_C \vec{F}\, d\vec{r} = f(A) - f(A) = 0$.

One can use Green's Theorem to find the area of an enclosed region by integrating along its boundary. Let $C$ be a closed curve, enclosing the region $R$, parametrized by $\vec{r}(t) = \langle f(t), g(t) \rangle$. We know the area of $R$ is computed by the double integral $\iint_R dA$, where the integrand is 1. By creating a field $\vec{F}$ where curl $\vec{F} = 1$, we can employ Green's Theorem to compute the area of $R$ as $\oint_C \vec{F} \cdot d\vec{r}$.

One is free to choose any field $\vec{F}$ to use as long as curl $\vec{F} = 1$. Common choices are $\vec{F} = \langle 0, x \rangle$, $\vec{F} = \langle -y, 0 \rangle$ and $\vec{F} = \langle -y/2, x/2 \rangle$. We demonstrate this below.

### Example 14.4.4  Using Green's Theorem to find area

Let $C$ be the closed curve parametrized by $\vec{r}(t) = \langle t - t^3, t^2 \rangle$ on $-1 \leq t \leq 1$, enclosing the region $R$, as shown in Figure 14.4.6. Find the area of $R$.

**SOLUTION** We can choose any field $\vec{F}$, as long as curl $\vec{F} = 1$. We choose $\vec{F} = \langle -y, 0 \rangle$. We also confirm (left to the reader) that $\vec{r}(t)$ traverses the region $R$ in a counterclockwise fashion. Thus

$$\text{Area of } R = \iint_R dA$$
$$= \oint_C \vec{F} \cdot d\vec{r}$$
$$= \int_{-1}^{1} \langle -t^2, 0 \rangle \cdot \langle 1 - 3t^2, 2t \rangle\, dt$$
$$= \int_{-1}^{1} (-t^2)(1 - 3t^2)\, dt$$
$$= \frac{8}{15}.$$

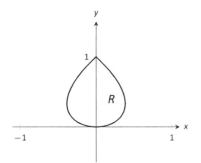

Figure 14.4.6: The region $R$, whose area is found in Example 14.4.4.

Notes:

## 14.4 Flow, Flux, Green's Theorem and the Divergence Theorem

### The Divergence Theorem

Green's Theorem makes a connection between the circulation around a closed region R and the sum of the curls over R. The Divergence Theorem makes a somewhat "opposite" connection: the total flux across the boundary of R is equal to the sum of the divergences over R.

> **Theorem 14.4.2**     **The Divergence Theorem (in the plane)**
>
> Let R be a closed, bounded region of the plane whose boundary C is composed of finitely many smooth curves, let $\vec{r}(t)$ be a counterclockwise parametrization of C, and let $\vec{F} = \langle M, N \rangle$ where $M_x$ and $N_y$ are continuous over R. Then
> $$\oint_C \vec{F} \cdot \vec{n}\, ds = \iint_R \operatorname{div} \vec{F}\, dA.$$

**Example 14.4.5**     **Confirming the Divergence Theorem**
Let $\vec{F} = \langle x - y, x + y \rangle$, let C be the circle of radius 2 centered at the origin and define R to be the interior of that circle, as shown in Figure 14.4.7. Verify the Divergence Theorem; that is, find the flux across C and show it is equal to the double integral of $\operatorname{div} \vec{F}$ over R.

**SOLUTION**     We parametrize the circle in the usual way, with $\vec{r}(t) = \langle 2\cos t, 2\sin t \rangle$, $0 \leq t \leq 2\pi$. The flux across C is

$$\oint_C \vec{F} \cdot \vec{n}\, ds = \oint_C \left(Mg'(t) - Nf'(t)\right) dt$$
$$= \int_0^{2\pi} \left((2\cos t - 2\sin t)(2\cos t) - (2\cos t + 2\sin t)(-2\sin t)\right) dt$$
$$= \int_0^{2\pi} 4\, dt = 8\pi.$$

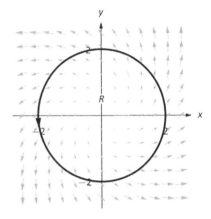

Figure 14.4.7: The region R used in Example 14.4.5.

We compute the divergence of $\vec{F}$ as $\operatorname{div} \vec{F} = M_x + N_y = 2$. Since the divergence is constant, we can compute the following double integral easily:

$$\iint_R \operatorname{div} \vec{F}\, dA = \iint_R 2\, dA = 2\iint_R dA = 2(\text{area of } R) = 8\pi,$$

which matches our previous result.

**Example 14.4.6**     **Flux when $\operatorname{div} \vec{F} = 0$**
Let $\vec{F}$ be any field where $\operatorname{div} \vec{F} = 0$, and let $C_1$ and $C_2$ be any two nonintersecting

Notes:

# Chapter 14 Vector Analysis

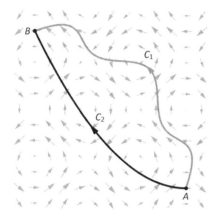

Figure 14.4.8: As used in Example 14.4.6, the vector field has a divergence of 0 and the two paths only intersect at their initial and terminal points.

paths, except that each begin at point A and end at point B (see Figure 14.4.8). Show why the flux across $C_1$ and $C_2$ is the same.

**SOLUTION** By referencing Figure 14.4.8, we see we can make a closed path C that combines $C_1$ with $C_2$, where $C_2$ is traversed with its opposite orientation. We label the enclosed region R. Since div $\vec{F} = 0$, the Divergence Theorem states that

$$\oint_C \vec{F} \cdot \vec{n}\, ds = \iint_R \text{div}\, \vec{F}\, dA = \iint_R 0\, dA = 0.$$

Using the properties and notation given in Theorem 14.3.1, consider:

$$0 = \oint_C \vec{F} \cdot \vec{n}\, ds$$
$$= \int_{C_1} \vec{F} \cdot \vec{n}\, ds + \int_{C_2^*} \vec{F} \cdot \vec{n}\, ds$$

(where $C_2^*$ is the path $C_2$ traversed with opposite orientation)

$$= \int_{C_1} \vec{F} \cdot \vec{n}\, ds - \int_{C_2} \vec{F} \cdot \vec{n}\, ds.$$
$$\int_{C_2} \vec{F} \cdot \vec{n}\, ds = \int_{C_1} \vec{F} \cdot \vec{n}\, ds.$$

Thus the flux across each path is equal.

In this section, we have investigated flow and flux, quantities that measure interactions between a vector field and a planar curve. We can also measure flow along spatial curves, though as mentioned before, it does not make sense to measure flux across spatial curves.

It does, however, make sense to measure the amount of a vector field that passes across a surface in space – i.e, the flux across a surface. We will study this, though in the next section we first learn about a more powerful way to describe surfaces than using functions of the form $z = f(x, y)$.

Notes:

# Exercises 14.4

## Terms and Concepts

1. Let $\vec{F}$ be a vector field and let $C$ be a curve. *Flow* is a measure of the amount of $\vec{F}$ going _____ $C$; *flux* is a measure of the amount of $\vec{F}$ going _____ $C$.

2. What is circulation?

3. Green's Theorem states, informally, that the circulation around a closed curve that bounds a region $R$ is equal to the sum of _____ across $R$.

4. The Divergence Theorem states, informally, that the outward flux across a closed curve that bounds a region $R$ is equal to the sum of _____ across $R$.

5. Let $\vec{F}$ be a vector field and let $C_1$ and $C_2$ be any nonintersecting paths except that each starts at point $A$ and ends at point $B$. If _____ $= 0$, then $\int_{C_1} \vec{F} \cdot \vec{T}\, ds = \int_{C_2} \vec{F} \cdot \vec{T}\, ds$.

6. Let $\vec{F}$ be a vector field and let $C_1$ and $C_2$ be any nonintersecting paths except that each starts at point $A$ and ends at point $B$. If _____ $= 0$, then $\int_{C_1} \vec{F} \cdot \vec{n}\, ds = \int_{C_2} \vec{F} \cdot \vec{n}\, ds$.

## Problems

In Exercises 7 – 12, a vector field $\vec{F}$ and a curve $C$ are given. Evaluate $\int_C \vec{F} \cdot \vec{n}\, ds$, the flux of $\vec{F}$ over $C$.

7. $\vec{F} = \langle x+y, x-y \rangle$; $C$ is the curve with initial and terminal points $(3,-2)$ and $(3,2)$, respectively, parametrized by $\vec{r}(t) = \langle 3t^2, 2t \rangle$ on $-1 \leq t \leq 1$.

8. $\vec{F} = \langle x+y, x-y \rangle$; $C$ is the curve with initial and terminal points $(3,-2)$ and $(3,2)$, respectively, parametrized by $\vec{r}(t) = \langle 3, t \rangle$ on $-2 \leq t \leq 2$.

9. $\vec{F} = \langle x^2, y+1 \rangle$; $C$ is line segment from $(0,0)$ to $(2,4)$.

10. $\vec{F} = \langle x^2, y+1 \rangle$; $C$ is the portion of the parabola $y = x^2$ from $(0,0)$ to $(2,4)$.

11. $\vec{F} = \langle y, 0 \rangle$; $C$ is the line segment from $(0,0)$ to $(0,1)$.

12. $\vec{F} = \langle y, 0 \rangle$; $C$ is the line segment from $(0,0)$ to $(1,1)$.

In Exercises 13 – 16, a vector field $\vec{F}$ and a closed curve $C$, enclosing a region $R$, are given. Verify Green's Theorem by evaluating $\oint_C \vec{F} \cdot d\vec{r}$ and $\iint_R \operatorname{curl} \vec{F}\, dA$, showing they are equal.

13. $\vec{F} = \langle x-y, x+y \rangle$; $C$ is the closed curve composed of the parabola $y = x^2$ on $0 \leq x \leq 2$ followed by the line segment from $(2,4)$ to $(0,0)$.

14. $\vec{F} = \langle -y, x \rangle$; $C$ is the unit circle.

15. $\vec{F} = \langle 0, x^2 \rangle$; $C$ the triangle with corners at $(0,0)$, $(2,0)$ and $(1,1)$.

16. $\vec{F} = \langle x+y, 2x \rangle$; $C$ the curve that starts at $(0,1)$, follows the parabola $y = (x-1)^2$ to $(3,4)$, then follows a line back to $(0,1)$.

In Exercises 17 – 20, a closed curve $C$ enclosing a region $R$ is given. Find the area of $R$ by computing $\oint_C \vec{F} \cdot d\vec{r}$ for an appropriate choice of vector field $\vec{F}$.

17. $C$ is the ellipse parametrized by $\vec{r}(t) = \langle 4\cos t, 3\sin t \rangle$ on $0 \leq t \leq 2\pi$.

18. $C$ is the curve parametrized by $\vec{r}(t) = \langle \cos t, \sin(2t) \rangle$ on $-\pi/2 \leq t \leq \pi/2$.

19. $C$ is the curve parametrized by $\vec{r}(t) = \langle \cos t, \sin(2t) \rangle$ on $0 \leq t \leq 2$.

20. $C$ is the curve parametrized by $\vec{r}(t) = \langle 2\cos t + \frac{1}{10}\cos(10t), 2\sin t + \frac{1}{10}\sin(10t) \rangle$ on $0 \leq t \leq 2\pi$.

In Exercises 21 – 24, a vector field $\vec{F}$ and a closed curve $C$, enclosing a region $R$, are given. Verify the Divergence Theorem by evaluating $\oint_C \vec{F} \cdot \vec{n}\, ds$ and $\iint_R \operatorname{div} \vec{F}\, dA$, showing they are equal.

21. $\vec{F} = \langle x-y, x+y \rangle$; $C$ is the closed curve composed of the parabola $y = x^2$ on $0 \leq x \leq 2$ followed by the line segment from $(2,4)$ to $(0,0)$.

22. $\vec{F} = \langle -y, x \rangle$; $C$ is the unit circle.

23. $\vec{F} = \langle 0, y^2 \rangle$; $C$ the triangle with corners at $(0,0)$, $(2,0)$ and $(1,1)$.

24. $\vec{F} = \langle x^2/2, y^2/2 \rangle$; $C$ the curve that starts at $(0,1)$, follows the parabola $y = (x-1)^2$ to $(3,4)$, then follows a line back to $(0,1)$.

## 14.5 Parametrized Surfaces and Surface Area

Thus far we have focused mostly on 2-dimensional vector fields, measuring flow and flux along/across curves in the plane. Both Green's Theorem and the Divergence Theorem make connections between planar regions and their boundaries. We now move our attention to 3-dimensional vector fields, considering both curves and surfaces in space.

We are accustomed to describing surfaces as functions of two variables, usually written as $z = f(x, y)$. For our coming needs, this method of describing surfaces will prove to be insufficient. Instead, we will *parametrize* our surfaces, describing them as the set of terminal points of some vector–valued function $\vec{r}(u, v) = \langle f(u, v), g(u, v), h(u, v) \rangle$. The bulk of this section is spent practicing the skill of describing a surface $\mathcal{S}$ using a vector-valued function. Once this skill is developed, we'll show how to find the surface area $S$ of a parametrically-defined surface $\mathcal{S}$, a skill needed in the remaining sections of this chapter.

**Note:** We use the letter $S$ to denote Surface Area. This section begins a study into surfaces, and it is natural to label a surface with the letter "S". We distinguish a surface from its surface area by using a calligraphic S to denote a surface: $\mathcal{S}$. When writing this letter by hand, it may be useful to add serifs to the letter, such as: $\mathcal{S}$

---

**Definition 14.5.1**     **Parametrized Surface**

Let $\vec{r}(u, v) = \langle f(u, v), g(u, v), h(u, v) \rangle$ be a vector–valued function that is continuous and one to one on the interior of its domain $R$ in the $u$-$v$ plane. The set of all terminal points of $\vec{r}$ (i.e., the *range* of $\vec{r}$) is the **surface** $\mathcal{S}$, and $\vec{r}$ along with its domain $R$ form a **parametrization** of $\mathcal{S}$.

This parametrization is **smooth** on $R$ if $\vec{r}_u$ and $\vec{r}_v$ are continuous and $\vec{r}_u \times \vec{r}_v$ is never $\vec{0}$ on the interior of $R$.

---

**Note:** A function is *one to one* on its domain if the function never repeats an output value over the domain. In the case of $\vec{r}(u, v)$, $\vec{r}$ is one to one if $\vec{r}(u_1, v_1) \neq \vec{r}(u_2, v_2)$ for all points $(u_1, v_1) \neq (u_2, v_2)$ in the domain of $\vec{r}$.

Given a point $(u_0, v_0)$ in the domain of a vector–valued function $\vec{r}$, the vectors $\vec{r}_u(u_0, v_0)$ and $\vec{r}_v(u_0, v_0)$ are tangent to the surface $\mathcal{S}$ at $\vec{r}(u_0, v_0)$ (a proof of this is developed later in this section). The definition of smoothness dictates that $\vec{r}_u \times \vec{r}_v \neq \vec{0}$; this ensures that neither $\vec{r}_u$ nor $\vec{r}_v$ are $\vec{0}$, nor are they ever parallel. Therefore smoothness guarantees that $\vec{r}_u$ and $\vec{r}_v$ determine a plane that is tangent to $\mathcal{S}$.

A surface $\mathcal{S}$ is said to be **orientable** if a field of normal vectors can be defined on $\mathcal{S}$ that vary continuously along $\mathcal{S}$. This definition may be hard to understand; it may help to know that orientable surfaces are often called "two sided." A sphere is an orientable surface, and one can easily envision an "inside" and "outside" of the sphere. A paraboloid is orientable, where again one can generally envision "inside" and "outside" sides (or "top" and "bottom" sides) to this surface. Just about every surface that one can imagine is orientable, and we'll assume all surfaces we deal with in this text are orientable.

It is enlightening to examine a classic non-orientable surface: the Möbius

Notes:

band, shown in Figure 14.5.1. Vectors normal to the surface are given, starting at the point indicated in the figure. These normal vectors "vary continuously" as they move along the surface. Letting each vector indicate the "top" side of the band, we can easily see near any vector which side is the "top".

However, if as we progress along the band, we recognize that we are labeling "both sides" of the band as the top; in fact, there are not two "sides" to this band, but one. The Möbius band is a non-orientable surface.

We now practice parameterizing surfaces.

**Example 14.5.1  Parameterizing a surface over a rectangle**
Parametrize the surface $z = x^2 + 2y^2$ over the rectangular region $R$ defined by $-3 \leq x \leq 3, -1 \leq y \leq 1$.

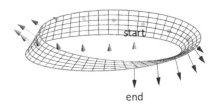

Figure 14.5.1: A Möbius band, a non-orientable surface.

**SOLUTION** There is a straightforward way to parametrize a surface of the form $z = f(x, y)$ over a rectangular domain. We let $x = u$ and $y = v$, and let $\vec{r}(u, v) = \langle u, v, f(u, v) \rangle$. In this instance, we have $\vec{r}(u, v) = \langle u, v, u^2 + 2v^2 \rangle$, for $-3 \leq u \leq 3, -1 \leq v \leq 1$. This surface is graphed in Figure 14.5.2.

**Example 14.5.2  Parameterizing a surface over a circular disk**
Parametrize the surface $z = x^2 + 2y^2$ over the circular region $R$ enclosed by the circle of radius 2 that is centered at the origin.

**SOLUTION** We can parametrize the circular boundary of $R$ with the vector-valued function $\langle 2 \cos u, 2 \sin u \rangle$, where $0 \leq u \leq 2\pi$. We can obtain the interior of $R$ by scaling this function by a variable amount, i.e., by multiplying by $v$: $\langle 2v \cos u, 2v \sin u \rangle$, where $0 \leq v \leq 1$.

It is important to understand the role of $v$ in the above function. When $v = 1$, we get the boundary of $R$, a circle of radius 2. When $v = 0$, we simply get the point $(0, 0)$, the center of $R$ (which can be thought of as a circle with radius of 0). When $v = 1/2$, we get the circle of radius 1 that is centered at the origin, which is the circle *halfway* between the boundary and the center. As $v$ varies from 0 to 1, we create a series of concentric circles that fill out all of $R$.

Thus far, we have determined the $x$ and $y$ components of our parametrization of the surface: $x = 2v \cos u$ and $y = 2v \sin u$. We find the $z$ component simply by using $z = f(x, y) = x^2 + 2y^2$:

$$z = (2v \cos u)^2 + 2(2v \sin u)^2 = 4v^2 \cos^2 u + 8v^2 \sin^2 u.$$

Thus $\vec{r}(u, v) = \langle 2v \cos u, 2v \sin u, 4v^2 \cos^2 u + 8v^2 \sin^2 u \rangle$, $0 \leq u \leq 2\pi$, $0 \leq v \leq 1$, which is graphed in Figure 14.5.3. The way that this graphic was generated highlights how the surface was parametrized. When viewing from above, one can see lines emanating from the origin; they represent different values of $u$ as

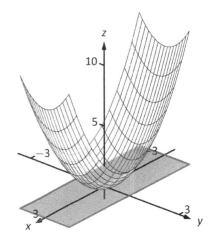

Figure 14.5.2: The surface parametrized in Example 14.5.1.

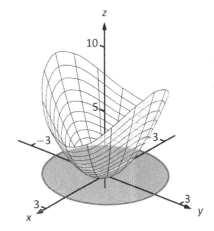

Figure 14.5.3: The surface parametrized in Example 14.5.2.

Notes:

## Chapter 14 Vector Analysis

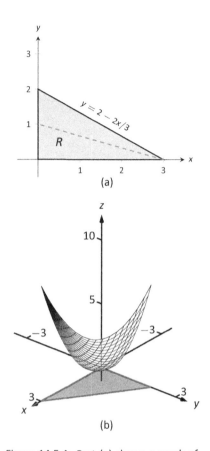

Figure 14.5.4: Part (a) shows a graph of the region $R$, and part (b) shows the surface over $R$, as defined in Example 14.5.3.

$u$ sweeps from an angle of 0 up to $2\pi$. One can also see concentric circles, each corresponding to a different value of $v$.

Examples 14.5.1 and 14.5.2 demonstrate an important principle when parameterizing surfaces given in the form $z = f(x, y)$ over a region $R$: if one can determine $x$ and $y$ in terms of $u$ and $v$, then $z$ follows directly as $z = f(x, y)$.

In the following two examples, we parametrize the same surface over triangular regions. Each will use $v$ as a "scaling factor" as done in Example 14.5.2.

**Example 14.5.3    Parameterizing a surface over a triangle**
Parametrize the surface $z = x^2 + 2y^2$ over the triangular region $R$ enclosed by the coordinate axes and the line $y = 2 - 2x/3$, as shown in Figure 14.5.4(a).

**SOLUTION**    We may begin by letting $x = u$, $0 \leq u \leq 3$, and $y = 2-2u/3$. This gives only the line on the "upper" side of the triangle. To get all of the region $R$, we can once again scale $y$ by a variable factor, $v$.

Still letting $x = u$, $0 \leq u \leq 3$, we let $y = v(2 - 2u/3)$, $0 \leq v \leq 1$. When $v = 0$, all $y$-values are 0, and we get the portion of the $x$-axis between $x = 0$ and $x = 3$. When $v = 1$, we get the upper side of the triangle. When $v = 1/2$, we get the line $y = 1/2(2 - 2u/3) = 1 - u/3$, which is the line "halfway up" the triangle, shown in the figure with a dashed line.

Letting $z = f(x, y) = x^2 + 2y^2$, we have $\vec{r}(u, v) = \langle u, v(2 - 2u/3), u^2 + 2(v(2 - 2u/3))^2 \rangle$, $0 \leq u \leq 3$, $0 \leq v \leq 1$. This surface is graphed in Figure 14.5.4(b). Again, when one looks from above, we can see the scaling effects of $v$: the series of lines that run to the point $(3, 0)$ each represent a different value of $v$.

Another common way to parametrize the surface is to begin with $y = u$, $0 \leq u \leq 2$. Solving the equation of the line $y = 2 - 2x/3$ for $x$, we have $x = 3 - 3y/2$, leading to using $x = v(3 - 3u/2)$, $0 \leq v \leq 1$. With $z = x^2 + 2y^2$, we have $\vec{r}(u, v) = \langle v(3-3u/2), u, (v(3-3u/2))^2 + 2v^2 \rangle$, $0 \leq u \leq 2$, $0 \leq v \leq 1$.

**Example 14.5.4    Parameterizing a surface over a triangle**
Parametrize the surface $z = x^2 + 2y^2$ over the triangular region $R$ enclosed by the lines $y = 3 - 2x/3$, $y = 1$ and $x = 0$ as shown in Figure 14.5.5(a).

**SOLUTION**    While the region $R$ in this example is very similar to the region $R$ in the previous example, and our method of parameterizing the surface is fundamentally the same, it will feel as though our answer is much different than before.

We begin with letting $x = u$, $0 \leq u \leq 3$. We may be tempted to let $y = v(3 - 2u/3)$, $0 \leq v \leq 1$, but this is incorrect. When $v = 1$, we obtain the upper

Notes:

line of the triangle as desired. However, when $v = 0$, the $y$-value is 0, which does not lie in the region $R$.

We will describe the general method of proceeding following this example. For now, consider $y = 1 + v(2 - 2u/3)$, $0 \leq v \leq 1$. Note that when $v = 1$, we have $y = 3 - 2u/3$, the upper line of the boundary of $R$. Also, when $v = 0$, we have $y = 1$, which is the lower boundary of $R$. With $z = x^2 + 2y^2$, we determine $\vec{r}(u, v) = \langle u, 1 + v(2 - 2u/3), u^2 + 2(1 + v(2 - 2u/3))^2 \rangle$, $0 \leq u \leq 3$, $0 \leq v \leq 1$.

The surface is graphed in Figure 14.5.5(b).

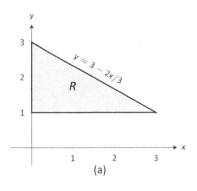

Given a surface of the form $z = f(x, y)$, one can often determine a parametrization of the surface over a region $R$ in a manner similar to determining bounds of integration over a region $R$. Using the techniques of Section 13.1, suppose a region $R$ can be described by $a \leq x \leq b$, $g_1(x) \leq y \leq g_2(x)$, i.e., the area of $R$ can be found using the iterated integral

$$\int_a^b \int_{g_1(x)}^{g_2(x)} dy\, dx.$$

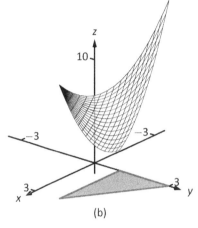

Figure 14.5.5: Part (a) shows a graph of the region $R$, and part (b) shows the surface over $R$, as defined in Example 14.5.4.

When parameterizing the surface, we can let $x = u$, $a \leq u \leq b$, and we can let $y = g_1(u) + v(g_2(u) - g_1(u))$, $0 \leq v \leq 1$. The parametrization of $x$ is straightforward, but look closely at how $y$ is determined. When $v = 0$, $y = g_1(u) = g_1(x)$. When $v = 1$, $y = g_2(u) = g_2(x)$.

As a specific example, consider the triangular region $R$ from Example 14.5.4, shown in Figure 14.5.5(a). Using the techniques of Section 13.1, we can find the area of $R$ as

$$\int_0^3 \int_1^{3-2x/3} dy\, dx.$$

Following the above discussion, we can set $x = u$, where $0 \leq u \leq 3$, and set $y = 1 + v(3 - 2u/3 - 1) = 1 + v(2 - 2u/3)$, $0 \leq v \leq 1$, as used in that example.

One can do a similar thing if $R$ is bounded by $c \leq y \leq d$, $h_1(y) \leq x \leq h_2(y)$, but for the sake of simplicity we leave it to the reader to flesh out those details. The principles outlined above are given in the following Key Idea for reference.

Notes:

## Chapter 14 Vector Analysis

> **Key Idea 14.5.1  Parameterizing Surfaces**
>
> Let a surface $S$ be the graph of a function $z = f(x, y)$, where the domain of $f$ is a closed, bounded region $R$ in the x-y plane. Let $R$ be bounded by $a \leq x \leq b$, $g_1(x) \leq y \leq g_2(x)$, i.e., the area of $R$ can be found using the iterated integral $\int_a^b \int_{g_1(x)}^{g_2(x)} dy\, dx$, and let $h(u, v) = g_1(u) + v(g_2(u) - g_1(u))$.
> $S$ can be parametrized as
> $$\vec{r}(u,v) = \langle u, h(u,v), f(u, h(u,v)) \rangle, \quad a \leq u \leq b,\; 0 \leq v \leq 1.$$

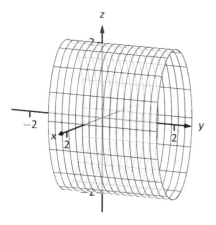

Figure 14.5.6: The cylinder parametrized in Example 14.5.5.

**Example 14.5.5  Parameterizing a cylinderical surface**
Find a parametrization of the cylinder $x^2 + z^2/4 = 1$, where $-1 \leq y \leq 2$, as shown in Figure 14.5.6.

**SOLUTION**  The equation $x^2 + z^2/4 = 1$ can be envisioned to describe an ellipse in the x-z plane; as the equation lacks a y-term, the equation describes a cylinder (recall Definition 10.1.2) that extends without bound parallel to the y-axis. This ellipse has a vertical major axis of length 4, a horizontal minor axis of length 2, and is centered at the origin. We can parametrize this ellipse using sines and cosines; our parametrization can begin with

$$\vec{r}(u,v) = \langle \cos u, \text{???}, 2 \sin u \rangle, \quad 0 \leq u \leq 2\pi,$$

where we still need to determine the y component.

While the cylinder $x^2 + z^2/4 = 1$ is satisfied by any y value, the problem states that all y values are to be between $y = -1$ and $y = 2$. Since the value of y does not depend at all on the values of x or z, we can use another variable, v, to describe y. Our final answer is

$$\vec{r}(u,v) = \langle \cos u, v, 2 \sin u \rangle, \quad 0 \leq u \leq 2\pi,\; -1 \leq v \leq 2.$$

**Example 14.5.6  Parameterizing an elliptic cone**
Find a parametrization of the elliptic cone $z^2 = \frac{x^2}{4} + \frac{y^2}{9}$, where $-2 \leq z \leq 3$, as shown in Figure 14.5.7.

**SOLUTION**  One way to parametrize this cone is to recognize that given a z value, the cross section of the cone at that z value is an ellipse with equation $\frac{x^2}{(2z)^2} + \frac{y^2}{(3z)^2} = 1$. We can let $z = v$, for $-2 \leq v \leq 3$ and then parametrize the above ellipses using sines, cosines and v.

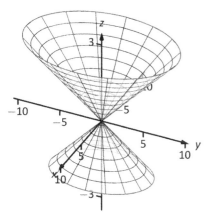

Figure 14.5.7: The elliptic cone as described in Example 14.5.6.

Notes:

## 14.5 Parametrized Surfaces and Surface Area

We can parametrize the x component of our surface with $x = 2z\cos u$ and the y component with $y = 3z\sin u$, where $0 \leq u \leq 2\pi$. Putting all components together, we have

$$\vec{r}(u,v) = \langle 2v\cos u, 3v\sin u, v\rangle, \quad 0 \leq u \leq 2\pi, \quad -2 \leq v \leq 3.$$

When $v$ takes on negative values, the radii of the cross–sectional ellipses become "negative," which can lead to some surprising results. Consider Figure 14.5.8, where the cone is graphed for $0 \leq u \leq \pi$. Because $v$ is negative below the x-y plane, the radii of the cross–sectional ellipses are negative, and the opposite side of the cone is sketched below the x-y plane.

**Example 14.5.7   Parameterizing an ellipsoid**
Find a parametrization of the ellipsoid $\frac{x^2}{25} + y^2 + \frac{z^2}{4} = 1$ as shown in Figure 14.5.9(a).

**Solution**   Recall Key Idea 10.2.1 from Section 10.2, which states that all unit vectors in space have the form $\langle \sin\theta\cos\varphi, \sin\theta\sin\varphi, \cos\theta\rangle$ for some angles $\theta$ and $\varphi$. If we choose our angles appropriately, this allows us to draw the unit sphere. To get an ellipsoid, we need only scale each component of the sphere appropriately.

The x-radius of the given ellipsoid is 5, the y-radius is 1 and the z-radius is 2. Substituting $u$ for $\theta$ and $v$ for $\varphi$, we have $\vec{r}(u,v) = \langle 5\sin u\cos v, \sin u\sin v, 2\cos u\rangle$, where we still need to determine the ranges of $u$ and $v$.

Note how the x and y components of $\vec{r}$ have cos $v$ and sin $v$ terms, respectively. This hints at the fact that ellipses are drawn parallel to the x-y plane as $v$ varies, which implies we should have $v$ range from 0 to $2\pi$.

One may be tempted to let $0 \leq u \leq 2\pi$ as well, but note how the z component is $2\cos u$. We only need cos $u$ to take on values between $-1$ and 1 once, therefore we can restrict $u$ to $0 \leq u \leq \pi$.

The final parametrization is thus

$$\vec{r}(u,v) = \langle 5\sin u\cos v, \sin u\sin v, 2\cos u\rangle, \quad 0 \leq u \leq \pi, \quad 0 \leq v \leq 2\pi.$$

In Figure 14.5.9(b), the ellipsoid is graphed on $\frac{\pi}{4} \leq u \leq \frac{2\pi}{3}, \frac{\pi}{4} \leq v \leq \frac{3\pi}{2}$ to demonstrate how each variable affects the surface.

Parametrization is a powerful way to represent surfaces. One of the advantages of the methods of parametrization described in this section is that the domain of $\vec{r}(u,v)$ is always a rectangle; that is, the bounds on $u$ and $v$ are constants. This will make some of our future computations easier to evaluate.

Just as we could parametrize curves in more than one way, there will always be multiple ways to parametrize a surface. Some ways will be more "natural"

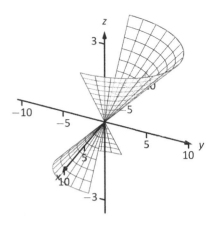

Figure 14.5.8: The elliptic cone as described in Example 14.5.6 with restricted domain.

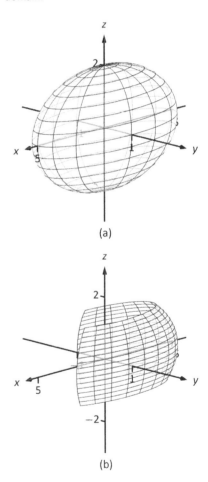

Figure 14.5.9: An ellipsoid in (a), drawn again in (b) with its domain restricted, as described in Example 14.5.7.

# Chapter 14 Vector Analysis

than others, but these other ways are not incorrect. Because technology is often readily available, it is often a good idea to check one's work by graphing a parametrization of a surface to check if it indeed represents what it was intended to.

## Surface Area

It will become important in the following sections to be able to compute the surface area of a surface $S$ given a smooth parametrization $\vec{r}(u,v)$, $a \leq u \leq b$, $c \leq v \leq d$. Following the principles given in the integration review at the beginning of this chapter, we can say that

$$\text{Surface Area of } S = S = \iint_S dS,$$

where $dS$ represents a small amount of surface area. That is, to compute total surface area $S$, add up lots of small amounts of surface area $dS$ across the entire surface $S$. The key to finding surface area is knowing how to compute $dS$. We begin by approximating.

In Section 13.5 we used the area of a plane to approximate the surface area of a small portion of a surface. We will do the same here.

Let $R$ be the region of the $u$-$v$ plane bounded by $a \leq u \leq b$, $c \leq v \leq d$ as shown in Figure 14.5.10(a). Partition $R$ into rectangles of width $\Delta u = \frac{b-a}{n}$ and height $\Delta v = \frac{d-c}{n}$, for some $n$. Let $p = (u_0, v_0)$ be the lower left corner of some rectangle in the partition, and let $m$ and $q$ be neighboring corners as shown.

The point $p$ maps to a point $P = \vec{r}(u_0, v_0)$ on the surface $S$, and the rectangle with corners $p$, $m$ and $q$ maps to some region (probably not rectangular) on the surface as shown in Figure 14.5.10(b), where $M = \vec{r}(m)$ and $Q = \vec{r}(q)$. We wish to approximate the surface area of this mapped region.

Let $\vec{u} = M - P$ and $\vec{v} = Q - P$. These two vectors form a parallelogram, illustrated in Figure 14.5.10(c), whose area *approximates* the surface area we seek. In this particular illustration, we can see that parallelogram does not particularly match well the region we wish to approximate, but that is acceptable; by increasing the number of partitions of $R$, $\Delta u$ and $\Delta v$ shrink and our approximations will become better.

From Section 10.4 we know the area of this parallelogram is $||\vec{u} \times \vec{v}||$. If we repeat this approximation process for each rectangle in the partition of $R$, we can sum the areas of all the parallelograms to get an approximation of the surface area $S$:

$$\text{Surface area of } S = S \approx \sum_{j=1}^{n}\sum_{i=1}^{n} ||\vec{u}_{i,j} \times \vec{v}_{i,j}||,$$

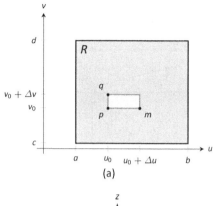

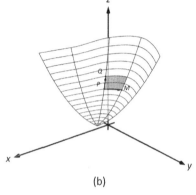

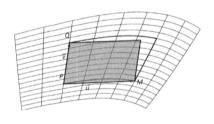

Figure 14.5.10: Illustrating the process of finding surface area by approximating with planes.

Notes:

## 14.5 Parametrized Surfaces and Surface Area

where $\vec{u}_{i,j} = \vec{r}(u_i + \Delta u, v_j) - \vec{r}(u_i, v_j)$ and $\vec{v}_{i,j} = \vec{r}(u_i, v_j + \Delta v) - \vec{r}(u_i, v_j)$.

From our previous calculus experience, we expect that taking a limit as $n \to \infty$ will result in the exact surface area. However, the current form of the above double sum makes it difficult to realize what the result of that limit is. The following rewriting of the double summation will be helpful:

$$\sum_{j=1}^{n} \sum_{i=1}^{n} || \vec{u}_{i,j} \times \vec{v}_{i,j} || =$$

$$\sum_{j=1}^{n} \sum_{i=1}^{n} || \left(\vec{r}(u_i + \Delta u, v_j) - \vec{r}(u_i, v_j)\right) \times \left(\vec{r}(u_i, v_j + \Delta v) - \vec{r}(u_i, v_j)\right) || =$$

$$\sum_{j=1}^{n} \sum_{i=1}^{n} \left\| \frac{\vec{r}(u_i + \Delta u, v_j) - \vec{r}(u_i, v_j)}{\Delta u} \times \frac{\vec{r}(u_i, v_j + \Delta v) - \vec{r}(u_i, v_j)}{\Delta v} \right\| \Delta u \Delta v.$$

We now take the limit as $n \to \infty$, forcing $\Delta u$ and $\Delta v$ to 0. As $\Delta u \to 0$,

$$\frac{\vec{r}(u_i + \Delta u, v_j) - \vec{r}(u_i, v_j)}{\Delta u} \to \vec{r}_u(u_i, v_j) \quad \text{and}$$

$$\frac{\vec{r}(u_i, v_j + \Delta v) - \vec{r}(u_i, v_j)}{\Delta v} \to \vec{r}_v(u_i, v_j).$$

(This limit process also demonstrates that $\vec{r}_u(u, v)$ and $\vec{r}_v(u, v)$ are tangent to the surface $\mathcal{S}$ at $\vec{r}(u, v)$. We don't need this fact now, but it will be important in the next section.)

Thus, in the limit, the double sum leads to a double integral:

$$\lim_{n \to \infty} \sum_{j=1}^{n} \sum_{i=1}^{n} || \vec{u}_{i,j} \times \vec{v}_{i,j} || = \int_{c}^{d} \int_{a}^{b} || \vec{r}_u \times \vec{r}_v || \, du \, dv.$$

---

**Theorem 14.5.1**    **Surface Area of Parametrically Defined Surfaces**

Let $\vec{r}(u, v)$ be a smooth parametrization of a surface $\mathcal{S}$ over a closed, bounded region $R$ of the $u$-$v$ plane.

- The surface area differential $dS$ is: $dS = || \vec{r}_u \times \vec{r}_v || \, dA$.

- The surface area $S$ of $\mathcal{S}$ is

$$S = \iint_{\mathcal{S}} dS = \iint_{R} || \vec{r}_u \times \vec{r}_v || \, dA.$$

---

Notes:

**Example 14.5.8  Finding the surface area of a parametrized surface**

Using the parametrization found in Example 14.5.2, find the surface area of $z = x^2 + 2y^2$ over the circular disk of radius 2, centered at the origin.

**SOLUTION**   In Example 14.5.2, we parametrized the surface as $\vec{r}(u,v) = \langle 2v\cos u, 2v\sin u, 4v^2\cos^2 u + 8v^2\sin^2 u\rangle$, for $0 \leq u \leq 2\pi$, $0 \leq v \leq 1$. To find the surface area using Theorem 14.5.1, we need $\|\vec{r}_u \times \vec{r}_v\|$. We find:

$$\vec{r}_u = \langle -2v\sin u, 2v\cos u, 8v^2\cos u \sin u\rangle$$
$$\vec{r}_v = \langle 2\cos u, 2\sin v, 8v\cos^2 u + 16v\sin^2 u\rangle$$
$$\vec{r}_u \times \vec{r}_v = \langle 16v^2\cos u, 32v^2\sin u, -4v\rangle$$
$$\|\vec{r}_u \times \vec{r}_v\| = \sqrt{256v^4\cos^2 u + 1024v^4\sin^2 u + 16v^2}.$$

Thus the surface area is

$$S = \iint_S dS = \iint_R \|\vec{r}_u \times \vec{r}_v\|\, dA$$
$$= \int_0^1 \int_0^{2\pi} \sqrt{256v^4\cos^2 u + 1024v^4\sin^2 u + 16v^2}\, du\, dv \approx 53.59.$$

There is a lot of tedious work in the above calculations and the final integral is nontrivial. The use of a computer-algebra system is highly recommended.

In Section 14.1, we recalled the arc length differential $ds = \|\vec{r}'(t)\|\, dt$. In subsequent sections, we used that differential, but in most applications the "$\|\vec{r}'(t)\|$" part of the differential canceled out of the integrand (to our benefit, as integrating the square roots of functions is generally difficult). We will find a similar thing happens when we use the surface area differential $dS$ in the following sections. That is, our main goal is not to be able to compute surface area; rather, surface area is a tool to obtain other quantities that are more important and useful. In our applications, we will use $dS$, but most of the time the "$\|\vec{r}_u \times \vec{r}_v\|$" part will cancel out of the integrand, making the subsequent integration easier to compute.

Notes:

# Exercises 14.5

## Terms and Concepts

1. In your own words, describe what an orientable surface is.

2. Give an example of a non-orientable surface.

## Problems

In Exercises 3 – 4, parametrize the surface defined by the function $z = f(x,y)$ over each of the given regions $R$ of the x-y plane.

3. $z = 3x^2 y$;

    (a) $R$ is the rectangle bounded by $-1 \leq x \leq 1$ and $0 \leq y \leq 2$.

    (b) $R$ is the circle of radius 3, centered at $(1, 2)$.

    (c) $R$ is the triangle with vertices $(0,0)$, $(1,0)$ and $(0,2)$.

    (d) $R$ is the region bounded by the x-axis and the graph of $y = 1 - x^2$.

4. $z = 4x + 2y^2$;

    (a) $R$ is the rectangle bounded by $1 \leq x \leq 4$ and $5 \leq y \leq 7$.

    (b) $R$ is the ellipse with major axis of length 8 parallel to the x-axis, and minor axis of length 6 parallel to the y-axis, centered at the origin.

    (c) $R$ is the triangle with vertices $(0,0)$, $(2,2)$ and $(0,4)$.

    (d) $R$ is the annulus bounded between the circles, centered at the origin, with radius 2 and radius 5.

In Exercises 5 – 8, a surface $S$ in space is described that cannot be defined in terms of a function $z = f(x,y)$. Give a parametrization of $S$.

5. $S$ is the rectangle in space with corners at $(0,0,0)$, $(0,2,0)$, $(0,2,1)$ and $(0,0,1)$.

6. $S$ is the triangle in space with corners at $(1,0,0)$, $(1,0,1)$ and $(0,0,1)$.

7. $S$ is the ellipsoid $\dfrac{x^2}{9} + \dfrac{y^2}{4} + \dfrac{z^2}{16} = 1$.

8. $S$ is the elliptic cone $y^2 = x^2 + \dfrac{z^2}{16}$, for $-1 \leq y \leq 5$.

In Exercises 9 – 16, a domain $D$ in space is given. Parametrize each of the bounding surfaces of $D$.

9. $D$ is the domain bounded by the planes $z = \frac{1}{2}(3-x)$, $x = 1$, $y = 0$, $y = 2$ and $z = 0$.

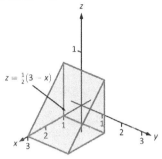

10. $D$ is the domain bounded by the planes $z = 2x + 4y - 4$, $x = 2$, $y = 1$ and $z = 0$.

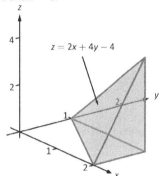

11. $D$ is the domain bounded by $z = 2y$, $y = 4 - x^2$ and $z = 0$.

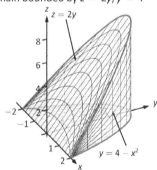

12. $D$ is the domain bounded by $y = 1 - z^2$, $y = 1 - x^2$, $x = 0$, $y = 0$ and $z = 0$.

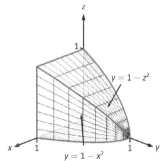

13. $D$ is the domain bounded by the cylinder $x + y^2/9 = 1$ and the planes $z = 1$ and $z = 3$.

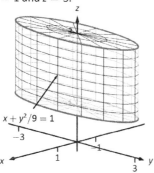

14. $D$ is the domain bounded by the cone $x^2 + y^2 = (z-1)^2$ and the plane $z = 0$.

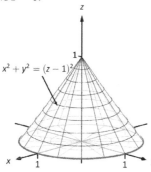

15. $D$ is the domain bounded by the cylinder $z = 1 - x^2$ and the planes $y = -1$, $y = 2$ and $z = 0$.

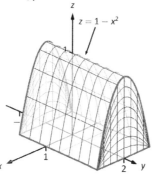

16. $D$ is the domain bounded by the paraboloid $z = 4 - x^2 - 4y^2$ and the plane $z = 0$.

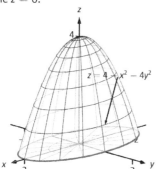

**In Exercises 17 – 20, find the surface area $S$ of the given surface $\mathcal{S}$. (The associated integrals are computable without the assistance of technology.)**

17. $\mathcal{S}$ is the plane $z = 2x + 3y$ over the rectangle $-1 \leq x \leq 1$, $2 \leq y \leq 3$.

18. $\mathcal{S}$ is the plane $z = x + 2y$ over the triangle with vertices at $(0,0)$, $(1,0)$ and $(0,1)$.

19. $\mathcal{S}$ is the plane $z = x + y$ over the circular disk, centered at the origin, with radius 2.

20. $\mathcal{S}$ is the plane $z = x + y$ over the annulus bounded by the circles, centered at the origin, with radius 1 and radius 2.

**In Exercises 21 – 24, set up the double integral that finds the surface area $S$ of the given surface $\mathcal{S}$, then use technology to approximate its value.**

21. $\mathcal{S}$ is the paraboloid $z = x^2 + y^2$ over the circular disk of radius 3 centered at the origin.

22. $\mathcal{S}$ is the paraboloid $z = x^2 + y^2$ over the triangle with vertices at $(0,0)$, $(0,1)$ and $(1,1)$.

23. $\mathcal{S}$ is the plane $z = 5x - y$ over the region enclosed by the parabola $y = 1 - x^2$ and the $x$-axis.

24. $\mathcal{S}$ is the hyperbolic paraboloid $z = x^2 - y^2$ over the circular disk of radius 1 centered at the origin.

## 14.6 Surface Integrals

Consider a smooth surface $\mathcal{S}$ that represents a thin sheet of metal. How could we find the mass of this metallic object?

If the density of this object is constant, then we can find mass via "mass= density $\times$ surface area," and we could compute the surface area using the techniques of the previous section.

What if the density were not constant, but variable, described by a function $\delta(x, y, z)$? We can describe the mass using our general integration techniques as

$$\text{mass} = \iint_\mathcal{S} dm,$$

where $dm$ represents "a little bit of mass." That is, to find the total mass of the object, sum up lots of little masses over the surface.

How do we find the "little bit of mass" $dm$? On a small portion of the surface with surface area $\Delta S$, the density is approximately constant, hence $dm \approx \delta(x, y, z)\Delta S$. As we use limits to shrink the size of $\Delta S$ to 0, we get $dm = \delta(x, y, z)dS$; that is, a little bit of mass is equal to a density times a small amount of surface area. Thus the total mass of the thin sheet is

$$\text{mass} = \iint_\mathcal{S} \delta(x, y, z)\, dS. \tag{14.3}$$

To evaluate the above integral, we would seek $\vec{r}(u, v)$, a smooth parametrization of $\mathcal{S}$ over a region $R$ of the $u$-$v$ plane. The density would become a function of $u$ and $v$, and we would integrate $\iint_R \delta(u, v)\, \|\vec{r}_u \times \vec{r}_v\|\, dA$.

The integral in Equation (14.3) is a specific example of a more general construction defined below.

---

**Definition 14.6.1**     **Surface Integral**

Let $G(x, y, z)$ be a continuous function defined on a surface $\mathcal{S}$. The **surface integral of $G$ on $\mathcal{S}$** is

$$\iint_\mathcal{S} G(x, y, z)\, dS.$$

---

Surface integrals can be used to measure a variety of quantities beyond mass. If $G(x, y, z)$ measures the static charge density at a point, then the surface integral will compute the total static charge of the sheet. If $G$ measures the amount of fluid passing through a screen (represented by $\mathcal{S}$) at a point, then the surface integral gives the total amount of fluid going through the screen.

Notes:

## Chapter 14 Vector Analysis

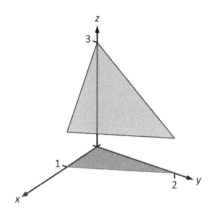

Figure 14.6.1: The surface whose mass is computed in Example 14.6.1.

**Example 14.6.1  Finding the mass of a thin sheet**

Find the mass of a thin sheet modeled by the plane $2x + y + z = 3$ over the triangular region of the x-y plane bounded by the coordinate axes and the line $y = 2 - 2x$, as shown in Figure 14.6.1, with density function $\delta(x, y, z) = x^2 + 5y + z$, where all distances are measured in cm and the density is given as gm/cm$^2$.

**SOLUTION** We begin by parameterizing the planar surface $\mathcal{S}$. Using the techniques of the previous section, we can let $x = u$ and $y = v(2 - 2u)$, where $0 \leq u \leq 1$ and $0 \leq v \leq 1$. Solving for $z$ in the equation of the plane, we have $z = 3 - 2x - y$, hence $z = 3 - 2u - v(2 - 2u)$, giving the parametrization $\vec{r}(u, v) = \langle u, v(2 - 2u), 3 - 2u - v(2 - 2u) \rangle$.

We need $dS = \|\vec{r}_u \times \vec{r}_v\| \, dA$, so we need to compute $\vec{r}_u$, $\vec{r}_v$ and the norm of their cross product. We leave it to the reader to confirm the following:

$$\vec{r}_u = \langle 1, -2v, 2v - 2 \rangle, \quad \vec{r}_v = \langle 0, 2 - 2u, 2u - 2 \rangle,$$

$$\vec{r}_u \times \vec{r}_v = \langle 4 - 4u, 2 - 2u, 2 - 2u \rangle \quad \text{and} \quad \|\vec{r}_u \times \vec{r}_v\| = 2\sqrt{6}\sqrt{(u-1)^2}.$$

We need to be careful to not "simplify" $\|\vec{r}_u \times \vec{r}_v\| = 2\sqrt{6}\sqrt{(u-1)^2}$ as $2\sqrt{6}(u - 1)$; rather, it is $2\sqrt{6}|u - 1|$. In this example, $u$ is bounded by $0 \leq u \leq 1$, and on this interval $|u - 1| = 1 - u$. Thus $dS = 2\sqrt{6}(1 - u)dA$.

The density is given as a function of x, y and z, for which we'll substitute the corresponding components of $\vec{r}$ (with the slight abuse of notation that we used in previous sections):

$$\delta(x, y, z) = \delta(\vec{r}(u, v))$$
$$= u^2 + 5v(2 - 2u) + 3 - 2u - v(2 - 2u)$$
$$= u^2 - 8uv - 2u + 8v + 3.$$

Thus the mass of the sheet is:

$$M = \iint_{\mathcal{S}} dm$$
$$= \iint_{R} \delta(\vec{r}(u,v)) \|\vec{r}_u \times \vec{r}_v\| \, dA$$
$$= \int_0^1 \int_0^1 (u^2 - 8uv - 2u + 8v + 3)(2\sqrt{6}(1 - u)) \, du \, dv$$
$$= \frac{31}{\sqrt{6}} \approx 12.66 \text{ gm}.$$

Notes:

## Flux

Let a surface $\mathcal{S}$ lie within a vector field $\vec{F}$. One is often interested in measuring the *flux* of $\vec{F}$ across $\mathcal{S}$; that is, measuring "how much of the vector field passes across $\mathcal{S}$." For instance, if $\vec{F}$ represents the velocity field of moving air and $\mathcal{S}$ represents the shape of an air filter, the flux will measure how much air is passing through the filter per unit time.

As flux measures the amount of $\vec{F}$ passing across $\mathcal{S}$, we need to find the "amount of $\vec{F}$ orthogonal to $\mathcal{S}$." Similar to our measure of flux in the plane, this is equal to $\vec{F} \cdot \vec{n}$, where $\vec{n}$ is a unit vector normal to $\mathcal{S}$ at a point. We now consider how to find $\vec{n}$.

Given a smooth parametrization $\vec{r}(u, v)$ of $\mathcal{S}$, the work in the previous section showing the development of our method of computing surface area also shows that $\vec{r}_u(u, v)$ and $\vec{r}_v(u, v)$ are tangent to $\mathcal{S}$ at $\vec{r}(u, v)$. Thus $\vec{r}_u \times \vec{r}_v$ is orthogonal to $\mathcal{S}$, and we let

$$\vec{n} = \frac{\vec{r}_u \times \vec{r}_v}{\|\vec{r}_u \times \vec{r}_v\|},$$

which is a unit vector normal to $\mathcal{S}$ at $\vec{r}(u, v)$.

The measurement of flux across a surface is a surface integral; that is, to measure total flux we sum the product of $\vec{F} \cdot \vec{n}$ times a small amount of surface area: $\vec{F} \cdot \vec{n}\, dS$.

A nice thing happens with the actual computation of flux: the $\|\vec{r}_u \times \vec{r}_v\|$ terms go away. Consider:

$$\begin{aligned}
\text{Flux} &= \iint_{\mathcal{S}} \vec{F} \cdot \vec{n}\, dS \\
&= \iint_R \vec{F} \cdot \frac{\vec{r}_u \times \vec{r}_v}{\|\vec{r}_u \times \vec{r}_v\|} \|\vec{r}_u \times \vec{r}_v\|\, dA \\
&= \iint_R \vec{F} \cdot (\vec{r}_u \times \vec{r}_v)\, dA.
\end{aligned}$$

The above only makes sense if $\mathcal{S}$ is orientable; the normal vectors $\vec{n}$ must vary continuously across $\mathcal{S}$. We assume that $\vec{n}$ does vary continuously. (If the parametrization $\vec{r}$ of $\mathcal{S}$ is smooth, then our above definition of $\vec{n}$ will vary continuously.)

Notes:

## Chapter 14 Vector Analysis

> **Definition 14.6.2**     **Flux over a surface**
>
> Let $\vec{F}$ be a vector field with continuous components defined on an orientable surface $\mathcal{S}$ with normal vector $\vec{n}$. The **flux** of $\vec{F}$ across $\mathcal{S}$ is
>
> $$\text{Flux} = \iint_{\mathcal{S}} \vec{F} \cdot \vec{n}\, dS.$$
>
> If $\mathcal{S}$ is parametrized by $\vec{r}(u,v)$, which is smooth on its domain $R$, then
>
> $$\text{Flux} = \iint_R \vec{F}\bigl(\vec{r}(u,v)\bigr) \cdot (\vec{r}_u \times \vec{r}_v)\, dA.$$

Since $\mathcal{S}$ is orientable, we adopt the convention of saying one passes from the "back" side of $\mathcal{S}$ to the "front" side when moving across the surface parallel to the direction of $\vec{n}$. Also, when $\mathcal{S}$ is closed, it is natural to speak of the regions of space "inside" and "outside" $\mathcal{S}$. We also adopt the convention that when $\mathcal{S}$ is a closed surface, $\vec{n}$ should point to the outside of $\mathcal{S}$. If $\vec{n} = \vec{r}_u \times \vec{r}_v$ points inside $\mathcal{S}$, use $\vec{n} = \vec{r}_v \times \vec{r}_u$ instead.

When the computation of flux is positive, it means that the field is moving from the back side of $\mathcal{S}$ to the front side; when flux is negative, it means the field is moving opposite the direction of $\vec{n}$, and is moving from the front of $\mathcal{S}$ to the back. When $\mathcal{S}$ is not closed, there is not a "right" and "wrong" direction in which $\vec{n}$ should point, but one should be mindful of its direction to make full sense of the flux computation.

We demonstrate the computation of flux, and its interpretation, in the following examples.

**Example 14.6.2**     **Finding flux across a surface**

Let $\mathcal{S}$ be the surface given in Example 14.6.1, where $\mathcal{S}$ is parametrized by $\vec{r}(u,v) = \langle u, v(2-2u), 3-2u-v(2-2u)\rangle$ on $0 \leq u \leq 1$, $0 \leq v \leq 1$, and let $\vec{F} = \langle 1, x, -y\rangle$, as shown in Figure 14.6.2. Find the flux of $\vec{F}$ across $\mathcal{S}$.

**Solution**     Using our work from the previous example, we have $\vec{n} = \vec{r}_u \times \vec{r}_v = \langle 4-4u, 2-2u, 2-2u\rangle$. We also need $\vec{F}(\vec{r}(u,v)) = \langle 1, u, -v(2-2u)\rangle$.

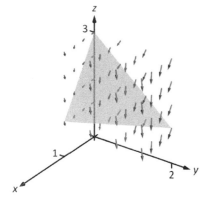

Figure 14.6.2: The surface and vector field used in Example 14.6.2.

Notes:

## 14.6 Surface Integrals

Thus the flux of $\vec{F}$ across $\mathcal{S}$ is:

$$\text{Flux} = \iint_{\mathcal{S}} \vec{F} \cdot \vec{n}\, dS$$
$$= \iint_R \langle 1, u, -v(2-2u)\rangle \cdot \langle 4-4u, 2-2u, 2-2u\rangle \, dA$$
$$= \int_0^1 \int_0^1 \left(-4u^2v - 2u^2 + 8uv - 2u - 4v + 4\right) du\, dv$$
$$= 5/3.$$

To make full use of this numeric answer, we need to know the direction in which the field is passing across $\mathcal{S}$. The graph in Figure 14.6.2 helps, but we need a method that is not dependent on a graph.

Pick a point $(u, v)$ in the interior of $R$ and consider $\vec{n}(u, v)$. For instance, choose $(1/2, 1/2)$ and look at $\vec{n}(1/2, 1/2) = \langle 2, 1, 1\rangle/\sqrt{6}$. This vector has positive *x, y* and *z* components. Generally speaking, one has *some* idea of what the surface $\mathcal{S}$ looks like, as that surface is for some reason important. In our case, we know $\mathcal{S}$ is a plane with *z*-intercept of $z = 3$. Knowing $\vec{n}$ and the flux measurement of positive $5/3$, we know that the field must be passing from "behind" $\mathcal{S}$, i.e., the side the origin is on, to the "front" of $\mathcal{S}$.

**Example 14.6.3    Flux across surfaces with shared boundaries**
Let $\mathcal{S}_1$ be the unit disk in the *x-y* plane, and let $\mathcal{S}_2$ be the paraboloid $z = 1 - x^2 - y^2$, for $z \geq 0$, as graphed in Figure 14.6.3. Note how these two surfaces each have the unit circle as a boundary.

Let $\vec{F}_1 = \langle 0, 0, 1\rangle$ and $\vec{F}_2 = \langle 0, 0, z\rangle$. Using normal vectors for each surface that point "upward," i.e., with a postive *z*-component, find the flux of each field across each surface.

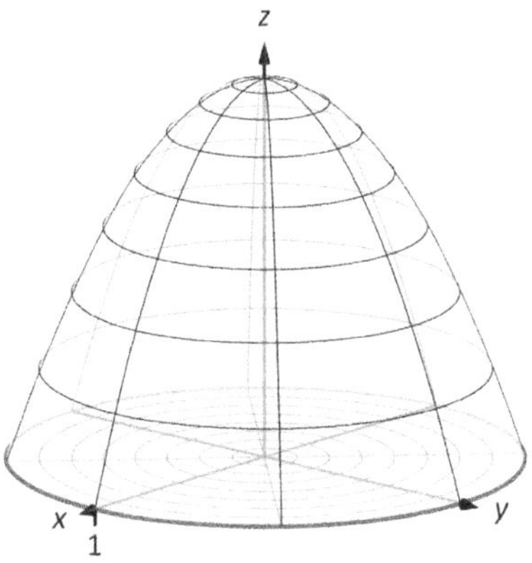

**SOLUTION**    We begin by parameterizing each surface.
The boundary of the unit disk in the *x-y* plane is the unit circle, which can be described with $\langle \cos u, \sin u, 0\rangle$, $0 \leq u \leq 2\pi$. To obtain the interior of the circle as well, we can scale by *v*, giving

$$\vec{r}_1(u, v) = \langle v\cos u, v\sin u, 0\rangle, \quad 0 \leq u \leq 2\pi \quad 0 \leq v \leq 1.$$

As the boundary of $\mathcal{S}_2$ is also the unit circle, the *x* and *y* components of $\vec{r}_2$ will be the same as those of $\vec{r}_1$; we just need a different *z* component. With $z = 1 - x^2 - y^2$, we have

$$\vec{r}_2(u, v) = \langle v\cos u, v\sin u, 1 - v^2\cos^2 u - v^2\sin^2 u\rangle = \langle v\cos u, v\sin u, 1 - v^2\rangle,$$

where $0 \leq u \leq 2\pi$ and $0 \leq v \leq 1$.

Figure 14.6.3: The surfaces used in Example 14.6.3.

Notes:

We now compute the normal vectors $\vec{n}_1$ and $\vec{n}_2$.
For $\vec{n}_1$: $\vec{r}_{1u} = \langle -v\sin u, v\cos u, 0\rangle$, $\vec{r}_{1v} = \langle \cos u, \sin u, 0\rangle$, so

$$\vec{n}_1 = \vec{r}_{1u} \times \vec{r}_{1v} = \langle 0, 0, -v\rangle.$$

As this vector has a negative $z$-component, we instead use

$$\vec{n}_1 = \vec{r}_{1v} \times \vec{r}_{1u} = \langle 0, 0, v\rangle.$$

Similarly, $\vec{n}_2$: $\vec{r}_{2u} = \langle -v\sin u, v\cos u, 0\rangle$, $\vec{r}_{2v} = \langle \cos u, \sin u, -2v\rangle$, so

$$\vec{n}_2 = \vec{r}_{2u} \times \vec{r}_{2v} = \langle -2v^2\cos u, -2v^2\sin u, -v\rangle.$$

Again, this normal vector has a negative $z$-component so we use

$$\vec{n}_2 = \vec{r}_{2v} \times \vec{r}_{2u} = \langle 2v^2\cos u, 2v^2\sin u, v\rangle.$$

We are now set to compute flux. Over field $\vec{F}_1 = \langle 0, 0, 1\rangle$:

$$\begin{aligned}
\text{Flux across } \mathcal{S}_1 &= \iint_{\mathcal{S}_1} \vec{F}_1 \cdot \vec{n}_1 \, dS \\
&= \iint_R \langle 0,0,1\rangle \cdot \langle 0,0,v\rangle \, dA \\
&= \int_0^1 \int_0^{2\pi} (v) \, du \, dv \\
&= \pi.
\end{aligned}$$

$$\begin{aligned}
\text{Flux across } \mathcal{S}_2 &= \iint_{\mathcal{S}_2} \vec{F}_1 \cdot \vec{n}_2 \, dS \\
&= \iint_R \langle 0,0,1\rangle \cdot \langle 2v^2\cos u, 2v^2\sin u, v\rangle \, dA \\
&= \int_0^1 \int_0^{2\pi} (v) \, du \, dv \\
&= \pi.
\end{aligned}$$

These two results are equal and positive. Each are positive because both normal vectors are pointing in the positive $z$-directions, as does $\vec{F}_1$. As the field passes through each surface in the direction of their normal vectors, the flux is measured as positive.

We can also intuitively understand why the results are equal. Consider $\vec{F}_1$ to represent the flow of air, and let each surface represent a filter. Since $\vec{F}_1$ is

Notes:

constant, and moving "straight up," it makes sense that all air passing through $S_1$ also passes through $S_2$, and vice–versa.

If we treated the surfaces as creating one piecewise–smooth surface $S$, we would find the total flux across $S$ by finding the flux across each piece, being sure that each normal vector pointed to the outside of the closed surface. Above, $\vec{n}_1$ does not point outside the surface, though $\vec{n}_2$ does. We would instead want to use $-\vec{n}_1$ in our computation. We would then find that the flux across $S_1$ is $-\pi$, and hence the total flux across $S$ is $-\pi + \pi = 0$. (As 0 is a special number, we should wonder if this answer has special significance. It does, which is briefly discussed following this example and will be more fully developed in the next section.)

We now compute the flux across each surface with $\vec{F}_2 = \langle 0, 0, z \rangle$:

$$\text{Flux across } S_1 = \iint_{S_1} \vec{F}_2 \cdot \vec{n}_1 \, dS.$$

Over $S_1$, $\vec{F}_2 = \vec{F}_2(\vec{r}_2(u,v)) = \langle 0, 0, 0 \rangle$. Therefore,

$$= \iint_R \langle 0, 0, 0 \rangle \cdot \langle 0, 0, v \rangle \, dA$$
$$= \int_0^1 \int_0^{2\pi} (0) \, du \, dv$$
$$= 0.$$

$$\text{Flux across } S_2 = \iint_{S_2} \vec{F}_2 \cdot \vec{n}_2 \, dS.$$

Over $S_2$, $\vec{F}_2 = \vec{F}_2(\vec{r}_2(u,v)) = \langle 0, 0, 1-v^2 \rangle$. Therefore,

$$= \iint_R \langle 0, 0, 1-v^2 \rangle \cdot \langle 2v^2 \cos u, 2v^2 \sin u, v \rangle \, dA$$
$$= \int_0^1 \int_0^{2\pi} (v^3 - v) \, du \, dv$$
$$= \pi/2.$$

This time the measurements of flux differ. Over $S_1$, the field $\vec{F}_2$ is just $\vec{0}$, hence there is no flux. Over $S_2$, the flux is again positive as $\vec{F}_2$ points in the positive $z$ direction over $S_2$, as does $\vec{n}_2$.

Notes:

In the previous example, the surfaces $S_1$ and $S_2$ form a closed surface that is piecewise smooth. That the measurement of flux across each surface was the same for some fields (and not for others) is reminiscent of a result from Section 14.4, where we measured flux across curves. The quick answer to why the flux was the same when considering $\vec{F}_1$ is that div $\vec{F}_1 = 0$. In the next section, we'll see the second part of the Divergence Theorem which will more fully explain this occurrence. We will also explore Stokes' Theorem, the spatial analogue to Green's Theorem.

Notes:

# Exercises 14.6

## Terms and Concepts

1. In the plane, flux is a measurement of how much of the vector field passes across a _____; in space, flux is a measurement of how much of the vector field passes across a _____.

2. When computing flux, what does it mean when the result is a negative number?

3. When $S$ is a closed surface, we choose the normal vector so that it points to the _____ of the surface.

4. If $S$ is a plane, and $\vec{F}$ is always parallel to $S$, then the flux of $\vec{F}$ across $S$ will be _____.

## Problems

In Exercises 5 – 6, a surface $S$ that represents a thin sheet of material with density $\delta$ is given. Find the mass of each thin sheet.

5. $S$ is the plane $f(x,y) = x + y$ on $-2 \leq x \leq 2$, $-3 \leq y \leq 3$, with $\delta(x, y, z) = z$.

6. $S$ is the unit sphere, with $\delta(x, y, z) = x + y + z + 10$.

In Exercises 7 – 14, a surface $S$ and a vector field $\vec{F}$ are given. Compute the flux of $\vec{F}$ across $S$. (If $S$ is not a closed surface, choose $\vec{n}$ so that it has a positive z-component, unless otherwise indicated.)

7. $S$ is the plane $f(x, y) = 3x + y$ on $0 \leq x \leq 1$, $1 \leq y \leq 4$; $\vec{F} = \langle x^2, -z, 2y \rangle$.

8. $S$ is the plane $f(x, y) = 8 - x - y$ over the triangle with vertices at $(0, 0)$, $(1, 0)$ and $(1, 5)$; $\vec{F} = \langle 3, 1, 2 \rangle$.

9. $S$ is the paraboloid $f(x, y) = x^2 + y^2$ over the unit disk; $\vec{F} = \langle 1, 0, 0 \rangle$.

10. $S$ is the unit sphere; $\vec{F} = \langle y - z, z - x, x - y \rangle$.

11. $S$ is the square in space with corners at $(0, 0, 0)$, $(1, 0, 0)$, $(1, 0, 1)$ and $(0, 0, 1)$ (choose $\vec{n}$ such that it has a positive y-component); $\vec{F} = \langle 0, -z, y \rangle$.

12. $S$ is the disk in the y-z plane with radius 1, centered at $(0, 1, 1)$ (choose $\vec{n}$ such that it has a positive x-component); $\vec{F} = \langle y, z, x \rangle$.

13. $S$ is the closed surface composed of $S_1$, whose boundary is the ellipse in the x-y plane described by $\frac{x^2}{25} + \frac{y^2}{9} = 1$ and $S_2$, part of the elliptical paraboloid $f(x, y) = 1 - \frac{x^2}{25} - \frac{y^2}{9}$ (see graph); $\vec{F} = \langle 5, 2, 3 \rangle$.

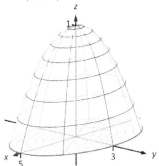

14. $S$ is the closed surface composed of $S_1$, part of the unit sphere and $S_2$, part of the plane $z = 1/2$ (see graph); $\vec{F} = \langle x, -y, z \rangle$.

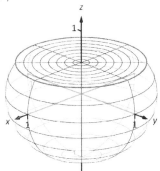

## 14.7 The Divergence Theorem and Stokes' Theorem

### The Divergence Theorem

Theorem 14.4.2 gives the Divergence Theorem in the plane, which states that the flux of a vector field across a closed *curve* equals the sum of the divergences over the region enclosed by the curve. Recall that the flux was measured via a line integral, and the sum of the divergences was measured through a double integral.

We now consider the three-dimensional version of the Divergence Theorem. It states, in words, that the flux across a closed *surface* equals the sum of the divergences over the domain enclosed by the surface. Since we are in space (versus the plane), we measure flux via a surface integral, and the sums of divergences will be measured through a triple integral.

**Note:** the term "outer unit normal vector" used in Theorem 14.7.1 means $\vec{n}$ points to the outside of $S$.

---

**Theorem 14.7.1**     **The Divergence Theorem (in space)**

Let $D$ be a closed domain in space whose boundary is an orientable, piecewise smooth surface $S$ with outer unit normal vector $\vec{n}$, and let $\vec{F}$ be a vector field whose components are differentiable on $D$. Then

$$\iint_S \vec{F} \cdot \vec{n}\, dS = \iiint_D \operatorname{div} \vec{F}\, dV.$$

---

**Example 14.7.1**     **Using the Divergence Theorem in space**

Let $D$ be the domain in space bounded by the planes $z = 0$ and $z = 2x$, along with the cylinder $x = 1 - y^2$, as graphed in Figure 14.7.1, let $S$ be the boundary of $D$, and let $\vec{F} = \langle x + y, y^2, 2z \rangle$.

Verify the Divergence Theorem by finding the total outward flux of $\vec{F}$ across $S$, and show this is equal to $\iiint_D \operatorname{div} \vec{F}\, dV$.

**SOLUTION**     The surface $S$ is piecewise smooth, comprising surfaces $S_1$, which is part of the plane $z = 2x$, surface $S_2$, which is part of the cylinder $x = 1 - y^2$, and surface $S_3$, which is part of the plane $z = 0$. To find the total outward flux across $S$, we need to compute the outward flux across each of these three surfaces.

We leave it to the reader to confirm that surfaces $S_1$, $S_2$ and $S_3$ can be pa-

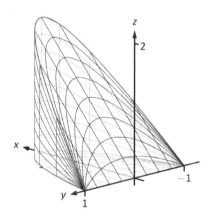

Figure 14.7.1: The surfaces used in Example 14.7.1.

Notes:

rameterized by $\vec{r}_1$, $\vec{r}_2$ and $\vec{r}_3$ respectively as

$$\vec{r}_1(u,v) = \langle v(1-u^2), u, 2v(1-u^2) \rangle,$$
$$\vec{r}_2(u,v) = \langle (1-u^2), u, 2v(1-u^2) \rangle,$$
$$\vec{r}_3(u,v) = \langle v(1-u^2), u, 0 \rangle,$$

where $-1 \leq u \leq 1$ and $0 \leq v \leq 1$ for all three functions.

We compute a unit normal vector $\vec{n}$ for each as $\frac{\vec{r}_u \times \vec{r}_v}{||\vec{r}_u \times \vec{r}_v||}$, though recall that as we are integrating $\vec{F} \cdot \vec{n}\, dS$, we actually only use $\vec{r}_u \times \vec{r}_v$. Finally, in previous flux computations, it did not matter which direction $\vec{n}$ pointed as long as we made note of its direction. When using the Divergence Theorem, we need $\vec{n}$ to point to the outside of the closed surface, so in practice this means we'll either use $\vec{r}_u \times \vec{r}_v$ or $\vec{r}_v \times \vec{r}_u$, depending on which points outside of the closed surface $\mathcal{S}$.

We leave it to the reader to confirm the following cross products and integrations are correct.

For $\mathcal{S}_1$, we need to use $\vec{r}_{1v} \times \vec{r}_{1u} = \langle 2(u^2-1), 0, 1-u^2 \rangle$. (Note the z-component is nonnegative as $u \leq 1$, therefore this vector always points up, meaning *to the outside*, of $\mathcal{S}$.) The flux across $\mathcal{S}_1$ is:

$$\text{Flux across } \mathcal{S}_1 := \iint_{\mathcal{S}_1} \vec{F} \cdot \vec{n}_1\, dS$$
$$= \int_0^1 \int_{-1}^1 \vec{F}(\vec{r}_1(u,v)) \cdot (\vec{r}_{1v} \times \vec{r}_{1u})\, du\, dv$$
$$= \int_0^1 \int_{-1}^1 \langle v(1-u^2)+u, u^2, 4v(1-u^2) \rangle \cdot \langle 2(u^2-1), 0, 1-u^2 \rangle\, du\, dv$$
$$= \int_0^1 \int_{-1}^1 (2u^4 v + 2u^3 - 4u^2 v - 2u + 2v)\, du\, dv$$
$$= \frac{16}{15}.$$

For $\mathcal{S}_2$, we use $\vec{r}_{2u} \times \vec{r}_{2v} = \langle 2(1-u^2), 4u(1-u^2), 0 \rangle$. (Note the x-component is always nonnegative, meaning this vector points outside $\mathcal{S}$.) The flux across $\mathcal{S}_2$

Notes:

is:

$$\text{Flux across } S_2 := \iint_{S_2} \vec{F} \cdot \vec{n}_2 \, dS$$

$$= \int_0^1 \int_{-1}^1 \vec{F}(\vec{r}_2(u,v)) \cdot (\vec{r}_{2u} \times \vec{r}_{2v}) \, du \, dv$$

$$= \int_0^1 \int_{-1}^1 \langle 1 - u^2 + u, u^2, 4v(1-u^2) \rangle \cdot \langle 2(1-u^2), 4u(1-u^2), 0 \rangle \, du\, dv$$

$$= \int_0^1 \int_{-1}^1 \left( 4u^5 - 2u^4 - 2u^3 + 4u^2 - 2u - 2 \right) du\, dv$$

$$= \frac{32}{15}.$$

For $S_3$, we use $\vec{r}_{3u} \times \vec{r}_{3v} = \langle 0, 0, u^2 - 1 \rangle$. (Note the z-component is never positive, meaning this vector points down, outside of $S$.) The flux across $S_3$ is:

$$\text{Flux across } S_3 := \iint_{S_3} \vec{F} \cdot \vec{n}_3 \, dS$$

$$= \int_0^1 \int_{-1}^1 \vec{F}(\vec{r}_3(u,v)) \cdot (\vec{r}_{3u} \times \vec{r}_{3v}) \, du\, dv$$

$$= \int_0^1 \int_{-1}^1 \langle v(1-u^2) + u, u^2, 0 \rangle \cdot \langle 0, 0, u^2 - 1 \rangle \, du\, dv$$

$$= \int_0^1 \int_{-1}^1 0 \, du\, dv$$

$$= 0.$$

Thus the total outward flux, measured by surface integrals across all three component surfaces of $S$, is $16/15 + 32/15 + 0 = 48/15 = 16/5 = 3.2$. We now find the total outward flux by integrating div $\vec{F}$ over $D$.

Following the steps outlined in Section 13.6, we see the bounds of $x$, $y$ and $z$ can be set as (thinking "surface to surface, curve to curve, point to point"):

$$0 \leq z \leq 2x; \quad 0 \leq x \leq 1 - y^2; \quad -1 \leq y \leq 1.$$

With div $\vec{F} = 1 + 2y + 2 = 2y + 3$, we find the total outward flux of $\vec{F}$ over $S$ as:

$$\text{Flux} = \iiint_D \text{div}\, \vec{F} \, dV = \int_{-1}^1 \int_0^{1-y^2} \int_0^{2x} (2y + 3) \, dz\, dx\, dy = 16/5,$$

the same result we obtained previously.

Notes:

## 14.7 The Divergence Theorem and Stokes' Theorem

In Example 14.7.1 we see that the total outward flux of a vector field across a closed surface can be found two different ways because of the Divergence Theorem. One computation took far less work to obtain. In that particular case, since $S$ was comprised of three separate surfaces, it was far simpler to compute one triple integral than three surface integrals (each of which required partial derivatives and a cross product). In practice, if outward flux needs to be measured, one would choose only one method. We will use both methods in this section simply to reinforce the truth of the Divergence Theorem.

We practice again in the following example.

### Example 14.7.2 Using the Divergence Theorem in space

Let $S$ be the surface formed by the paraboloid $z = 1 - x^2 - y^2$, $z \geq 0$, and the unit disk centered at the origin in the $x$-$y$ plane, graphed in Figure 14.7.2, and let $\vec{F} = \langle 0, 0, z \rangle$. (This surface and vector field were used in Example 14.6.3.)

Verify the Divergence Theorem; find the total outward flux across $S$ and evaluate the triple integral of div $\vec{F}$, showing that these two quantities are equal.

**SOLUTION**   We find the flux across $S$ first. As $S$ is piecewise–smooth, we decompose it into smooth components $S_1$, the disk, and $S_2$, the paraboloid, and find the flux across each.

In Example 14.6.3, we found the flux across $S_1$ is 0. We also found that the flux across $S_2$ is $\pi/2$. (In that example, the normal vector had a positive $z$ component hence was an outer normal.) Thus the total outward flux is $0 + \pi/2 = \pi/2$.

We now compute $\iiint_D \text{div} \, \vec{F} \, dV$. We can describe $D$ as the domain bounded by (think "surface to surface, curve to curve, point to point"):

$$0 \leq z \leq 1 - x^2 - y^2, \quad -\sqrt{1-x^2} \leq y \leq \sqrt{1-x^2}, \quad -1 \leq x \leq 1.$$

This description of $D$ is not very easy to integrate. With polar, we can do better. Let $R$ represent the unit disk, which can be described in polar simply as $r$, where $0 \leq r \leq 1$ and $0 \leq \theta \leq 2\pi$. With $x = r\cos\theta$ and $y = r\sin\theta$, the surface $S_2$ becomes

$$z = 1 - x^2 - y^2 \Rightarrow 1 - (r\cos\theta)^2 - (r\sin\theta)^2 \Rightarrow 1 - r^2.$$

Thus $D$ can be described as the domain bounded by:

$$0 \leq z \leq 1 - r^2, \quad 0 \leq r \leq 1, \quad 0 \leq \theta \leq 2\pi.$$

With div $\vec{F} = 1$, we can integrate, recalling that $dV = r \, dz \, dr \, d\theta$:

$$\iiint_D \text{div} \, \vec{F} \, dV = \int_0^{2\pi} \int_0^1 \int_0^{1-r^2} r \, dz \, dr \, d\theta = \frac{\pi}{2},$$

which matches our flux computation above.

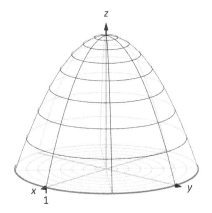

Figure 14.7.2: The surfaces used in Example 14.7.2.

Notes:

## Chapter 14 Vector Analysis

**Example 14.7.3    A "paradox" of the Divergence Theorem and Gauss's Law**

The magnitude of many physical quantities (such as light intensity or electromagnetic and gravitational forces) follow an "inverse square law": the magnitude of the quantity at a point is inversely proportional to the square of the distance to the source of the quantity.

Let a point light source be placed at the origin and let $\vec{F}$ be the vector field which describes the intensity and direction of the emanating light. At a point $(x, y, z)$, the unit vector describing the direction of the light passing through that point is $\langle x, y, z \rangle / \sqrt{x^2 + y^2 + z^2}$. As the intensity of light follows the inverse square law, the magnitude of $\vec{F}$ at $(x, y, z)$ is $k/(x^2 + y^2 + z^2)$ for some constant $k$. Taken together,

$$\vec{F}(x, y, z) = \frac{k}{(x^2 + y^2 + z^2)^{3/2}} \langle x, y, z \rangle.$$

Consider the cube, centered at the origin, with sides of length $2a$ for some $a > 0$ (hence corners of the cube lie at $(a, a, a)$, $(-a, -a, -a)$, etc., as shown in Figure 14.7.3). Find the flux across the six faces of the cube and compare this to $\iiint_D \text{div}\,\vec{F}\,dV$.

**SOLUTION**    Let $\mathcal{S}_1$ be the "top" face of the cube, which can be parametrized by $\vec{r}(u, v) = \langle u, v, a \rangle$ for $-a \leq u \leq a$, $-a \leq v \leq a$. We leave it to the reader to confirm that $\vec{r}_u \times \vec{r}_v = \langle 0, 0, 1 \rangle$, which points outside of the cube.

The flux across this face is:

$$\text{Flux} = \iint_{\mathcal{S}_1} \vec{F} \cdot \vec{n}\,dS$$

$$= \int_{-a}^{a} \int_{-a}^{a} \vec{F}(\vec{r}(u, v)) \cdot (\vec{r}_u \times \vec{r}_v)\,du\,dv$$

$$= \int_{-a}^{a} \int_{-a}^{a} \frac{k\,a}{(u^2 + v^2 + a^2)^{3/2}}\,du\,dv.$$

This double integral is *not* trivial to compute, requiring multiple trigonometric substitutions. This example is not meant to stress integration techniques, so we leave it to the reader to confirm the result is

$$= \frac{2k\pi}{3}.$$

Note how the result is independent of $a$; no matter the size of the cube, the flux through the top surface is always $2k\pi/3$.

An argument of symmetry shows that the flux through each of the six faces is $2k\pi/3$, thus the total flux through the faces of the cube is $6 \times 2k\pi/3 = 4k\pi$.

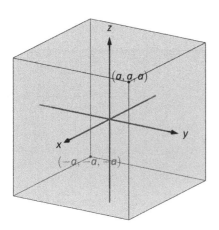

Figure 14.7.3: The cube used in Example 14.7.3.

Notes:

## 14.7 The Divergence Theorem and Stokes' Theorem

It takes a bit of algebra, but we can show that div $\vec{F} = 0$. Thus the Divergence Theorem would seem to imply that the total flux through the faces of the cube should be

$$\text{Flux} = \iiint_D \text{div}\, \vec{F}\, dV = \iiint_D 0\, dV = 0,$$

but clearly this does not match the result from above. What went wrong?

Revisit the statement of the Divergence Theorem. One of the conditions is that the components of $\vec{F}$ must be differentiable on the domain enclosed by the surface. In our case, $\vec{F}$ is *not* differentiable at the origin – it is not even defined! As $\vec{F}$ does not satisfy the conditions of the Divergence Theorem, it does not apply, and we cannot expect $\iint_S \vec{F} \cdot \vec{n}\, dA = \iiint_D \text{div}\, \vec{F}\, dV$.

Since $\vec{F}$ is differentiable everywhere except the origin, the Divergence Theorem does apply over any domain that does not include the origin. Let $S_2$ be any surface that encloses the cube used before, and let $\hat{D}$ be the domain *between* the cube and $S_2$; note how $\hat{D}$ does not include the origin and so the Divergence Theorem does apply over this domain. The total outward flux over $\hat{D}$ is thus $\iiint_{\hat{D}} \text{div}\, \vec{F}\, dV = 0$, which means the amount of flux coming out of $S_2$ is the same as the amount of flux coming out of the cube. The conclusion: the flux across *any* surface enclosing the origin will be $4k\pi$.

This has an important consequence in electrodynamics. Let $q$ be a point charge at the origin. The electric field generated by this point charge is

$$\vec{E} = \frac{q}{4\pi\varepsilon_0} \frac{\langle x, y, z\rangle}{(x^2 + y^2 + z^2)^{3/2}},$$

i.e., it is $\vec{F}$ with $k = q/(4\pi\varepsilon_0)$, where $\varepsilon_0$ is a physical constant (the "permittivity of free space"). Gauss's Law states that the outward flux of $\vec{E}$ across any surface enclosing the origin is $q/\varepsilon_0$.

Our interest in the Divergence Theorem is twofold. First, it's truth alone is interesting: to study the behavior of a vector field across a closed surface, one can examine properties of that field within the surface. Secondly, it offers an alternative way of computing flux. When there are multiple methods of computing a desired quantity, one has power to select the easiest computation as illustrated next.

**Example 14.7.4    Using the Divergence Theorem to compute flux**
Let $S$ be the cube bounded by the planes $x = \pm 1, y = \pm 1, z = \pm 1$, and let $\vec{F} = \langle x^2 y, 2yz, x^2 z^3 \rangle$. Compute the outward flux of $\vec{F}$ over $S$.

**Solution**    We compute div $\vec{F} = 2xy + 2z + 3x^2 z^2$. By the Divergence Theorem, the outward flux is the triple integral over the domain $D$ enclosed by

Notes:

$S$:

$$\text{Outward flux:} \quad \int_{-1}^{1}\int_{-1}^{1}\int_{-1}^{1} (2xy + 2z + 3x^2z^2)\, dz\, dy\, dx = \frac{8}{3}.$$

The direct flux computation requires six surface integrals, one for each face of the cube. The Divergence Theorem offers a much more simple computation.

## Stokes' Theorem

Just as the spatial Divergence Theorem of this section is an extension of the planar Divergence Theorem, Stokes' Theorem is the spatial extension of Green's Theorem. Recall that Green's Theorem states that the circulation of a vector field around a closed curve in the plane is equal to the sum of the curl of the field over the region enclosed by the curve. Stokes' Theorem effectively makes the same statement: given a closed curve that lies on a surface $S$, the circulation of a vector field around that curve is the same as the sum of "the curl of the field" across the enclosed surface. We use quotes around "the curl of the field" to signify that this statement is not quite correct, as we do not sum curl $\vec{F}$, but curl $\vec{F} \cdot \vec{n}$, where $\vec{n}$ is a unit vector normal to $S$. That is, we sum the portion of curl $\vec{F}$ that is orthogonal to $S$ at a point.

Green's Theorem dictated that the curve was to be traversed counterclockwise when measuring circulation. Stokes' Theorem will follow a right hand rule: when the thumb of one's right hand points in the direction of $\vec{n}$, the path $C$ will be traversed in the direction of the curling fingers of the hand (this is equivalent to traversing counterclockwise in the plane).

---

**Theorem 14.7.2     Stokes' Theorem**

Let $S$ be a piecewise smooth, orientable surface whose boundary is a piecewise smooth curve $C$, let $\vec{n}$ be a unit vector normal to $S$, let $C$ be traversed with respect to $\vec{n}$ according to the right hand rule, and let the components of $\vec{F}$ have continuous first partial derivatives over $S$. Then

$$\oint_C \vec{F} \cdot d\vec{r} = \iint_S (\operatorname{curl} \vec{F}) \cdot \vec{n}\, dS.$$

---

In general, the best approach to evaluating the surface integral in Stokes' Theorem is to parametrize the surface $S$ with a function $\vec{r}(u, v)$. We can find a unit normal vector $\vec{n}$ as

$$\vec{n} = \frac{\vec{r}_u \times \vec{r}_v}{\|\vec{r}_u \times \vec{r}_v\|}.$$

---

Notes:

Since $dS = \|\vec{r}_u \times \vec{r}_v\| \, dA$, the surface integral in practice is evaluated as

$$\iint_{\mathcal{S}} (\text{curl } \vec{F}) \cdot (\vec{r}_u \times \vec{r}_v) \, dA,$$

where $\vec{r}_u \times \vec{r}_v$ may be replaced by $\vec{r}_v \times \vec{r}_u$ to properly match the direction of this vector with the orientation of the parameterization of C.

**Example 14.7.5   Verifying Stokes' Theorem**
Considering the planar surface $f(x, y) = 7 - 2x - 2y$, let C be the curve in space that lies on this surface above the circle of radius 1 and centered at $(1, 1)$ in the x-y plane, let $\mathcal{S}$ be the planar region enclosed by C, as illustrated in Figure 14.7.4, and let $\vec{F} = \langle x+y, 2y, y^2 \rangle$. Verify Stoke's Theorem by showing $\oint_C \vec{F} \cdot d\vec{r} = \iint_{\mathcal{S}} (\text{curl } \vec{F}) \cdot \vec{n} \, dS$.

**SOLUTION**    We begin by parameterizing C and then find the circulation. A unit circle centered at $(1, 1)$ can be parametrized with $x = \cos t + 1$, $y = \sin t + 1$ on $0 \leq t \leq 2\pi$; to put this curve on the surface $f$, make the z component equal $f(x, y)$: $z = 7 - 2(\cos t + 1) - 2(\sin t + 1) = 3 - 2\cos t - 2\sin t$. All together, we parametrize C with $\vec{r}(t) = \langle \cos t + 1, \sin t + 1, 3 - 2\cos t - 2\sin t \rangle$.

The circulation of $\vec{F}$ around C is

$$\oint_C \vec{F} \cdot d\vec{r} = \int_0^{2\pi} \vec{F}(\vec{r}(t)) \cdot \vec{r}'(t) \, dt$$
$$= \int_0^{2\pi} \left( 2\sin^3 t - 2\cos t \sin^2 t + 3\sin^2 t - 3\cos t \sin t \right) dt$$
$$= 3\pi.$$

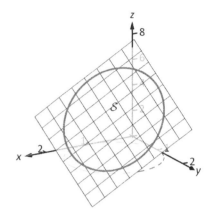

Figure 14.7.4: As given in Example 14.7.5, the surface $\mathcal{S}$ is the portion of the plane bounded by the curve.

We now parametrize $\mathcal{S}$. (We reuse the letter "r" for our surface as this is our custom.) Based on the parametrization of C above, we describe $\mathcal{S}$ with $\vec{r}(u, v) = \langle v\cos u + 1, v\sin u + 1, 3 - 2v\cos u - 2v\sin u \rangle$, where $0 \leq u \leq 2\pi$ and $0 \leq v \leq 1$.

We leave it to the reader to confirm that $\vec{r}_u \times \vec{r}_v = \langle 2v, 2v, v \rangle$. As $0 \leq v \leq 1$, this vector always has a non-negative z-component, which the right–hand rule requires given the orientation of C used above. We also leave it to the reader to confirm $\text{curl } \vec{F} = \langle 2y, 0, -1 \rangle$.

The surface integral of Stokes' Theorem is thus

$$\iint_{\mathcal{S}} (\text{curl } \vec{F}) \cdot \vec{n} \, dS = \iint_{\mathcal{S}} (\text{curl } \vec{F}) \cdot (\vec{r}_u \times \vec{r}_v) \, dA$$
$$= \int_0^1 \int_0^{2\pi} \langle 2v\sin u + 2, 0, -1 \rangle \cdot \langle 2v, 2v, v \rangle \, du \, dv$$
$$= 3\pi,$$

which matches our previous result.

Notes:

## Chapter 14 Vector Analysis

One of the interesting results of Stokes' Theorem is that if two surfaces $S_1$ and $S_2$ share the same boundary, then $\iint_{S_1}(\text{curl }\vec{F})\cdot\vec{n}\,dS = \iint_{S_2}(\text{curl }\vec{F})\cdot\vec{n}\,dS$. That is, the value of these two surface integrals is somehow independent of the interior of the surface. We demonstrate this principle in the next example.

**Example 14.7.6    Stokes' Theorem and surfaces that share a boundary**
Let $C$ be the curve given in Example 14.7.5 and note that it lies on the surface $z = 6 - x^2 - y^2$. Let $S$ be the region of this surface bounded by $C$, and let $\vec{F} = \langle x + y, 2y, y^2\rangle$ as in the previous example. Compute $\iint_S(\text{curl }\vec{F})\cdot\vec{n}\,dS$ to show it equals the result found in the previous example.

**Solution**    We begin by demonstrating that $C$ lies on the surface $z = 6 - x^2 - y^2$. We can parametrize the $x$ and $y$ components of $C$ with $x = \cos t + 1$, $y = \sin t + 1$ as before. Lifting these components to the surface $f$ gives the $z$ component as $z = 6 - x^2 - y^2 = 6 - (\cos t + 1)^2 - (\sin t + 1)^2 = 3 - 2\cos t - 2\sin t$, which is the same $z$ component as found in Example 14.7.5. Thus the curve $C$ lies on the surface $z = 6 - x^2 - y^2$, as illustrated in Figure 14.7.5.

Since $C$ and $\vec{F}$ are the same as in the previous example, we already know that $\oint_C \vec{F}\cdot d\vec{r} = 3\pi$. We confirm that this is also the value of $\iint_S(\text{curl }\vec{F})\cdot\vec{n}\,dS$. We parametrize $S$ with

$$\vec{r}(u,v) = \langle v\cos u + 1, v\sin u + 1, 6 - (v\cos u + 1)^2 - (v\sin u + 1)^2\rangle,$$

where $0 \leq u \leq 2\pi$ and $0 \leq v \leq 1$, and leave it to the reader to confirm that

$$\vec{r}_u \times \vec{r}_v = \langle 2v(v\cos u + 1), 2v(v\sin u + 1), v\rangle,$$

which also conforms to the right-hand rule with regard to the orientation of $C$. With curl $\vec{F} = \langle 2y, 0, -1\rangle$ as before, we have

$$\iint_S (\text{curl }\vec{F})\cdot\vec{n}\,dS =$$
$$\int_0^1 \int_0^{2\pi} \langle 2v\sin u + 2, 0, -1\rangle \cdot \langle 2v(v\cos u + 1), 2v(v\sin u + 1), v\rangle\,du\,dv =$$
$$3\pi.$$

Even though the surfaces used in this example and in Example 14.7.5 are very different, because they share the same boundary, Stokes' Theorem guarantees they have equal "sum of curls" across their respective surfaces.

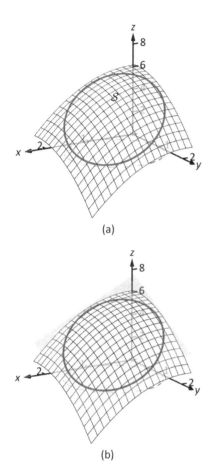

(a)

(b)

Figure 14.7.5: As given in Example 14.7.6, the surface $S$ is the portion of the plane bounded by the curve.

Notes:

## 14.7 The Divergence Theorem and Stokes' Theorem

### A Common Thread of Calculus

We have threefold interest in each of the major theorems of this chapter: the Fundamental Theorem of Line Integrals, Green's, Stokes' and the Divergence Theorems. First, we find the beauty of their truth interesting. Second, each provides two methods of computing a desired quantity, sometimes offering a simpler method of computation.

There is yet one more reason of interest in the major theorems of this chapter. These important theorems also all share an important principle with the Fundamental Theorem of Calculus, introduced in Chapter 5.

Revisit this fundamental theorem, adopting the notation used heavily in this chapter. Let $I$ be the interval $[a, b]$ and let $y = F(x)$ be differentiable on $I$, with $F'(x) = f(x)$. The Fundamental Theorem of Calculus states that

$$\int_I f(x)\, dx = F(b) - F(a).$$

That is, the sum of the rates of change of a function $F$ over an interval $I$ can also be calculated with a certain sum of $F$ itself on the boundary of $I$ (in this case, at the points $x = a$ and $x = b$).

Each of the named theorems above can be expressed in similar terms. Consider the Fundamental Theorem of Line Integrals: given a function $z = f(x, y)$, the gradient $\nabla f$ is a type of rate of change of $f$. Given a curve $C$ with initial and terminal points $A$ and $B$, respectively, this fundamental theorem states that

$$\int_C \nabla f\, ds = f(B) - f(A),$$

where again the sum of a rate of change of $f$ along a curve $C$ can also be evaluated by a certain sum of $f$ at the boundary of $C$ (i.e., the points $A$ and $B$).

Green's Theorem is essentially a special case of Stokes' Theorem, so we consider just Stokes' Theorem here. Recalling that the curl of a vector field $\vec{F}$ is a measure of a rate of change of $\vec{F}$, Stokes' Theorem states that over a surface $\mathcal{S}$ bounded by a closed curve $C$,

$$\iint_\mathcal{S} (\operatorname{curl} \vec{F}) \cdot \vec{n}\, dS = \oint_C \vec{F} \cdot d\vec{r},$$

i.e., the sum of a rate of change of $\vec{F}$ can be calculated with a certain sum of $\vec{F}$ itself over the boundary of $\mathcal{S}$. In this case, the latter sum is also an infinite sum, requiring an integral.

Finally, the Divergence Theorems state that the sum of divergences of a vector field (another measure of a rate of change of $\vec{F}$) over a region can also be computed with a certain sum of $\vec{F}$ over the boundary of that region. When the

---

Notes:

region is planar, the latter sum of $\vec{F}$ is an integral; when the region is spatial, the latter sum of $\vec{F}$ is a double integral.

The common thread among these theorems: the sum of a rate of change of a function over a region can be computed as another sum of the function itself on the boundary of the region. While very general, this is a very powerful and important statement.

Notes:

# Exercises 14.7

## Terms and Concepts

1. What are the differences between the Divergence Theorems of Section 14.4 and this section?

2. What property of a vector field does the Divergence Theorem relate to flux?

3. What property of a vector field does Stokes' Theorem relate to circulation?

4. Stokes' Theorem is the spatial version of what other theorem?

## Problems

**In Exercises 5 – 8, a closed surface $S$ enclosing a domain $D$ and a vector field $\vec{F}$ are given. Verify the Divergence Theorem on $S$; that is, show $\iint_S \vec{F} \cdot \vec{n}\, dS = \iiint_D \operatorname{div} \vec{F}\, dV$.**

5. $S$ is the surface bounding the domain $D$ enclosed by the plane $z = 2 - x/2 - 2y/3$ and the coordinate planes in the first octant; $\vec{F} = \langle x^2, y^2, x \rangle$.

6. $S$ is the surface bounding the domain $D$ enclosed by the cylinder $x^2 + y^2 = 1$ and the planes $z = -3$ and $z = 3$; $\vec{F} = \langle -x, y, z \rangle$.

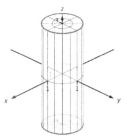

7. $S$ is the surface bounding the domain $D$ enclosed by $z = xy(3-x)(3-y)$ and the plane $z = 0$; $\vec{F} = \langle 3x, 4y, 5z+1 \rangle$.

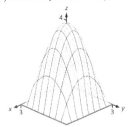

8. $S$ is the surface composed of $S_1$, the paraboloid $z = 4 - x^2 - y^2$ for $z \geq 0$, and $S_2$, the disk of radius 2 centered at the origin; $\vec{F} = \langle x, y, z^2 \rangle$.

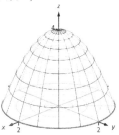

**In Exercises 9 – 12, a closed curve $C$ that is the boundary of a surface $S$ is given along with a vector field $\vec{F}$. Verify Stokes' Theorem on $C$; that is, show $\oint_C \vec{F} \cdot d\vec{r} = \iint_S (\operatorname{curl} \vec{F}) \cdot \vec{n}\, dS$.**

9. $C$ is the curve parametrized by $\vec{r}(t) = \langle \cos t, \sin t, 1 \rangle$ and $S$ is the portion of $z = x^2 + y^2$ enclosed by $C$; $\vec{F} = \langle z, -x, y \rangle$.

10. $C$ is the curve parametrized by $\vec{r}(t) = \langle \cos t, \sin t, e^{-1} \rangle$ and $S$ is the portion of $z = e^{-x^2-y^2}$ enclosed by $C$; $\vec{F} = \langle -y, x, 1 \rangle$.

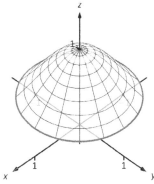

11. $C$ is the curve that follows the triangle with vertices at $(0,0,2)$, $(4,0,0)$ and $(0,3,0)$, traversing the vertices in that order and returning to $(0,0,2)$, and $S$ is the portion of the plane $z = 2 - x/2 - 2y/3$ enclosed by $C$; $\vec{F} = \langle y, -z, y \rangle$.

12. $C$ is the curve whose $x$ and $y$ coordinates follow the parabola $y = 1 - x^2$ from $x = 1$ to $x = -1$, then follow the line from $(-1, 0)$ back to $(1, 0)$, where the $z$ coordinates of $C$ are determined by $f(x, y) = 2x^2 + y^2$, and $S$ is the portion of $z = 2x^2 + y^2$ enclosed by $C$; $\vec{F} = \langle y^2 + z, x, x^2 - y \rangle$.

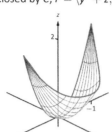

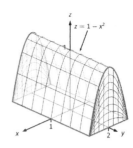

**In Exercises 13 – 16, a closed surface $S$ and a vector field $\vec{F}$ are given. Find the outward flux of $\vec{F}$ over $S$ either through direct computation or through the Divergence Theorem.**

13. $S$ is the surface formed by the intersections of $z = 0$ and $z = (x^2 - 1)(y^2 - 1)$; $\vec{F} = \langle x^2 + 1, yz, xz^2 \rangle$.

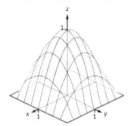

14. $S$ is the surface formed by the intersections of the planes $z = \frac{1}{2}(3 - x)$, $x = 1$, $y = 0$, $y = 2$ and $z = 0$; $\vec{F} = \langle x, y^2, z \rangle$.

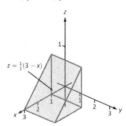

15. $S$ is the surface formed by the intersections of the planes $z = 2y$, $y = 4 - x^2$ and $z = 0$; $\vec{F} = \langle xz, 0, xz \rangle$.

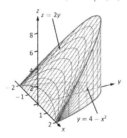

16. $S$ is the surface formed by the intersections of the cylinder $z = 1 - x^2$ and the planes $y = -2$, $y = 2$ and $z = 0$; $\vec{F} = \langle 0, y^3, 0 \rangle$.

**In Exercises 17 – 20, a closed curve $C$ that is the boundary of a surface $S$ is given along with a vector field $\vec{F}$. Find the circulation of $\vec{F}$ around $C$ either through direct computation or through Stokes' Theorem.**

17. $C$ is the curve whose $x$- and $y$-values are determined by the three sides of a triangle with vertices at $(-1, 0)$, $(1, 0)$ and $(0, 1)$, traversed in that order, and the $z$-values are determined by the function $z = xy$; $\vec{F} = \langle z - y^2, x, z \rangle$.

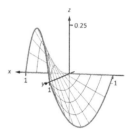

18. $C$ is the curve whose $x$- and $y$-values are given by $\vec{r}(t) = \langle 2\cos t, 2\sin t \rangle$ and the $z$-values are determined by the function $z = x^2 + y^3 - 3y + 1$; $\vec{F} = \langle -y, x, z \rangle$.

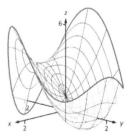

19. $C$ is the curve whose $x$- and $y$-values are given by $\vec{r}(t) = \langle \cos t, 3\sin t \rangle$ and the $z$-values are determined by the function $z = 5 - 2x - y$; $\vec{F} = \langle -\frac{1}{3}y, 3x, \frac{2}{3}y - 3x \rangle$.

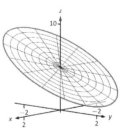

20. C is the curve whose x- and y-values are sides of the square with vertices at $(1, 1)$, $(-1, 1)$, $(-1, -1)$ and $(1, -1)$, traversed in that order, and the z-values are determined by the function $z = 10 - 5x - 2y$; $\vec{F} = \langle 5y^2, 2y^2, y^2 \rangle$.

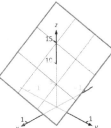

**Exercises 21 – 24 are designed to challenge your understanding and require no computation.**

21. Let $S$ be any closed surface enclosing a domain $D$. Consider $\vec{F}_1 = \langle x, 0, 0 \rangle$ and $\vec{F}_2 = \langle y, y^2, z - 2yz \rangle$.
    These fields are clearly very different. Why is it that the total outward flux of each field across $S$ is the same?

22. (a) Green's Theorem can be used to find the area of a region enclosed by a curve by evaluating a line integral with the appropriate choice of vector field $\vec{F}$. What condition on $\vec{F}$ makes this possible?

    (b) Likewise, Stokes' Theorem can be used to find the surface area of a region enclosed by a curve in space by evaluating a line integral with the appropriate choice of vector field $\vec{F}$. What condition on $\vec{F}$ makes this possible?

23. The Divergence Theorem establishes equality between a particular double integral and a particular triple integral. What types of circumstances would lead one to choose to evaluate the triple integral over the double integral?

24. Stokes' Theorem establishes equality between a particular line integral and a particular double integral. What types of circumstances would lead one to choose to evaluate the double integral over the line integral?

# A: Solutions To Selected Problems

## Chapter 9

### Section 9.1

1. When defining the conics as the intersections of a plane and a double napped cone, degenerate conics are created when the plane intersects the tips of the cones (usually taken as the origin). Nondegenerate conics are formed when this plane does not contain the origin.

3. Hyperbola

5. With a horizontal transverse axis, the $x^2$ term has a positive coefficient; with a vertical transverse axis, the $y^2$ term has a positive coefficient.

7. $y = \frac{1}{2}(x-3)^2 + \frac{3}{2}$

9. $x = -\frac{1}{4}(y-5)^2 + 2$

11. $y = -\frac{1}{4}(x-1)^2 + 2$

13. $y = 4x^2$

15. focus: $(0, 1)$; directrix: $y = -1$. The point $P$ is 2 units from each.

17.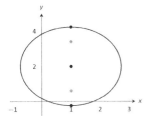

19. $\frac{(x+1)^2}{9} + \frac{(y-2)^2}{4} = 1$; foci at $(-1 \pm \sqrt{5}, 2)$; $e = \sqrt{5}/3$

21. $\frac{x^2}{9} + \frac{y^2}{5} = 1$

23. $\frac{(x-2)^2}{45} + \frac{y^2}{49} = 1$

25. $\frac{(x-1)^2}{2} + (y-2)^2 = 1$

27. $\frac{x^2}{4} + \frac{(y-3)^2}{6} = 1$

29. $x^2 - \frac{y^2}{3} = 1$

31. $\frac{(y-3)^2}{4} - \frac{(x-1)^2}{9} = 1$

33.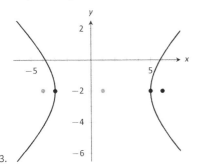

35. $\frac{x^2}{4} - \frac{y^2}{5} = 1$

37. $\frac{(x-3)^2}{16} - \frac{(y-3)^2}{9} = 1$

39. $\frac{x^2}{4} - \frac{y^2}{3} = 1$

41. $(y-2)^2 - \frac{x^2}{10} = 1$

43. (a) $c = \sqrt{12-4} = 2\sqrt{2}$.

    (b) The sum of distances for each point is $2\sqrt{12} \approx 6.9282$.

45. The sound originated from a point approximately 31m to the left of $B$ and 1340m above it.

### Section 9.2

1. T

3. rectangular

5.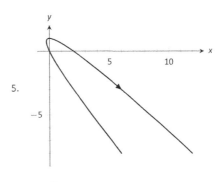

7.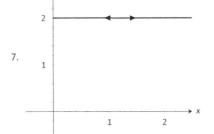

9.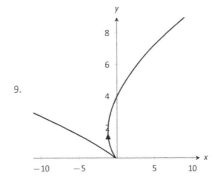

11.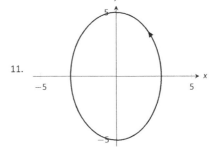

13.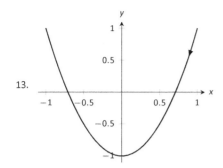

15.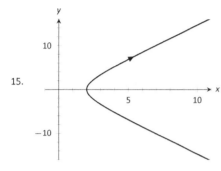

17.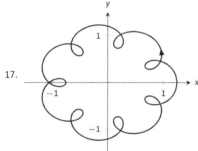

19. (a) Traces the parabola $y = x^2$, moves from left to right.
    (b) Traces the parabola $y = x^2$, but only from $-1 \leq x \leq 1$; traces this portion back and forth infinitely.
    (c) Traces the parabola $y = x^2$, but only for $0 < x$. Moves left to right.
    (d) Traces the parabola $y = x^2$, moves from right to left.

21. $y = -1.5x + 8.5$

23. $\frac{(x-1)^2}{16} + \frac{(y+2)^2}{9} = 1$

25. $y = 2x + 3$

27. $y = e^{2x} - 1$

29. $x^2 - y^2 = 1$

31. $y = \frac{b}{a}(x - x_0) + y_0$; line through $(x_0, y_0)$ with slope $b/a$.

33. $\frac{(x-h)^2}{a^2} + \frac{(y-k)^2}{b^2} = 1$; ellipse centered at $(h, k)$ with horizontal axis of length $2a$ and vertical axis of length $2b$.

35. $x = (t+11)/6, y = (t^2 - 97)/12$. At $t = 1, x = 2, y = -8$. $y' = 6x - 11$; when $x = 2, y' = 1$.

37. $x = \cos^{-1} t, y = \sqrt{1-t^2}$. At $t = 1, x = 0, y = 0$. $y' = \cos x$; when $x = 0, y' = 1$.

39. $t = \pm 1$

41. $t = \pi/2, 3\pi/2$

43. $t = -1$

45. $t = \ldots \pi/2, 3\pi/2, 5\pi/2, \ldots$

47. $x = 4t, y = -16t^2 + 64t$

49. $x = 10t, y = -16t^2 + 320t$

51. $x = 3\cos(2\pi t) + 1, y = 3\sin(2\pi t) + 1$; other answers possible

53. $x = 5\cos t, y = \sqrt{24} \sin t$; other answers possible

55. $x = 2\tan t, y = \pm 6 \sec t$; other answers possible

## Section 9.3

1. F

3. F

5. (a) $\frac{dy}{dx} = 2t$
   (b) Tangent line: $y = 2(x-1) + 1$; normal line: $y = -1/2(x-1) + 1$

7. (a) $\frac{dy}{dx} = \frac{2t+1}{2t-1}$
   (b) Tangent line: $y = 3x + 2$; normal line: $y = -1/3x + 2$

9. (a) $\frac{dy}{dx} = \csc t$
   (b) $t = \pi/4$: Tangent line: $y = \sqrt{2}(x - \sqrt{2}) + 1$; normal line: $y = -1/\sqrt{2}(x - \sqrt{2}) + 1$

11. (a) $\frac{dy}{dx} = \frac{\cos t \sin(2t) + \sin t \cos(2t)}{-\sin t \sin(2t) + 2\cos t \cos(2t)}$
    (b) Tangent line: $y = x - \sqrt{2}$; normal line: $y = -x - \sqrt{2}$

13. $t = 0$

15. $t = -1/2$

17. The graph does not have a horizontal tangent line.

19. The solution is non-trivial; use identities $\sin(2t) = 2\sin t \cos t$ and $\cos(2t) = \cos^2 t - \sin^2 t$ to rewrite $g'(t) = 2\sin t(2\cos^2 t - \sin^2 t)$. On $[0, 2\pi]$, $\sin t = 0$ when $t = 0, \pi, 2\pi$, and $2\cos^2 t - \sin^2 t = 0$ when $t = \tan^{-1}(\sqrt{2}), \pi \pm \tan^{-1}(\sqrt{2}), 2\pi - \tan^{-1}(\sqrt{2})$.

21. $t_0 = 0$; $\lim_{t \to 0} \frac{dy}{dx} = 0$.

23. $t_0 = 1$; $\lim_{t \to 1} \frac{dy}{dx} = \infty$.

25. $\frac{d^2 y}{dx^2} = 2$; always concave up

27. $\frac{d^2 y}{dx^2} = -\frac{4}{(2t-1)^3}$; concave up on $(-\infty, 1/2)$; concave down on $(1/2, \infty)$.

29. $\frac{d^2 y}{dx^2} = -\cot^3 t$; concave up on $(-\infty, 0)$; concave down on $(0, \infty)$.

31. $\frac{d^2 y}{dx^2} = \frac{4(13 + 3\cos(4t))}{(\cos t + 3\cos(3t))^3}$, obtained with a computer algebra system; concave up on $(-\tan^{-1}(\sqrt{2}/2), \tan^{-1}(\sqrt{2}/2))$, concave down on $(-\pi/2, -\tan^{-1}(\sqrt{2}/2)) \cup (\tan^{-1}(\sqrt{2}/2), \pi/2)$

33. $L = 6\pi$

35. $L = 2\sqrt{34}$

37. $L \approx 2.4416$ (actual value: $L = 2.42211$)

39. $L \approx 4.19216$ (actual value: $L = 4.18308$)

41. The answer is $16\pi$ for both (of course), but the integrals are different.

43. $SA \approx 8.50101$ (actual value $SA = 8.02851$)

## Section 9.4

1. Answers will vary.

3. T

5.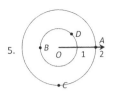

7. $A = P(2.5, \pi/4)$ and $P(-2.5, 5\pi/4)$;
$B = P(-1, 5\pi/6)$ and $P(1, 11\pi/6)$;
$C = P(3, 4\pi/3)$ and $P(-3, \pi/3)$;
$D = P(1.5, 2\pi/3)$ and $P(-1.5, 5\pi/3)$;

9. $A = (\sqrt{2}, \sqrt{2})$
$B = (\sqrt{2}, -\sqrt{2})$
$C = P(\sqrt{5}, -0.46)$
$D = P(\sqrt{5}, 2.68)$

11.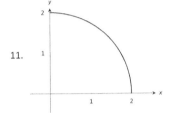

13.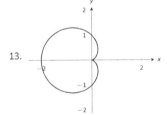

15.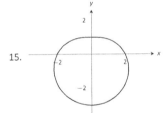

17.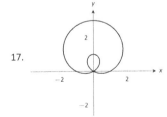

19.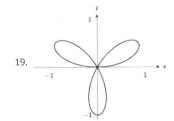

21.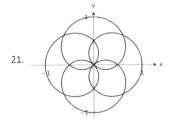

23.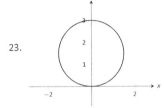

25.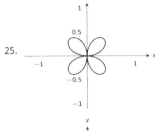

27.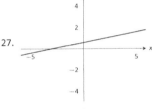

29.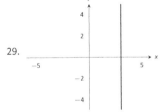

31. $(x-3)^2 + y^2 = 3$
33. $(x-1/2)^2 + (y-1/2)^2 = 1/2$
35. $x = 3$
37. $x^4 + x^2 y^2 - y^2 = 0$
39. $x^2 + y^2 = 4$
41. $\theta = \pi/4$
43. $r = 5 \sec\theta$
45. $r = \cos\theta / \sin^2\theta$
47. $r = \sqrt{7}$
49. $P(\sqrt{3}/2, \pi/6), P(0, \pi/2), P(-\sqrt{3}/2, 5\pi/6)$
51. $P(0,0) = P(0, \pi/2), P(\sqrt{2}, \pi/4)$
53. $P(\sqrt{2}/2, \pi/12), P(-\sqrt{2}/2, 5\pi/12), P(\sqrt{2}/2, 3\pi/4)$
55. For all points, $r = 1$; $\theta = \pi/12, 5\pi/12, 7\pi/12, 11\pi/12, 13\pi/12, 17\pi/12, 19\pi/12, 23\pi/12$.
57. Answers will vary. If $m$ and $n$ do not have any common factors, then an interval of $2n\pi$ is needed to sketch the entire graph.

**Section 9.5**

1. Using $x = r\cos\theta$ and $y = r\sin\theta$, we can write $x = f(\theta)\cos\theta$, $y = f(\theta)\sin\theta$.

3. (a) $\frac{dy}{dx} = -\cot\theta$

   (b) tangent line: $y = -(x - \sqrt{2}/2) + \sqrt{2}/2$; normal line: $y = x$

5. (a) $\frac{dy}{dx} = \frac{\cos\theta(1+2\sin\theta)}{\cos^2\theta - \sin\theta(1+\sin\theta)}$

   (b) tangent line: $x = 3\sqrt{3}/4$; normal line: $y = 3/4$

7. (a) $\frac{dy}{dx} = \frac{\theta\cos\theta + \sin\theta}{\cos\theta - \theta\sin\theta}$

   (b) tangent line: $y = -2/\pi x + \pi/2$; normal line: $y = \pi/2 x + \pi/2$

9. (a) $\frac{dy}{dx} = \frac{4\sin(\theta)\cos(4\theta) + \sin(4\theta)\cos(\theta)}{4\cos(\theta)\cos(4\theta) - \sin(\theta)\sin(4\theta)}$

   (b) tangent line: $y = 5\sqrt{3}(x + \sqrt{3}/4) - 3/4$; normal line: $y = -1/5\sqrt{3}(x + \sqrt{3}/4) - 3/4$

11. horizontal: $\theta = \pi/2, 3\pi/2$; vertical: $\theta = 0, \pi, 2\pi$

13. horizontal: $\theta = \tan^{-1}(1/\sqrt{5})$, $\pi/2$, $\pi - \tan^{-1}(1/\sqrt{5})$, $\pi + \tan^{-1}(1/\sqrt{5})$, $3\pi/2$, $2\pi - \tan^{-1}(1/\sqrt{5})$; vertical: $\theta = 0$, $\tan^{-1}(\sqrt{5})$, $\pi - \tan^{-1}(\sqrt{5})$, $\pi$, $\pi + \tan^{-1}(\sqrt{5})$, $2\pi - \tan^{-1}(\sqrt{5})$

15. In polar: $\theta = 0 \cong \theta = \pi$
    In rectangular: $y = 0$

17. area = $4\pi$

19. area = $\pi/12$

21. area = $3\pi/2$

23. area = $2\pi + 3\sqrt{3}/2$

25. area = 1

27. area = $\frac{1}{32}(4\pi - 3\sqrt{3})$

29. $4\pi$

31. area = $\sqrt{2}\pi$

33. $L \approx 2.2592$; (actual value $L = 2.22748$)

35. $SA = 16\pi$

37. $SA = 32\pi/5$

39. $SA = 36\pi$

# Chapter 10

### Section 10.1

1. right hand

3. curve (a parabola); surface (a cylinder)

5. a hyperboloid of two sheets

7. $||\overline{AB}|| = \sqrt{6}$; $||\overline{BC}|| = \sqrt{17}$; $||\overline{AC}|| = \sqrt{11}$. Yes, it is a right triangle as $||\overline{AB}||^2 + ||\overline{AC}||^2 = ||\overline{BC}||^2$.

9. Center at $(4, -1, 0)$; radius = 3

11. Interior of a sphere with radius 1 centered at the origin.

13. The first octant of space; all points $(x, y, z)$ where each of $x$, $y$ and $z$ are non-negative. (Analogous to the first quadrant in the plane.)

15.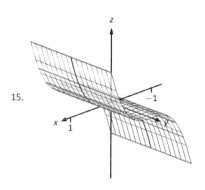

17.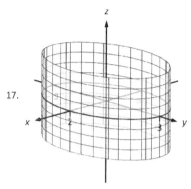

19. $x^2 + z^2 = \frac{1}{(1+y^2)^2}$

21. $z = (\sqrt{x^2+y^2})^2 = x^2 + y^2$

23. (a) $x = y^2 + \frac{z^2}{9}$

25. (b) $x^2 + \frac{y^2}{9} + \frac{z^2}{4} = 1$

27.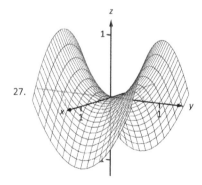

29.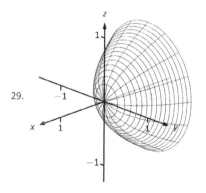

31.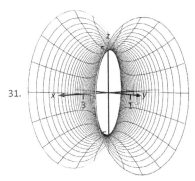

## Section 10.2

1. Answers will vary.
3. A vector with magnitude 1.
5. Their respective unit vectors are parallel; unit vectors $\vec{u}_1$ and $\vec{u}_2$ are parallel if $\vec{u}_1 = \pm\vec{u}_2$.
7. $\vec{PQ} = \langle 1, 6 \rangle = 1\vec{i} + 6\vec{j}$
9. $\vec{PQ} = \langle 6, -1, 6 \rangle = 6\vec{i} - \vec{j} + 6\vec{k}$
11. (a) $\vec{u} + \vec{v} = \langle 2, -1 \rangle; \vec{u} - \vec{v} = \langle 0, -3 \rangle; 2\vec{u} - 3\vec{v} = \langle -1, -7 \rangle$.
    (c) $\vec{x} = \langle 1/2, 2 \rangle$.
13.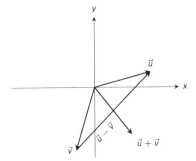
15.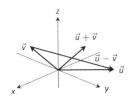
17. $\|\vec{u}\| = \sqrt{5}, \|\vec{v}\| = \sqrt{13}, \|\vec{u}+\vec{v}\| = \sqrt{26}, \|\vec{u}-\vec{v}\| = \sqrt{10}$
19. $\|\vec{u}\| = \sqrt{5}, \|\vec{v}\| = 3\sqrt{5}, \|\vec{u}+\vec{v}\| = 2\sqrt{5}, \|\vec{u}-\vec{v}\| = 4\sqrt{5}$
21. When $\vec{u}$ and $\vec{v}$ have the same direction. (Note: parallel is not enough.)
23. $\vec{u} = \langle 0.6, 0.8 \rangle$
25. $\vec{u} = \langle 1/\sqrt{3}, -1/\sqrt{3}, 1/\sqrt{3} \rangle$
27. $\vec{u} = \langle \cos 120°, \sin 120° \rangle = \langle -1/2, \sqrt{3}/2 \rangle$.
29. The force on each chain is $100/\sqrt{3} \approx 57.735$lb.
31. The force on the chain with angle $\theta$ is approx. 45.124lb; the force on the chain with angle $\varphi$ is approx. 59.629lb.
33. $\theta = 45°$; the weight is lifted 0.29 ft (about 3.5in).
35. $\theta = 45°$; the weight is lifted 2.93 ft.

## Section 10.3

1. Scalar
3. By considering the sign of the dot product of the two vectors. If the dot product is positive, the angle is acute; if the dot product is negative, the angle is obtuse.
5. $-22$
7. 3
9. not defined
11. Answers will vary.
13. $\theta = 0.3218 \approx 18.43°$
15. $\theta = \pi/4 = 45°$
17. Answers will vary; two possible answers are $\langle -7, 4 \rangle$ and $\langle 14, -8 \rangle$.
19. Answers will vary; two possible answers are $\langle 1, 0, -1 \rangle$ and $\langle 4, 5, -9 \rangle$.
21. $\text{proj}_{\vec{v}}\vec{u} = \langle -1/2, 3/2 \rangle$.
23. $\text{proj}_{\vec{v}}\vec{u} = \langle -1/2, -1/2 \rangle$.
25. $\text{proj}_{\vec{v}}\vec{u} = \langle 1, 2, 3 \rangle$.
27. $\vec{u} = \langle -1/2, 3/2 \rangle + \langle 3/2, 1/2 \rangle$.
29. $\vec{u} = \langle -1/2, -1/2 \rangle + \langle -5/2, 5/2 \rangle$.
31. $\vec{u} = \langle 1, 2, 3 \rangle + \langle 0, 3, -2 \rangle$.
33. 1.96lb
35. 141.42ft–lb
37. 500ft–lb
39. 500ft–lb

## Section 10.4

1. vector
3. "Perpendicular" is one answer.
5. Torque
7. $\vec{u} \times \vec{v} = \langle 12, -15, 3 \rangle$
9. $\vec{u} \times \vec{v} = \langle -5, -31, 27 \rangle$
11. $\vec{u} \times \vec{v} = \langle 0, -2, 0 \rangle$
13. $\vec{u} \times \vec{v} = \langle 0, 0, ad - bc \rangle$
15. $\vec{i} \times \vec{k} = -\vec{j}$
17. Answers will vary.
19. 5
21. 0
23. $\sqrt{14}$
25. 3
27. $5\sqrt{2}/2$
29. 1
31. 7
33. 2
35. $\pm\frac{1}{\sqrt{6}}\langle 1, 1, -2 \rangle$
37. $\langle 0, \pm 1, 0 \rangle$
39. 87.5ft–lb
41. $200/3 \approx 66.67$ft–lb
43. With $\vec{u} = \langle u_1, u_2, u_3 \rangle$ and $\vec{v} = \langle v_1, v_2, v_3 \rangle$, we have
$$\vec{u} \cdot (\vec{u} \times \vec{v}) = \langle u_1, u_2, u_3 \rangle \cdot (\langle u_2v_3 - u_3v_2, -(u_1v_3 - u_3v_1), u_1v_2 - u_2v_1 \rangle)$$
$$= u_1(u_2v_3 - u_3v_2) - u_2(u_1v_3 - u_3v_1) + u_3(u_1v_2 - u_2v_1)$$
$$= 0.$$

## Section 10.5

1. A point on the line and the direction of the line.

3. parallel, skew

5. vector: $\ell(t) = \langle 2, -4, 1 \rangle + t \langle 9, 2, 5 \rangle$
   parametric: $x = 2 + 9t, y = -4 + 2t, z = 1 + 5t$
   symmetric: $(x-2)/9 = (y+4)/2 = (z-1)/5$

7. Answers can vary: vector: $\ell(t) = \langle 2, 1, 5 \rangle + t \langle 5, -3, -1 \rangle$
   parametric: $x = 2 + 5t, y = 1 - 3t, z = 5 - t$
   symmetric: $(x-2)/5 = -(y-1)/3 = -(z-5)$

9. Answers can vary; here the direction is given by $\vec{d}_1 \times \vec{d}_2$: vector:
   $\ell(t) = \langle 0, 1, 2 \rangle + t \langle -10, 43, 9 \rangle$
   parametric: $x = -10t, y = 1 + 43t, z = 2 + 9t$
   symmetric: $-x/10 = (y-1)/43 = (z-2)/9$

11. Answers can vary; here the direction is given by $\vec{d}_1 \times \vec{d}_2$: vector:
    $\ell(t) = \langle 7, 2, -1 \rangle + t \langle 1, -1, 2 \rangle$
    parametric: $x = 7 + t, y = 2 - t, z = -1 + 2t$
    symmetric: $x - 7 = 2 - y = (z+1)/2$

13. vector: $\ell(t) = \langle 1, 1 \rangle + t \langle 2, 3 \rangle$
    parametric: $x = 1 + 2t, y = 1 + 3t$
    symmetric: $(x-1)/2 = (y-1)/3$

15. parallel

17. intersecting; $\vec{\ell}_1(3) = \vec{\ell}_2(4) = \langle 9, -5, 13 \rangle$

19. skew

21. same

23. $\sqrt{41}/3$

25. $5\sqrt{2}/2$

27. $3/\sqrt{2}$

29. Since both $P$ and $Q$ are on the line, $\overrightarrow{PQ}$ is parallel to $\vec{d}$. Thus $\overrightarrow{PQ} \times \vec{d} = \vec{0}$, giving a distance of 0.

31. (a) The distance formula cannot be used because since $\vec{d}_1$ and $\vec{d}_2$ are parallel, $\vec{c}$ is $\vec{0}$ and we cannot divide by $\|\vec{0}\|$.

    (b) Since $\vec{d}_1$ and $\vec{d}_2$ are parallel, $\overrightarrow{P_1P_2}$ lies in the plane formed by the two lines. Thus $\overrightarrow{P_1P_2} \times \vec{d}_2$ is orthogonal to this plane, and $\vec{c} = (\overrightarrow{P_1P_2} \times \vec{d}_2) \times \vec{d}_2$ is parallel to the plane, but still orthogonal to both $\vec{d}_1$ and $\vec{d}_2$. We desire the length of the projection of $\overrightarrow{P_1P_2}$ onto $\vec{c}$, which is what the formula provides.

    (c) Since the lines are parallel, one can measure the distance between the lines at any location on either line (just as to find the distance between straight railroad tracks, one can use a measuring tape anywhere along the track, not just at one specific place.) Let $P = P_1$ and $Q = P_2$ as given by the equations of the lines, and apply the formula for distance between a point and a line.

## Section 10.6

1. A point in the plane and a normal vector (i.e., a direction orthogonal to the plane).

3. Answers will vary.

5. Answers will vary.

7. Standard form: $3(x-2) - (y-3) + 7(z-4) = 0$
   general form: $3x - y + 7z = 31$

9. Answers may vary;
   Standard form: $8(x-1) + 4(y-2) - 4(z-3) = 0$
   general form: $8x + 4y - 4z = 4$

11. Answers may vary;
    Standard form: $-7(x-2) + 2(y-1) + (z-2) = 0$
    general form: $-7x + 2y + z = -10$

13. Answers may vary;
    Standard form: $2(x-1) - (y-1) = 0$
    general form: $2x - y = 1$

15. Answers may vary;
    Standard form: $2(x-2) - (y+6) - 4(z-1) = 0$
    general form: $2x - y - 4z = 6$

17. Answers may vary;
    Standard form: $(x-5) + (y-7) + (z-3) = 0$
    general form: $x + y + z = 15$

19. Answers may vary;
    Standard form: $3(x+4) + 8(y-7) - 10(z-2) = 0$
    general form: $3x + 8y - 10z = 24$

21. Answers may vary:
    $\ell = \begin{cases} x = 14t \\ y = -1 - 10t \\ z = 2 - 8t \end{cases}$

23. $(-3, -7, -5)$

25. No point of intersection; the plane and line are parallel.

27. $\sqrt{5/7}$

29. $1/\sqrt{3}$

31. If $P$ is any point in the plane, and $Q$ is also in the plane, then $\overrightarrow{PQ}$ lies parallel to the plane and is orthogonal to $\vec{n}$, the normal vector. Thus $\vec{n} \cdot \overrightarrow{PQ} = 0$, giving the distance as 0.

# Chapter 11

## Section 11.1

1. parametric equations

3. displacement

5.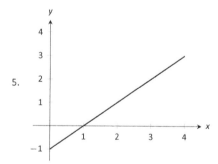

7.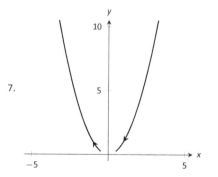

9.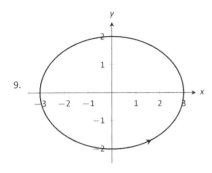

11.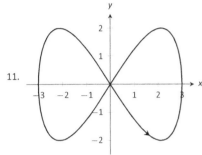

13.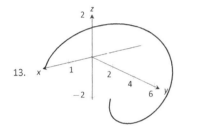

15.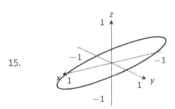

17. $\|\vec{r}(t)\| = \sqrt{t^2 + t^4} = |t|\sqrt{t^2 + 1}$.

19. $\|\vec{r}(t)\| = \sqrt{4\cos^2 t + 4\sin^2 t + t^2} = \sqrt{t^2 + 4}$.

21. Answers may vary, though most direct solution is
$\vec{r}(t) = \langle 2\cos t + 1, 2\sin t + 2\rangle$.

23. Answers may vary, though most direct solution is
$\vec{r}(t) = \langle 1.5\cos t, 5\sin t\rangle$.

25. Answers may vary, though most direct solutions are
$\vec{r}(t) = \langle t, 5(t-2) + 3\rangle$ and
$\vec{r}(t) = \langle t+2, 5t+3\rangle$.

27. Specific forms may vary, though most direct solutions are
$\vec{r}(t) = \langle 1, 2, 3\rangle + t\langle 3, 3, 3\rangle$ and
$\vec{r}(t) = \langle 3t+1, 3t+2, 3t+3\rangle$.

29. Answers may vary, though most direct solution is
$\vec{r}(t) = \langle 2\cos t, 2\sin t, 2t\rangle$.

31. $\langle 1, 0\rangle$

33. $\langle 0, 0, 1\rangle$

## Section 11.2

1. component

3. It is difficult to identify the points on the graphs of $\vec{r}(t)$ and $\vec{r}\,'(t)$ that correspond to each other.

5. $\langle 11, 74, \sin 5\rangle$

7. $\langle 1, e\rangle$

9. $(-\infty, 0) \cup (0, \infty)$

11. $\vec{r}\,'(t) = \langle -\sin t, e^t, 1/t\rangle$

13. $\vec{r}\,'(t) = (2t)\langle \sin t, 2t+5\rangle + (t^2)\langle \cos t, 2\rangle = \langle 2t\sin t + t^2 \cos t, 6t^2 + 10t\rangle$

15. $\vec{r}\,'(t) = \langle 2t, 1, 0\rangle \times \langle \sin t, 2t+5, 1\rangle + \langle t^2+1, t-1, 1\rangle \times \langle \cos t, 2, 0\rangle = \langle -1, \cos t - 2t, 6t^2 + 10t + 2 + \cos t - \sin t - t\cos t\rangle$

17.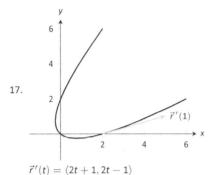
$\vec{r}\,'(t) = \langle 2t+1, 2t-1\rangle$

19.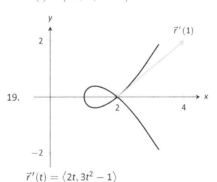
$\vec{r}\,'(t) = \langle 2t, 3t^2 - 1\rangle$

21. $\ell(t) = \langle 2, 0\rangle + t\langle 3, 1\rangle$

23. $\ell(t) = \langle -3, 0, \pi\rangle + t\langle 0, -3, 1\rangle$

25. $t = 2n\pi$, where $n$ is an integer; so
$t = \ldots -4\pi, -2\pi, 0, 2\pi, 4\pi, \ldots$

27. $\vec{r}(t)$ is not smooth at $t = 3\pi/4 + n\pi$, where $n$ is an integer

29. Both derivatives return $\langle 5t^4, 4t^3 - 3t^2, 3t^2\rangle$.

31. Both derivatives return
$\langle 2t - e^t - 1, \cos t - 3t^2, (t^2 + 2t)e^t - (t-1)\cos t - \sin t\rangle$.

33. $\langle \frac{1}{4}t^4, \sin t, te^t - e^t\rangle + \vec{C}$

35. $\langle -2, 0\rangle$

37. $\vec{r}(t) = \langle \frac{1}{2}t^2 + 2, -\cos t + 3\rangle$

39. $\vec{r}(t) = \langle t^4/12 + t + 4, t^3/6 + 2t + 5, t^2/2 + 3t + 6\rangle$

41. $2\sqrt{13}\pi$

43. $\frac{1}{54}\left((22)^{3/2} - 8\right)$

45. As $\vec{r}(t)$ has constant length, $\vec{r}(t) \cdot \vec{r}(t) = c^2$ for some constant $c$. Thus
$$\vec{r}(t) \cdot \vec{r}(t) = c^2$$
$$\frac{d}{dt}\left(\vec{r}(t) \cdot \vec{r}(t)\right) = \frac{d}{dt}(c^2)$$
$$\vec{r}\,'(t) \cdot \vec{r}(t) + \vec{r}(t) \cdot \vec{r}\,'(t) = 0$$
$$2\vec{r}(t) \cdot \vec{r}\,'(t) = 0$$
$$\vec{r}(t) \cdot \vec{r}\,'(t) = 0.$$

## Section 11.3

1. Velocity is a vector, indicating an objects direction of travel and its rate of distance change (i.e., its speed). Speed is a scalar.

3. The average velocity is found by dividing the displacement by the time traveled – it is a vector. The average speed is found by dividing the distance traveled by the time traveled – it is a scalar.

5. One example is traveling at a constant speed $s$ in a circle, ending at the starting position. Since the displacement is $\vec{0}$, the average velocity is $\vec{0}$, hence $||\vec{0}|| = 0$. But traveling at constant speed $s$ means the average speed is also $s > 0$.

7. $\vec{v}(t) = \langle 2, 5, 0 \rangle$, $\vec{a}(t) = \langle 0, 0, 0 \rangle$

9. $\vec{v}(t) = \langle -\sin t, \cos t\rangle$, $\vec{a}(t) = \langle -\cos t, -\sin t\rangle$

11. $\vec{v}(t) = \langle 1, \cos t\rangle$, $\vec{a}(t) = \langle 0, -\sin t\rangle$

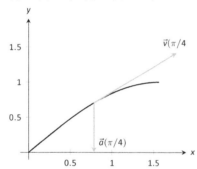

13. $\vec{v}(t) = \langle 2t+1, -2t+2\rangle$, $\vec{a}(t) = \langle 2, -2\rangle$

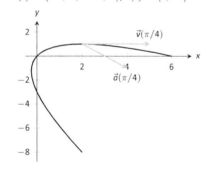

15. $||\vec{v}(t)|| = \sqrt{4t^2+1}$.
Min at $t = 0$; Max at $t = \pm 1$.

17. $||\vec{v}(t)|| = 5$.
Speed is constant, so there is no difference between min/max

19. $||\vec{v}(t)|| = |\sec t|\sqrt{\tan^2 t + \sec^2 t}$.
min: $t = 0$; max: $t = \pi/4$

21. $||\vec{v}(t)|| = 13$.
speed is constant, so there is no difference between min/max

23. $||\vec{v}(t)|| = \sqrt{4t^2+1+t^2/(1-t^2)}$.
min: $t = 0$; max: there is no max; speed approaches $\infty$ as $t \to \pm 1$

25. (a) $\vec{r}_1(1) = \langle 1,1\rangle$; $\vec{r}_2(1) = \langle 1,1\rangle$
    (b) $\vec{v}_1(1) = \langle 1,2\rangle$; $||\vec{v}_1(1)|| = \sqrt{5}$; $\vec{a}_1(1) = \langle 0,2\rangle$
    $\vec{v}_2(1) = \langle 2,4\rangle$; $||\vec{v}_2(1)|| = 2\sqrt{5}$; $\vec{a}_2(1) = \langle 2,12\rangle$

27. (a) $\vec{r}_1(2) = \langle 6,4\rangle$; $\vec{r}_2(2) = \langle 6,4\rangle$
    (b) $\vec{v}_1(2) = \langle 3,2\rangle$; $||\vec{v}_1(2)|| = \sqrt{13}$; $\vec{a}_1(2) = \langle 0,0\rangle$
    $\vec{v}_2(2) = \langle 6,4\rangle$; $||\vec{v}_2(2)|| = 2\sqrt{13}$; $\vec{a}_2(2) = \langle 0,0\rangle$

29. $\vec{v}(t) = \langle 2t+1, 3t+2\rangle$, $\vec{r}(t) = \langle t^2+t+5, 3t^2/2+2t-2\rangle$

31. $\vec{v}(t) = \langle \sin t, \cos t\rangle$, $\vec{r}(t) = \langle 1-\cos t, \sin t\rangle$

33. Displacement: $\langle 0,0,6\pi\rangle$; distance traveled: $2\sqrt{13}\pi \approx 22.65$ft; average velocity: $\langle 0,0,3\rangle$; average speed: $\sqrt{13} \approx 3.61$ft/s

35. Displacement: $\langle 0,0\rangle$; distance traveled: $2\pi \approx 6.28$ft; average velocity: $\langle 0,0\rangle$; average speed: 1ft/s

37. At $t$-values of $\sin^{-1}(9/30)/(4\pi) + n/2 \approx 0.024 + n/2$ seconds, where $n$ is an integer.

39. (a) Holding the crossbow at an angle of 0.013 radians, $\approx 0.745°$ will hit the target 0.4s later. (Another solution exists, with an angle of $89°$, landing 18.75s later, but this is impractical.)
    (b) In the .4 seconds the arrow travels, a deer, traveling at 20mph or 29.33ft/s, can travel 11.7ft. So she needs to lead the deer by 11.7ft.

41. The position function is $\vec{r}(t) = \langle 220t, -16t^2+1000\rangle$. The $y$-component is 0 when $t = 7.9$; $\vec{r}(7.9) = \langle 1739.25, 0\rangle$, meaning the box will travel about 1740ft horizontally before it lands.

## Section 11.4

1. 1

3. $\vec{T}(t)$ and $\vec{N}(t)$.

5. $\vec{T}(t) = \left\langle \frac{4t}{\sqrt{20t^2-4t+1}}, \frac{2t-1}{\sqrt{20t^2-4t+1}} \right\rangle$; $\vec{T}(1) = \langle 4/\sqrt{17}, 1/\sqrt{17}\rangle$

7. $\vec{T}(t) = \frac{\cos t \sin t}{\sqrt{\cos^2 t \sin^2 t}} \langle -\cos t, \sin t\rangle$. (Be careful; this cannot be simplified as just $\langle -\cos t, \sin t\rangle$ as $\sqrt{\cos^2 t \sin^2 t} \neq \cos t \sin t$, but rather $|\cos t \sin t|$.) $\vec{T}(\pi/4) = \langle -\sqrt{2}/2, \sqrt{2}/2\rangle$

9. $\ell(t) = \langle 2, 0\rangle + t\langle 4/\sqrt{17}, 1/\sqrt{17}\rangle$; in parametric form,
$\ell(t) = \begin{cases} x &= 2 + 4t/\sqrt{17} \\ y &= t/\sqrt{17} \end{cases}$

11. $\ell(t) = \langle \sqrt{2}/4, \sqrt{2}/4\rangle + t\langle -\sqrt{2}/2, \sqrt{2}/2\rangle$; in parametric form,
$\ell(t) = \begin{cases} x &= \sqrt{2}/4 - \sqrt{2}t/2 \\ y &= \sqrt{2}/4 + \sqrt{2}t/2 \end{cases}$

13. $\vec{T}(t) = \langle -\sin t, \cos t\rangle$; $\vec{N}(t) = \langle -\cos t, -\sin t\rangle$

15. $\vec{T}(t) = \left\langle -\frac{\sin t}{\sqrt{4\cos^2 t + \sin^2 t}}, \frac{2\cos t}{\sqrt{4\cos^2 t + \sin^2 t}} \right\rangle$;
$\vec{N}(t) = \left\langle -\frac{2\cos t}{\sqrt{4\cos^2 t + \sin^2 t}}, -\frac{\sin t}{\sqrt{4\cos^2 t + \sin^2 t}} \right\rangle$

17. (a) Be sure to show work
    (b) $\vec{N}(\pi/4) = \langle -5/\sqrt{34}, -3/\sqrt{34}\rangle$

19. (a) Be sure to show work
    (b) $\vec{N}(0) = \left\langle -\frac{1}{\sqrt{5}}, \frac{2}{\sqrt{5}} \right\rangle$

21. $\vec{T}(t) = \frac{1}{\sqrt{5}}\langle 2, \cos t, -\sin t\rangle$; $\vec{N}(t) = \langle 0, -\sin t, -\cos t\rangle$

23. $\vec{T}(t) = \frac{1}{\sqrt{a^2+b^2}}\langle -a\sin t, a\cos t, b\rangle$; $\vec{N}(t) = \langle -\cos t, -\sin t, 0\rangle$

25. $a_T = \frac{4t}{\sqrt{1+4t^2}}$ and $a_N = \sqrt{4 - \frac{16t^2}{1+4t^2}}$
At $t = 0$, $a_T = 0$ and $a_N = 2$;
At $t = 1$, $a_T = 4/\sqrt{5}$ and $a_N = 2/\sqrt{5}$.
At $t = 0$, all acceleration comes in the form of changing the direction of velocity and not the speed; at $t = 1$, more acceleration comes in changing the speed than in changing direction.

27. $a_T = 0$ and $a_N = 2$
    At $t = 0$, $a_T = 0$ and $a_N = 2$;
    At $t = \pi/2$, $a_T = 0$ and $a_N = 2$.
    The object moves at constant speed, so all acceleration comes from changing direction, hence $a_T = 0$. $\vec{a}(t)$ is always parallel to $\vec{N}(t)$, but twice as long, hence $a_N = 2$.

29. $a_T = 0$ and $a_N = a$
    At $t = 0$, $a_T = 0$ and $a_N = a$;
    At $t = \pi/2$, $a_T = 0$ and $a_N = a$.
    The object moves at constant speed, meaning that $a_T$ is always 0. The object "rises" along the $z$-axis at a constant rate, so all acceleration comes in the form of changing direction circling the $z$-axis. The greater the radius of this circle the greater the acceleration, hence $a_N = a$.

## Section 11.5

1. time and/or distance

3. Answers may include lines, circles, helixes

5. $\kappa$

7. $s = 3t$, so $\vec{r}(s) = \langle 2s/3, s/3, -2s/3 \rangle$

9. $s = \sqrt{13}t$, so $\vec{r}(s) = \langle 3\cos(s/\sqrt{13}), 3\sin(s/\sqrt{13}), 2s/\sqrt{13} \rangle$

11. $\kappa = \frac{|6x|}{(1+(3x^2-1)^2)^{3/2}}$;
    $\kappa(0) = 0$, $\kappa(1/2) = \frac{192}{17\sqrt{17}} \approx 2.74$.

13. $\kappa = \frac{|\cos x|}{(1+\sin^2 x)^{3/2}}$;
    $\kappa(0) = 1$, $\kappa(\pi/2) = 0$

15. $\kappa = \frac{|2\cos t \cos(2t) + 4\sin t \sin(2t)|}{(4\cos^2(2t)+\sin^2 t)^{3/2}}$;
    $\kappa(0) = 1/4$, $\kappa(\pi/4) = 8$

17. $\kappa = \frac{|6t^2+2|}{(4t^2+(3t^2-1)^2)^{3/2}}$;
    $\kappa(0) = 2$, $\kappa(5) = \frac{19}{1394\sqrt{1394}} \approx 0.0004$

19. $\kappa = 0$;
    $\kappa(0) = 0$, $\kappa(1) = 0$

21. $\kappa = \frac{3}{13}$;
    $\kappa(0) = 3/13$, $\kappa(\pi/2) = 3/13$

23. maximized at $x = \pm\frac{\sqrt{2}}{\sqrt[4]{5}}$

25. maximized at $t = 1/4$

27. radius of curvature is $5\sqrt{5}/4$.

29. radius of curvature is 9.

31. $x^2 + (y - 1/2)^2 = 1/4$, or $\vec{c}(t) = \langle 1/2 \cos t, 1/2 \sin t + 1/2 \rangle$

33. $x^2 + (y + 8)^2 = 81$, or $\vec{c}(t) = \langle 9\cos t, 9\sin t - 8 \rangle$

# Chapter 12

## Section 12.1

1. Answers will vary.

3. topographical

5. surface

7. domain: $\mathbb{R}^2$
   range: $z \geq 2$

9. domain: $\mathbb{R}^2$
   range: $\mathbb{R}$

11. domain: $\mathbb{R}^2$
    range: $0 < z \leq 1$

13. domain: $\{(x,y) \mid x^2 + y^2 \leq 9\}$, i.e., the domain is the circle and interior of a circle centered at the origin with radius 3.
    range: $0 \leq z \leq 3$

15. Level curves are lines $y = (3/2)x - c/2$.

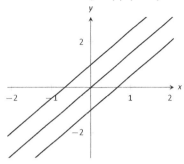

17. Level curves are parabolas $x = y^2 + c$.

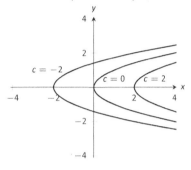

19. When $c \neq 0$, the level curves are circles, centered at $(1/c, -1/c)$ with radius $\sqrt{2/c^2 - 1}$. When $c = 0$, the level curve is the line $y = x$.

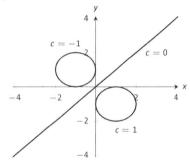

21. Level curves are ellipses of the form $\frac{x^2}{c^2} + \frac{y^2}{c^2/4} = 1$, i.e., $a = c$ and $b = c/2$.

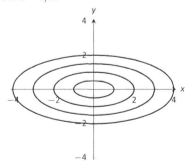

23. domain: $x + 2y - 4z \neq 0$; the set of points in $\mathbb{R}^3$ NOT in the domain form a plane through the origin.
    range: $\mathbb{R}$

25. domain: $z \geq x^2 - y^2$; the set of points in $\mathbb{R}^3$ above (and including) the hyperbolic paraboloid $z = x^2 - y^2$.
    range: $[0, \infty)$

27. The level surfaces are spheres, centered at the origin, with radius $\sqrt{c}$.

29. The level surfaces are paraboloids of the form $z = \frac{x^2}{c} + \frac{y^2}{c}$; the larger $c$, the "wider" the paraboloid.

31. The level curves for each surface are similar; for $z = \sqrt{x^2 + 4y^2}$ the level curves are ellipses of the form $\frac{x^2}{c^2} + \frac{y^2}{c^2/4} = 1$, i.e., $a = c$ and $b = c/2$; whereas for $z = x^2 + 4y^2$ the level curves are ellipses of the form $\frac{x^2}{c} + \frac{y^2}{c/4} = 1$, i.e., $a = \sqrt{c}$ and $b = \sqrt{c}/2$. The first set of ellipses are spaced evenly apart, meaning the function grows at a constant rate; the second set of ellipses are more closely spaced together as $c$ grows, meaning the function grows faster and faster as $c$ increases.
The function $z = \sqrt{x^2 + 4y^2}$ can be rewritten as $z^2 = x^2 + 4y^2$, an elliptic cone; the function $z = x^2 + 4y^2$ is a paraboloid, each matching the description above.

## Section 12.2

1. Answers will vary.

3. Answers will vary.
   One possible answer: $\{(x, y)|x^2 + y^2 \leq 1\}$

5. Answers will vary.
   One possible answer: $\{(x, y)|x^2 + y^2 < 1\}$

7. (a) Answers will vary.
       interior point: $(1, 3)$
       boundary point: $(3, 3)$
   (b) $S$ is a closed set
   (c) $S$ is bounded

9. (a) Answers will vary.
       interior point: none
       boundary point: $(0, -1)$
   (b) $S$ is a closed set, consisting only of boundary points
   (c) $S$ is bounded

11. (a) $D = \{(x, y) \,|\, 9 - x^2 - y^2 \geq 0\}$.
    (b) $D$ is a closed set.
    (c) $D$ is bounded.

13. (a) $D = \{(x, y) \,|\, y > x^2\}$.
    (b) $D$ is an open set.
    (c) $D$ is unbounded.

15. (a) Along $y = 0$, the limit is 1.
    (b) Along $x = 0$, the limit is $-1$.
    Since the above limits are not equal, the limit does not exist.

17. (a) Along $y = mx$, the limit is $\frac{mx(1-m)}{m^2x + 1} = 0$ for all $m$.
    (b) Along $x = 0$, the limit is $-1$.
    Since the above limits are not equal, the limit does not exist.

19. (a) Along $y = 2$, the limit is:
    $$\lim_{(x,y)\to(1,2)} \frac{x + y - 3}{x^2 - 1} = \lim_{x\to 1} \frac{x - 1}{x^2 - 1}$$
    $$= \lim_{x\to 1} \frac{1}{x + 1}$$
    $$= 1/2.$$
    (b) Along $y = x + 1$, the limit is:
    $$\lim_{(x,y)\to(1,2)} \frac{x + y - 3}{x^2 - 1} = \lim_{x\to 1} \frac{2(x - 1)}{x^2 - 1}$$
    $$= \lim_{x\to 1} \frac{2}{x + 1}$$
    $$= 1.$$
    Since the limits along the lines $y = 2$ and $y = x + 1$ differ, the overall limit does not exist.

## Section 12.3

1. A constant is a number that is added or subtracted in an expression; a coefficient is a number that is being multiplied by a nonconstant function.

3. $f_x$

5. $f_x = 2xy - 1, f_y = x^2 + 2$
   $f_x(1, 2) = 3, f_y(1, 2) = 3$

7. $f_x = -\sin x \sin y, f_y = \cos x \cos y$
   $f_x(\pi/3, \pi/3) = -3/4, f_y(\pi/3, \pi/3) = 1/4$

9. $f_x = 2xy + 6x, f_y = x^2 + 4$
   $f_{xx} = 2y + 6, f_{yy} = 0$
   $f_{xy} = 2x, f_{yx} = 2x$

11. $f_x = 1/y, f_y = -x/y^2$
    $f_{xx} = 0, f_{yy} = 2x/y^3$
    $f_{xy} = -1/y^2, f_{yx} = -1/y^2$

13. $f_x = 2xe^{x^2+y^2}, f_y = 2ye^{x^2+y^2}$
    $f_{xx} = 2e^{x^2+y^2} + 4x^2e^{x^2+y^2}, f_{yy} = 2e^{x^2+y^2} + 4y^2e^{x^2+y^2}$
    $f_{xy} = 4xye^{x^2+y^2}, f_{yx} = 4xye^{x^2+y^2}$

15. $f_x = \cos x \cos y, f_y = -\sin x \sin y$
    $f_{xx} = -\sin x \cos y, f_{yy} = -\sin x \cos y$
    $f_{xy} = -\sin y \cos x, f_{yx} = -\sin y \cos x$

17. $f_x = -5y^3 \sin(5xy^3), f_y = -15xy^2 \sin(5xy^3)$
    $f_{xx} = -25y^6 \cos(5xy^3),$
    $f_{yy} = -225x^2y^4 \cos(5xy^3) - 30xy \sin(5xy^3)$
    $f_{xy} = -75xy^5 \cos(5xy^3) - 15y^2 \sin(5xy^3),$
    $f_{yx} = -75xy^5 \cos(5xy^3) - 15y^2 \sin(5xy^3)$

19. $f_x = \frac{2y^2}{\sqrt{4xy^2+1}}, f_y = \frac{4xy}{\sqrt{4xy^2+1}}$
    $f_{xx} = -\frac{4y^4}{\sqrt{4xy^2+1}^3}, f_{yy} = -\frac{16x^2y^2}{\sqrt{4xy^2+1}^3} + \frac{4x}{\sqrt{4xy^2+1}}$
    $f_{xy} = -\frac{8xy^3}{\sqrt{4xy^2+1}^3} + \frac{4y}{\sqrt{4xy^2+1}}, f_{yx} = -\frac{8xy^3}{\sqrt{4xy^2+1}^3} + \frac{4y}{\sqrt{4xy^2+1}}$

21. $f_x = -\frac{2x}{(x^2+y^2+1)^2}, f_y = -\frac{2y}{(x^2+y^2+1)^2}$
    $f_{xx} = \frac{8x^2}{(x^2+y^2+1)^3} - \frac{2}{(x^2+y^2+1)^2}, f_{yy} = \frac{8y^2}{(x^2+y^2+1)^3} - \frac{2}{(x^2+y^2+1)^2}$
    $f_{xy} = \frac{8xy}{(x^2+y^2+1)^3}, f_{yx} = \frac{8xy}{(x^2+y^2+1)^3}$

23. $f_x = 6x, f_y = 0$
    $f_{xx} = 6, f_{yy} = 0$
    $f_{xy} = 0, f_{yx} = 0$

25. $f_x = \frac{1}{4xy}, f_y = -\frac{\ln x}{4y^2}$
    $f_{xx} = -\frac{1}{4x^2y}, f_{yy} = \frac{\ln x}{2y^3}$
    $f_{xy} = -\frac{1}{4xy^2}, f_{yx} = -\frac{1}{4xy^2}$

27. $f(x, y) = x \sin y + x + C$, where $C$ is any constant.

29. $f(x, y) = 3x^2y - 4xy^2 + 2y + C$, where $C$ is any constant.

31. $f_x = 2xe^{2y-3z}, f_y = 2x^2e^{2y-3z}, f_z = -3x^2e^{2y-3z}$
    $f_{yz} = -6x^2e^{2y-3z}, f_{zy} = -6x^2e^{2y-3z}$

33. $f_x = \frac{3}{7y^2z}, f_y = -\frac{6x}{7y^3z}, f_z = -\frac{3x}{7y^2z^2}$
    $f_{yz} = \frac{6x}{7y^3z^2}, f_{zy} = \frac{6x}{7y^3z^2}$

## Section 12.4

1. T

3. T

5. $dz = (\sin y + 2x)dx + (x \cos y)dy$

7. $dz = 5dx - 7dy$

9. $dz = \frac{x}{\sqrt{x^2+y}}dx + \frac{1}{2\sqrt{x^2+y}}dy$, with $dx = -0.05$ and $dy = .1$. At $(3,7)$, $dz = 3/4(-0.05) + 1/8(.1) = -0.025$, so $f(2.95, 7.1) \approx -0.025 + 4 = 3.975$.

11. $dz = (2xy - y^2)dx + (x^2 - 2xy)dy$, with $dx = 0.04$ and $dy = 0.06$. At $(2,3)$, $dz = 3(0.04) + (-8)(0.06) = -0.36$, so $f(2.04, 3.06) \approx -0.36 - 6 = -6.36$.

13. The total differential of volume is $dV = 4\pi dr + \pi dh$. The coefficient of $dr$ is greater than the coefficient of $dh$, so the volume is more sensitive to changes in the radius.

15. Using trigonometry, $\ell = x \tan \theta$, so $d\ell = \tan \theta dx + x\sec^2\theta d\theta$. With $\theta = 85°$ and $x = 30$, we have $d\ell = 11.43dx + 3949.38d\theta$. The measured length of the wall is much more sensitive to errors in $\theta$ than in $x$. While it can be difficult to compare sensitivities between measuring feet and measuring degrees (it is somewhat like "comparing apples to oranges"), here the coefficients are so different that the result is clear: a small error in degree has a much greater impact than a small error in distance.

17. $dw = 2xyz^3\,dx + x^2z^3\,dy + 3x^2yz^2\,dz$

19. $dx = 0.05$, $dy = -0.1$. $dz = 9(.05) + (-2)(-0.1) = 0.65$. So $f(3.05, 0.9) \approx 7 + 0.65 = 7.65$.

21. $dx = 0.5$, $dy = 0.1$, $dz = -0.2$. $dw = 2(0.5) + (-3)(0.1) + 3.7(-0.2) = -0.04$, so $f(2.5, 4.1, 4.8) \approx -1 - 0.04 = -1.04$.

## Section 12.5

1. Because the parametric equations describe a level curve, $z$ is constant for all $t$. Therefore $\frac{dz}{dt} = 0$.

3. $\frac{dx}{dt}$, and $\frac{\partial f}{\partial y}$

5. F

7. (a) $\frac{dz}{dt} = 3(2t) + 4(2) = 6t + 8$.
   (b) At $t = 1$, $\frac{dz}{dt} = 14$.

9. (a) $\frac{dz}{dt} = 5(-2\sin t) + 2(\cos t) = -10\sin t + 2\cos t$
   (b) At $t = \pi/4$, $\frac{dz}{dt} = -4\sqrt{2}$.

11. (a) $\frac{dz}{dt} = 2x(\cos t) + 4y(3\cos t)$.
    (b) At $t = \pi/4$, $x = \sqrt{2}/2$, $y = 3\sqrt{2}/2$, and $\frac{dz}{dt} = 19$.

13. $t = -4/3$; this corresponds to a minimum

15. $t = \tan^{-1}(1/5) + n\pi$, where $n$ is an integer

17. We find that
$$\frac{dz}{dt} = 38\cos t \sin t.$$
Thus $\frac{dz}{dt} = 0$ when $t = \pi n$ or $\pi n + \pi/2$, where $n$ is any integer.

19. (a) $\frac{\partial z}{\partial s} = 2xy(1) + x^2(2) = 2xy + 2x^2$;
    $\frac{\partial z}{\partial t} = 2xy(-1) + x^2(4) = -2xy + 4x^2$
    (b) With $s = 1$, $t = 0$, $x = 1$ and $y = 2$. Thus $\frac{\partial z}{\partial s} = 6$ and $\frac{\partial z}{\partial t} = 0$

21. (a) $\frac{\partial z}{\partial s} = 2x(\cos t) + 2y(\sin t) = 2x\cos t + 2y\sin t$;
    $\frac{\partial z}{\partial t} = 2x(-s\sin t) + 2y(s\cos t) = -2xs\sin t + 2ys\cos t$
    (b) With $s = 2$, $t = \pi/4$, $x = \sqrt{2}$ and $y = \sqrt{2}$. Thus $\frac{\partial z}{\partial s} = 4$ and $\frac{\partial z}{\partial t} = 0$

23. $f_x = 2x\tan y$, $f_y = x^2\sec^2 y$;
$\frac{dy}{dx} = -\frac{2\tan y}{x\sec^2 y}$

25. $f_x = \frac{(x+y^2)(2x) - (x^2+y)(1)}{(x+y^2)^2}$,
$f_y = \frac{(x+y^2)(1) - (x^2+y)(2y)}{(x+y^2)^2}$;
$\frac{dy}{dx} = -\frac{2x(x+y^2) - (x^2+y)}{x+y^2 - 2y(x^2+y)}$

27. $\frac{dz}{dt} = 2(4) + 1(-5) = 3$.

29. $\frac{\partial z}{\partial s} = -4(5) + 9(-2) = -38$,
$\frac{\partial z}{\partial t} = -4(7) + 9(6) = 26$.

## Section 12.6

1. A partial derivative is essentially a special case of a directional derivative; it is the directional derivative in the direction of $x$ or $y$, i.e., $\langle 1, 0\rangle$ or $\langle 0, 1\rangle$.

3. $\vec{u} = \langle 0, 1\rangle$

5. maximal, or greatest

7. $\nabla f = \langle -2xy + y^2 + y, -x^2 + 2xy + x\rangle$

9. $\nabla f = \left\langle \frac{-2x}{(x^2+y^2+1)^2}, \frac{-2y}{(x^2+y^2+1)^2}\right\rangle$

11. $\nabla f = \langle 2x - y - 7, 4y - x\rangle$

13. $\nabla f = \langle -2xy + y^2 + y, -x^2 + 2xy + x\rangle$; $\nabla f(2,1) = \langle -2, 2\rangle$. Be sure to change all directions to unit vectors.
    (a) $2/5$ ($\vec{u} = \langle 3/5, 4/5\rangle$)
    (b) $-2/\sqrt{5}$ ($\vec{u} = \langle -1/\sqrt{5}, -2/\sqrt{5}\rangle$)

15. $\nabla f = \left\langle \frac{-2x}{(x^2+y^2+1)^2}, \frac{-2y}{(x^2+y^2+1)^2}\right\rangle$; $\nabla f(1,1) = \langle -2/9, -2/9\rangle$. Be sure to change all directions to unit vectors.
    (a) $0$ ($\vec{u} = \langle 1/\sqrt{2}, -1/\sqrt{2}\rangle$)
    (b) $2\sqrt{2}/9$ ($\vec{u} = \langle -1/\sqrt{2}, -1/\sqrt{2}\rangle$)

17. $\nabla f = \langle 2x - y - 7, 4y - x\rangle$; $\nabla f(4,1) = \langle 0, 0\rangle$.
    (a) 0
    (b) 0

19. $\nabla f = \langle -2xy + y^2 + y, -x^2 + 2xy + x\rangle$
    (a) $\nabla f(2,1) = \langle -2, 2\rangle$
    (b) $\|\nabla f(2,1)\| = \|\langle -2, 2\rangle\| = \sqrt{8}$
    (c) $\langle 2, -2\rangle$
    (d) $\langle 1/\sqrt{2}, 1/\sqrt{2}\rangle$

21. $\nabla f = \left\langle \frac{-2x}{(x^2+y^2+1)^2}, \frac{-2y}{(x^2+y^2+1)^2}\right\rangle$
    (a) $\nabla f(1,1) = \langle -2/9, -2/9\rangle$.
    (b) $\|\nabla f(1,1)\| = \|\langle -2/9, -2/9\rangle\| = 2\sqrt{2}/9$
    (c) $\langle 2/9, 2/9\rangle$
    (d) $\langle 1/\sqrt{2}, -1/\sqrt{2}\rangle$

23. $\nabla f = \langle 2x - y - 7, 4y - x\rangle$
    (a) $\nabla f(4,1) = \langle 0, 0\rangle$
    (b) 0
    (c) $\langle 0, 0\rangle$
    (d) All directions give a directional derivative of 0.

25. (a) $\nabla F(x,y,z) = \langle 6xz^3 + 4y, 4x, 9x^2z^2 - 6z\rangle$
    (b) $113/\sqrt{3}$

27. (a) $\nabla F(x,y,z) = \langle 2xy^2, 2y(x^2 - z^2), -2y^2z\rangle$
    (b) 0

## Section 12.7

A.11

1. Answers will vary. The displacement of the vector is one unit in the x-direction and 3 units in the z-direction, with no change in y. Thus along a line parallel to $\vec{v}$, the change in z is 3 times the change in x – i.e., a "slope" of 3. Specifically, the line in the x-z plane parallel to z has a slope of 3.

3. T

5. (a) $\ell_x(t) = \begin{cases} x = 2 + t \\ y = 3 \\ z = -48 - 12t \end{cases}$

   (b) $\ell_y(t) = \begin{cases} x = 2 \\ y = 3 + t \\ z = -48 - 40t \end{cases}$

   (c) $\ell_{\vec{u}}(t) = \begin{cases} x = 2 + t/\sqrt{10} \\ y = 3 + 3t/\sqrt{10} \\ z = -48 - 66\sqrt{2/5}t \end{cases}$

7. (a) $\ell_x(t) = \begin{cases} x = 4 + t \\ y = 2 \\ z = 2 + 3t \end{cases}$

   (b) $\ell_y(t) = \begin{cases} x = 4 \\ y = 2 + t \\ z = 2 - 5t \end{cases}$

   (c) $\ell_{\vec{u}}(t) = \begin{cases} x = 4 + t/\sqrt{2} \\ y = 2 + t/\sqrt{2} \\ z = 2 - \sqrt{2}t \end{cases}$

9. $\ell_{\vec{n}}(t) = \begin{cases} x = 2 - 12t \\ y = 3 - 40t \\ z = -48 - t \end{cases}$

11. $\ell_{\vec{n}}(t) = \begin{cases} x = 4 + 3t \\ y = 2 - 5t \\ z = 2 - t \end{cases}$

13. $(1.425, 1.085, -48.078)$, $(2.575, 4.915, -47.952)$

15. $(5.014, 0.31, 1.662)$ and $(2.986, 3.690, 2.338)$

17. $-12(x - 2) - 40(y - 3) - (z + 48) = 0$

19. $3(x - 4) - 5(y - 2) - (z - 2) = 0$ (Note that this tangent plane is the same as the original function, a plane.)

21. $\nabla F = \langle x/4, y/2, z/8 \rangle$; at $P$, $\nabla F = \langle 1/4, \sqrt{2}/2, \sqrt{6}/8 \rangle$

    (a) $\ell_{\vec{n}}(t) = \begin{cases} x = 1 + t/4 \\ y = \sqrt{2} + \sqrt{2}t/2 \\ z = \sqrt{6} + \sqrt{6}t/8 \end{cases}$

    (b) $\frac{1}{4}(x - 1) + \frac{\sqrt{2}}{2}(y - \sqrt{2}) + \frac{\sqrt{6}}{8}(z - \sqrt{6}) = 0$.

23. $\nabla F = \langle y^2 - z^2, 2xy, -2xz \rangle$; at $P$, $\nabla F = \langle 0, 4, 4 \rangle$

    (a) $\ell_{\vec{n}}(t) = \begin{cases} x = 2 \\ y = 1 + 4t \\ z = -1 + 4t \end{cases}$

    (b) $4(y - 1) + 4(z + 1) = 0$.

### Section 12.8

1. F; it is the "other way around."

3. T

5. One critical point at $(-4, 2)$; $f_{xx} = 1$ and $D = 4$, so this point corresponds to a relative minimum.

7. One critical point at $(6, -3)$; $D = -4$, so this point corresponds to a saddle point.

9. Two critical points: at $(0, -1)$; $f_{xx} = 2$ and $D = -12$, so this point corresponds to a saddle point; at $(0, 1)$, $f_{xx} = 2$ and $D = 12$, so this corresponds to a relative minimum.

11. There are infinite critical points, whenever $x = 0$ or $y = 0$. With $D = -12x^2y^2$, at each critical point $D = 0$ and the test is inconclusive. (Some elementary thought shows that each is an absolute minimum.)

13. One critical point: $f_x = 0$ when $x = 3$; $f_y = 0$ when $y = 0$, so one critical point at $(3, 0)$, which is a relative maximum, where
$f_{xx} = \frac{y^2 - 16}{(16 - (x-3)^2 - y^2)^{3/2}}$ and $D = \frac{16}{(16 - (x-3)^2 - y^2)^2}$.
Both $f_x$ and $f_y$ are undefined along the circle $(x - 3)^2 + y^2 = 16$; at any point along this curve, $f(x, y) = 0$, the absolute minimum of the function.

15. The triangle is bound by the lines $y = -1$, $y = 2x + 1$ and $y = -2x + 1$.
Along $y = -1$, there is a critical point at $(0, -1)$.
Along $y = 2x + 1$, there is a critical point at $(-3/5, -1/5)$.
Along $y = -2x + 1$, there is a critical point at $(3/5, -1/5)$.
The function $f$ has one critical point, irrespective of the constraint, at $(0, -1/2)$.
Checking the value of $f$ at these four points, along with the three vertices of the triangle, we find the absolute maximum is at $(0, 1, 3)$ and the absolute minimum is at $(0, -1/2, 3/4)$.

17. The region has no "corners" or "vertices," just a smooth edge. To find critical points along the circle $x^2 + y^2 = 4$, we solve for $y^2$: $y^2 = 4 - x^2$. We can go further and state $y = \pm\sqrt{4 - x^2}$.
We can rewrite $f$ as
$f(x) = x^2 + 2x + (4 - x^2) + 2\sqrt{4 - x^2} = 2x + 4 + 2\sqrt{4 - x^2}$.
(We will return and use $-\sqrt{4 - x^2}$ later.) Solving $f'(x) = 0$, we get $x = \sqrt{2} \Rightarrow y = \sqrt{2}$. $f'(x)$ is also undefined at $x = \pm 2$, where $y = 0$.
Using $y = -\sqrt{4 - x^2}$, we rewrite $f(x, y)$ as
$f(x) = 2x + 4 - 2\sqrt{4 - x^2}$. Solving $f'(x) = 0$, we get $x = -\sqrt{2}$, $y = -\sqrt{2}$. Again, $f'(x)$ is undefined at $x = \pm 2$.
The function $z = f(x, y)$ itself has a critical point at $(-1, -1)$.
Checking the value of $f$ at $(-1, -1)$, $(\sqrt{2}, \sqrt{2})$, $(-\sqrt{2}, -\sqrt{2})$, $(2, 0)$ and $(-2, 0)$, we find the absolute maximum is at $(\sqrt{2}, \sqrt{2}, 4 + 4\sqrt{2})$ and the absolute minimum is at $(-1, -1, -2)$.

## Chapter 13

### Section 13.1

1. $C(y)$, meaning that instead of being just a constant, like the number 5, it is a function of $y$, which acts like a constant when taking derivatives with respect to $x$.

3. curve to curve, then from point to point

5. (a) $18x^2 + 42x - 117$

   (b) $-108$

7. (a) $x^4/2 - x^2 + 2x - 3/2$

   (b) $23/15$

9. (a) $\sin^2 y$

   (b) $\pi/2$

11. $\int_1^4 \int_{-2}^1 dy\, dx$ and $\int_{-2}^1 \int_1^4 dx\, dy$.
area of $R = 9u^2$

13. $\int_2^4 \int_{x-1}^{7-x} dy\, dx$. The order $dx\, dy$ needs two iterated integrals as $x$ is bounded above by two different functions. This gives:
$\int_1^3 \int_2^{y+1} dx\, dy + \int_3^5 \int_2^{7-y} dx\, dy$.
area of $R = 4u^2$

15. $\int_0^1 \int_{x^4}^{\sqrt{x}} dy\, dx$ and $\int_0^1 \int_{y^2}^{\sqrt[4]{y}} dx\, dy$

    area of $R = 7/15 u^2$

17.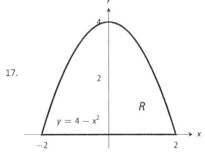

    area of $R = \int_0^4 \int_{-\sqrt{4-y}}^{\sqrt{4-y}} dx\, dy$

19.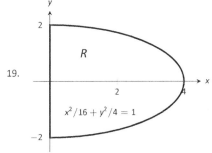

    area of $R = \int_0^4 \int_{-\sqrt{4-x^2/4}}^{\sqrt{4-x^2/4}} dy\, dx$

21.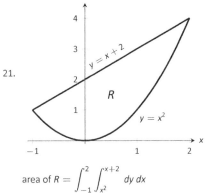

    area of $R = \int_{-1}^2 \int_{x^2}^{x+2} dy\, dx$

### Section 13.2

1. volume

3. The double integral gives the signed volume under the surface. Since the surface is always positive, it is always above the x-y plane and hence produces only "positive" volume.

5. $6;\ \int_{-1}^1 \int_1^2 \left(\dfrac{x}{y} + 3\right) dy\, dx$

7. $112/3;\ \int_0^2 \int_0^{4-2y} (3x^2 - y + 2)\, dx\, dy$

9. $16/5;\ \int_{-1}^1 \int_0^{1-x^2} (x + y + 2)\, dy\, dx$

11. (a)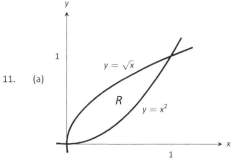

    (b) $\int_0^1 \int_{x^2}^{\sqrt{x}} x^2 y\, dy\, dx = \int_0^1 \int_{y^2}^{\sqrt{y}} x^2 y\, dx\, dy$.

    (c) $\dfrac{3}{56}$

13. (a)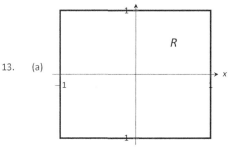

    (b) $\int_{-1}^1 \int_{-1}^1 x^2 - y^2\, dy\, dx = \int_{-1}^1 \int_{-1}^1 x^2 - y^2\, dx\, dy$.

    (c) 0

15. (a)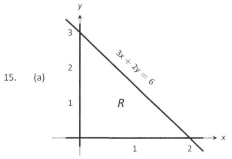

    (b)

    (c) $\int_0^2 \int_0^{3-3/2x} (6 - 3x - 2y)\, dy\, dx = \int_0^3 \int_0^{2-2/3y} (6 - 3x - 2y)\, dx\, dy$.

    (d) 6

17. (a)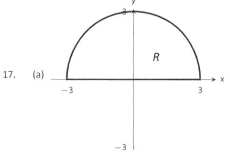

    (b) $\int_{-3}^3 \int_0^{\sqrt{9-x^2}} (x^3 y - x)\, dy\, dx = \int_0^3 \int_{-\sqrt{9-y^2}}^{\sqrt{9-y^2}} (x^3 y - x)\, dx\, dy$.

(c) 0

19. Integrating $e^{x^2}$ with respect to $x$ is not possible in terms of elementary functions. $\int_0^2 \int_0^{2x} e^{x^2}\,dy\,dx = e^4 - 1$.

21. Integrating $\int_y^1 \frac{2y}{x^2+y^2}\,dx$ gives $\tan^{-1}(1/y) - \pi/4$; integrating $\tan^{-1}(1/y)$ is hard.
$\int_0^1 \int_0^x \frac{2y}{x^2+y^2}\,dy\,dx = \ln 2$.

23. average value of $f = 6/2 = 3$

25. average value of $f = \frac{112/3}{4} = 28/3$

### Section 13.3

1. $f(r\cos\theta, r\sin\theta), r\,dr\,d\theta$

3. $\int_0^{2\pi} \int_0^1 (3r\cos\theta - r\sin\theta + 4)r\,dr\,d\theta = 4\pi$

5. $\int_0^\pi \int_{\cos\theta}^{3\cos\theta} (8 - r\sin\theta)r\,dr\,d\theta = 16\pi$

7. $\int_0^{2\pi} \int_1^2 (\ln(r^2))r\,dr\,d\theta = 2\pi(\ln 16 - 3/2)$

9. $\int_{-\pi/2}^{\pi/2} \int_0^6 (r^2\cos^2\theta - r^2\sin^2\theta)r\,dr\,d\theta =$
$\int_{-\pi/2}^{\pi/2} \int_0^6 (r^2\cos(2\theta))r\,dr\,d\theta = 0$

11. $\int_{-\pi/2}^{\pi/2} \int_0^5 (r^2)\,dr\,d\theta = 125\pi/3$

13. $\int_0^{\pi/4} \int_0^{\sqrt{8}} (r\cos\theta + r\sin\theta)r\,dr\,d\theta = 16\sqrt{2}/3$

15. (a) This is impossible to integrate with rectangular coordinates as $e^{-(x^2+y^2)}$ does not have an antiderivative in terms of elementary functions.

(b) $\int_0^{2\pi} \int_0^a re^{-r^2}\,dr\,d\theta = \pi(1 - e^{-a^2})$.

(c) $\lim_{a\to\infty} \pi(1 - e^{-a^2}) = \pi$. This implies that there is a finite volume under the surface $e^{-(x^2+y^2)}$ over the entire x-y plane.

### Section 13.4

1. Because they are scalar multiples of each other.

3. "little masses"

5. $M_x$ measures the moment about the x-axis, meaning we need to measure distance from the x-axis. Such measurements are measures in the y-direction.

7. $\bar{x} = 5.25$

9. $(\bar{x}, \bar{y}) = (0, 3)$

11. $M = 150$gm;

13. $M = 2$lb

15. $M = 16\pi \approx 50.27$kg

17. $M = 54\pi \approx 169.65$lb

19. $M = 150$gm; $M_y = 600$; $M_x = -75$; $(\bar{x}, \bar{y}) = (4, -0.5)$

21. $M = 2$lb; $M_y = 0$; $M_x = 2/3$; $(\bar{x}, \bar{y}) = (0, 1/3)$

23. $M = 16\pi \approx 50.27$kg; $M_y = 4\pi$; $M_x = 4\pi$; $(\bar{x}, \bar{y}) = (1/4, 1/4)$

25. $M = 54\pi \approx 169.65$lb; $M_y = 0$; $M_x = 504$; $(\bar{x}, \bar{y}) = (0, 2.97)$

27. $I_x = 64/3$; $I_y = 64/3$; $I_O = 128/3$

29. $I_x = 16/3$; $I_y = 64/3$; $I_O = 80/3$

### Section 13.5

1. arc length

3. surface areas

5. Intuitively, adding $h$ to $f$ only shifts $f$ up (i.e., parallel to the z-axis) and does not change its shape. Therefore it will not change the surface area over $R$.

Analytically, $f_x = g_x$ and $f_y = g_y$; therefore, the surface area of each is computed with identical double integrals.

7. $SA = \int_0^{2\pi} \int_0^{2\pi} \sqrt{1 + \cos^2 x \cos^2 y + \sin^2 x \sin^2 y}\,dx\,dy$

9. $SA = \int_{-1}^1 \int_{-1}^1 \sqrt{1 + 4x^2 + 4y^2}\,dx\,dy$

11. $SA = \int_0^3 \int_{-1}^1 \sqrt{1 + 9 + 49}\,dx\,dy = 6\sqrt{59} \approx 46.09$

13. This is easier in polar:
$SA = \int_0^{2\pi} \int_0^4 r\sqrt{1 + 4r^2\cos^2 t + 4r^2\sin^2 t}\,dr\,d\theta$
$= \int_0^{2\pi} \int_0^4 r\sqrt{1 + 4r^2}\,dr\,d\theta$
$= \frac{\pi}{6}(65\sqrt{65} - 1) \approx 273.87$

15. 
$SA = \int_0^2 \int_0^{2x} \sqrt{1 + 1 + 4x^2}\,dy\,dx$
$= \int_0^2 (2x\sqrt{2 + 4x^2})\,dx$
$= \frac{26}{3}\sqrt{2} \approx 12.26$

17. This is easier in polar:
$SA = \int_0^{2\pi} \int_0^5 r\sqrt{1 + \frac{4r^2\cos^2\theta + 4r^2\sin^2\theta}{r^2\sin^2\theta + r^2\cos^2\theta}}\,dr\,d\theta$
$= \int_0^{2\pi} \int_0^5 r\sqrt{5}\,dr\,d\theta$
$= 25\pi\sqrt{5} \approx 175.62$

19. Integrating in polar is easiest considering $R$:
$SA = \int_0^{2\pi} \int_0^1 r\sqrt{1 + c^2 + d^2}\,dr\,d\theta$
$= \int_0^{2\pi} \frac{1}{2}\left(\sqrt{1 + c^2 + d^2}\right) d\theta$
$= \pi\sqrt{1 + c^2 + d^2}$.

The value of $h$ does not matter as it only shifts the plane vertically (i.e., parallel to the z-axis). Different values of $h$ do not create different ellipses in the plane.

### Section 13.6

1. surface to surface, curve to curve and point to point

3. Answers can vary. From this section we used triple integration to find the volume of a solid region, the mass of a solid, and the center of mass of a solid.

5. $V = \int_{-1}^1 \int_{-1}^1 (8 - x^2 - y^2 - (2x + y))\,dx\,dy = 88/3$

7. $V = \int_0^\pi \int_0^x (\cos x \sin y + 2 - \sin x \cos y)\,dy\,dx = \pi^2 - \pi \approx 6.728$

9. $dz\,dy\,dx$: $\int_0^3 \int_0^{1-x/3} \int_0^{2-2x/3-2y} dz\,dy\,dx$

$dz\,dx\,dy$: $\int_0^1 \int_0^{3-3y} \int_0^{2-2x/3-2y} dz\,dx\,dy$

$dy\,dz\,dx$: $\int_0^3 \int_0^{2-2x/3} \int_0^{1-x/3-z/2} dy\,dz\,dx$

$dy\,dx\,dz$: $\int_0^2 \int_0^{3-3z/2} \int_0^{1-x/3-z/2} dy\,dx\,dz$

$dx\,dz\,dy$: $\int_0^1 \int_0^{2-2y} \int_0^{3-3y-3z/2} dx\,dz\,dy$

$dx\,dy\,dz$: $\int_0^2 \int_0^{1-z/2} \int_0^{3-3y-3z/2} dx\,dy\,dz$

$V = \int_0^3 \int_0^{1-x/3} \int_0^{2-2x/3-2y} dz\,dy\,dx = 1$.

11. $dz\,dy\,dx$: $\int_0^2 \int_{-2}^0 \int_{y^2/2}^{-y} dz\,dy\,dx$

$dz\,dx\,dy$: $\int_{-2}^0 \int_0^2 \int_{y^2/2}^{-y} dz\,dx\,dy$

$dy\,dz\,dx$: $\int_0^2 \int_0^2 \int_{-\sqrt{2z}}^{-z} dy\,dz\,dx$

$dy\,dx\,dz$: $\int_0^2 \int_0^2 \int_{-\sqrt{2z}}^{-z} dy\,dx\,dz$

$dx\,dz\,dy$: $\int_{-2}^0 \int_{y^2/2}^{-y} \int_0^2 dx\,dz\,dy$

$dx\,dy\,dz$: $\int_0^2 \int_{-\sqrt{2z}}^{-z} \int_0^2 dx\,dy\,dz$

$V = \int_0^2 \int_0^2 \int_{-\sqrt{2z}}^{-z} dy\,dz\,dx = 4/3$.

13. $dz\,dy\,dx$: $\int_0^2 \int_{1-x/2}^1 \int_0^{2x+4y-4} dz\,dy\,dx$

$dz\,dx\,dy$: $\int_0^1 \int_{2-2y}^2 \int_0^{2x+4y-4} dz\,dx\,dy$

$dy\,dz\,dx$: $\int_0^2 \int_0^{2x} \int_{z/4-x/2+1}^1 dy\,dz\,dx$

$dy\,dx\,dz$: $\int_0^4 \int_{z/2}^2 \int_{z/4-x/2+1}^1 dy\,dx\,dz$

$dx\,dz\,dy$: $\int_0^1 \int_0^{4y} \int_{z/2-2y+2}^2 dx\,dz\,dy$

$dx\,dy\,dz$: $\int_0^4 \int_{z/4}^1 \int_{z/2-2y+2}^2 dx\,dy\,dz$

$V = \int_0^4 \int_{z/4}^1 \int_{z/2-2y+2}^2 dx\,dy\,dz = 4/3$.

15. $dz\,dy\,dx$: $\int_0^1 \int_0^{1-x^2} \int_0^{\sqrt{1-y}} dz\,dy\,dx$

$dz\,dx\,dy$: $\int_0^1 \int_0^{\sqrt{1-y}} \int_0^{\sqrt{1-y}} dz\,dx\,dy$

$dy\,dz\,dx$: $\int_0^1 \int_0^x \int_0^{1-x^2} dy\,dz\,dx + \int_0^1 \int_x^1 \int_0^{1-z^2} dy\,dz\,dx$

$dy\,dx\,dz$: $\int_0^1 \int_0^z \int_0^{1-z^2} dy\,dx\,dz + \int_0^1 \int_z^1 \int_0^{1-x^2} dy\,dx\,dz$

$dx\,dz\,dy$: $\int_0^1 \int_0^{\sqrt{1-y}} \int_0^{\sqrt{1-y}} dx\,dz\,dy$

$dx\,dy\,dz$: $\int_0^1 \int_0^{1-z^2} \int_0^{\sqrt{1-y}} dx\,dy\,dz$

Answers will vary. Neither order is particularly "hard." The order $dz\,dy\,dx$ requires integrating a square root, so powers can be messy; the order $dy\,dz\,dx$ requires two triple integrals, but each uses only polynomials.

17. 8

19. $\pi$

21. $M = 10$, $M_{yz} = 15/2$, $M_{xz} = 5/2$, $M_{xy} = 5$;
$(\bar{x}, \bar{y}, \bar{z}) = (3/4, 1/4, 1/2)$

23. $M = 16/5$, $M_{yz} = 16/3$, $M_{xz} = 104/45$, $M_{xy} = 32/9$;
$(\bar{x}, \bar{y}, \bar{z}) = (5/3, 13/18, 10/9) \approx (1.67, 0.72, 1.11)$

### Section 13.7

1. In cylindrical, $r$ determines how far from the origin one goes in the x-y plane before considering the z-component. Equivalently, if on projects a point in cylindrical coordinates onto the x-y plane, $r$ will be the distance of this projection from the origin.

   In spherical, $\rho$ is the distance from the origin to the point.

3. Cylinders (tubes) centered at the origin, parallel to the z-axis; planes parallel to the z-axis that intersect the z-axis; planes parallel to the x-y plane.

5. (a) Cylindrical: $(2\sqrt{2}, \pi/4, 1)$ and $(2, 5\pi/6, 0)$
   Spherical: $(3, \pi/4, \cos^{-1}(1/3))$ and $(2, 5\pi/6, \pi/2)$

   (b) Rectangular: $(\sqrt{2}, \sqrt{2}, 2)$ and $(0, -3, -4)$
   Spherical: $(2\sqrt{2}, \pi/4, \pi/4)$ and $(5, 3\pi/2, \pi - \tan^{-1}(3/4))$

   (c) Rectangular: $(1, 1, \sqrt{2})$ and $(0, 0, 1)$
   Cylindrical: $(\sqrt{2}, \pi/4, \sqrt{2})$ and $(0, 0, 1)$

7. (a) A cylindrical surface or tube, centered along the z-axis of radius 1, extending from the x-y plane up to the plane $z = 1$ (i.e., the tube has a length of 1).

   (b) This is a region of space, being half of a tube with "thick" walls of inner radius 1 and outer radius 2, centered along the z-axis with a length of 1, where the half "below" the x-z plane is removed.

   (c) This is upper half of the sphere of radius 3 centered at the origin (i.e., the upper hemisphere).

   (d) This is a region of space, where the ball of radius 2, centered at the origin, is removed from the ball of radius 3, centered at the origin.

9. $\int_{\theta_1}^{\theta_2} \int_{r_1}^{r_2} \int_{z_1}^{z_2} h(r, \theta, z) r\,dz\,dr\,d\theta$

11. The region in space is bounded between the planes $z = 0$ and $z = 2$, inside of the cylinder $x^2 + y^2 = 4$, and the planes $\theta = 0$ and $\theta = \pi/2$: describes a "wedge" of a cylinder of height 2 and radius 2; the angle of the wedge is $\pi/2$, or $90°$.

13. Bounded between the plane $z = 1$ and the cone $z = 1 - \sqrt{x^2 + y^2}$: describes an inverted cone, with height of 1, point at $(0, 0, 1)$ and base radius of 1.

15. Describes a quarter of a ball of radius 3, centered at the origin; the quarter resides above the x-y plane and above the x-z plane.

17. Describes the portion of the unit ball that resides in the first octant.

19. Bounded above the cone $z = \sqrt{x^2 + y^2}$ and below the sphere $x^2 + y^2 + z^2 = 4$: describes a shape that is somewhat "diamond"-like; some think of it as looking like an ice cream cone (see Figure 13.7.8). It describes a cone, where the side makes an angle of $\pi/4$ with the positive z-axis, topped by the portion of the ball of radius 2, centered at the origin.

21. The region in space is bounded below by the cone $z = \sqrt{3}\sqrt{x^2 + y^2}$ and above by the plane $z = 1$: it describes a cone, with point at the origin, centered along the positive z-axis, with height of 1 and base radius of $\tan(\pi/6) = 1/\sqrt{3}$.

23. In cylindrical coordinates, the density is $\delta(r, \theta, z) = r + 1$. Thus mass is

$$\int_0^{2\pi} \int_0^2 \int_0^4 (r+1) r\,dz\,dr\,d\theta = 112\pi/3.$$

25. In cylindrical coordinates, the density is $\delta(r,\theta,z) = 1$. Thus mass is
$$\int_0^\pi \int_0^1 \int_0^{4-r\sin\theta} r\, dz\, dr\, d\theta = 2\pi - 2/3 \approx 5.617.$$

27. In cylindrical coordinates, the density is $\delta(r,\theta,z) = r+1$. Thus mass is
$$M = \int_0^{2\pi} \int_0^2 \int_0^4 (r+1)r\, dz\, dr\, d\theta = 112\pi/3.$$
We find $M_{yz} = 0$, $M_{xz} = 0$, and $M_{xy} = 224\pi/3$, placing the center of mass at $(0,0,2)$.

29. In cylindrical coordinates, the density is $\delta(r,\theta,z) = 1$. Thus mass is
$$\int_0^\pi \int_0^1 \int_0^{4-r\sin\theta} r\, dz\, dr\, d\theta = 2\pi - 2/3 \approx 5.617.$$
We find $M_{yz} = 0$, $M_{xz} = 8/3 - \pi/8$, and $M_{xy} = 65\pi/16 - 8/3$, placing the center of mass at $\approx (0, 0.405, 1.80)$.

31. In spherical coordinates, the density is $\delta(\rho,\theta,\varphi) = 1$. Thus mass is
$$\int_0^{\pi/2} \int_0^{2\pi} \int_0^1 \rho^2 \sin(\varphi)\, d\rho\, d\theta\, d\varphi = 2\pi/3.$$

33. In spherical coordinates, the density is $\delta(\rho,\theta,\varphi) = \rho\cos\varphi$. Thus mass is
$$\int_0^{\pi/4} \int_0^{2\pi} \int_0^1 (\rho\cos(\varphi))\rho^2 \sin(\varphi)\, d\rho\, d\theta\, d\varphi = \pi/8.$$

35. In spherical coordinates, the density is $\delta(\rho,\theta,\varphi) = 1$. Thus mass is
$$\int_0^{\pi/2} \int_0^{2\pi} \int_0^1 \rho^2 \sin(\varphi)\, d\rho\, d\theta\, d\varphi = 2\pi/3.$$
We find $M_{yz} = 0$, $M_{xz} = 0$, and $M_{xy} = \pi/4$, placing the center of mass at $(0, 0, 3/8)$.

37. In spherical coordinates, the density is $\delta(\rho,\theta,\varphi) = \rho\cos\varphi$. Thus mass is
$$\int_0^{\pi/4} \int_0^{2\pi} \int_0^1 (\rho\cos(\varphi))\rho^2 \sin(\varphi)\, d\rho\, d\theta\, d\varphi = \pi/8.$$
We find $M_{yz} = 0$, $M_{xz} = 0$, and $M_{xy} = (4-\sqrt{2})\pi/30$, placing the center of mass at $(0, 0, 4(4-\sqrt{2})/15)$.

39. Rectangular: $\int_{-1}^1 \int_{-\sqrt{1-x^2}}^{\sqrt{1-x^2}} \int_{-\sqrt{1-x^2-y^2}}^{\sqrt{1-x^2-y^2}} dz\, dy\, dx$
Cylindrical: $\int_0^{2\pi} \int_0^1 \int_{-\sqrt{1-r^2}}^{\sqrt{1-r^2}} r\, dz\, dr\, d\theta$
Spherical: $\int_0^\pi \int_0^{2\pi} \int_0^1 \rho^2 \sin(\varphi)\, d\rho\, d\theta\, d\varphi$
Spherical appears simplest, avoiding the integration of square-roots and using techniques such as Substitution; all bounds are constants.

41. Rectangular: $\int_{-1}^1 \int_{-\sqrt{1-x^2}}^{\sqrt{1-x^2}} \int_{\sqrt{x^2+y^2}}^1 dz\, dy\, dx$
Cylindrical: $\int_0^{2\pi} \int_0^1 \int_r^1 r\, dz\, dr\, d\theta$
Spherical: $\int_0^{\pi/4} \int_0^{2\pi} \int_0^{\sec\varphi} \rho^2 \sin(\varphi)\, d\rho\, d\theta\, d\varphi$
Cylindrical appears simplest, avoiding the integration of square-roots that rectangular uses. Spherical is not difficult, though it requires Substitution, an extra step.

# Chapter 14

## Section 14.1

1. When $C$ is a curve in the plane and $f$ is a surface defined over $C$, then $\int_C f(s)\, ds$ describes the area under the spatial curve that lies on $f$, over $C$.

3. The variable $s$ denotes the arc-length parameter, which is generally difficult to use. The Key Idea allows one to parametrize a curve using another, ideally easier-to-use, parameter.

5. $12\sqrt{2}$

7. $40\pi$

9. Over the first subcurve of $C$, the line integral has a value of $3/2$; over the second subcurve, the line integral has a value of $4/3$. The total value of the line integral is thus $17/6$.

11. $\int_0^1 (5t^2 + 2t + 2)\sqrt{(4t+1)^2 + 1}\, dt \approx 17.071$

13. $\oint_0^{2\pi} (10 - 4\cos^2 t - \sin^2 t)\sqrt{\cos^2 t + 4\sin^2 t}\, dt \approx 74.986$

15. $7\sqrt{26}/3$

17. $8\pi^3$

19. $M = 8\sqrt{2}\pi^2$; center of mass is $(0, -1/(2\pi), 8\pi/3)$.

## Section 14.2

1. Answers will vary. Appropriate answers include velocities of moving particles (air, water, etc.); gravitational or electromagnetic forces.

3. Specific answers will vary, though should relate to the idea that the vector field is spinning clockwise at that point.

5. Correct answers should look similar to

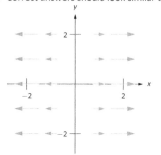

7. Correct answers should look similar to

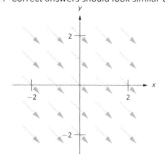

9. div $\vec{F} = 1 + 2y$
curl $\vec{F} = 0$

11. div $\vec{F} = x\cos(xy) - y\sin(xy)$
curl $\vec{F} = y\cos(xy) + x\sin(xy)$

13. div $\vec{F} = 3$
curl $\vec{F} = \langle -1, -1, -1 \rangle$

15. div $\vec{F} = 1 + 2y$
curl $\vec{F} = 0$

17. div $\vec{F} = 2y - \sin z$
curl $\vec{F} = \vec{0}$

## Section 14.3

1. False. It is true for line integrals over scalar fields, though.

3. True.

5. We can conclude that $\vec{F}$ is conservative.
7. $11/6$. (One parametrization for C is $\vec{r}(t) = \langle 3t, t \rangle$ on $0 \leq t \leq 1$.)
9. 0. (One parametrization for C is $\vec{r}(t) = \langle \cos t, \sin t \rangle$ on $0 \leq t \leq \pi$.)
11. 12. (One parametrization for C is $\vec{r}(t) = \langle 1, 2, 3 \rangle + t \langle 3, 1, -1 \rangle$ on $0 \leq t \leq 1$.)
13. 5/6 joules. (One parametrization for C is $\vec{r}(t) = \langle t, t \rangle$ on $0 \leq t \leq 1$.)
15. 24 ft-lbs.
17. (a) $f(x,y) = xy + x$
    (b) curl $\vec{F} = 0$.
    (c) 1. (One parametrization for C is $\vec{r}(t) = \langle t, -1t \rangle$ on $0 \leq t \leq 1$.)
    (d) 1 (with $A = (0,1)$ and $B = (1,0)$, $f(B) - f(A) = 1$.)
19. (a) $f(x,y) = x^2 yz$
    (b) curl $\vec{F} = \vec{0}$.
    (c) 250.
    (d) 250 (with $A = (1, -1, 0)$ and $B = (5, 5, 2)$, $f(B) - f(A) = 250$.)
21. Since $\vec{F}$ is conservative, it is the gradient of some potential function. That is, $\nabla f = \langle f_x, f_y, f_z \rangle = \vec{F} = \langle M, N, P \rangle$. In particular, $M = f_x$, $N = f_y$ and $P = f_z$.
    Note that
    curl $\vec{F} = \langle P_y - N_z, M_z - P_x, N_x - M_y \rangle = \langle f_{zy} - f_{yz}, f_{xz} - f_{zx}, f_{yx} - f_{xy} \rangle$, which, by Theorem 12.3.1, is $\langle 0, 0, 0 \rangle$.

### Section 14.4

1. along, across
3. the curl of $\vec{F}$, or curl $\vec{F}$
5. curl $\vec{F}$
7. 12
9. $-2/3$
11. $1/2$
13. The line integral $\oint_C \vec{F} \cdot d\vec{r}$, over the parabola, is 38/3; over the line, it is $-10$. The total line integral is thus $38/3 - 10 = 8/3$. The double integral of curl $\vec{F} = 2$ over R also has value $8/3$.
15. Three line integrals need to be computed to compute $\oint_C \vec{F} \cdot d\vec{r}$. It does not matter which corner one starts from first, but be sure to proceed around the triangle in a counterclockwise fashion.
    From $(0,0)$ to $(2,0)$, the line integral has a value of 0. From $(2,0)$ to $(1,1)$ the integral has a value of $7/3$. From $(1,1)$ to $(0,0)$ the line integral has a value of $-1/3$. Total value is 2.
    The double integral of curl $\vec{F}$ over R also has value 2.
17. Any choice of $\vec{F}$ is appropriate as long as curl $\vec{F} = 1$. When $\vec{F} = \langle -y/2, x/2 \rangle$, the integrand of the line integral is simply 6. The area of R is $12\pi$.
19. Any choice of $\vec{F}$ is appropriate as long as curl $\vec{F} = 1$. The choices of $\vec{F} = \langle -y, 0 \rangle$, $\langle 0, x \rangle$ and $\langle -y/2, x/2 \rangle$ each lead to reasonable integrands. The area of R is $16/15$.
21. The line integral $\oint_C \vec{F} \cdot \vec{n} \, ds$, over the parabola, is $-22/3$; over the line, it is 10. The total line integral is thus $-22/3 + 10 = 8/3$. The double integral of div $\vec{F} = 2$ over R also has value $8/3$.
23. Three line integrals need to be computed to compute $\oint_C \vec{F} \cdot \vec{n} \, ds$. It does not matter which corner one starts from first, but be sure to proceed around the triangle in a counterclockwise fashion.
    From $(0,0)$ to $(2,0)$, the line integral has a value of 0. From $(2,0)$ to $(1,1)$ the integral has a value of $1/3$. From $(1,1)$ to $(0,0)$ the line integral has a value of $1/3$. Total value is $2/3$.
    The double integral of div $\vec{F}$ over R also has value $2/3$.

### Section 14.5

1. Answers will vary, though generally should meaningfully include terms like "two sided".
3. (a) $\vec{r}(u,v) = \langle u, v, 3u^2 v \rangle$ on $-1 \leq u \leq 1, 0 \leq v \leq 2$.
   (b) $\vec{r}(u,v) = \langle 3v \cos u + 1, 3v \sin u + 2, 3(3v \cos u + 1)^2 (3v \sin u + 2) \rangle$, on $0 \leq u \leq 2\pi, 0 \leq v \leq 1$.
   (c) $\vec{r}(u,v) = \langle u, v(2 - 2u), 3u^2 v(2 - 2u) \rangle$ on $0 \leq u, v \leq 1$.
   (d) $\vec{r}(u,v) = \langle u, v(1 - u^2), 3u^2 v(1 - u^2) \rangle$ on $-1 \leq u \leq 1$, $0 \leq v \leq 1$.
5. $\vec{r}(u,v) = \langle 0, u, v \rangle$ with $0 \leq u \leq 2, 0 \leq v \leq 1$.
7. $\vec{r}(u,v) = \langle 3 \sin u \cos v, 2 \sin u \sin v, 4 \cos u \rangle$ with $0 \leq u \leq \pi$, $0 \leq v \leq 2\pi$.
9. Answers may vary.
   For $z = \frac{1}{2}(3 - x)$: $\vec{r}(u,v) = \langle u, v, \frac{1}{2}(3 - u) \rangle$, with $1 \leq u \leq 3$ and $0 \leq v \leq 2$.
   For $x = 1$: $\vec{r}(u,v) = \langle 0, u, v \rangle$, with $0 \leq u \leq 2, 0 \leq v \leq 1$
   For $y = 0$: $\vec{r}(u,v) = \langle u, 0, v/2(3 - u) \rangle$, with $1 \leq u \leq 3$, $0 \leq v \leq 1$
   For $y = 2$: $\vec{r}(u,v) = \langle u, 2, v/2(3 - u) \rangle$, with $1 \leq u \leq 3$, $0 \leq v \leq 1$
   For $z = 0$: $\vec{r}(u,v) = \langle u, v, 0 \rangle$, with $1 \leq u \leq 3, 0 \leq v \leq 2$
11. Answers may vary.
    For $z = 2y$: $\vec{r}(u,v) = \langle u, v(4 - u^2), 2v(4 - u^2) \rangle$ with $-2 \leq u \leq 2$ and $0 \leq v \leq 1$.
    For $y = 4 - x^2$: $\vec{r}(u,v) = \langle u, 4 - u^2, 2v(4 - u^2) \rangle$ with $-2 \leq u \leq 2$ and $0 \leq v \leq 1$.
    For $z = 0$: $\vec{r}(u,v) = \langle u, v(4 - u^2), 0 \rangle$ with $-2 \leq u \leq 2$ and $0 \leq v \leq 1$.
13. Answers may vary.
    For $x + y^2/9 = 1$: $\vec{r}(u,v) = \langle \cos u, 3 \sin u, v \rangle$ with $0 \leq u \leq 2\pi$ and $1 \leq v \leq 3$.
    For $z = 1$: $\vec{r}(u,v) = \langle v \cos u, 3v \sin u, 1 \rangle$ with $0 \leq u \leq 2\pi$ and $0 \leq v \leq 1$.
    For $z = 3$: $\vec{r}(u,v) = \langle v \cos u, 3v \sin u, 3 \rangle$ with $0 \leq u \leq 2\pi$ and $0 \leq v \leq 1$.
15. Answers may vary.
    For $z = 1 - x^2$: $\vec{r}(u,v) = \langle u, v, 1 - u^2 \rangle$ with $-1 \leq u \leq 1$ and $-1 \leq v \leq 2$.
    For $y = -1$: $\vec{r}(u,v) = \langle u, -1, v(1 - u^2) \rangle$ with $-1 \leq u \leq 1$ and $0 \leq v \leq 1$.
    For $y = 2$: $\vec{r}(u,v) = \langle u, 2, v(1 - u^2) \rangle$ with $-1 \leq u \leq 1$ and $0 \leq v \leq 1$.
    For $z = 0$: $\vec{r}(u,v) = \langle u, v, 0 \rangle$ with $-1 \leq u \leq 1$ and $-1 \leq v \leq 2$.
17. $S = 2\sqrt{14}$.
19. $S = 4\sqrt{3}\pi$.
21. $S = \int_0^3 \int_0^{2\pi} \sqrt{v^2 + 4v^4} \, du \, dv = (37\sqrt{37} - 1)\pi/6 \approx 117.319$.
23. $S = \int_0^1 \int_{-1}^1 \sqrt{(5u^2 - 2uv - 5)^2 + u^4 + (1 - u^2)^2} \, du \, dv \approx 7.084$.

### Section 14.6

1. curve; surface
3. outside
5. $240\sqrt{3}$
7. 24

9. 0

11. $-1/2$

13. 0; the flux over $S_1$ is $-45\pi$ and the flux over $S_2$ is $45\pi$.

### Section 14.7

1. Answers will vary; in Section 14.4, the Divergence Theorem connects outward flux over a closed curve in the plane to the divergence of the vector field, whereas in this section the Divergence Theorem connects outward flux over a closed surface in space to the divergence of the vector field.

3. Curl.

5. Outward flux across the plane $z = 2 - x/2 - 2y/3$ is 14; across the plane $z = 0$ the outward flux is $-8$; across the planes $x = 0$ and $y = 0$ the outward flux is 0.
   Total outward flux: 14.
   $\iint_D \text{div } \vec{F} \, dV = \int_0^4 \int_0^{3-3x/4} \int_0^{2-x/2-2y/3} (2x + 2y) \, dz \, dy \, dx = 14$.

7. Outward flux across the surface $z = xy(3-x)(3-y)$ is 252; across the plane $z = 0$ the outward flux is $-9$.
   Total outward flux: 243.
   $\iint_D \text{div } \vec{F} \, dV = \int_0^3 \int_0^3 \int_0^{xy(3-x)(3-y)} 12 \, dz \, dy \, dx = 243$.

9. Circulation on $C$: $\oint_C \vec{F} \cdot d\vec{r} = \pi$
   $\iint_S (\text{curl } \vec{F}) \cdot \vec{n} \, dS = \pi$.

11. Circulation on $C$: The flow along the line from $(0,0,2)$ to $(4,0,0)$ is 0; from $(4,0,0)$ to $(0,3,0)$ it is $-6$, and from $(0,3,0)$ to $(0,0,2)$ it is 6. The total circulation is $0 + (-6) + 6 = 0$.
    $\iint_S (\text{curl } \vec{F}) \cdot \vec{n} \, dS = \iint_S 0 \, dS = 0$.

13. $128/225$

15. $8192/105 \approx 78.019$

17. $5/3$

19. $23\pi$

21. Each field has a divergence of 1; by the Divergence Theorem, the total outward flux across $S$ is $\iint_D 1 \, dS$ for each field.

23. Answers will vary. Often the closed surface $S$ is composed of several smooth surfaces. To measure total outward flux, this may require evaluating multiple double integrals. Each double integral requires the parametrization of a surface and the computation of the cross product of partial derivatives. One triple integral may require less work, especially as the divergence of a vector field is generally easy to compute.

# Index

!, 405
Absolute Convergence Theorem, 456
absolute maximum, 129
absolute minimum, 129
Absolute Value Theorem, 410
acceleration, 77, 651
Alternating Harmonic Series, 427, 454, 467
Alternating Series Test, 450
$a_N$, 669, 679
analytic function, 488
angle of elevation, 656
antiderivative, 197
    of vector–valued function, 646
arc length, 379, 527, 553, 648, 673
arc length parameter, 673, 675
asymptote
    horizontal, 50
    vertical, 48
$a_T$, 669, 679
average rate of change, 635
average value of a function, 777
average value of function, 244

Binomial Series, 489
Bisection Method, 42
boundary point, 690
bounded sequence, 412
    convergence, 413
bounded set, 690

center of mass, 791–793, 795, 822
Chain Rule, 101
    multivariable, 721, 724
    notation, 107
circle of curvature, 678
circulation, 870
closed, 690
closed disk, 690
concave down, 151
concave up, 151
concavity, 151, 524
    inflection point, 152
    test for, 152
conic sections, 498
    degenerate, 498
    ellipse, 501
    hyperbola, 504
    parabola, 498
connected, 865
    simply, 865
conservative field, 865, 866, 868

Constant Multiple Rule
    of derivatives, 84
    of integration, 201
    of series, 427
constrained optimization, 754
continuous function, 37, 696
    properties, 40, 697
    vector–valued, 638
contour lines, 684
convergence
    absolute, 454, 456
    Alternating Series Test, 450
    conditional, 454
    Direct Comparison Test, 437
        for integration, 347
    Integral Test, 434
    interval of, 462
    Limit Comparison Test, 438
        for integration, 349
    $n^{\text{th}}$–term test, 429
    of geometric series, 422
    of improper int., 342, 347, 349
    of monotonic sequences, 416
    of $p$-series, 423
    of power series, 461
    of sequence, 408, 413
    of series, 419
    radius of, 462
    Ratio Comparison Test, 443
    Root Comparison Test, 446
coordinates
    cylindrical, 828
    polar, 533
    spherical, 831
critical number, 131
critical point, 131, 749–751
cross product
    and derivatives, 643
    applications, 605
        area of parallelogram, 606
        torque, 608
        volume of parallelepiped, 607
    definition, 601
    properties, 603, 604
curl, 853
    of conservative fields, 868
curvature, 675
    and motion, 679
    equations for, 677
    of circle, 677, 678
    radius of, 678

curve
    parametrically defined, 511
    rectangular equation, 511
    smooth, 517
curve sketching, 159
cusp, 517
cycloid, 633
cylinder, 563
cylindrical coordinates, 828

decreasing function, 142
    finding intervals, 143
definite integral, 209
    and substitution, 278
    of vector–valued function, 646
    properties, 211
del operator, 851
derivative
    acceleration, 78
    as a function, 66
    at a point, 62
    basic rules, 82
    Chain Rule, 101, 107, 721, 724
    Constant Multiple Rule, 84
    Constant Rule, 82
    differential, 189
    directional, 729, 731, 732, 735, 736
    exponential functions, 107
    First Deriv. Test, 145
    Generalized Power Rule, 102
    higher order, 85
        interpretation, 86
    hyperbolic funct., 324
    implicit, 111, 726
    interpretation, 75
    inverse function, 122
    inverse hyper., 327
    inverse trig., 125
    logarithmic differentiation, 118
    Mean Value Theorem, 138
    mixed partial, 704
    motion, 78
    multivariable differentiability, 713, 718
    normal line, 63
    notation, 66, 85
    parametric equations, 521
    partial, 700, 708
    Power Rule, 82, 95, 116
    power series, 465
    Product Rule, 89
    Quotient Rule, 92
    second, 85
    Second Deriv. Test, 155
    Sum/Difference Rule, 84
    tangent line, 62
    third, 85
    trigonometric functions, 94
    vector–valued functions, 639, 640, 643
    velocity, 78
differentiable, 62, 713, 718

differential, 189
    notation, 189
Direct Comparison Test
    for integration, 347
    for series, 437
directional derivative, 729, 731, 732, 735, 736
directrix, 498, 563
Disk Method, 364
displacement, 238, 634, 648
distance
    between lines, 619
    between point and line, 619
    between point and plane, 628
    between points in space, 560
    traveled, 659
divergence, 852, 853
    Alternating Series Test, 450
    Direct Comparison Test, 437
        for integration, 347
    Integral Test, 434
    Limit Comparison Test, 438
        for integration, 349
    $n^{\text{th}}$–term test, 429
    of geometric series, 422
    of improper int., 342, 347, 349
    of $p$-series, 423
    of sequence, 408
    of series, 419
    Ratio Comparison Test, 443
    Root Comparison Test, 446
Divergence Theorem
    in space, 900
    in the plane, 877
dot product
    and derivatives, 643
    definition, 588
    properties, 589, 590
double integral, 770, 771
    in polar, 781
    properties, 774

eccentricity, 503, 507
elementary function, 248
ellipse
    definition, 501
    eccentricity, 503
    parametric equations, 517
    reflective property, 504
    standard equation, 502
extrema
    absolute, 129, 749
    and First Deriv. Test, 145
    and Second Deriv. Test, 155
    finding, 132
    relative, 130, 749, 750
Extreme Value Theorem, 130, 754
extreme values, 129

factorial, 405
First Derivative Test, 145

first octant, 560
floor function, 38
flow, 870, 872
fluid pressure/force, 397, 399
flux, 870, 872, 893, 894
focus, 498, 501, 504
Fubini's Theorem, 771
function
    of three variables, 687
    of two variables, 683
    vector–valued, 631
Fundamental Theorem of Calculus, 236, 237
    and Chain Rule, 240
Fundamental Theorem of Line Integrals, 864, 866

Gabriel's Horn, 384
Gauss's Law, 904
Generalized Power Rule, 102
geometric series, 421, 422
gradient, 731, 732, 735, 736, 746
    and level curves, 732
    and level surfaces, 746
Green's Theorem, 874

Harmonic Series, 427
Head To Tail Rule, 578
Hooke's Law, 390
hyperbola
    definition, 504
    eccentricity, 507
    parametric equations, 517
    reflective property, 507
    standard equation, 505
hyperbolic function
    definition, 321
    derivatives, 324
    identities, 324
    integrals, 324
    inverse, 325
        derivative, 327
        integration, 327
        logarithmic def., 326

implicit differentiation, 111, 726
improper integration, 342, 345
incompressible vector field, 852
increasing function, 142
    finding intervals, 143
indefinite integral, 197
    of vector–valued function, 646
indeterminate form, 2, 49, 335, 336
inflection point, 152
initial point, 574
initial value problem, 202
Integral Test, 434
integration
    arc length, 379
    area, 209, 762, 763
    area between curves, 241, 354
    average value, 244
    by parts, 283
    by substitution, 265
    definite, 209
        and substitution, 278
        properties, 211
        Riemann Sums, 232
    displacement, 238
    distance traveled, 659
    double, 770
    fluid force, 397, 399
    Fun. Thm. of Calc., 236, 237
    general application technique, 353
    hyperbolic funct., 324
    improper, 342, 345, 347, 349
    indefinite, 197
    inverse hyper., 327
    iterated, 761
    Mean Value Theorem, 243
    multiple, 761
    notation, 198, 209, 237, 761
    numerical, 248
        Left/Right Hand Rule, 248, 255
        Simpson's Rule, 253, 255, 256
        Trapezoidal Rule, 251, 255, 256
    of multivariable functions, 759
    of power series, 465
    of trig. functions, 271
    of trig. powers, 294, 299
    of vector–valued function, 646
    of vector–valued functions, 646
    partial fraction decomp., 314
    Power Rule, 202
    Sum/Difference Rule, 202
    surface area, 383, 529, 554
    trig. subst., 305
    triple, 808, 819–821
    volume
        cross-sectional area, 362
        Disk Method, 364
        Shell Method, 371, 375
        Washer Method, 366, 375
    with cylindrical coordinates, 829
    with spherical coordinates, 833
    work, 387
interior point, 690
Intermediate Value Theorem, 42
interval of convergence, 462
iterated integration, 761, 770, 771, 808, 819–821
    changing order, 765
    properties, 774, 814

L'Hôpital's Rule, 332, 334
lamina, 787
Left Hand Rule, 218, 223, 248
Left/Right Hand Rule, 255
level curves, 684, 732
level surface, 688, 746
limit
    Absolute Value Theorem, 410
    at infinity, 50
    definition, 10

difference quotient, 6
does not exist, 4, 32
indeterminate form, 2, 49, 335, 336
L'Hôpital's Rule, 332, 334
left handed, 30
of infinity, 46
of multivariable function, 691, 692, 698
of sequence, 408
of vector–valued functions, 637
one sided, 30
properties, 18, 692
pseudo-definition, 2
right handed, 30
Squeeze Theorem, 22
Limit Comparison Test
for integration, 349
for series, 438
line integral
Fundamental Theorem, 864, 866
over scalar field, 841, 843, 859
over vector field, 860
path independent, 865, 866
properties over a scalar field, 846
properties over a vector field, 863
lines, 612
distances between, 619
equations for, 614
intersecting, 615
parallel, 615
skew, 615
logarithmic differentiation, 118

Möbius band, 881
Maclaurin Polynomial, *see* Taylor Polynomial
definition, 474
Maclaurin Series, *see* Taylor Series
definition, 485
magnitude of vector, 574
mass, 787, 788, 822, 847
center of, 791, 847
maximum
absolute, 129, 749
and First Deriv. Test, 145
and Second Deriv. Test, 155
relative/local, 130, 749, 752
Mean Value Theorem
of differentiation, 138
of integration, 243
Midpoint Rule, 218, 223
minimum
absolute, 129, 749
and First Deriv. Test, 145, 155
relative/local, 130, 749, 752
moment, 793, 795, 822
monotonic sequence, 414
multiple integration, *see* iterated integration
multivariable function, 683, 687
continuity, 696–698, 714, 719
differentiability, 713, 714, 718, 719
domain, 683, 687

level curves, 684
level surface, 688
limit, 691, 692, 698
range, 683, 687

Newton's Method, 168
norm, 574
normal line, 63, 521, 742
normal vector, 623
$n^{\text{th}}$–term test, 429
numerical integration, 248
Left/Right Hand Rule, 248, 255
Simpson's Rule, 253, 255
error bounds, 256
Trapezoidal Rule, 251, 255
error bounds, 256

octant
first, 560
one to one, 880
open, 690
open ball, 698
open disk, 690
optimization, 181
constrained, 754
orientable, 880
orthogonal, 592, 742
decomposition, 596
orthogonal decomposition of vectors, 596
orthogonal projection, 594
osculating circle, 678
outer unit normal vector, 900

*p*-series, 423
parabola
definition, 498
general equation, 499
reflective property, 501
parallel vectors, 582
Parallelogram Law, 578
parametric equations
arc length, 527
concavity, 524
definition, 511
finding $\frac{d^2y}{dx^2}$, 525
finding $\frac{dy}{dx}$, 521
normal line, 521
of a surface, 880
surface area, 529
tangent line, 521
parametrized surface, 880
partial derivative, 700, 708
high order, 708
meaning, 702
mixed, 704
second derivative, 704
total differential, 712, 718
partition, 225
size of, 225
path independent, 865, 866

perpendicular, *see* orthogonal
piecewise smooth curve, 846
planes
    coordinate plane, 562
    distance between point and plane, 628
    equations of, 624
    introduction, 562
    normal vector, 623
    tangent, 745
point of inflection, 152
polar
    coordinates, 533
    function
        arc length, 553
        gallery of graphs, 540
        surface area, 554
    functions, 536
        area, 549
        area between curves, 551
        finding $\frac{dy}{dx}$, 546
        graphing, 536
polar coordinates, 533
    plotting points, 533
potential function, 857, 866
Power Rule
    differentiation, 82, 89, 95, 116
    integration, 202
power series, 460
    algebra of, 491
    convergence, 461
    derivatives and integrals, 465
projectile motion, 656, 657, 670

quadric surface
    definition, 566
    ellipsoid, 568
    elliptic cone, 567
    elliptic paraboloid, 567
    gallery, 567–569
    hyperbolic paraboloid, 569
    hyperboloid of one sheet, 568
    hyperboloid of two sheets, 569
    sphere, 568
    trace, 566
Quotient Rule, 92

$\mathbb{R}$, 574
radius of convergence, 462
radius of curvature, 678
Ratio Comparison Test
    for series, 443
rearrangements of series, 455, 456
related rates, 174
Riemann Sum, 218, 222, 225
    and definite integral, 232
Right Hand Rule, 218, 223, 248
right hand rule
    of Cartesian coordinates, 560
    of the cross product, 605
Rolle's Theorem, 138

Root Comparison Test
    for series, 446

saddle point, 751, 752
Second Derivative Test, 155, 752
sensitivity analysis, 717
sequence
    Absolute Value Theorem, 410
    positive, 437
sequences
    boundedness, 412
    convergent, 408, 413, 416
    definition, 405
    divergent, 408
    limit, 408
    limit properties, 411
    monotonic, 414
series
    absolute convergence, 454
    Absolute Convergence Theorem, 456
    alternating, 449
        Approximation Theorem, 452
    Alternating Series Test, 450
    Binomial, 489
    conditional convergence, 454
    convergent, 419
    definition, 419
    Direct Comparison Test, 437
    divergent, 419
    geometric, 421, 422
    Integral Test, 434
    interval of convergence, 462
    Limit Comparison Test, 438
    Maclaurin, 485
    $n^{\text{th}}$–term test, 429
    $p$-series, 423
    partial sums, 419
    power, 460, 461
        derivatives and integrals, 465
    properties, 427
    radius of convergence, 462
    Ratio Comparison Test, 443
    rearrangements, 455, 456
    Root Comparison Test, 446
    Taylor, 485
    telescoping, 424, 425
Shell Method, 371, 375
signed area, 209
signed volume, 770, 771
simple curve, 865
simply connected, 865
Simpson's Rule, 253, 255
    error bounds, 256
smooth, 642
    curve, 517
    surface, 880
smooth curve
    piecewise, 846
speed, 651
sphere, 561

spherical coordinates, 831
Squeeze Theorem, 22
Stokes' Theorem, 906
Sum/Difference Rule
    of derivatives, 84
    of integration, 202
    of series, 427
summation
    notation, 219
    properties, 221
surface, 880
    smooth, 880
surface area, 800
    of parametrized surface, 886, 887
    solid of revolution, 383, 529, 554
surface integral, 891
surface of revolution, 564, 565

tangent line, 62, 521, 546, 641
    directional, 739
tangent plane, 745
Taylor Polynomial
    definition, 474
    Taylor's Theorem, 477
Taylor Series
    common series, 491
    definition, 485
    equality with generating function, 487
Taylor's Theorem, 477
telescoping series, 424, 425
terminal point, 574
torque, 608
total differential, 712, 718
    sensitivity analysis, 717
total signed area, 209
trace, 566
Trapezoidal Rule, 251, 255
    error bounds, 256
triple integral, 808, 819–821
    properties, 814

unbounded sequence, 412
unbounded set, 690
unit normal vector
    $a_N$, 669
    and acceleration, 668, 669
    and curvature, 679
    definition, 666
    in $\mathbb{R}^2$, 668
unit tangent vector
    and acceleration, 668, 669
    and curvature, 675, 679
    $a_T$, 669
    definition, 664
    in $\mathbb{R}^2$, 668
unit vector, 580
    properties, 582
    standard unit vector, 584
    unit normal vector, 666
    unit tangent vector, 664

vector field, 850
    conservative, 865, 866
    curl of, 853
    divergence of, 852, 853
    over vector field, 860
    potential function of, 857, 866
vector–valued function
    algebra of, 632
    arc length, 648
    average rate of change, 635
    continuity, 638
    definition, 631
    derivatives, 639, 640, 643
    describing motion, 651
    displacement, 634
    distance traveled, 659
    graphing, 631
    integration, 646
    limits, 637
    of constant length, 645, 655, 656, 665
    projectile motion, 656, 657
    smooth, 642
    tangent line, 641
vectors, 574
    algebra of, 577
    algebraic properties, 580
    component form, 575
    cross product, 601, 603, 604
    definition, 574
    dot product, 588–590
    Head To Tail Rule, 578
    magnitude, 574
    norm, 574
    normal vector, 623
    orthogonal, 592
    orthogonal decomposition, 596
    orthogonal projection, 594
    parallel, 582
    Parallelogram Law, 578
    resultant, 578
    standard unit vector, 584
    unit vector, 580, 582
    zero vector, 578
velocity, 77, 651
volume, 770, 771, 806

Washer Method, 366, 375
work, 387, 599

# Differentiation Rules

1. $\dfrac{d}{dx}(cx) = c$
2. $\dfrac{d}{dx}(u \pm v) = u' \pm v'$
3. $\dfrac{d}{dx}(u \cdot v) = uv' + u'v$
4. $\dfrac{d}{dx}\left(\dfrac{u}{v}\right) = \dfrac{vu' - uv'}{v^2}$
5. $\dfrac{d}{dx}(u(v)) = u'(v)v'$
6. $\dfrac{d}{dx}(c) = 0$
7. $\dfrac{d}{dx}(x) = 1$
8. $\dfrac{d}{dx}(x^n) = nx^{n-1}$
9. $\dfrac{d}{dx}(e^x) = e^x$
10. $\dfrac{d}{dx}(a^x) = \ln a \cdot a^x$
11. $\dfrac{d}{dx}(\ln x) = \dfrac{1}{x}$
12. $\dfrac{d}{dx}(\log_a x) = \dfrac{1}{\ln a} \cdot \dfrac{1}{x}$
13. $\dfrac{d}{dx}(\sin x) = \cos x$
14. $\dfrac{d}{dx}(\cos x) = -\sin x$
15. $\dfrac{d}{dx}(\csc x) = -\csc x \cot x$
16. $\dfrac{d}{dx}(\sec x) = \sec x \tan x$
17. $\dfrac{d}{dx}(\tan x) = \sec^2 x$
18. $\dfrac{d}{dx}(\cot x) = -\csc^2 x$
19. $\dfrac{d}{dx}(\sin^{-1} x) = \dfrac{1}{\sqrt{1-x^2}}$
20. $\dfrac{d}{dx}(\cos^{-1} x) = \dfrac{-1}{\sqrt{1-x^2}}$
21. $\dfrac{d}{dx}(\csc^{-1} x) = \dfrac{-1}{|x|\sqrt{x^2-1}}$
22. $\dfrac{d}{dx}(\sec^{-1} x) = \dfrac{1}{|x|\sqrt{x^2-1}}$
23. $\dfrac{d}{dx}(\tan^{-1} x) = \dfrac{1}{1+x^2}$
24. $\dfrac{d}{dx}(\cot^{-1} x) = \dfrac{-1}{1+x^2}$
25. $\dfrac{d}{dx}(\cosh x) = \sinh x$
26. $\dfrac{d}{dx}(\sinh x) = \cosh x$
27. $\dfrac{d}{dx}(\tanh x) = \text{sech}^2 x$
28. $\dfrac{d}{dx}(\text{sech } x) = -\text{sech } x \tanh x$
29. $\dfrac{d}{dx}(\text{csch } x) = -\text{csch } x \coth x$
30. $\dfrac{d}{dx}(\coth x) = -\text{csch}^2 x$
31. $\dfrac{d}{dx}(\cosh^{-1} x) = \dfrac{1}{\sqrt{x^2-1}}$
32. $\dfrac{d}{dx}(\sinh^{-1} x) = \dfrac{1}{\sqrt{x^2+1}}$
33. $\dfrac{d}{dx}(\text{sech}^{-1} x) = \dfrac{-1}{x\sqrt{1-x^2}}$
34. $\dfrac{d}{dx}(\text{csch}^{-1} x) = \dfrac{-1}{|x|\sqrt{1+x^2}}$
35. $\dfrac{d}{dx}(\tanh^{-1} x) = \dfrac{1}{1-x^2}$
36. $\dfrac{d}{dx}(\coth^{-1} x) = \dfrac{1}{1-x^2}$

# Integration Rules

1. $\displaystyle\int c \cdot f(x)\, dx = c \int f(x)\, dx$
2. $\displaystyle\int f(x) \pm g(x)\, dx = \int f(x)\, dx \pm \int g(x)\, dx$
3. $\displaystyle\int 0\, dx = C$
4. $\displaystyle\int 1\, dx = x + C$
5. $\displaystyle\int x^n\, dx = \dfrac{1}{n+1}x^{n+1} + C,\ n \neq -1$
6. $\displaystyle\int e^x\, dx = e^x + C$
7. $\displaystyle\int \ln x\, dx = x \ln x - x + C$
8. $\displaystyle\int a^x\, dx = \dfrac{1}{\ln a} \cdot a^x + C$
9. $\displaystyle\int \dfrac{1}{x}\, dx = \ln|x| + C$
10. $\displaystyle\int \cos x\, dx = \sin x + C$
11. $\displaystyle\int \sin x\, dx = -\cos x + C$
12. $\displaystyle\int \tan x\, dx = -\ln|\cos x| + C$
13. $\displaystyle\int \sec x\, dx = \ln|\sec x + \tan x| + C$
14. $\displaystyle\int \csc x\, dx = -\ln|\csc x + \cot x| + C$
15. $\displaystyle\int \cot x\, dx = \ln|\sin x| + C$
16. $\displaystyle\int \sec^2 x\, dx = \tan x + C$
17. $\displaystyle\int \csc^2 x\, dx = -\cot x + C$
18. $\displaystyle\int \sec x \tan x\, dx = \sec x + C$
19. $\displaystyle\int \csc x \cot x\, dx = -\csc x + C$
20. $\displaystyle\int \cos^2 x\, dx = \dfrac{1}{2}x + \dfrac{1}{4}\sin(2x) + C$
21. $\displaystyle\int \sin^2 x\, dx = \dfrac{1}{2}x - \dfrac{1}{4}\sin(2x) + C$
22. $\displaystyle\int \dfrac{1}{x^2+a^2}\, dx = \dfrac{1}{a}\tan^{-1}\left(\dfrac{x}{a}\right) + C$
23. $\displaystyle\int \dfrac{1}{\sqrt{a^2-x^2}}\, dx = \sin^{-1}\left(\dfrac{x}{a}\right) + C$
24. $\displaystyle\int \dfrac{1}{x\sqrt{x^2-a^2}}\, dx = \dfrac{1}{a}\sec^{-1}\left(\dfrac{|x|}{a}\right) + C$
25. $\displaystyle\int \cosh x\, dx = \sinh x + C$
26. $\displaystyle\int \sinh x\, dx = \cosh x + C$
27. $\displaystyle\int \tanh x\, dx = \ln(\cosh x) + C$
28. $\displaystyle\int \coth x\, dx = \ln|\sinh x| + C$
29. $\displaystyle\int \dfrac{1}{\sqrt{x^2-a^2}}\, dx = \ln\left|x + \sqrt{x^2-a^2}\right| + C$
30. $\displaystyle\int \dfrac{1}{\sqrt{x^2+a^2}}\, dx = \ln\left|x + \sqrt{x^2+a^2}\right| + C$
31. $\displaystyle\int \dfrac{1}{a^2-x^2}\, dx = \dfrac{1}{2a}\ln\left|\dfrac{a+x}{a-x}\right| + C$
32. $\displaystyle\int \dfrac{1}{x\sqrt{a^2-x^2}}\, dx = \dfrac{1}{a}\ln\left(\dfrac{x}{a+\sqrt{a^2-x^2}}\right) + C$
33. $\displaystyle\int \dfrac{1}{x\sqrt{x^2+a^2}}\, dx = \dfrac{1}{a}\ln\left|\dfrac{x}{a+\sqrt{x^2+a^2}}\right| + C$

## The Unit Circle

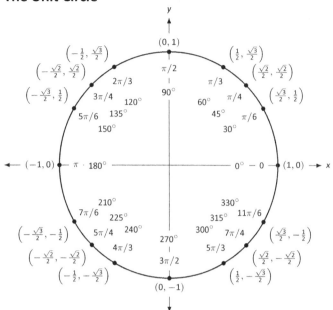

## Definitions of the Trigonometric Functions

### Unit Circle Definition

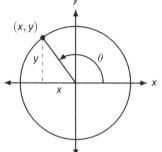

$$\sin\theta = y \qquad \cos\theta = x$$
$$\csc\theta = \frac{1}{y} \qquad \sec\theta = \frac{1}{x}$$
$$\tan\theta = \frac{y}{x} \qquad \cot\theta = \frac{x}{y}$$

### Right Triangle Definition

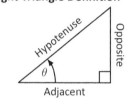

$$\sin\theta = \frac{O}{H} \qquad \csc\theta = \frac{H}{O}$$
$$\cos\theta = \frac{A}{H} \qquad \sec\theta = \frac{H}{A}$$
$$\tan\theta = \frac{O}{A} \qquad \cot\theta = \frac{A}{O}$$

## Common Trigonometric Identities

**Pythagorean Identities**

$\sin^2 x + \cos^2 x = 1$

$\tan^2 x + 1 = \sec^2 x$

$1 + \cot^2 x = \csc^2 x$

**Cofunction Identities**

$\sin\left(\dfrac{\pi}{2} - x\right) = \cos x \qquad \csc\left(\dfrac{\pi}{2} - x\right) = \sec x$

$\cos\left(\dfrac{\pi}{2} - x\right) = \sin x \qquad \sec\left(\dfrac{\pi}{2} - x\right) = \csc x$

$\tan\left(\dfrac{\pi}{2} - x\right) = \cot x \qquad \cot\left(\dfrac{\pi}{2} - x\right) = \tan x$

**Double Angle Formulas**

$\sin 2x = 2\sin x \cos x$

$\cos 2x = \cos^2 x - \sin^2 x$

$\qquad = 2\cos^2 x - 1$

$\qquad = 1 - 2\sin^2 x$

$\tan 2x = \dfrac{2\tan x}{1 - \tan^2 x}$

**Sum to Product Formulas**

$\sin x + \sin y = 2\sin\left(\dfrac{x+y}{2}\right)\cos\left(\dfrac{x-y}{2}\right)$

$\sin x - \sin y = 2\sin\left(\dfrac{x-y}{2}\right)\cos\left(\dfrac{x+y}{2}\right)$

$\cos x + \cos y = 2\cos\left(\dfrac{x+y}{2}\right)\cos\left(\dfrac{x-y}{2}\right)$

$\cos x - \cos y = -2\sin\left(\dfrac{x+y}{2}\right)\sin\left(\dfrac{x-y}{2}\right)$

**Power–Reducing Formulas**

$\sin^2 x = \dfrac{1 - \cos 2x}{2}$

$\cos^2 x = \dfrac{1 + \cos 2x}{2}$

$\tan^2 x = \dfrac{1 - \cos 2x}{1 + \cos 2x}$

**Even/Odd Identities**

$\sin(-x) = -\sin x$

$\cos(-x) = \cos x$

$\tan(-x) = -\tan x$

$\csc(-x) = -\csc x$

$\sec(-x) = \sec x$

$\cot(-x) = -\cot x$

**Product to Sum Formulas**

$\sin x \sin y = \dfrac{1}{2}\big(\cos(x-y) - \cos(x+y)\big)$

$\cos x \cos y = \dfrac{1}{2}\big(\cos(x-y) + \cos(x+y)\big)$

$\sin x \cos y = \dfrac{1}{2}\big(\sin(x+y) + \sin(x-y)\big)$

**Angle Sum/Difference Formulas**

$\sin(x \pm y) = \sin x \cos y \pm \cos x \sin y$

$\cos(x \pm y) = \cos x \cos y \mp \sin x \sin y$

$\tan(x \pm y) = \dfrac{\tan x \pm \tan y}{1 \mp \tan x \tan y}$

# Areas and Volumes

## Triangles

$h = a \sin \theta$

Area = $\frac{1}{2}bh$

Law of Cosines:
$c^2 = a^2 + b^2 - 2ab \cos \theta$

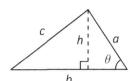

## Right Circular Cone

Volume = $\frac{1}{3}\pi r^2 h$

Surface Area =
$\pi r \sqrt{r^2 + h^2} + \pi r^2$

## Parallelograms

Area = $bh$

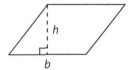

## Right Circular Cylinder

Volume = $\pi r^2 h$

Surface Area =
$2\pi rh + 2\pi r^2$

## Trapezoids

Area = $\frac{1}{2}(a + b)h$

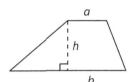

## Sphere

Volume = $\frac{4}{3}\pi r^3$

Surface Area = $4\pi r^2$

## Circles

Area = $\pi r^2$

Circumference = $2\pi r$

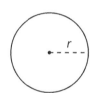

## General Cone

Area of Base = $A$

Volume = $\frac{1}{3}Ah$

## Sectors of Circles

$\theta$ in radians

Area = $\frac{1}{2}\theta r^2$

$s = r\theta$

## General Right Cylinder

Area of Base = $A$

Volume = $Ah$

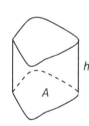

# Algebra

## Factors and Zeros of Polynomials
Let $p(x) = a_n x^n + a_{n-1} x^{n-1} + \cdots + a_1 x + a_0$ be a polynomial. If $p(a) = 0$, then $a$ is a *zero* of the polynomial and a solution of the equation $p(x) = 0$. Furthermore, $(x - a)$ is a *factor* of the polynomial.

## Fundamental Theorem of Algebra
An *n*th degree polynomial has *n* (not necessarily distinct) zeros. Although all of these zeros may be imaginary, a real polynomial of odd degree must have at least one real zero.

## Quadratic Formula
If $p(x) = ax^2 + bx + c$, and $0 \leq b^2 - 4ac$, then the real zeros of $p$ are $x = (-b \pm \sqrt{b^2 - 4ac})/2a$

## Special Factors
$$x^2 - a^2 = (x - a)(x + a) \qquad x^3 - a^3 = (x - a)(x^2 + ax + a^2)$$
$$x^3 + a^3 = (x + a)(x^2 - ax + a^2) \qquad x^4 - a^4 = (x^2 - a^2)(x^2 + a^2)$$
$$(x + y)^n = x^n + nx^{n-1}y + \frac{n(n-1)}{2!}x^{n-2}y^2 + \cdots + nxy^{n-1} + y^n$$
$$(x - y)^n = x^n - nx^{n-1}y + \frac{n(n-1)}{2!}x^{n-2}y^2 - \cdots \pm nxy^{n-1} \mp y^n$$

## Binomial Theorem
$$(x + y)^2 = x^2 + 2xy + y^2 \qquad (x - y)^2 = x^2 - 2xy + y^2$$
$$(x + y)^3 = x^3 + 3x^2 y + 3xy^2 + y^3 \qquad (x - y)^3 = x^3 - 3x^2 y + 3xy^2 - y^3$$
$$(x + y)^4 = x^4 + 4x^3 y + 6x^2 y^2 + 4xy^3 + y^4 \qquad (x - y)^4 = x^4 - 4x^3 y + 6x^2 y^2 - 4xy^3 + y^4$$

## Rational Zero Theorem
If $p(x) = a_n x^n + a_{n-1} x^{n-1} + \cdots + a_1 x + a_0$ has integer coefficients, then every *rational zero* of $p$ is of the form $x = r/s$, where $r$ is a factor of $a_0$ and $s$ is a factor of $a_n$.

## Factoring by Grouping
$$acx^3 + adx^2 + bcx + bd = ax^2(cs + d) + b(cx + d) = (ax^2 + b)(cx + d)$$

## Arithmetic Operations
$$ab + ac = a(b + c) \qquad \frac{a}{b} + \frac{c}{d} = \frac{ad + bc}{bd} \qquad \frac{a + b}{c} = \frac{a}{c} + \frac{b}{c}$$

$$\frac{\left(\dfrac{a}{b}\right)}{\left(\dfrac{c}{d}\right)} = \left(\frac{a}{b}\right)\left(\frac{d}{c}\right) = \frac{ad}{bc} \qquad \frac{\left(\dfrac{a}{b}\right)}{c} = \frac{a}{bc} \qquad \frac{a}{\left(\dfrac{b}{c}\right)} = \frac{ac}{b}$$

$$a\left(\frac{b}{c}\right) = \frac{ab}{c} \qquad \frac{a - b}{c - d} = \frac{b - a}{d - c} \qquad \frac{ab + ac}{a} = b + c$$

## Exponents and Radicals
$$a^0 = 1, \ a \neq 0 \qquad (ab)^x = a^x b^x \qquad a^x a^y = a^{x+y} \qquad \sqrt{a} = a^{1/2} \qquad \frac{a^x}{a^y} = a^{x-y} \qquad \sqrt[n]{a} = a^{1/n}$$

$$\left(\frac{a}{b}\right)^x = \frac{a^x}{b^x} \qquad \sqrt[n]{a^m} = a^{m/n} \qquad a^{-x} = \frac{1}{a^x} \qquad \sqrt[n]{ab} = \sqrt[n]{a}\sqrt[n]{b} \qquad (a^x)^y = a^{xy} \qquad \sqrt[n]{\frac{a}{b}} = \frac{\sqrt[n]{a}}{\sqrt[n]{b}}$$

# Additional Formulas

## Summation Formulas:

$$\sum_{i=1}^{n} c = cn \qquad \sum_{i=1}^{n} i = \frac{n(n+1)}{2}$$

$$\sum_{i=1}^{n} i^2 = \frac{n(n+1)(2n+1)}{6} \qquad \sum_{i=1}^{n} i^3 = \left(\frac{n(n+1)}{2}\right)^2$$

## Trapezoidal Rule:

$$\int_a^b f(x)\,dx \approx \frac{\Delta x}{2}\left[f(x_1) + 2f(x_2) + 2f(x_3) + \ldots + 2f(x_n) + f(x_{n+1})\right]$$

with Error $\leq \dfrac{(b-a)^3}{12n^2}\left[\max |f''(x)|\right]$

## Simpson's Rule:

$$\int_a^b f(x)\,dx \approx \frac{\Delta x}{3}\left[f(x_1) + 4f(x_2) + 2f(x_3) + 4f(x_4) + \ldots + 2f(x_{n-1}) + 4f(x_n) + f(x_{n+1})\right]$$

with Error $\leq \dfrac{(b-a)^5}{180n^4}\left[\max |f^{(4)}(x)|\right]$

## Arc Length:

$$L = \int_a^b \sqrt{1 + f'(x)^2}\,dx$$

## Surface of Revolution:

$$S = 2\pi \int_a^b f(x)\sqrt{1 + f'(x)^2}\,dx$$

(where $f(x) \geq 0$)

$$S = 2\pi \int_a^b x\sqrt{1 + f'(x)^2}\,dx$$

(where $a, b \geq 0$)

## Work Done by a Variable Force:

$$W = \int_a^b F(x)\,dx$$

## Force Exerted by a Fluid:

$$F = \int_a^b w\,d(y)\,\ell(y)\,dy$$

## Taylor Series Expansion for $f(x)$:

$$p_n(x) = f(c) + f'(c)(x-c) + \frac{f''(c)}{2!}(x-c)^2 + \frac{f'''(c)}{3!}(x-c)^3 + \ldots + \frac{f^{(n)}(c)}{n!}(x-c)^n$$

## Maclaurin Series Expansion for $f(x)$, where $c = 0$:

$$p_n(x) = f(0) + f'(0)x + \frac{f''(0)}{2!}x^2 + \frac{f'''(0)}{3!}x^3 + \ldots + \frac{f^{(n)}(0)}{n!}x^n$$

# Summary of Tests for Series:

| Test | Series | Condition(s) of Convergence | Condition(s) of Divergence | Comment |
|---|---|---|---|---|
| $n$th-Term | $\sum_{n=1}^{\infty} a_n$ | | $\lim_{n\to\infty} a_n \neq 0$ | This test cannot be used to show convergence. |
| Geometric Series | $\sum_{n=0}^{\infty} r^n$ | $\|r\| < 1$ | $\|r\| \geq 1$ | Sum $= \dfrac{1}{1-r}$ |
| Telescoping Series | $\sum_{n=1}^{\infty} (b_n - b_{n+a})$ | $\lim_{n\to\infty} b_n = L$ | | Sum $= \left(\sum_{n=1}^{a} b_n\right) - L$ |
| $p$-Series | $\sum_{n=1}^{\infty} \dfrac{1}{(an+b)^p}$ | $p > 1$ | $p \leq 1$ | |
| Integral Test | $\sum_{n=0}^{\infty} a_n$ | $\int_1^{\infty} a(n)\,dn$ is convergent | $\int_1^{\infty} a(n)\,dn$ is divergent | $a_n = a(n)$ must be continuous |
| Direct Comparison | $\sum_{n=0}^{\infty} a_n$ | $\sum_{n=0}^{\infty} b_n$ converges and $0 \leq a_n \leq b_n$ | $\sum_{n=0}^{\infty} b_n$ diverges and $0 \leq b_n \leq a_n$ | |
| Limit Comparison | $\sum_{n=0}^{\infty} a_n$ | $\sum_{n=0}^{\infty} b_n$ converges and $\lim_{n\to\infty} a_n/b_n \geq 0$ | $\sum_{n=0}^{\infty} b_n$ diverges and $\lim_{n\to\infty} a_n/b_n > 0$ | Also diverges if $\lim_{n\to\infty} a_n/b_n = \infty$ |
| Ratio Test | $\sum_{n=0}^{\infty} a_n$ | $\lim_{n\to\infty} \dfrac{a_{n+1}}{a_n} < 1$ | $\lim_{n\to\infty} \dfrac{a_{n+1}}{a_n} > 1$ | $\{a_n\}$ must be positive. Also diverges if $\lim_{n\to\infty} a_{n+1}/a_n = \infty$ |
| Root Test | $\sum_{n=0}^{\infty} a_n$ | $\lim_{n\to\infty} (a_n)^{1/n} < 1$ | $\lim_{n\to\infty} (a_n)^{1/n} > 1$ | $\{a_n\}$ must be positive. Also diverges if $\lim_{n\to\infty} (a_n)^{1/n} = \infty$ |

Made in the USA
Monee, IL
31 January 2022